Springer Series in Reliability Engineering

Series editor

Hoang Pham, Piscataway, USA

More information about this series at http://www.springer.com/series/6917

Xiao-Sheng Si · Zheng-Xin Zhang
Chang-Hua Hu

Data-Driven Remaining Useful Life Prognosis Techniques

Stochastic Models, Methods and Applications

国防工業出版社
National Defense Industry Press

Springer

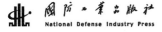

Xiao-Sheng Si
Department of Automation
Xi'an Institute of High-Technology
Xi'an, Shaanxi
China

Chang-Hua Hu
Department of Automation
Xi'an Institute of High-Technology
Xi'an, Shaanxi
China

Zheng-Xin Zhang
Department of Automation
Xi'an Institute of High-Technology
Xi'an, Shaanxi
China

ISSN 1614-7839 ISSN 2196-999X (electronic)
Springer Series in Reliability Engineering
ISBN 978-3-662-54028-2 ISBN 978-3-662-54030-5 (eBook)
DOI 10.1007/978-3-662-54030-5

Jointly published with National Defense Industry Press, Beijing, China

Library of Congress Control Number: 2016961678

Printed on acid-free paper

This Springer imprint is published by Springer Nature
The registered company is Springer-Verlag GmbH Germany
The registered company address is: Heidelberger Platz 3, 14197 Berlin, Germany

Preface

The remaining useful life (RUL) of a system is defined as the length from the current time to the end of the useful life. The concept of the RUL has been widely used in operational research, reliability, and statistics literature with important applications in other fields such as materials science, biostatistics, and econometrics. However, there are many definitions as what is regarded as the useful life. In 'Businessdictionary.com,' it defines the useful life "the period during which an asset or property is expected to be usable for the purpose it was acquired'. However, in accounting, it is defined as 'the expected period of time during which a depreciating asset will be productive." The keyword here is 'usable' or 'productive' which is again upon individual explanations. Clearly the definition of the useful life depends on the context and operational characteristics. In this book we will assume that the definition of the useful life is known to the owner of the asset and the main interest is to investigate the modeling methods for RUL estimation given condition and health monitoring information.

In conventional data-based approaches, estimating the RUL is achieved by evaluating the conditional lifetime distribution given that a system has survived up to a specific time. The obtained RUL distributions from these approaches are generally based on the life characteristics of a population of identical systems and lifetime data are required. However, such data are scarce in reality or even nonexistent at all for systems which are costly or time-consuming to collect the life data. With the advances in CM technologies, degradation data can be obtained from routine CM as feasible and low-cost alternatives to estimate the RUL. These data are usually correlated with the underlying physical degradation process. If they are properly modeled, degradation data can be used to predict unexpected failures and accurately estimate the lifetime of gradually degraded systems. In many situations, such as the drift degradation of an inertial navigation system used in the aerospace industry, it is natural to view the failure event of interest as the result of a stochastic degradation process crossing a threshold level, i.e., to model the hitting time of the degradation as a time-dependent stochastic process. On the other hand, dynamic environments induce changes in the physics of failure.

RUL prognosis is one of the key factors in condition-based maintenance (CBM), and prognostics and health management. It is critically important to assess the RUL of an asset while in use since it has impacts on the planning of maintenance activities, spare parts provision, operational performance, and profitability of the owner of an asset. RUL estimation has also an important role in the management of product reuse and recycle which has strategic impacts on energy consumption, raw material use, pollution, and landfill. The reused products must have sufficient long lives left among others to be able to be reused. This puts the importance of the estimation of RUL beyond CBM and prognostics and health management because of the green issues associated. As a consequence, developing RUL prognosis methods is much desired for health management of degrading systems to prevent sudden failure and reduce the safety risk. In the past four decades, valuable contributions to prognostics in reliability field have been made. This book is intended to summarize the research results studied mainly by the authors in the past decade.

This book introduces the main ideas of data-driven remaining useful life prognosis techniques, with an emphasis on stochastic models, methods, and applications. It gives a thorough survey of new methods that have been developed in the recent years and demonstrates them with examples. To the knowledge of the authors, all major aspects of RUL prognosis are treated for the first time in a single book from a common viewpoint. With the presentation of RUL prognosis methods for degrading systems, the book provides novel materials that have not yet been described in monographs or textbooks.

This monograph consists of four parts:

- **Part I: Introduction, Degradation Data Acquisition and Evaluation**. Advances in data-driven RUL prognosis techniques are reviewed. As fundamental issues for data-driven RUL prognosis, methods of how to acquiring the degradation data and how to evaluate the usability of the acquired data are presented.
- **Part II: Prognostic Techniques for Linear Degrading Systems**. Methods for adaptive RUL prognosis, exact RUL prognosis solution, RUL prognosis with multiple kinds of variability for linear degrading systems are presented and the methods are demonstrated by case studies.
- **Part III: Prognostic Techniques for Nonlinear Degrading Systems**. Methods for nonlinear degradation modeling, adaptive RUL prognosis, nonlinear RUL prognosis under multiple sources of variability, residual storage life prognosis with switching systems for nonlinear degrading systems are presented and the methods are demonstrated by case studies.
- **Part IV: Applications of Prognostic Information**. This part discusses the applications of prognostic information such as mission reliability estimation, condition-based replacement, spare parts forecasting, and joint optimization of spare part ordering and replacement.

As each of the models used requires its own mathematical background and the methods based on these models follow different lines of thinking, the book cannot present the methods for all details. The aim is to give the readers a broad view of the field and provide them with bibliographical notes for further reading. A further reason for the different depth with which the chapters tackle the RUL prognosis problems is given by the status of research. In the introductory parts of all chapters, the problems to be solved are posed in a framework that is familiar to practicing engineers. They describe the new ideas and concepts of RUL prognosis in an intuitive way, before these ideas are brought into a strict mathematical form. Examples illustrate the applicability of the methods. Bibliographical notes at the end of each chapter point to the origins of the presented ideas and the current research lines. The evaluation of the methods and the application studies should help the readers to assess the available methods and the limits of the present knowledge about RUL prognosis with respect to their particular field of application.

Together with four parts, the book is composed of 16 chapters. Chapter 1 is devoted to an introduction to advances in data-driven RUL prognosis techniques. Chapter 2 considers the problem of planning repeated degradation test for degrading products with three-source variability. In Chap. 3, the attention is paid to specifying measurement errors for required lifetime estimation performance so as to evaluate the data usability. A linear degradation model with a recursive filter algorithm and Bayesian updating is presented to estimate the PDF of the RUL in Chap. 4. Chapter 5 derives the exact and closed-form solution of RUL prognosis for linear degrading systems. Chapter 6 presents a Wiener-process-based degradation modeling framework for RUL estimation with three-source variability. In Chap. 7, a diffusion process-based model was presented to characterize the dynamics and nonlinearity of degradation processes, and the corresponding RUL distribution is formulated. The results in Chap. 7 are further extended to an age- and state-dependent case in Chap. 8. In Chap. 9, an adaptive and nonlinear prognostic model is presented to estimate the RUL using the history of the observed data to date. Chapter 10 develops a real-time RUL estimation method based on a state space model considering that the degradation process is hidden and nonlinear. Chapter 11 presents a general nonlinear diffusion process-based model to estimate the RUL with the temporal variability, unit-to-unit variability, and measurement variability. In Chap. 12, the problem of predicting RSL for a class of systems with operation state switches is concerned. Chapter 13 applies the prognostic information to reliability estimation of phased-mission systems. In Chap. 14, a real-time variable cost-based maintenance model is presented based on nonlinear prognostic information. Chapter 15 presents an adaptive spare parts demand forecasting method based on degradation modeling of the CM data. In Chap. 16, a new sequential maintenance and inventory model is developed to consider the effects of both expectation of the maintenance cost and its variability under prognostic information.

In preparing the book, efforts have been made to maintain a balance between the required theoretical and mathematical rigor in the exposition of the methods and the clarity in the illustration of the numerical examples and practical applications. For this reason, this book can serve well as a reference to both reliability and risk analysis researchers and engineers. Furthermore, sufficient references leading to further studies are cited at the end of each chapter. This book will serve as a textbook and reference book for graduate students and researchers in reliability and maintenance. Although the book is self-explanatory, a standard background in probability theory, mathematical statistics, and stochastic processes is recommended.

Finally, we wish to thank Profs. Wenbin Wang, Donghua Zhou, and Michael Pecht for their cooperation and valuable discussions. In addition, it is with sincere appreciation that we thank the support by National Nature Science Foundation of China under Grant 61174030, 61374126, 61473094, 61573076, 61573366, and the NSF of Shaanxi Province of China under grant 2015JQ6235.

Xi'an, China
July 2016

Xiao-Sheng Si
Zheng-Xin Zhang
Chang-Hua Hu

Contents

Part I Introduction, Degradation Data Acquisition and Evaluation

1 Advances in Data-Driven RUL Prognosis Techniques 3
1.1 Introduction 3
1.2 Methods Considering Unit-to-Unit Variability 5
1.2.1 Random Coefficients Regression Models. 6
1.2.2 Stochastic Process Models with Random Coefficients 6
1.3 Methods Considering Impact of Heterogeneity in Working Environment 8
1.3.1 Methods Based on Stochastic Filtering 8
1.3.2 Multi-stage Degradation Models 9
1.3.3 Covariate Hazards Model 10
1.3.4 Degradation Models Involving Random Shocks 11
1.4 Methods Considering the Impact of Tasks and Workloads. 13
1.4.1 Degradation Modeling for Systems with Dynamic Workloads. 13
1.4.2 Degradation Modeling for System with Maintenances. 14
1.5 Future Research Directions. 15
References. 17

2 Planning Repeated Degradation Testing for Degrading Products. 23
2.1 Introduction 23
2.2 Degradation Modeling with Three-Source Variability. 25
2.3 Parameter Estimation and Information Matrix. 27

2.4 Estimating the Degradation Distribution and Lifetime Distribution . . . 28
2.4.1 The Quantiles of Degradation Distribution and Its Variance . . . 28
2.4.2 The Lifetime Distribution . . . 29
2.5 Degradation Test Planning . . . 32
2.6 An Illustrativc Examplc . . . 33
References . . . 36

3 Specifying Measurement Errors for Required Lifetime Estimation Performance . . . 39
3.1 Introduction . . . 39
3.2 Properties of the WPDM . . . 42
3.3 Properties of the WPDM with the ME . . . 43
3.4 Permissible ME Parameters for Lifetime Estimation . . . 45
3.4.1 Performance Measures to Quantify the Difference in Lifetime Estimation with Versus Without the ME . . . 45
3.4.2 Permissible ME Parameters Using the Relative Increase Ratio of the CV . . . 46
3.4.3 Permissible ME Parameters Using the Relative Increase Ratio of the Variance . . . 49
3.5 Effect of Lifetime Estimation with or Without ME on an Age-Based Replacement Decision . . . 50
3.6 Experimental Studies . . . 52
3.6.1 A Numerical Illustration . . . 52
3.6.2 The Case Study . . . 55
Appendix . . . 62
References . . . 67

Part II Prognostic Techniques for Linear Degrading Systems

4 An Adaptive Remaining Useful Life Estimation Approach with a Recursive Filter . . . 73
4.1 Introduction . . . 73
4.2 Wiener-Process-Based Degradation Modeling and RUL Estimation . . . 76
4.2.1 An Outline of Wiener-Process-Based Degradation Model for Lifetime Analysis . . . 76
4.2.2 Wiener-Process-Based Degradation Modeling . . . 79
4.2.3 Real-Time Updating of the RUL Distribution . . . 81
4.3 Parameter Estimation . . . 84
4.3.1 EM Algorithm . . . 84
4.3.2 The Implementation of EM Algorithm for the Proposed Model . . . 86

4.3.3 Convergence Analysis of Adaptive Model Parameter Estimation Algorithm 90
4.4 A Practical Case Study 92
4.4.1 Problem Description 92
4.4.2 The Implementation of Our Model for RUL Estimation of the INS 94
4.4.3 Comparative Studies 96
References 100

5 An Exact and Closed-Form Solution to Degradation Path-Dependent RUL Estimation 103
5.1 Introduction 103
5.2 A Degradation Path-Dependent Approach for Adaptive RUL Estimation 107
5.2.1 A General Description of Stochastic Process Based Degradation Models 107
5.2.2 A Degradation Path-Dependent Approach for Adaptive RUL Estimation via Real-Time CM Data 108
5.3 Linear Model 110
5.4 Exponential Model 123
5.5 Experimental Studies 130
5.5.1 Numerical Example 131
5.5.2 A Practical Case Study of the Developed Approach in Condition-Based Replacement 136
References 141

6 Estimating RUL with Three-Source Variability in Degradation Modeling 143
6.1 Introduction 143
6.1.1 Motivation 143
6.1.2 Related Works 145
6.1.3 Main Works of This Chapter 146
6.2 Description of Degradation Modeling with Three-Source Variability for RUL Estimation 147
6.3 RUL Estimation with Three-Source Variability 149
6.3.1 RUL Estimation with Temporal Variability and Unit-to-Unit Variability 149
6.3.2 RUL Estimation with Temporal Variability and Uncertain Measurements 154
6.3.3 RUL Estimation with Three-Source Variability 158
6.4 Parameter Estimation 167

6.5 Experimental Studies . . . 170
6.5.1 Problem Description . . . 171
6.5.2 Comparisons for Model Fitting . . . 174
6.5.3 Comparisons for the Estimated RUL . . . 175
References . . . 178

Part III Prognostic Techniques for Nonlinear Degrading Systems

7 RUL Estimation Based on a Nonlinear Diffusion Degradation Process . . . 183
7.1 Introduction . . . 183
7.2 Literature Review . . . 186
7.3 Motivating Examples and RUL Modeling Principle . . . 187
7.4 Lifetime Distribution and Parameter Estimation of the Proposed Degradation Model . . . 191
7.4.1 Derivation of the Lifetime Distribution . . . 191
7.4.2 Lifetime Distribution Under Random Effects . . . 198
7.4.3 The Distribution of the RUL Estimation . . . 200
7.5 Parameters Estimation . . . 202
7.6 Examples of the Applications of the Models . . . 205
7.6.1 Laser Data . . . 206
7.6.2 Drift Degradation Data of INS . . . 208
7.6.3 Fatigue Crack Data of 2017-T4 . . . 211
References . . . 213

8 Prognostics for Age- and State-Dependent Nonlinear Degrading Systems . . . 217
8.1 Introduction . . . 217
8.2 Problem Formulation . . . 219
8.3 RUL Estimation by Degradation Modeling . . . 221
8.4 Model Parameter Estimation Framework . . . 226
8.5 An Illustrative Example . . . 228
8.5.1 Degradation Model and Lifetime Estimation . . . 228
8.5.2 Parameters Estimation . . . 230
8.5.3 Verifying the Accuracy of the Proposed Method . . . 231
8.6 Case Study . . . 232
References . . . 243

9 Adaptive Prognostic Approach via Nonlinear Degradation Modeling . . . 247
9.1 Introduction . . . 247
9.2 Nonlinear Model Description and RUL Estimation . . . 250
9.2.1 Modeling Description . . . 250
9.2.2 Derivation of the RUL Distribution . . . 251
9.2.3 Adaptive RUL Estimation . . . 253

9.3 Adaptive Parameter Estimation 256
9.4 An Illustrative Example 260
9.5 Numerical Example and Case Study 261
9.5.1 Numerical Example 261
9.5.2 Lithium-Ion Battery Life Prognosis 265
References 268

10 Prognostics for Hidden and Age-Dependent Nonlinear Degrading Systems 273
10.1 Introduction 273
10.1.1 Motivation 273
10.1.2 Related Works 274
10.1.3 Main Works of This Chapter 276
10.2 Problem Formulation and RUL Estimation 277
10.2.1 Problem Formulation 277
10.2.2 RUL Estimation 279
10.2.3 Comparative Discussions 284
10.3 Parameter Estimation 287
10.4 Illustrative Examples 291
10.4.1 The Derivation of the RUL for Three Cases 291
10.4.2 The Derivation of Parameter Estimation Algorithm for Three Cases 293
10.5 Simulation Study 297
10.6 Case Study 305
10.6.1 The Data and State-Space-Based Degradation Model 305
10.6.2 Results and Discussions 307
References 309

11 Prognostics for Nonlinear Degrading Systems with Three-Source Variability 313
11.1 Introduction 313
11.2 Nonlinear Prognostic Model Description 315
11.3 RUL Estimate Method with Three-Source Variability 317
11.3.1 RUL Estimate Only with the Temporal Variability 317
11.3.2 RUL Estimate with the Temporal Variability and the Unit-to-Unit Variability 318
11.3.3 RUL Estimate with the Temporal Variability and the Measurement Variability 321
11.3.4 RUL Estimate with Three-Source Variability 324
11.3.5 Parameter Estimation 328

11.4 Experimental Studies 328
11.4.1 Simulation Study 329
11.4.2 Case Study 332
References 335

12 RSL Prediction Approach for Systems with Operation State Switches 337
12.1 Introduction 337
12.2 Literature Review 339
12.3 Problem Description for RSL Estimation 340
12.4 Model Formulation for Transitions Between the Operating State and Storage State 342
12.4.1 Randomly Varying System Operation Process 342
12.4.2 Bayesian Estimation for Parameters in the System's Operation Process 344
12.5 Model Formulation of the System Degradation Process to Predict the RSL 346
12.5.1 Predicting the RSL Conditional on the Model Parameters and Fixed System Operation Process 346
12.5.2 Bayesian Estimation for Parameters in the Degradation Process 349
12.5.3 RSL Prediction Considering the Future Transitions and Updated Parameters 350
12.6 Case Study 353
12.6.1 Background and Data Description 353
12.6.2 Results and Discussions 357
References 359

Part IV Applications of Prognostic Information

13 Reliability Estimation Approach for PMS 363
13.1 Introduction 363
13.2 Assumptions and Problem Description 366
13.2.1 Problem Description 366
13.2.2 Assumptions 367
13.3 Mission Process to Estimate the Mission Time 368
13.4 System Degradation Process to Estimate the Lifetime 374
13.4.1 Model Description 374
13.4.2 Bayesian Updating of Model Parameters 376
13.4.3 Estimating the RUL of PMS 377
13.5 Reliability Estimation for PMS 382

13.6 Experimental Studies 383
13.6.1 Numerical Simulations. 383
13.6.2 Case Study 388
References. 390

14 A Real-Time Variable Cost-Based Maintenance Model 393
14.1 Introduction 393
14.2 Degradation Modeling for Prognostics 395
14.2.1 Degradation Modeling. 395
14.2.2 RUL Estimation. 398
14.3 Replacement Decision Modeling. 399
14.4 A Case Study 401
References. 403

15 An Adaptive Spare Parts Demand Forecasting Method Based on Degradation Modeling 405
15.1 Introduction 405
15.2 Degradation Modeling Description 407
15.3 Adaptive Lifetime Estimation 408
15.4 Adaptively Forecasting Spare Parts Demand. 410
15.5 Adaptive Parameter Estimation. 412
15.6 Case Study 413
References. 416

16 Variable Cost-Based Maintenance and Inventory Model 419
16.1 Introduction 419
16.2 Degradation Modeling for Prognostics 420
16.2.1 Degradation Modeling. 421
16.2.2 RUL Estimation. 421
16.3 Parameter Estimation 422
16.4 Replacement and Inventory Decision Modeling 424
16.5 Case Study 427
References. 430

Acronyms

AIC	Akaike information criterion
BM	Brownian motion
CBM	Condition-based maintenance
CDF	Cumulative distribution function
CM	Condition monitoring
CTMC	Continuous-time Markov chain
EKF	Extended Kalman filter
EKS	Extended Kalman smoother
FHT	First hitting time
FPK	Fokker–Planck–Kolmogorov
FPT	First passage time
HMM	Hidden Markov model
HSMM	Hidden semi-Markov model
INS	Inertial navigation system
KF	Kalman filter
ME	Measurement error
MLE	Maximum likelihood estimation
MSE	Mean squared error
MTTF	Mean time to failure
PDF	Probability density function
PHM	Prognostics and health management
PMS	Phased-mission system
RE	Relative error
RSL	Residual storage life
RTS	Rauch–Tung–Striebel
RUL	Remaining useful life
STF	Strong tracking filter
TMSE	Total MSE

Part I
Introduction, Degradation Data Acquisition and Evaluation

Chapter 1
Advances in Data-Driven RUL Prognosis Techniques

1.1 Introduction

Prognosis and health management (PHM) has drawn increasing attention and gained deepening recognition and widening applications during the past decades [1–4]. Actually, the initial health and usage inspection system was fist equipped in the early helicopters of US military and the synthetically health management philosophy was presented for spacecraft in the 1970s. Recently, the comprehensive solution for system performance prognosis and maintenance has been achieved in the Joint Strike Fighter F-35 project [5]. Further, the ability of PHM has already been listed by the Department of Defense (DOD) of United states as one of the essential norms for weapon system purchasing. This shows the significant implication of PHM in military fields. On the other side, industrial practice indicates that PHM technology can effectively reduce the maintenance cost, improve the reliability and guarantee the completion of tasks of the system [6, 7]. Research institutes including NASA [8], University of Maryland [6] and George's University [9], as well as commercial companies such as Boeing have launched a great deal of theoretical and applied research works about PHM technology. The PHM conference has been successfully organized and held by IEEE Reliability Society in Shenzhen, Macau, Beijing, Rome, and Zhangjiajie respectively, Beijing in six consecutive years since 2010.

Remaining useful life (RUL) estimation, offering guidance for sequential management involving inspection schedule, maintenance, replacement and spare parts ordering, has been considered as the kernel technology of PHM, and the focus of current research in the field of reliability also. According to Petch's classical monograph about PHM technology [6], methods for RUL estimation can be classified into three kinds: namely physical model-based methods, data-driven methods and their combinations. However, with the development of industry and the continuing extension of human exploring activities, the complexity of a system, together with the diversity and uncertainty of its operating environments, continues to increase, which results in extreme difficulties in constructing physical models capturing the system and its operating circumvents. Meanwhile, data-driven methods, including

X.-S. Si et al., *Data-Driven Remaining Useful Life Prognosis Techniques*,
Springer Series in Reliability Engineering, DOI 10.1007/978-3-662-54030-5_1

artificial intelligence-based methods and statistical data-based methods have become an effective avenue to evaluate reliability and estimate RUL, especially for vital systems with high reliability and long lifetime. Artificial intelligence based methods can hardly provide a probability density function (PDF) estimate capturing stochastic and uncertain characteristics of the RUL, while this desire is a natural result for stochastic data-driven methods [10]. To address the uncertainty of prognosis, we mean statistic data-driven methods as data-driven methods throughout this chapter. According to the observability of underlying degradation process, Si et al. provided an review on data-driven methods for both direct and indirect observed degradation data, introducing many common methods including Gamma processes, Wiener processes, Hidden (semi-) Markov models, stochastic regression models, stochastic filtering-based models and covariate hazard-based models, from the perspectives of applying procedure, merits and drawbacks [11]. While being satisfactory for RUL estimation under each specific applying condition, these methods exhibit some limits in cases with heterogeneity from the inner states or the external operating conditions of systems.

Heterogeneity is widespread in the inner states of the system and the related working environments. Examples involve that a weapon system may experience various operating conditions, saying storage, inspection, transport, and maintenance during its life cycle due to different tasks; that a manufacture system produces different products under different workloads; and that even systems from the same category may exhibit various degrading paths in the same environment. The performance degradation of a system is a result of interactions of both inner deterioration and working environment of the system, indicating a need for incorporating the heterogeneity into degradation modeling, to achieve a more accurate RUL estimation. For particular heterogeneity, such as the unit-to-unit variability, changing working conditions and periodic tasks, many recent advances in RUL estimation have appeared. However, to the best of the authors' knowledge, there is still no review regarding degradation modeling and RUL estimation for systems with heterogeneity. Therefore, this chapter tries the best to fill this gap.

Toward the end of this chapter, three kinds of heterogeneity are considered consecutively: the unit-to-unit variability for systems from the same category, the variability in time-varying operating conditions, and the diversity of tasks and workloads of system during their life cycles. The first kind of variability describes the differences in degradation processes of units from the same category, while the second represents noninform working conditions related to the degradation, such as the time-varying, multi-state and stochastic working environments or random shocks. The third kind of heterogeneity captures the influence of changes in tasks and management activities involving inspection, maintenance, etc. Accordingly, this chapter classifies methods addressing degradation modeling and RUL estimation with heterogeneity into three kinds, each of which considers one kind of heterogeneity introduced above and consists of some subclassifications. The taxonomy of RUL estimation approaches for systems with heterogeneity is illustrated in Fig. 1.1.

The remainder of the chapter is structured as follows: Section 1.2 summarizes methods considering unit-to-unit variability, saying degradation models with random

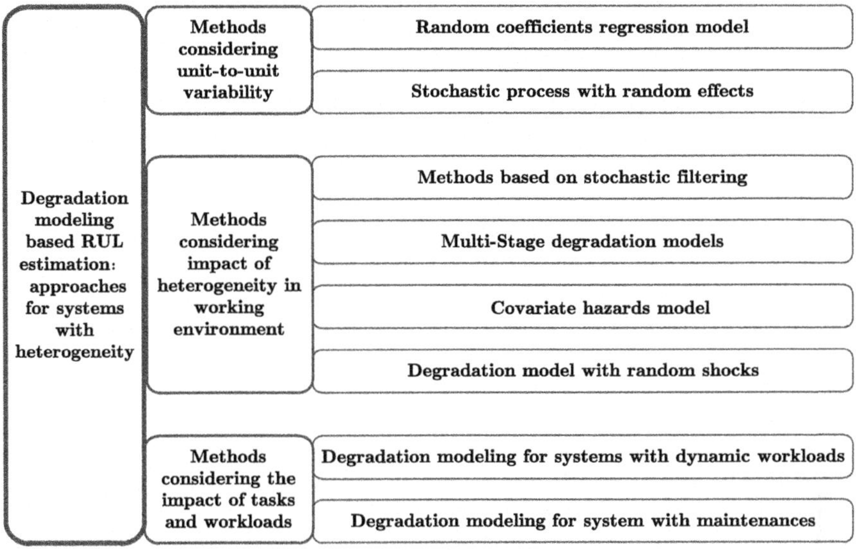

Fig. 1.1 Taxonomy of RUL estimation approaches for system under heterogeneity

effect In Sect. 1.3 methods considering the impact of the working conditions are provided. Methods for incorporating the influence of diversity in tasks and workloads are reviewed in Sect. 1.4. Section 1.5 concludes the chapter and provides several possible directions for future studies.

1.2 Methods Considering Unit-to-Unit Variability

A large number of experiments and engineering phenomena show that systems of the same category, even from one batch degrade differently from one another in performance. This kind of difference in degradation is usually defined as the unit-to-unit variability, due to the variability in inner structures of the considered systems, as well as the diversity in their working environment. Commonly, models with random effects are employed to capture the unit-to-unit variability, when we model the degradation process and estimate the RUL. The most typical way to do so is to specify some parameters of the model as random variables governed by distributions with computing convenience, presenting the individuality in degradation processes from different units and leave the rest of parameters as constants describing the universality in degradation of systems from the same category or batch. In the following, random coefficients regression models and stochastic process models with random coefficients of this kind are discussed, respectively.

1.2.1 Random Coefficients Regression Models

Random effects first appeared in random coefficients regression models. In the most frequently cited paper about degradation modeling and RUL estimate [12], Lu and Meeker described the random coefficients regression model in a general form as

$$X(t_{ij}) = g(t_{ij}; \boldsymbol{\phi}, \boldsymbol{\theta}) + \varepsilon,$$

where $X(t_{ij})$ is the amount of the degradation of the ith device in the jth inspecting time t_{ij}, the fixed coefficients $\boldsymbol{\phi}$ and the random coefficients $\boldsymbol{\theta}$ are, respectively, used to characterize the universality and individuality in degradation of different systems, and is the random noise.

Random coefficients regression models have been extended, developed and applied widely in many areas, in which a series of extended works presented by Gebraeel et al. are included [13–15] Son et al. compared various kinds of RUL estimation method based on random coefficients regression models [16]. Suk and Paul proposed a nonlinear random coefficients regression method for degradation data [17], and applied the model to the degradation of the vacuum fluorescent tube display. To improve the accuracy of parameter estimation, Weaver and Meeker also studied the optimal design of repeated measures degradation studies, and the method to design accelerated repeated degradation studies [18, 19]. A procedure deciding the minimum sample size and the minimum times of systematic sampling for each item to achieve an anticipant accuracy of estimation (large sample approximate variance) has been provided in their works.

However, according to Wang's analysis in [20], the assumptions of random coefficient regression models result in several limitations, involving the need for more historical degradation data from different systems of the same category, the difficulty in capturing the time-varying dynamics of systems and the independency between random noise with time.

1.2.2 Stochastic Process Models with Random Coefficients

Incorporating random coefficients into stochastic degradation process-based models enables both considerations of time-varying dynamics of an individual system, and description of unit-to-unit variability, and thus has been favored by many researchers. Suppose that the degradation of a system is modeled by a stochastic process $\{X(t); t \leq 0, \boldsymbol{\theta}, \boldsymbol{\vartheta}\}$, with constant parameters $\boldsymbol{\theta}$ and random parameters $\boldsymbol{\vartheta}$. Under the concept of first passage time (FPT), the RUL of the system conditional on the observation $X(t_k)$ at time t_k is defined as

$$L_k := \inf \{l_k : X(t_k + l_k) \geq \omega | X(t_k) < \omega\},$$

where is a preset constant failure threshold.

Lawless proposed a Gamma process-based model containing the covariates and random effects, and applied it to degradation modeling and RUL estimation [21]. When fitting the semi-parametric Gamma process to degradation data, Ye et al. also took the random effects into consideration. Further, the unit-to-unit variability was captured using random parameters following some particular distributions in recent degradation models based on Inverse Gaussian process [22, 23]. The same specifications addressing differences in the degradation process of systems from the same category were used in the application of Inverse Gaussian process for systems with monotonous degradation by Wang [24] and Ye et al. [25]. For nonmonotonic degradation processes with fluctuations, Wang proposed a Wiener degradation model with random effects [26]. Si et al. presented a degradation path-based RUL estimation method with exact closed form solution of the estimated PDF of the RUL in linear and exponential cases, which also incorporated the random effects. Peng and Zeng analyzed the misspecification of linear degradation model in the framework of Wiener process with random drift coefficient [26]. Similarly, Si et al. [27] and Wang et al. [28] set some parameters in their methods as normally distributed random variables, when modeling nonlinear diffusion degradation process and additive hybrid degradation process, respectively. From results in the existing literature, stochastic process models with random effects can effectively improve the estimation accuracy and extend the applications of the initial degradation models, in both cases of monotonous and nonmonotonous degradation processes no matter linear or nonlinear.

In the industrial applications, the main flaw of degradation models with random effects is the complexity in computation. Therefore, the primary concern choosing the random parameters and their distribution models is the convenience of calculation. Normally distributed random variables are with high frequency in the related literature. For example, in Tseng and Yu [29], Lu and Meeker [12], Gebraeel [13], Si [30], all selected models with random variables following Normal distribution to characterize unit-to-unit variability. As for some particular degradation models, parameters subjected to special forms of distributions are preferred. Wang utilized Gamma distributions to mode the drift and diffusion coefficients in the Wiener degradation model [26], and Ye el al. also used gamma distributed parameters when constructing semi-parametric Gamma degradation process. These choices are made due to the purposes of computing convenience. The misspecifications of such distributions are considered by some researchers and some nonparametric distributions based on observations are recommended [31–34]. However, explicit results of the estimated RUL can hardly be derived when nonparametric distributions are used. Besides, the corresponding computation is always complicated, which makes it inadequate for real-time RUL estimation. Therefore, it is a challenge to reasonably choose random parameters and their distributions that cannot only capture the unit-to-unit variability but also benefit computation, when using degradation models with random effects.

1.3 Methods Considering Impact of Heterogeneity in Working Environment

1.3.1 Methods Based on Stochastic Filtering

As early as 1979, Sarma et al. estimated health state of aerospace engine using Kalman Filter (KF) technology, and achieved a maintenance decision optimization based on the estimated results [35]. Afterwards, Wang and Christer [], Batzel and Swanson [], proposed different state evaluation and RUL estimation methods, applied successfully to electromagnetic induction smelting furnaces, aeronautical batteries and other industrial systems, based on the construction of state-space models. As for the nonlinear and nonGaussian state-space models, Extended KF, Benes Filter, Multiple Model Filter and Particle Filter based methods for health state and RUL estimation, have been successively proposed [36–40]. When the impact of heterogeneity is incorporated into stochastic filter-based methods, two kinds of sub methods can be referred to, namely semi-stochastic filter based methods and adaptive parameter based methods.

Ability to handle unobservable degradation is an advantage of stochastic filter based methods, while the failure threshold of the unobservable degradation can hardly be specified. In this connection, the lifetime of a system is directly defined as a state in the state-space model by Wang and Christ, and the length of time interval between two consecutive inspections is treated as the decrease of the lifetime. As such, RUL estimation method based on stochastic filter was proposed in [20] through constructing a stochastic relationship between the condition monitoring data and the lifetime of the system. This original method has been extended to cases where the operating environments are considered, by establishing the stochastic relationship between lifetime with the condition monitoring information and the operating environments simultaneously [41].

In another class of approaches for degradation modeling, some important parameters are expanded as state of a state-space model, which is utilized to describe dynamics in parameters. These parameters are adaptive to the changing environmental variables and updated jointly online with the healthy state of the system. As a result, the updated states and parameters are substituted to obtain a new estimation of RUL. Wang and Mattgew set the drift coefficient in Wiener process as an adaptive parameter, which will be updated through KF technology once new observations are available [42]. Inspired by [42], Si et al. proposed a Wiener degradation model with nonlinear drift coefficient function, which also makes some parameter adaptive to the observed data [43, 44]. In this chapter concerning models for RUL estimation under three sources of variability, the drift coefficient was also treated as an adaptive parameter and expanded to a state in the state-space model describing the degradation of the system, and was updated with the degradation level on-line [45]. The successful application of this method in the RUL estimation of an inertial navigation system has shown much superiority of such method.

The best advantage of such stochastic filter based RUL estimation methods lies in that the parameters and the accordingly RUL estimation can be updated with the newly observed condition monitoring information. Furthermore, the dynamics in the degradation process and the probable measurement errors are taken into consideration, which makes it suitable for indirectly observable degradation process. However, these methods have a premise in common that an explicit state-space model must be constructed, which may be impossible in some cases. Another limitation is that the RUL estimation is obtained without consideration of the possible future changes in the degradation. In addition, the assumption in the semi-stochastic filter based method that there is a deterministic equal relationship between the reduction of lifetime and the inspection interval may not hold in many cases, especially when there are changes in the operating environments of the workloads of the system.

1.3.2 Multi-stage Degradation Models

Multi-stage degradation models are proposed to handle the period differences existing in the degradation process. In Wang's two stage degradation model, the degradation data after the defect point were used to estimate the parameters in the degradation model and predict the RUL of the system [46]. In order to evaluate the remaining storage lifetime of a system, Feng el al. proposed a multi-stage Wiener degradation model in [47], where some related works were listed. These works include the nonhomogeneous Poisson process which can be used to analyse time-varying failure rate of software, the nonlinear model with random coefficients which is applied to the multi-state nonmonotonic degradation process of hardware, the multi-stage linear regression model, the multi-stage linear stochastic process model, and so on. Li and Pham studied the reliability modeling problem of multi-state degrading systems, under the interaction of multi competing failure modes and random shocks [48]. The common shared by these models is the presence of change points, such as the defect point in two-stage model and the starting/finish points of each stage. Generally, the unknown locations of these change points have to be determined by selecting appropriate detection methods before model identification and RUL estimation. This problem has been considered as highlight but also aporia. Currently, the maximum likelihood estimation, stochastic filtering, and control charts are the most popular methods to estimate the change points in multi-stage degradation models. Thus, the accuracy of change points estimate has direct influence on the accuracy of the RUL estimation. Another popular multi-stage degradation process for RUL estimation is the Markovian model. To model the hidden degradation process, Hidden Markovian Model (HMM) was first introduced to RUL estimation and condition based maintenance (CBM) [49]. On this basis, Dong et al. proposed the RUL estimation framework by using Hidden semi-Markov Models (HsMM), which extends the exponential assumption of state sojourn time to more general situations [50]. He and Dong extended the work in [50] and obtained RUL estimation through a comprehensive consideration of sojourn time in each state which has been modeled by a single HsMM [51]. Prognosis of both performance and RUL were achieved in [52] by a combination of HsMM and AR model for time series data. Giorgio, Guida and Pulcini considered the age- and state-dependency of the degradation process in the framework of Markovian degradation model [53, 54]. A very good result was obtained when the proposed models were applied to the degradation process of marine engine cylinder.

The proposition of these models improves the accuracy of degradation modeling, and enriches the selections of models for different degradation processes. However, almost all multi-stage degradation models face the problem of determining the number of degradation stages, and a large amount of training data as well as a complex computation procedure are needed for parameter estimation. Further, instead of a derivable analytical solution to the PDF estimate of the RUL, a time-consuming simulation based methods have to be executed. In addition, the RUL estimation in multi-stage degradation models is based on the information since the latest change point. Such an estimation of RUL is accurate if there will be no change occur in the future time of the system. In more practical situations with possible change points in the future, severe bias will be introduced into the RUL estimation if using such estimation mechanism, i.e., ignoring the possible change points in the future. To tackle this problem, the possible change points in the future should be considered in multi-stage degradation processing modeling and RUL estimation.

1.3.3 Covariate Hazards Model

Factors that affect degradation in performance of systems are defined as covariates in engineering pactive. The classical model for lifetime analysis, named proportion hazards model, is the most widely used ones in the fields of RUL estimation, reliability analysis/evaluation, decision-making and optimization on maintenances, etc. The existing works related to hazards models have been reviewed in [55]. The description of the system failure rate is the core of the proportional hazards model, and also the key for reliability assessment and RUL estimation. Failure rate in the proportional hazards model usually consists of the product of a reference failure rate function $h_0(t)$ and the covariate function $\psi(\boldsymbol{\beta} z(t))$,

$$h(t|z(t)) = h_0(t)\psi(\boldsymbol{\beta} z(t)),$$

where $z(t)$ are the covariate variable, $\boldsymbol{\beta}$ are the regression coefficients which can be estimated using historical lifetime data or censored lifetime data of the system from the same category. Proportional intensities model and proportional covariates model, developed from the proportional hazards model, are also popular models for RUL estimation [].

In heterogeneous working environments, the failure rate of the system will be affected. Ye et al. studied the influence of heterogeneity in the working conditions on the estimation of the RUL, based on the analysis of accelerated life test [56].

A system may experience fixed, time-varying, and even stochastic environmental conditions and the corresponding covariates may also be constants, time-varying or even random variables. In order to characterize the influence of random covariates on the failure rates in the proportional hazard models, researchers have considered using some stochastic process to model the time-varying covariates, and incorporating the modeled covariates into the proportion hazard models. For example, Markov chains, which can naturally describe the operating process of a system, are the most frequently used process to model the changing procedure of covariates [57–59]. A HMM with a known state transition law was utilized to model the stochastic degradation process, and the formula to calculate the mean RUL was derived in [60]. Lu and Liu studied the relationship between failure rates and the dynamic working environment [61]. In their research, the changing covariates were modeled by a two states (normal/severe) Markov Chain, as such, failure rate functions are changing with operation function, and the lifetime of the system can then be determined by its working conditions.

Although their strong explanatory property makes covariate hazard models dominant both in theory and application researches, this kind of method does have some inborn limitations, which have been summarized by Si et al. in [11]. Furthermore, some difficulties should be solved before the practical applications of this kind of methods. First, with the development of high reliability and small amount systems, the lifetime data required for estimating parameters $\boldsymbol{\beta}$ and reference failure rate function $h_0(t)$ are difficult and expensive to obtain. Second, it is hard to determine the form of covariate function when systems become complex.

1.3.4 Degradation Models Involving Random Shocks

During the degradation process, a system may suffer various kinds of shocks, which will impact the performance of the system as well as its underlying degradation process. Typically, there are five different types of random shock models existing in the literature [62]: (i) extreme shock model: the system fails when the size of a shock is beyond a specified threshold value; (ii) cumulative shocks model: a system fails when the accumulated damage of shocks is beyond a critical level; (iii) m-shock model: a system failures after suffering m shocks whose sizes are greater than a critical level; (iv) run shock model: failure occurs when there is a series of n consecutive shocks that are greater than a threshold; and (v) shock model: a system experiences failure when the inter-arrival time of two sequential shocks is less than a threshold. As for reliability modeling and RUL estimation under random shocks, there have been a number of studies including [63, 64] to which we can refer. Random shocks, whose influences on the performance of a system are addressed in this section, are regarded as heterogeneity in the working environment. In general, failure is a result from the interaction and competition of the performance degradation and external random shocks [65–67]. Poison process (homogeneous/nonhomogeneous) [68], Markov Chains [69], and the phase-type distribution [70, 71] have all been used to describe the arriving process of random shocks. Models for degradation processes with random shocks can be divided into two categories here, according to the existing of interdependency between the continuous degradation processes with random shocks.

Degradation processes and random shocks are supposed to be independent of the first type of models. Klutke and Yang first proposed an availability model for the system under interaction of degradation and random shocks [65]. Afterward, Huang and Askin analyzed and constructed a reliability model for systems under the competing impact of the degradation process and sudden failure [72]. Li and Pham proposed a reliability model for a system suffering two types of degradation and a type of random shocks [73]. Chen and Li assumed that from the external environment the degrading system may experience two types of random shocks, i.e., fatal or nonfatal [74]. An optimal maintenance strategy was proposed under a further assumption that system's tolerance of the total number of nonfatal shocks decreases subjected to the times of maintenance. A common underlying assumption of the works in [65, 72–74] is that degradation processes causing softer failures and random shocks leading to hard failures are independent from each other, and no mutual influence between degradation and shocks exists.

The interactions between shocks and degradations are considered in the second type of models. When studying the reliability and maintenance model for the system under competing degradation process and random shocks, Wang and Pham supposed that fatal shocks caused a direct failure of the system while nonfatal shocks resulted in abrupt increases in the degradation level [75]. The interdependency of soft failure caused by degradation and the hard failure caused by random shocks was included in Peng's work about reliability modeling [76]. Liu et al. considered the relationships between the failure rate of a system with age, degradation level and their interactions in the degradation model [64]. Recently, Koosha studied the influence of various types of random shocks on the degradation processes of the system and supposed that the level of degradation process jumped once a shock came while the degradation rate changed only after a particular type random shock [62]. A reliability model for dependent competing failure processes with changing degradation rate was then proposed based on this dependency of degradation process on random shocks.

The primary drawbacks of using such kind of methods are the following: (1) Lots of existing works incorporated random shocks to the linear degradation process, while degradation processes of actual systems are often nonlinear. To be more practical, the influence of random shocks on nonlinear degradation processes should be considered, which has seldom been done except [76, 77]. (2) As for discretely inspected system, the time and influence amplitude of random shocks can hardly be measured directly, which may introduce extreme difficulties in model identification and parameter estimation. (3) In cases where the dependency between degradation processes with random shocks is considered, attention has been focused on the influence of random rocks on (levels and rates) degradation process, while researches about impact of degradation on random shocks and the interdependency between each other are rarely reported except for [77, 78]. (4) Random shocks in degradation models are assumed to be negative, causing the increase in degradation level and even the failure of the system. However, there exist some shocks improving the system's performance, e.g., the state-of-health regeneration phenomena in lithium-ion battery systems. As such, this kind of shock should be further considered into the degradation process in the future.

1.4 Methods Considering the Impact of Tasks and Workloads

1.4.1 Degradation Modeling for Systems with Dynamic Workloads

Due to the diversity and randomness in the operating environments and workloads of systems, the characteristic of degradation varies with age throughout the whole lifetime cycle. If the dynamic operating model of the system's workload is constructed scientifically and incorporated into the degradation model reasonably, a more accurate estimation of the RUL will be achieved.

During the industrial applications, some systems change their working state in several different working modes, corresponding to which are the different workloads and various degradation processes. For example, a missile weapon system with an extremely long storage before being launched may experience different working states involving storage, transportation, inspection and maintenance during its service. Studies have shown that, due to influences of temperature, humidity and human factors in the storage conditions, the performance of gyroscopes installed in an inertial navigation system (INS) exhibit some decreasing trends, which will be accelerated by each electrifying inspection, after some time of storage [79]. Moreover, the states switching of the system is a stochastic process, because of the uncertainty in the coming of different tasks or missions A continuous-time Markov model (CTMM) with finite state is a natural selection to describe such a stochastic operating process [80, 81]. In literature, CTMM was used to capture the time-varying random working conditions of a system in Jeffrey and Steven's stochastic models for degradation-based reliability [82]. Si el al. also utilized two-state CTMM to represent the states switching process between storage and usage, and the operating model was successfully applied to estimate the remaining storage life (RSL) of gyroscopes in INS [83]. Hawkes proposed a reliability assessment model based on the CTMM modeling of working condition switches. Huynh modeled changing working conditions using CTMM and incorporated the results into the decision-making framework for adaptive CBM decisions [84]. Another focus when the dynamics of the workloads are of concern is to establish the relationship between the operating conditions and the degradation process of the system. This relationship is usually supposed to be totally known or at least particular functions with unknown parameters which can be estimated by using the observations of both operating conditions and degradation process. Jeffery pointed out that this relationship varies case-by-vase and should be determined according to the specific characteristics of the device [82]. When Wiener process was used as the degradation model, Si el al. assigned different drift coefficient values for the system in the state of usage and storage, respectively [83]. Besides, Arrhenius model and Eying model are frequently used to represent the relation between the degradation and working environments for electromechanical systems.

It is worth noting that the existing methods suffer some limitations. On one hand, the primary limit of CTMM is that the sojourn time in each state is exponentially distributed, which may be incompatible with the facts in practical applications in industry. To overcome this shortcoming, a semi-Markov model can be employed. Besides, when the operating information cannot be recorded directly, the according HMM and HsMM should be used to model the operation process of the system. On the other hand, with more and more complex structures of systems, the relationship between the degradation with operating conditions can be neither characterized by the simple existing laws, nor constructed through physical analysis, which may restrict the application of this kind of methods.

1.4.2 Degradation Modeling for System with Maintenances

Maintenance is an effective way to remove faults, reduce failure rates and improve the reliability throughout the lifetime of the system. Scientific and reasonable maintenance schedule can efficiently reduce the operating costs and the risks of the system, which also works for degrading systems [85]. Degradation modeling and maintenance activities of systems are closely related. On the one hand, the results of reliability evaluation and RUL estimation based on the degradation data offer the health evaluation information required for scheduling maintenance activities. On the other hand, maintenances improve the performance of the system and thus change the degradation path. To extend the application of the degradation model and improve the accuracy of RUL estimation, the influence of maintenance on degradation should be taken into consideration.

There are plenty of studies addressing preventive maintenance and optimal inspection based on degradation modeling [86–88], and some relate to the effects of maintenances on systems' performance [89, 90]. Popular models include the 'repair as new' model and the 'repair as old' model [91]. Both kinds of models assume that the performance of the system will be improved by maintenances, and the hazards functions are used to describe the effects of maintenance activities. In the 'repair as new' model, the system can be restored to the original state after a perfect maintenance, usually corresponding to the hazard increasing model [92]. The 'repair as old' models assume that maintenances on a system are imperfect so that the performance of the system recovers to a level worse than that of new system [93] Besides, the system degrades until the next maintenance or failure whichever comes first. Recently, Wang et al. employed a renewal-reward process perturbed by a diffusion, which is also defined as a Wiener process with random jumps elsewhere, to model the influence of maintenances [94]. The most inspiring idea in this work is that jumps with random arriving time and amplitude opposite to the trend of degradation are utilized. Under the concept of the FPT, the PDF of the RUL was provided based on simulation.

In consideration of degradation modeling under maintenance intervention, there are still many issues deserving further studies: (1) The time consumed by a maintenance which is ignored in most existing researches, is not negligible in some cases. Compared to occupying time of the degradation between two consecutive maintenances, the time consumed by maintenances is generally so short that it has always been ignored. The rationality of this assumption should be queried and studied when the maintenance time cannot be omitted. (2) The dependency of maintenance effects on the age of the system and time spent by maintenances should be considered. (3) Various maintenance activities such as minor repair, major repair, and replacement with different effects on the systems' degradation process exist in the maintenance policy of the system. The diversity of maintenance activities and the according influence on the degradation of systems should be treated discriminatingly and synthetically when estimating the RUL of the system.

1.5 Future Research Directions

Degradation modeling based RUL estimation, as the foundation and kernel technology of PHM, is now the focus of researches in reliability. With the development of complex systems, the extension of the sphere of human exploring activities, the diversification of workloads of the system and the increase in man–machine interaction, the system may experience more and more heterogeneity during its service. Thus, degradation modeling based RUL estimation approaches for systems with heterogeneity have been favored by many researchers and engineers. In this chapter, we reviewed the existing approaches in literature, from three aspects including methods considering unit-to-unit variability, methods considering the impact of heterogeneity in working environment and methods considering the impact of tasks and workloads. The pros and cons of these reviewed methods are discussed. From this survey, it is observed that the incorporation of heterogeneity into the degradation models enables more accurate RUL estimation under practical conditions, but plagues us with extra difficulties in the derivation, inference and computation of the PDF estimate of the RUL and the model parameters According to the current research results and the limitations of existing approaches, there are still a number of challenges and practical problems to be further studied. Specifically, it is concluded that the following research topics deserve the future studies and the fundamental theoretical problem behind these topics are still unsolved.

(1) It is desirable to develop age- and state-dependent prognosis models. There exist some industrial systems whose degradation processes are closely dependent on the age- and instantaneous degradation state of the degradation. However, the studies on age- and state-dependent degradation modeling are very limited in literature, as

opposed to a great deal of efforts made to age-dependent degradation models. Most recently, Giorgio et al. in [53, 54] made the first attempt along this direction and presented some Markov chain based degradation models whose transition probabilities between the process states depend on the current state and the current age of the system under study. It is worth pointing out that their developed models are only suitable to represent strictly monotonic degradation processes. However, in many industrial systems, a nonmonotonic degradation process, e.g., resulting from minor repair or reduced intensity of use, can provide a good description of the system's degradation signals. Besides, the continuous degradation process is approximated by a Markov chain with discrete degradation states in [53, 54]. This approximating process introduces many context-dependent parameters, which might pose difficulty in applications. Together with these discussions, it can be concluded that it is still challenging to develope age- and state-dependent degradation models for continuously degradation systems whose degradation progression might be nonmonotonic.

(2) The RUL estimating methods based on multi-source reliability data fusion are active demand for reliability engineers. After a period of service time of a system, various sources of reliability data can be collected, including lifetime data from accelerated lifetime test and the failed systems under industrial conditions, the degradation data from the accelerated degradation test and condition monitoring, and expert knowledge on system reliability, etc. These data, involving qualitative knowledge and quantitative information, contains abundant message reflecting the reliability of the system. However, effective and practical methods fusing these data to achieve a reasonable RUL estimation are still limited. The only exceptions are [95] where lifetime data from the accelerated lifetime test were used to assess lifetime distribution under working conditions and [96] where lifetime data of failed systems of the same category and degradation data are fused to estimate the RUL. Nevertheless, data fusion methods are still desired especially for degradation data from different working environments, and qualitative expert knowledge and degradation data of extremely expensive systems with high reliability.

(3) Another challenge is the RUL estimation for degrading systems with state changes. In engineering practice, a system may experience different working environments and changes of states in different modes. Si et al. introduced the state switches of an INS between usage and storage and the differences of degradation in each state [83]. The degradation process of the system in switching random environments was modeled in [97]. Li-ion batteries [98, 99] and OLED systems [100] alternate their states between the usage and storage. The capacitance/inner resistance of the Li-ion batteries decreases/increases with the cycles of charge and discharge, indicating the degradation of the performance in the usage state, while some recoveries reflected by increase/decrease of capacitance/inner occurred during the state of storage. Similar phenomena can also be observed in the degradation of OLEDs. To simplify computation, the state changes of the system and their influence are ignored in current studies, which pose biases and even errors to the estimated RUL. Though considering the state switches is difficult, it is of important theoretical significance and application prospect to take the state changes into account when the degradation process is modeled.

(4) To be more practical, it is desiderated to develop methods to estimate the RUL for systems with multiple failure modes and multiple components. The majority of engineering systems consist of many components, each of which may have several degradation modes that can lead to the failure of the whole system. Furthermore, there might be interrelationship among different components connected to each other. Whilst the failure of a degrading system is a result of interactions of inner complex structures and the external working conditions, most models are built for single-component systems with only one failure mode. Therefore, developing RUL estimation methods for multi-component systems with multi failure/degradation modes will speed up the application of theoretical methods to the practical systems.

References

1. Sandborn P, Pecht M (2007) Introduction to special section on electronic systems prognostics and health management. Microelectron Reliab 47(12):1847–1848
2. Dolev E (2009) Introduction to the special section on prognostics and health management. IEEE Trans Reliab 58(2):262–263
3. Lau D, Fong B (2011) Special issue on prognostics and health management. Microelectron Reliab 51(2):253–254
4. Wang W (2011) Special section on prognostics and systems health management (PHM), extended chapters from the PHM macau 2010 conference. IEEE Trans Reliab 60(1):2
5. Smith G, Schroeder J, Navarro S (1997) Development of a prognostics and health management capability for the Joint Strike Fighter. In: IEEE Autotestcon proceedings AUTOTESTCON, pp 676–682
6. Petch M (2008) Prognostics and health management of electronics. Wiley, New Jersey
7. Sun B, Zeng S, Kang R (2012) Benefits and challenges of system prognostics. IEEE Trans Reliab 61(2):323–335
8. Goebel K, Saha B, Saxena A (2008) Prognostics in battery health management. IEEE Instrum Meas Mag 11(4):33–40
9. Gebraeel N, Lawley M, Rong L (2005) Residual-life distributions from component degradation signals: a Bayesian approach. IIE Trans 37(6):543–557
10. Jardine K, Lin D, Banjevic D (2006) A review on machinery diagnostics and prognostics implementing condition-based maintenance. Mech Syst Signal Process 20(7):1483–1510
11. Si XS, Wang W, Hu CH, Zhou DH (2011) Remaining useful life estimation-a review on the statistical data driven approaches. Eur J Oper Res 213(1):1–14
12. Lu C, Meeker W (1993) Using degradation measures to estimate a time-to-failure distribution. Technometrics 35(2):161–174
13. Gebraeel N (2006) Sensory-updated residual life distributions for components with exponential degradation patterns. IEEE Trans Autom Sci Eng 3(4):382–393
14. Gebraeel N, Pan J (2008) Prognostic degradation models for computing and updating residual life distributions in a time-varying environment. IEEE Trans Reliab 57(4):539–550
15. Gebraeel N, Elwany A, Pan J (2009) Residual life predictions in the absence of prior degradation knowledge. IEEE Trans Reliab 58(1):106117
16. Son J, Zhou Q, Zhou S et al (2013) Evaluation and comparison of mixed effects model based prognosis for hard failure. IEEE Trans Reliab 62(2):379–794
17. Suk JB, Paul HK (2004) A nonlinear random-coefficients model for degradation testing. Technometrics 46(4):460469
18. Weaver BP, Meeker WQ, Escobar LA, Wendelberger J (2013) Method for planning repeated measures degradation studies. Technometrics 55(2):122–134

19. Weaver BP (2011) Methods for planning repeated measures degradation tests. Graduate Theses and Dissertations. Chapter 11967
20. Wang W, Christer A (2000) Towards a general condition based maintenance model for a stochastic dynamic system. J Oper Res Soc 51(4):145–155
21. Lawless J, Crowder M (2004) Covariates and random effects in a gamma process model with application to degradation and failure. Lifetime Data Anal 10(2):213–227
22. Peng WW, Li YF, Yang YJ, Huang HZ, Zuo MJ (2014) Inverse Gaussian process models for degradation analysis: a Bayesian perspective. Reliab Eng Syst Saf 130(1):175–189
23. Peng CY Inverse Gaussian Processes with Random Effects and Explanatory Variables for Degradation Data. Technometrics. http://dx.doi.org/10.1080/00401706.2013.879077
24. Wang X, Xu D (2010) An inverse Gaussian process model for degradation data. Technometrics 52(2):188197
25. Ye ZS, Chen N (2014) The inverse Gaussian process as a degradation model. Technometrics. doi:10.1080/00401706.2013.830074
26. Wang X (2010) Wiener processes with random effects for degradation data. J Multivar Anal 101(1):340–351
27. Si XS, Wang W, Hu CH, Zhou DH, Pecht M (2012) Remaining useful life estimation based on a nonlinear diffusion degradation process. IEEE Trans Reliab 61(1):5067
28. Wang ZQ, Wang W, Hu CH, Si XS (2014) An additive Wiener process-based prognostic model for hybrid deteriorating systems. IEEE Trans Reliab 63(1):208222
29. Peng CY, Tseng ST (2009) Mis-specification analysis of linear degradation models. IEEE Trans Reliab 58(3):444455
30. Si XS, Zhou DH (2013) A generalized result for degradation model-based reliability estimation. IEEE Trans Autom Sci Eng 11(2):632637
31. Huang X (2009) Diagnosis of random-effect model mis-specification in generalized linear mixed models for binary response. Biometrics 65(1):361368
32. McCulloch, John M, Neuhaus (2011) Mis-specifying the shape of a random effects distribution: why getting it wrong may not matter. Stat Sci 26(3):388402
33. Alonso A, Litiere S, Molenberghs G (2008) A family of tests to detect misspecifications in the random effects structure of generalized linear mixed models. Comput Stat Data Anal 52(1):44744486
34. Agrestia A, Brian C, Ohman-Stricklandc P (2004) Examples in which misspecification of a random effects distribution reduces efficiency, and possible remedies. Comput Stat Data Anal 47(1):639653
35. Sarma V, Kunhikrishnan K, Ramchand K (1979) A decision theory model for health monitoring of aero engines. J Aircr 16(3):222–224
36. Xu Z, Ji Y, Zhou D (2008) Real-time reliability prediction for a dynamic system based on the hidden degradation process identification. IEEE Trans Reliab 57(2):230–242
37. Luo J, Pattipati K, Qiao L et al (2008) Model-based prognostic techniques applied to a suspension system. IEEE Trans Syst Man Cybern Part A: Syst Hum 38(5):1156–1168
38. Tang L, DeCastro J, Kacprzynski G et al (2010) Filtering and prediction techniques for model-based prognosis and uncertainty management. In: IEEE prognostics and health management conference
39. Orchard M, Vachtsevanos G (2009) A particle-filtering approach for on-line fault diagnosis and failure prognosis. Trans Inst Meas Control 31(3–4):221–246
40. Phelps E, Willett P, Kirubarajan T et al (2007) Predicting time to failure using the IMM and excitable tests. IEEE Trans Syst Man Cybern Part A: Syst Hum 37(5):630–642
41. Wang W, Hussin B (2009) Plant residual time modelling based on observed variables in oil samples. J Oper Res Soc 60(6):789–796
42. Wang W, Matthew C, Xu WJ, Khairy K (2011) A model for residual life prediction based on Brownian motion with an adaptive drift. Microelectron Reliab 51(1):285–293
43. Si XS, Wang W, Hu CH, Chen MY, Zhou DH (2013) A Wiener process based degradation model with a recursive filter algorithm for remaining useful life estimation. Mech Syst Signal Process 35(1–2):219–237

44. Si, XS, Hu CH, Wang W, Chen MY (2011) An adaptive and nonlinear drift based Wiener process for remaining useful life estimation. In: Prognostics and system health management conference
45. Si XS, Wang W, Hu CH, Zhou DH (2014) Estimating remaining useful life with three-level variability in degradation modeling. IEEE Trans Reliab 63(1):167–190
46. Wang W (2007) A two-stage prognosis model in condition based maintenance. Eur J Oper Res 182(1):11771187
47. Feng J, Sun Q, Jin TD (2012) Storage life prediction for a high-performance capacitor using multi-phase Wiener degradation models. Commun Stat-Simul Comput 41(8):1317–1335
48. Li W, Pham H (2005) Reliability modeling of multi-state degraded system with multi competing failures and random shocks. IEEE Trans Reliab 54(2):297–303
49. Bunks, McCarthy D, Al-Ani T (2000) Condition-based maintenance of machines using hidden Markov models. Mech Syst Signal Process 14(4):597–612
50. Dong M, He D (2007) Hidden semi-Markov model-based methodology for multi-sensor equipment health diagnosis and prognosis. Eur J Oper Res 178(3):858–878
51. Dong M, He D (2007) A segmental hidden semi-Markov model (HSMM)-based diagnostics and prognostics framework and methodology. Mech Syst Signal Process 21(5):2248–2266
52. Dong M (2008) A novel approach to equipment health management based on autoregressive hidden semi-Markov model (AR-HSMM). Sci China Ser F: Inf Sci 51(9):1291–1304
53. Giorgio M, Guida M, Pulcini G (2011) An age- and state-dependent Markov model for degradation processes. IIE Trans 43(9):621–632
54. Giorgio M, Guida M, Pulcini G (2010) A parametric Markov chain to model age- and state-dependent wear processes. In: Pietro M, Piercesare S (eds) Complex data modelling and computationally intensive statistical methods. Springer, Milan
55. Gorjian N, Ma L, Mittinty M (2009) A review on reliability models with covariates
56. Ye ZS, Hong Y, Xie M (2013) How do heterogeneities in operating environments affect field failure predictions and test planning?. The Annals of Applied Statistics (to appear)
57. Banjevic D, Jardine A (2006) Calculation of reliability function and remaining useful life for a Markov failure time process. IMA J Manag Math 17(2):115–130
58. Elsayed E (2003) Mean residual life and optimal operating conditions for industrial furnace tubes. Case Studies in Reliability and Maintenance. Wiley, New Jersey, pp 497–515
59. Zhao X, Fouladirad M, Bérenguer C (2010) Condition-based inspection/replacement policies for nonmonotonous deteriorating systems with environmental covariates. Reliab Eng Syst Saf 95(8):921–934
60. Ghasemi A, Yacout S, Ouali M (2010) Evaluating the reliability function and the mean residual life for equipment with unobservable states. IEEE Trans Reliab 59(1):45–54
61. Lu XF, Liu M (2014) Hazard rate function in dynamic environment. Reliab Eng Syst Saf 130(1):50–60
62. Pafiee K, Feng QM, Coit DW (2014) Reliability modeling for dependent competing failure process with changing degradation rare. IIE Trans 46(1):483–496
63. Nakagawa T (2007) Shock and damage models in reliability theory. Springer, London
64. Liu Y, Huang HZ, Pham H (2006) Reliability evaluation of systems with degradation and random shocks. In: Proceedings of the annual reliability and maintainability symposium, pp 328–333
65. Klutke GA, Yang Y (2002) The availability of inspected systems subject to shocks and graceful degradation. IEEE Trans Reliab 51(3):371–374
66. Wang ZL, Du L, Huang HZ (2008) Reliability modeling for dependent competitive failure processes. In: Proceedings of the annual reliability and maintainability symposium, pp 278–282
67. Wang Z, Huang HZ, Li Y, Xiao NC (2011) An approach to reliability assessment under degradation and shock process. IEEE Trans Reliab 60(4):852–863
68. Cha W, Lee EY (2010) An extended stochastic failure model for a system subject to random shocks. Oper Res Lett 38(1):468–473

69. Murat K, Maillart LM (2009) Structured replacement policies for a Markov-modulated shock model. Oper Res Lett 38(1):280–284
70. Delia MC, Rafael PO, Segovia MC (2007) Survival probabilities for shock and wear models governed by phase-type distributions. Qual Technol Quant Manag 4(1):85–94
71. Javier RO, Mauricio SS, Raha AT (2014) Reliability analysis of shock-based deterioration using phase-type distributions. Probab Eng Mech 38(1):88–101
72. Huang W, Askin RG (2004) A generalized SSI reliability model considering stochastic loading and strength aging degradation. IEEE Trans Reliab 53(1):77–82
73. Li W, Pham H (2005) An inspection-maintenance model for systems with multiple competing processes. IEEE Trans Reliab 54(2):318–327
74. Chen YY, Li ZH (2008) An extended extreme shock maintenance model for a deteriorating system. Reliab Eng Syst Saf 93(1):1123–1129
75. Wang Y, Pham H (2011) A multi-objective optimization of imperfect preventive maintenance policy for dependent competing risk systems with hidden failure. IEEE Trans Reliab 60(4):770–781
76. Peng H, Feng Q, Coit DW (2011) Reliability and maintenance modeling for systems subject to multiple dependent competing failure processes. IIE Trans 43(1):12–22
77. Huynh KT, Barros A, Brenguer C et al (2011) A periodic inspection and replacement policy for system subject to degradation and traumatic events. Reliab Eng Syst Saf 96(4):497–508
78. Fan HD, Hu CH, Chen MY, Zhou DH (2011) Cooperative predictive maintenance of repairable systems with dependent failure modes and resource constraint. IEEE Trans Reliab 61(1):144–157
79. Perfetti G, Aubert T, Wildeboer W (2011) Influence of handling and storage conditions on morphological and mechanical properties of polymer-coated particles: characterization and modeling. Powder Technol 206(1):99–111
80. Kharoufeh J (2003) Explicit results for wear processes in a Markovian environment. Oper Res Lett 31(3):237–244
81. Kharoufeh J, Mixon D (2009) On a Markov-modulated shock and wear process. Nav Res Logist 56(6):563–576
82. Jeffrey P, Steven M (2005) Stochastic models for degradation-based reliability. IIE Trans 37(6):533–542
83. Si XS, Hu CH, Kong XY, Zhou DH (2014) A residual storage life prediction approach for systems with operation state switches. IEEE Trans Ind Electron. doi:10.1109/TIE.2014.2308135
84. Huynh K, Barros A, B'erenguer C (2012) Adaptive condition-based maintenance decision framework for deteriorating systems operating under variable environment and uncertain condition monitoring. Proc Inst Mech Eng Part O: J Risk Reliab 226(6):602–623
85. Scarf PA (2007) A Framework for condition monitoring and condition based maintenance. Qual Technol Quant Manag 4(2):301–312
86. Wang H (2002) A survey of maintenance policies of deteriorating systems. Eur J Oper Res 139(1):469–489
87. Wang W (2009) An inspection model for a process with types of inspections and repairs. Reliab Eng Syst Saf 94(1):526–533
88. Lam Y (1995) An optimal inspection-repair-replacement policy for standby systems. J Appl Probab 32(1):212–223
89. Endrenyi J, Anders GJ, Leite da Silva AM (1998) Probabilistic evaluation of the effect of maintenance on reliability-an application. IEEE Trans Power Syst 13(2):576–583
90. Vlok PJ, Wnek M, Zygmunt M (2004) Utilizing statistical residual life estimates of bearings to quantify the influence of preventive maintenance actions. Mech Syst Signal Process 18(1):833–847
91. Zhou XJ, Xi LF, Lee J (2007) Reliability-centered predictive maintenance scheduling for a continuously monitored system subject to degradation. Reliab Eng Syst Saf 92(1):530–534
92. Nakagawa T (1988) Sequential imperfect preventive maintenance policies. IEEE Trans Reliab 37(3):295–298
93. Mak M (1979) Reliable preventive maintenance policy. AIIE Trans 11(3):221–228

94. Wang ZQ, Hu CH, Wang W, Si XS (2014) A simulation-based remaining useful life prediction method considering the influence of maintenance activities. In: 2014 prognostics and system health management conference (PHM2014 Zhangjiajie), Zhangjiajie
95. Bagdonavicius, Nikulin MS (2002) Accelerated life models: modeling and statistical analysis. Chapman and Hall/CRC, Boca Raton
96. V. Couallier Some recent results on joint degradation and failure time modeling
97. Hawkes A, Cui L, Zheng Z (2011) Modeling the evolution of system reliability performance under alternative environments. IIE Trans 43(11):761–772
98. Eddahech A, Briat O, Vinassa M (2013) Lithium-Ion battery performance improvement based on capacity recovery exploitation. Electrochim Acta. doi:S0013-4686(13)02071-9
99. Liu DT, Pang JY, Zhou JB, Peng Y, Pecht M (2013) Prognostics for state of health estimation of lithium-ion batteries based on combination Gaussian process functional regression. Microelectron Reliab 53(1):832–839
100. Rao KS, Mohapatra YN (2014) Disentangling degradation and auto-recovery of luminescence in Alq3 based organic light emitting diodes. J Lumin 145(1):793–796

Chapter 2
Planning Repeated Degradation Testing for Degrading Products

2.1 Introduction

Degradation information regarding the system's health state, especially from highly reliable items, has been a useful alternative for the system's remaining useful life (RUL) estimation, as well as a valuable basis for condition-based maintenance (CBM). Once the degradation information of a system is available by the degradation test, one well-recognized method is to establish a stochastic degradation model to predict the distributions of the future degradation and the associated lifetime, based on the relationship between the degradation and failure time. However, the accuracy of the aforementioned degradation or lifetime distributions is heavily influenced by the accuracy of the parameter estimation, which is affected by the number of items and the sampling frequency of each item. Therefore, to achieve a satisfactory prognosis accuracy, engineers need to decide how many items should be measured and how often should the measurements be made, before the degradation studies are performed [1]. In addition, the degradation test is usually costly, particularly for highly valued systems or vital components. In this case, how to achieve a tradeoff between the limited fund and the required estimation accuracy for important statistics of interest is also an interesting problem deserving in-depth studies.

Researchers and engineers have paid much attention to the degradation test design, particularly in the field of accelerated degradation test planning. Meeker et al. in [2] discussed the modeling and analysis issues of accelerated degradation test. Tseng and Yu in [3, 4] proposed several appropriate termination rules for degradation experiments. The works in [5, 6] considered optimal step-stress accelerated degradation test design for Wiener process and Gamma process, respectively. In the works of [5, 6], the temporal variability in stochastic degradation characteristics is involved, while both the unit-to-unit diversity and the measurements variability are ignored. Shi et al. in [7] studied the test planning methods for accelerated destructive degradation, where only the measurement errors were taken into account. Recently, Bayesian methods for designing accelerated destructive degradation test have also been developed by [8]. In addition, Weaver et al. in [1] documented several useful methods for

X.-S. Si et al., *Data-Driven Remaining Useful Life Prognosis Techniques*,
Springer Series in Reliability Engineering, DOI 10.1007/978-3-662-54030-5_2

planning repeated degradation tests. In these two works, both diversities among units and measurement errors are considered, but the temporal variability is ignored in the process of stochastic degradation modeling.

In general, the degradation process of an item is stochastic. As a result, the lifetime and the degradation in the future are also random variables, resulting in the difficulty to predict the degradation in the future and estimate the lifetime with certainty. As summarized by [9], there are three sources of variability contributing to the uncertainties of degradation modeling and prognosis: (1) temporal variability; (2) unit-to-unite variability (usually modeled as random effect); and (3) measurement variability. The temporal variability is referred to as the inherent stochastic characteristics of the associated degradation process over time [10]. The unit-to-unit variability determines the heterogeneity among the degradation paths of different units [11]. The measurement variability describes the randomness in the measured data of the degradation, which might be contaminated by the uncertainty during the measurement process [12]. Therefore, a reasonable and appropriate degradation model for prognosis has to take into account these sources of variability. It has been found in [9] that degradation modeling with three-source variability shows great potentiality in improving the accuracy of the lifetime estimation.

By the above survey over recent advances in degradation modeling and degradation test design planning, it can be observed that, though a significant volume of research regarding planning degradation test has appeared so far, there is no literature addressing the problem of planning repeated degradation test for products whose degradation measurements exhibiting three-source variability. Such a method for planning repeated degradation test is useful and desired, particularly for the case that the concerned system is highly valued but with limited fund conducting extensive degradation tests.

In response to the above desire, this chapter considers the problem of planning repeated degradation test for degrading products with three-source variability. The degradation process is modeled as a Wiener process with a random linear drift coefficient and a constant volatility coefficient, while the measurement errors are described as additive zero-mean random variables. Based on the presented model, the lifetime distribution is formulated under the concept of the first passage time (FPT). After the model parameters are estimated by the expectation maximization (EM) algorithm, the large-sample approximate standard errors (ASE) of the maximum likelihood estimation (MLE) for the mean failure time and the quantile of degradation distribution are derived, respectively. Then, we take into account the relationship between the performance of the measurement errors and its cost, after which we propose a constrained optimum designing model by minimizing the test cost under the condition of a maximum acceptable ASE. An example is provided to illustrate the procedure and advantages of the proposed planning method.

In summary, the contributions of the chapter mainly include two aspects. On the one hand, based on a relatively general deteriorating model, we proposed a method for repeated degradation test planning for systems with three-source variability, which has not been considered before but such gap is filled by this work. On the other hand, we introduced the measurement error into the constrained optimization model for the first time. As such, the relationship between the accuracy and cost of measurement has been modeled and brought into the optimization of the test plans. Through such modeling, the proposed method can provide some practical guidance on repeated degradation tests panning for both researchers and engineers.

The remainder of this chapter is structured as follows. Section 2.2 describes the degradation model with three-source variability. The parameter estimation and information matrix derivation framework are presented in Sect. 2.3. Section 2.4 proposes the estimation methods for both the degradation distribution and the lifetime distribution. The method to choose an optimal degradation test plan is discussed in Sect. 2.5. Section 2.6 provided a numerical example for illustration.

2.2 Degradation Modeling with Three-Source Variability

In this chapter, the degradation process is modeled as a Wiener process with a linear drift, which has been widely used in the field of reliability. As one kind of stochastic-process-based degradation models, Wiener process has many favorable properties including the ability to handle nonmonotonous deteriorating process, an analytical result for the FPT distribution, and strong Markovian property, etc. In addition, it is very easy to extend Wiener process to nonlinear diffusion processes, and to incorporate the measurement errors, covariates as well as random effects. A comprehensive review of Wiener process as a degradation model can be founded in [13].

In this study, suppose that there are totally n units for the test and the i-th unit is measured m_i times. Let D_{ij} be the degradation state of the i-th unit at time t_{ij}, where $i = 1, \ldots, n$ and $j = 1, \ldots, m_i$. Then, based on the Wiener process, D_{ij} can be modeled as

$$D_{ij} = \lambda_{0i} + \lambda_{1i} t_{ij} + \sigma_B B(t_{ij}), \tag{2.1}$$

where λ_{0i} and λ_{1i} are the initial state (intercept) and the deteriorating rate (slope) of the ith unit, respectively; σ_B is the diffusion parameter which is a constant used to capture the level of the temporal variability, and $B(t)$ is a standard Brownian motion. Note that a linear drift $\lambda_{0i} + \lambda_{1i} t_{ij}$ is used here for the reason that many nonlinear degradation processes can be converted into an approximated linear form as Eq. (2.1) by some transformation techniques (see [14] for more details).

To describe the unit-to-unit variability, random effects are incorporated. Particularly, it is assumed that the intercept λ_{0i} and the slop λ_{1i} are treated as the random realizations following the bivariate-normal distribution $[\lambda_{0i}, \lambda_{1i}]^\tau \sim BVN(\boldsymbol{\lambda}, \boldsymbol{V})$, where $\boldsymbol{\lambda} = [\lambda_0, \lambda_1]^\tau$ represents the common properties of a specific kind of units in interpret and slope, $[\cdot]^\tau$ denotes the transpose of a vector or a matrix, and $\boldsymbol{V}$ is the covariance matrix characterizing the unit-to-unit variability with

$$\boldsymbol{V} = \begin{bmatrix} \sigma_{\lambda_0}^2 & \rho\sigma_{\lambda_0}\sigma_{\lambda_1} \\ \rho\sigma_{\lambda_0}\sigma_{\lambda_1} & \sigma_{\lambda_1}^2 \end{bmatrix}.$$

It is worth noting that, for convenience, the normally distributed random variables are used to depict the random effects. Actually, any other forms of distributions can be treated as random effects [11], but some difficulty in derivation may be involved.

In practice, the degradation state is usually partly observable. In other words, the actual observation of D_{ij}, denoted by y_{ij} is frequently influenced by the measurement errors. Thus, the degradation measurement model can be formulated as

$$y_{ij} = D_{ij} + \varepsilon_{ij}, \tag{2.2}$$

where ε_{ij} is also normally distributed with the mean 0 and variance σ^2, representing the measurement errors.

By now, the temporal variability, unit-to-unit variability, and measurement variability are all incorporated into the degradation model. For the notation convenience, we gather the observations of unit i into $\boldsymbol{Y}_i = [y_{i1}, y_{i2}, \ldots, y_{im_i}]^T$ and give an equivalent formulation of Eq. (2.2) as

$$\boldsymbol{Y}_i = \mathbf{Z}_i\boldsymbol{\lambda}_i^* + \boldsymbol{\xi}_i + \boldsymbol{\varepsilon}_i, \tag{2.3}$$

where $\boldsymbol{\lambda}_i^* = [\lambda_{0i}^*, \lambda_{1i}^*]^\tau$ is a random vector drawn from the bivariate distribution $BVN(\boldsymbol{\lambda}, \boldsymbol{V})$, $\boldsymbol{\xi}_i$ is another multivariate-normal distributed random vector following $MVN\left(\mathbf{0}, \sigma_B^2 \boldsymbol{T}_i\right)$ with

$$\boldsymbol{T}_i = \begin{bmatrix} t_{i1} & t_{i1} & \cdots & t_{i1} \\ t_{i1} & t_{i2} & \cdots & t_{i2} \\ \vdots & \vdots & \ddots & \vdots \\ t_{i1} & t_{i2} & \cdots & t_{im_i} \end{bmatrix}.$$

$\mathbf{Z}_i$ are the design matrix of the form

$$\mathbf{Z}_i = \begin{bmatrix} 1 & t_{i1} \\ \vdots & \vdots \\ 1 & t_{im_i} \end{bmatrix},$$

and $\boldsymbol{\varepsilon}_i = [\varepsilon_{i1}, \cdots, \varepsilon_{im_i}]^\tau$.

It is reasonable to assume that $\boldsymbol{\varepsilon}_i$, $\boldsymbol{\lambda}_i^*$ and $\boldsymbol{\xi}_i$ are mutually independent, and the components of $\boldsymbol{\varepsilon}_i$ are also independent and normally distributed with $\boldsymbol{\varepsilon}_i \sim MVN\left(\mathbf{0}, \sigma^2 \boldsymbol{I}_i\right)$, where $\boldsymbol{I}_i$ is a $m_i \times m_i$ identity matrix. As such, the observation $\boldsymbol{Y}_i$ follows a multivariate normal distribution $MVN\left(\mathbf{Z}_i\boldsymbol{\lambda}, \boldsymbol{\Sigma}_i\right)$ with

$$\boldsymbol{\Sigma}_i = \mathbf{Z}_i \boldsymbol{V} \mathbf{Z}_i^T + \sigma_B^2 \boldsymbol{T}_i + \sigma^2 \boldsymbol{I}_i.$$

Based on the above general model description, we elaborate the issues of the parameter estimation and the associated derivation of the information matrix in the following section.

2.3 Parameter Estimation and Information Matrix

Based on the structure of the presented degradation model involving three-source variability, the EM algorithm is chosen here to estimate the model parameters. Usually, the parameters with random effects are treated as unobservable variables. After an initial guess of the parameters, the expectation of the complete likelihood function for both observable and unobservable variables is computed conditional on the available observations in the E-Step, followed by the M-Step which updates the guess of the parameters by maximizing the conditional expectation. The specific algorithm for estimating the model parameters is omitted here due to the limited space. For the details, see [15–17].

Since the focus of this chapter is on test planning, we now check the variance-covariance matrix of the estimated parameters. With n independent observations $[\boldsymbol{y}_1, \ldots, \boldsymbol{y}_n]$ from $\boldsymbol{Y}_i$, the log-likelihood for unit i

$$\mathscr{L}_i = -\frac{1}{2}\log\left[\det \boldsymbol{\Sigma}_i\right] - \frac{1}{2}\left(\boldsymbol{y}_i - \mathbf{Z}_i\boldsymbol{\lambda}\right)^{\tau} \boldsymbol{\Sigma}_i^{-1}\left(\boldsymbol{y}_i - \mathbf{Z}_i\boldsymbol{\lambda}\right), \tag{2.4}$$

and the total log-likelihood for n units is

$$\begin{aligned}\mathscr{L} &= \sum_{i=1}^{n} \mathscr{L}_i \\ &= -\frac{1}{2}\sum_{i=1}^{n}\log\left[\det \boldsymbol{\Sigma}_i\right] - \frac{1}{2}\sum_{i=1}^{n}\left(\boldsymbol{y}_i - \mathbf{Z}_i\boldsymbol{\lambda}\right)^{\tau} \boldsymbol{\Sigma}_i^{-1}\left(\boldsymbol{y}_i - \mathbf{Z}_i\boldsymbol{\lambda}\right). \end{aligned} \tag{2.5}$$

Denote $\boldsymbol{\theta}^{\tau} = [\boldsymbol{\lambda}^{\tau}, \boldsymbol{\vartheta}^{\tau}]$ as the parameter vector to be estimated, where $\boldsymbol{\vartheta} = [\sigma_{\lambda_0}, \sigma_{\lambda_1}, \rho, \sigma_B, \sigma]^{\tau}$. From the large-sample theory, the large-sample approximate covariance matrix of the MLE can be formulated as

$$Avar(\hat{\boldsymbol{\theta}}) = [\mathscr{I}(\boldsymbol{\theta})]^{-1}, \tag{2.6}$$

where $\hat{\boldsymbol{\theta}}$ is the ML estimator of $\boldsymbol{\theta}$, and $\mathscr{I}(\boldsymbol{\theta}) = \sum_{i=1}^{n} \mathscr{I}_i$ is the Fisher information matrix with the definition

$$\mathscr{I}_i(\boldsymbol{\theta}) = E_Y\left[-\frac{\partial^2 \mathscr{L}_i}{\partial \boldsymbol{\theta}^2}\right]. \tag{2.7}$$

Setting $\boldsymbol{\mu}_i = \mathbf{Z}_i\boldsymbol{\lambda}$ for simplicity, the specific form of the information matrix can then be expressed as

$$\mathscr{I}_i(\boldsymbol{\lambda}, \boldsymbol{\vartheta}) = diag\left[\mathscr{I}_i(\boldsymbol{\lambda}), \mathscr{I}_i(\boldsymbol{\vartheta})\right], \tag{2.8}$$

where $diag(\cdot)$ is a diagonal matrix with blocks, and

$$\mathscr{I}_i(\boldsymbol{\lambda})_{j,k} = \frac{\partial \boldsymbol{\mu}_i^{\tau}}{\partial \boldsymbol{\lambda}_j} \boldsymbol{\Sigma}_i \frac{\partial \boldsymbol{\mu}_i}{\partial \boldsymbol{\lambda}_k}, \; 1 \le j, k \le 2,$$

$$\mathscr{I}_i(\boldsymbol{\vartheta})_{j,k} = \frac{1}{2} tr \left(\boldsymbol{\Sigma}_i^{-1} \frac{\partial \boldsymbol{\Sigma}_i}{\partial \boldsymbol{\vartheta}_j} \boldsymbol{\Sigma}_i^{-1} \frac{\partial \boldsymbol{\Sigma}_i}{\partial \boldsymbol{\vartheta}_k} \right), \; 1 \le j, k \le 5,$$

respectively.

An estimator of $Avar(\hat{\boldsymbol{\theta}})$ denoted by $\widehat{var}(\hat{\boldsymbol{\theta}})$ can be achieved by substituting ML estimator $\widehat{\boldsymbol{\theta}}$ into Eq. (2.6). It is worth pointing out here that $\hat{\boldsymbol{\theta}}$ is the MLE of the model parameters and such estimating process is achieved by the EM algorithm. $\widehat{var}(\hat{\boldsymbol{\theta}})$ is the foundation to evaluate the standard error of some other important quantiles, which will be shown in the following sections.

2.4 Estimating the Degradation Distribution and Lifetime Distribution

The degradation distribution and the lifetime distribution, without which many statistical inference will be impossible, are two most important concerns in prognosis. For the mean of lifetime and the quantile of the degradation distribution, a smaller standard error corresponds to a better degradation test plan. Therefore, we evaluate the degradation distribution and lifetime distribution, as well as some important quantiles which will be used as criteria of test planning.

2.4.1 The Quantiles of Degradation Distribution and Its Variance

We begin with the quantile of the degradation distribution. From Eq. (2.1), it follows that the degradation at time t is drawn from a normal distribution with mean $E[D(t)] = \lambda_0 + \lambda_1 t$ and variance $var[D(t)] = \sigma_{\lambda_0}^2 + \sigma_{\lambda_1}^2 t^2 + 2t\rho\sigma_{\lambda_0}\sigma_{\lambda_1} + t\sigma_B^2$. As a result, the p quantile of the degradation distribution at time t can be formulated by

$$d_p(t) = \lambda_0 + \lambda_1 t + \Phi^{-1}(p)\sqrt{\sigma_{\lambda_0}^2 + \sigma_{\lambda_1}^2 t^2 + 2t\rho\sigma_{\lambda_0}\sigma_{\lambda_1} + t\sigma_B^2}, \tag{2.9}$$

with the inverse standard normal cumulative distribution function $\Phi^{-1}(p)$.

Therefore, the ML estimator of $d_p(t)$, denoted by $\widehat{d_p}(t)$, can be accessed by computing Eq. (2.9) at the MLE $\hat{\boldsymbol{\theta}}$.

From Eq. (2.9), we note that the quantile $d_p(t)$ is a function of parameters $\boldsymbol{\theta}$, which means the formula for the approximated standard error (ASE) of $\widehat{d_p}(t)$ can be derived through the delta method. Thus, the large-sample approximate variance of $\widehat{d_p}(t)$ can be formulated as

$$Avar(\widehat{d_p}(t)) = \left[\frac{\partial d_p(t)}{\partial \boldsymbol{\theta}}\right]^{\tau} Avar(\hat{\boldsymbol{\theta}}) \left[\frac{\partial d_p(t)}{\partial \boldsymbol{\theta}}\right], \tag{2.10}$$

where $\frac{\partial d_p(t)}{\partial \boldsymbol{\theta}}$ is the partial deviation of $d_p(t)$ with respect to $\boldsymbol{\theta}$.

Equation (2.10) indicates that the standard error of $\widehat{d_p}(t)$ is $ASE_{\widehat{d_p}} = \sqrt{Avar(\widehat{d_p}(t))}$, which can be estimated by evaluating Eq. (2.10) at $\hat{\boldsymbol{\theta}}$ as $\widehat{SE}_{\widehat{d_p}} = \sqrt{\widehat{var}(\widehat{d_p})}$. Note that $\hat{\boldsymbol{\theta}}$ is the estimation for the standard error of $\widehat{d_p}(t)$ and $\widehat{var}(\widehat{d_p})$ is the estimation of $Avar(\widehat{d_p}(t))$. The specific forms of $\partial d_p(t)/\partial\theta_k$ are listed as

$$\begin{aligned}
&\frac{\partial d_p(t)}{\partial \lambda_0} = 1, \; \frac{\partial d_p(t)}{\partial \lambda_1} = t,\\
&\frac{\partial d_p(t)}{\partial \sigma_{\lambda_0}} = \frac{\Phi^{-1}(p)(\sigma_{\lambda_0} + t\rho\sigma_{\lambda_1})}{\xi},\\
&\frac{\partial d_p(t)}{\partial \sigma_{\lambda_1}} = \frac{\Phi^{-1}(p)(\sigma_{\lambda_1} + t\rho\sigma_{\lambda_0})}{\xi},\\
&\frac{\partial d_p(t)}{\partial \rho} = \frac{\Phi^{-1}(p)t\sigma_{\lambda_0}\sigma_{\lambda_1}}{\xi},\\
&\frac{\partial d_p(t)}{\partial \sigma_B} = \frac{\Phi^{-1}(p)\sigma_B t}{\xi}, \; \frac{\partial d_p(t)}{\partial \sigma} = 0,
\end{aligned}$$

with $\xi = \sqrt{\sigma_{\lambda_0}^2 + \sigma_{\lambda_1}^2 t^2 + 2t\rho\sigma_{\lambda_0}\sigma_{\lambda_1} + t\sigma_B^2}$.

Based on the availability for $\widehat{d_p}(t)$ and $\widehat{SE}_{\widehat{d_p}}$, a large-sample approximation $100(1-\alpha)\%$ confidence interval for $d_p(t)$ can be also achieved by

$$[\underline{d_p}(t), \overline{d_p}(t)] = \widehat{d_p}(t) \pm z_{(1-\alpha/2)}\widehat{SE}_{\widehat{d_p}}, \tag{2.11}$$

where $z_{(1-\alpha/2)}$ is the $(1-\alpha/2)$ standard normal quantile.

2.4.2 The Lifetime Distribution

For the degradation process with soft failure, the life of a unit ends when its performance degradation process hits a preset threshold ω, known as the failure threshold. Therefore, the lifetime is defined as the FPT of the degradation process crossing ω:

$$T = \inf\{t : D(t) \geq \omega | D(0) < \omega\}.$$

The work in [9] developed a useful method to derive the lifetime distribution for units with the degradation process described by Eq. (2.1). The probability density function (PDF) of lifetime $f_L(l)$ and the cumulative density function (CDF) $F_L(l)$

can be evaluated using the law of total probability, given the corresponding results of the degradation model ignoring the random effect in parameters.

Thus, we begin the discussion with a simplified model without considering random parameters. Then, the PDF and CDF of lifetime corresponding to the simplified model are respectively formulated as

$$f_{L|S}(l) = \frac{\omega - \lambda_0}{\sqrt{2\pi l^3 \sigma_B^2}} \exp\left(-\frac{(\omega - \lambda_0 - \lambda_1 l)^2}{2\sigma_B^2 l}\right), \tag{2.12}$$

and

$$F_{L|S}(l) = 1 - \Phi\left(\frac{\omega - \lambda_0 - \lambda_1 l}{\sigma_B \sqrt{l}}\right) + \exp\left(\frac{2\lambda_1(\omega - \lambda_0)}{\sigma_B^2}\right)\Phi\left(\frac{-\omega + \lambda_0 - \lambda_1 l}{\sigma_B \sqrt{l}}\right). \tag{2.13}$$

Equation (2.12) shows that the lifetime is an inverse Gaussian distribution with mean $\frac{\omega-\lambda_0}{\lambda_1}$ and covariance $\frac{(\omega-\lambda_0)\sigma_B^2}{\lambda_1^3}$. Due to the law of total probability, and the results in [17], when the random effect is considered, the PDF of lifetime $f_L(l)$ can be formulated as

$$f_L(l) = \frac{W\left(\sigma_B^2 l + \boldsymbol{b}^T \boldsymbol{V} \boldsymbol{b} - \boldsymbol{a}^T \boldsymbol{V} \boldsymbol{b}\right) - \sigma_B^2 l \boldsymbol{a}^T \boldsymbol{\lambda} - \boldsymbol{b}^T \boldsymbol{V} \boldsymbol{b} \boldsymbol{a}^T \boldsymbol{\lambda} + \boldsymbol{a}^T \boldsymbol{V} \boldsymbol{b} \boldsymbol{b}^T \boldsymbol{\lambda}}{\sqrt{2\pi\left(\sigma_B^2 l + \boldsymbol{b}^T \boldsymbol{V} \boldsymbol{b}\right)^3}} \times \exp\left[-\frac{\left(W - \boldsymbol{b}^T \boldsymbol{\lambda}\right)^2}{2\left(\sigma_B^2 l + \boldsymbol{b}^T \boldsymbol{V} \boldsymbol{b}\right)}\right], \tag{2.14}$$

where $\boldsymbol{a}^T = [1, 0]$ and $\boldsymbol{b}^T = [1, l]$ for brevity.

Accordingly, the CDF of the lifetime $F_L(l)$ can be obtained as

$$F_L(l) = 1 - \Phi\left(\frac{c_1 + \boldsymbol{d}_1^T \boldsymbol{\lambda}}{\sqrt{1 + \boldsymbol{d}_1^T \boldsymbol{V} \boldsymbol{d}_1}}\right) + \frac{1}{|\boldsymbol{A}\boldsymbol{V}|^{\frac{1}{2}}} \exp\left[\frac{2\boldsymbol{\lambda}^T \boldsymbol{B} \boldsymbol{\lambda} + 2\boldsymbol{\lambda}^T \boldsymbol{a}_2 + (2\boldsymbol{B}\boldsymbol{\lambda} + \boldsymbol{a}_2)^T \boldsymbol{a}_2^{-1} (2\boldsymbol{B}\boldsymbol{\lambda} + \boldsymbol{a}_2)}{2}\right] \times \Phi\left(\frac{c_2 + \boldsymbol{d}_2^T \boldsymbol{\lambda} + \boldsymbol{d}_2^T \boldsymbol{A}^{-1}(2\boldsymbol{B}\boldsymbol{\lambda} + \boldsymbol{a}_2)}{\sqrt{1 + \boldsymbol{d}_2^T \boldsymbol{A} \boldsymbol{d}_2}}\right), \tag{2.15}$$

with parameters $c_1 = \frac{W}{\sigma_B \sqrt{l}}$, $\boldsymbol{d}_1 = -\frac{1}{\sigma_B^2 \sqrt{l}}[1 \;\; l]$, $c_2 = -\frac{W}{\sigma_B \sqrt{l}}$, $\boldsymbol{a}_2 = [0 \;\; W]$, $\boldsymbol{d}_2 = -\frac{1}{\sigma_B^2 \sqrt{l}}[-1 \;\; l]$, $\boldsymbol{A} = \boldsymbol{V}^{-1} - 2\boldsymbol{B}$, and

$$\boldsymbol{B} = \begin{bmatrix} 0 & -1 \\ 0 & 0 \end{bmatrix}.$$

Even in the situation where the unit-to-unit variability is ignored, no analytical-form quantile of the lifetime distribution can be obtained, which leads to either the use of iterative algorithm or an approximation of the distribution itself [13]. By contrast, the mean of the lifetime, which is frequently used in statistical inference and decision-making, should be estimated with a requirement of the degradation test performance. In this chapter, the standard error of the MLE for the mean lifetime is considered in the planning of the degradation test.

Thanks to the law of total probability, we can obtain the mean lifetime as

$$E_L = \rho\sigma_{\lambda_0} + \frac{\sqrt{2}(\omega + \rho\lambda_1\sigma_{\lambda_0} - \lambda_0\sigma_{\lambda_1})}{\sigma_{\lambda_1}} G\left(\frac{\lambda_1}{\sqrt{2}\sigma_{\lambda_1}}\right), \tag{2.16}$$

where $G(x) = \exp(-x^2)\int_0^x \exp(u^2)du$ is the Dawson integral for all real x.

Furthermore, the delta method is employed to estimate the standard error for the MLE of the mean lifetime $\widehat{E_L}$. Then, the large-sample approximate variance of $\widehat{E_L}$ is formulated as

$$Avar(\widehat{E_L}) = \left[\frac{\partial E_L}{\partial \boldsymbol{\theta}}\right]^T Avar(\hat{\boldsymbol{\theta}}) \left[\frac{\partial E_L}{\partial \boldsymbol{\theta}}\right]. \tag{2.17}$$

The specific forms of $\frac{\partial E_L}{\partial \theta_k}$ are given as

$$\begin{aligned}
\frac{\partial E_L}{\partial \lambda_0} &= -\sqrt{2} G\left(\frac{\lambda_1}{\sqrt{2}\sigma_{\lambda_1}}\right), \\
\frac{\partial E_L}{\partial \lambda_1} &= \left[-\frac{\lambda_1}{\sigma_{\lambda_1}^2} G\left(\frac{\lambda_1}{\sqrt{2}\sigma_{\lambda_1}}\right) + G\left(\frac{\lambda_1}{\sqrt{2}\sigma_{\lambda_1}}\right) + \frac{1}{\sqrt{2}\sigma_{\lambda_1}}\right] \varXi \\
&\quad + \frac{\sqrt{2}\rho\sigma_{\lambda_0}}{\sigma_{\lambda_1}} G\left(\frac{\lambda_1}{\sqrt{2}\sigma_{\lambda_1}}\right), \\
\frac{\partial E_L}{\partial \sigma_{\lambda_0}} &= \frac{\sqrt{2}\rho\lambda_1}{\sigma_{\lambda_1}^2} G\left(\frac{\lambda_1}{\sqrt{2}\sigma_{\lambda_1}}\right) + \rho, \\
\frac{\partial E_L}{\partial \sigma_{\lambda_1}} &= -G\left(\frac{\lambda_1}{\sqrt{2}\sigma_{\lambda_1}}\right) \frac{\sqrt{2}(\omega + \rho\sigma_{\lambda_0})}{\sigma_{\lambda_1}^2} \\
&\quad + \varXi\left[G\left(\frac{\lambda_1}{\sqrt{2}\sigma_{\lambda_1}}\right)\frac{\lambda_1^2}{\sigma_{\lambda_1}^3} + G\left(\frac{\lambda_1}{\sqrt{2}\sigma_{\lambda_1}}\right) - \frac{\lambda_1}{\sqrt{2}\sigma_{\lambda_1}^2}\right], \\
\frac{\partial E_L}{\partial \rho} &= \frac{\sqrt{2}\sigma_{\lambda_0}\lambda_1}{\sigma_{\lambda_1}} G\left(\frac{\lambda_1}{\sqrt{2}\sigma_{\lambda_1}}\right) + \sigma_{\lambda_0},
\end{aligned}$$

$$\frac{\partial E_L}{\partial \sigma_B} = 0, \frac{\partial E_L}{\partial \sigma} = 0,$$

where $\Xi = \frac{\sqrt{2}(W+\rho\sigma_{\lambda_0}\sigma_{\lambda_1}-\lambda_0\sigma_{\lambda_1})}{\sigma_{\lambda_1}}$.

By the above derivations, the standard error of the MLE for $\widehat{E_L}$, saying $\widehat{SE}_{\widehat{E_L}}$, can be evaluated as $\widehat{SE}_{\widehat{E_L}} = \sqrt{\widehat{var}(\widehat{E_L})}$, where $\widehat{var}(\widehat{E_L})$ is the MLE of $Avar(\widehat{E_L})$ and can be achieved by calculating Eq. (2.17) at $\widehat{\boldsymbol{\theta}}$. Based on these results, we determine the degradation test planning in the following section.

2.5 Degradation Test Planning

The degradation test plan depends on the parameters of the model, which are not exactly known when we make the planning. Prior knowledge of the degradation model including previous experience (such as measurements of the degradation burn-in test, and degradation data from the field systems). In this case, expert knowledge and design specifications are used to choose a set of parameters as planning information before the planning begins, denoted by $\boldsymbol{\theta}^{\square}$. This set of parameters is known as the planning information.

Given the planning information $\boldsymbol{\theta}^{\square}$ and a test plan with specific number of units as well as measurement schedule of each unit, we can compute the corresponding fisher information matrix, which facilitates the evaluation of standard error for the mean of lifetime and the quantile of the degradation distribution. For simple test plans where all units are measured using the same schedule, a contour plot of the large-sample approximate standard error of the interesting statistical test-plan properties against the number of units n and the number of times each unit measured m is recommended for test selection in [1]. For test plans where units are measured using different schedules, the plots of standard errors against time are compared to choose a reasonable test plan.

The planning problem, when there is a constraint on standard error of the statistical test-plan properties and a desire to minimize the cost of degradation test, is often encountered. Generally speaking, the cost of a degradation test depends on the number of the tested units, the measurement times on each unit, and the used measurement method. Therefore, suppose that the relationship between the standard error of measurement and the cost of measurement method is formulated as

$$cost(\sigma) = C - \kappa \exp(\eta\sigma), \tag{2.18}$$

where C, κ, η are positive constant coefficients which can be determined based on experience.

From Eq. (2.18), it can be noted that a smaller σ corresponds to higher performance of the measurement method. Equation (2.18) indicates that higher the performance measure method is, the harder the performance of the measurement method can be further improved. This phenomenon is consistent with most of the industrial cases. When the performance level of the degradation measurement is low, the cost will increase rapidly to improve the precision of equipment. However, the rate of increase slows down as the performance of the measurement equipment increases.

Based on the above discussions, the cost of a test plan can be formulated as

$$cost(n, \boldsymbol{m}) = c_1 + c_2 n + \sum_{k=1}^{n} cost(\sigma) m_k, \tag{2.19}$$

where $\boldsymbol{m}$ is the collection of the number of measurements of each unit, i.e., $[m_1, \ldots, m_n]^{\tau}$. c_1, and c_2 denote the fixed cost of running the test and the cost of testing each unit, respectively. However, it is noted that other models describing the relationship between the standard error of measurement and the cost of measurement method can also be used, and the proposed framework in this chapter is not limited to the model as (2.18).

Then, the problem of selecting test plan can be formulated as the following constrained optimization problem:

$$\begin{aligned} [\boldsymbol{m}^*, n^*] &= \arg\min_{m,n} [cost(n, \boldsymbol{m})] \\ s.t. \quad & SE \leq \gamma, \end{aligned} \tag{2.20}$$

where γ is the maximum acceptable value for ASE and can be determined according to the specific yet practical requirement.

By optimizing (2.19), we can obtain the optimal test plan which gives the optimal number of the tested units and the testing times for each unit.

2.6 An Illustrative Example

To illustrate the procedure for planning repeated degradation test with there-source variability, we give an example of degradation test planning based on $d_{0.1}(t)$ in this section. For the purpose of comparison, the same planning information of shelf-life test design in example 3 from [1] together with a new setting of σ_B is used here. Since $d_{0.1}(t)$ depends on time t, we first give a plot of contourslice of $\widehat{SE}_{\widehat{d_{0.1}}}$ in t and the cost on the grid of m and n, where both large and small values of $\sigma^{\square}$ are considered. Suppose that $\sigma_B = 0.3$ for both cases. The individual costs of components in the test are set as $c_1 = 15000$, $c_2 = 1500$, $\kappa = 100$, and $\eta = 1.25$. The maximum acceptable γ is set as 0.8.

From Figs. 2.1 and 2.2, we can observe that the large-sample approximate standard error of $d_{0.1}(t)$, $\widehat{SE}_{\widehat{d_{0.1}}}$, cannot be decreased by taking more measurements when the error of measurement is small and there are only a few units, i.e., less than 5 units are measured. In addition, in the case of small measurement error, $\widehat{SE}_{\widehat{d_{0.1}}}$ cannot

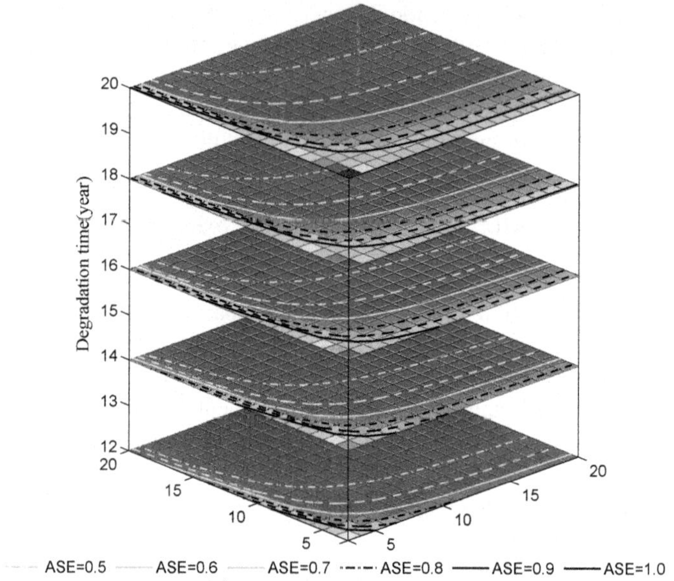

Fig. 2.1 Contourslice of the large-sample approximate standard error $\widehat{SE}_{\widehat{d_{0.1}}}$ when $\sigma^{\square} = 0.3$

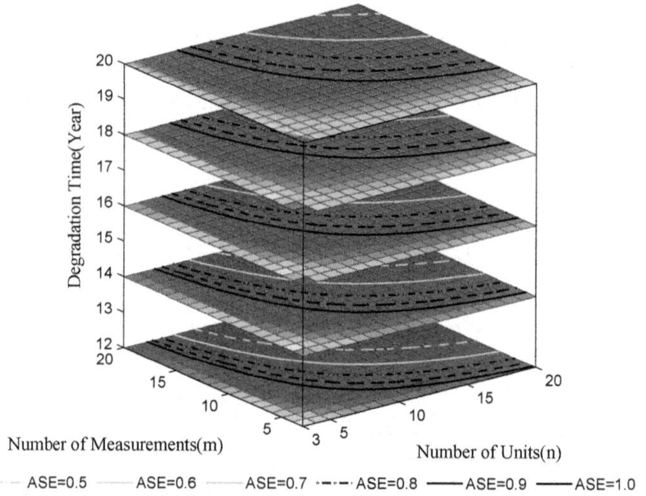

Fig. 2.2 Contourslice of the large-sample approximate standard error $\widehat{SE}_{\widehat{d_{0.1}}}$ when $\sigma^{\square} = 1.2$

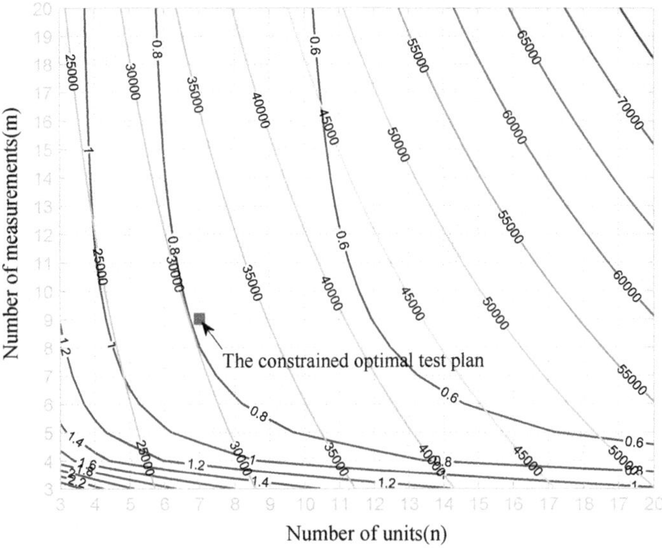

Fig. 2.3 the contour plot of cost and $\widehat{SE}_{\widehat{d_{0.1}}}$ when $\sigma^{\square} = 0.3$

be decreased efficiently by adding more units, when each unit are not measured sufficiently. In the case where the measurement error is large, $\widehat{SE}_{\widehat{d_{0.1}}}$ is more sensitive with m and n, which is indicated by Fig. 2.1. $\widehat{SE}_{\widehat{d_{0.1}}}$ is larger when the temporal variability is under consideration.

Figure 2.1 also shows that $\widehat{SE}_{\widehat{d_{0.1}}}$ increases with time t due to the temporal variability and the unit-to-unit variability in the parameter of the degradation model. Therefore, we choose the contourslice at a later time, say $t = 20$, to accomplish the test design. The top 1 slice of Fig. 2.1 is amplified in Fig. 2.2, which indicates a test plan where $n^* = 7$ items should be measured at $m^* = 9$ equally spaced times. For such a test plan, $\widehat{SE}_{\widehat{d_p}} = 0.7277$ and $cost(m^*, n^*) = 30696$.

To further investigate the influence of the measurement errors on the test planning, we check a series of contour plot of both $\widehat{SE}_{\widehat{d_p}}$ and the cost of test at different levels of measurement errors. An interesting result is found as follows. After setting $\sigma^{\square} = 0.8$, a test plan with $n = 8$, and $m = 18$ can achieve the precision level of $\widehat{SE}_{\widehat{d_{0.1}}} = 0.7863$ with a cost of 30856, which is very close to the constrained optimum design when $\sigma^{\square} = 0.3$. A comparison between Figs. 2.3 and 2.4 suggests that there are two strategies to achieve an acceptable level of $\widehat{SE}_{\widehat{d_p}}$ under the same constraint of budget. In other words, we can either spend less money on instruments with relatively worse performance but take more measurements on more items, or spend more money on instruments with better performance but cut down the number of items and measurements. These results are useful to design the degradation test plan for a practical product, particularly for vital products.

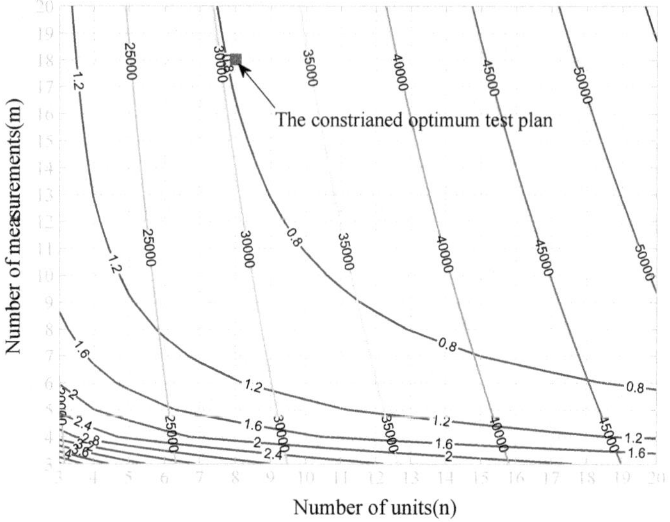

Fig. 2.4 the contour plot of cost and $\widehat{SE}_{\widehat{d_{0.1}}}$ when $\sigma^{\square} = 0.8$

References

1. Weaver BP, Meeker WQ, Escobar LA, Wendelberger J (2013) Method for planning repeated measures degradation studies. Technometrics 55(2):122–134
2. Meeker WQ, Escolar LA, Lu CJ (1998) Accelerated degradation tests: modeling and analysis. Technometrics 40(2):89–99
3. Tseng ST, Yu HF (1997) A termination rule for degradation experiments. IEEE Trans Reliab 49(1):130–133
4. Yu HF, Tseng ST (1998) On-line procedure for terminating an accelerated degradation test. Statistica Sinica 8:207–220
5. Liao CM, Tseng ST (2006) Optimal design for step-Stress accelerated degradation tests. IEEE Trans Reliab 55(1):59–66
6. Tseng ST, Balakrihnan N, Tsai CC (2009) Optimal step-stress accelerated degradation test plan for Gamma degradation process. IEEE Trans Reliab 58(4):611–618
7. Shi Y, Escobar LA, Meeker WQ (2009) Accelerated destructive degradation test planning. Technometrics 51(1):1–14
8. Shi Y, Meeker WQ (2012) Bayesian methods for accelerated destructive degradation test planning. IEEE Trans Reliab 61(1):245–253
9. Si XS, Wang WB, Hu CH, Zhou DH (2014) Estimating remaining useful life with three-source variability in degradation model. IEEE Trans Reliab 63(1):167–190
10. Tseng ST, Peng CY (2007) Stochastic diffusion modeling of degradation data. J Data Sci 5:315–333
11. Wang X (2010) Wiener processes with random effects for degradation data. J Multivar Anal 101:340–351
12. Ye ZS, Wang Y, Tsui KL, Pecht M (2013) Degradation data analysis using Wiener process with measurement errors. IEEE Trans Reliab 57(4):539–550
13. Ye ZS, Xie M (2015) Stochastic modelling and analysis of degradation for highly reliable products. Appl Stoch Models Bus Ind 31:16–32

14. Whitmore GA, Schenkelberg F (1997) Modelling accelerated degradation data using Wiener diffusion with a time scale transformation. Lifetime Data Anal 3:27–45
15. Dempster A, Laird N, Rubin D (1977) Maximum likelihood from incomplete data via the EM algorithm. J R Stat Soc Ser B 39(1):1–38
16. Schön TB (2009) An explanation of the expectation maximization algorithm, Division of Automatic Control, Linköping University, Linköping, Sweden, Technical Report LITH-ISY-R-2915
17. Si XS, Wang W, Hu CH, Chen MY, Zhou DH (2013) A Wiener-process-based degradation model with a recursive filter algorithm for remaining useful life estimation. Mech Syst Signal Process 35:219–237

Chapter 3
Specifying Measurement Errors for Required Lifetime Estimation Performance

3.1 Introduction

Reliable and accurate lifetime estimates for key engineering assets have long been a hot research topic attracting increasing attention in reliability and operational research communities and practices. Because estimating the lifetime is important and fundamental for maintenance schedules and logistic supports of assets, which can lead to the extension of the asset life and lifecycle cost reduction [1–5]. In particular, accurate lifetime estimation can lead to timely and efficient maintenance and logistic planning to reduce the extra costs due to unscheduled maintenance [6]. Therefore, the effectiveness of maintenance decisions and logistic planning relies heavily on the performance of the estimated lifetime of the asset.

Traditionally, if the past failure data of the assets from either fields or experiments are available, the lifetime can be estimated from the failure data by using likelihood-based inference methods [7, 8]. However, for expensive and highly reliable assets, failure data are scarce or limited. In practice, most failures of assets arise from a degradation mechanism at work and there are measurable characteristics that can be observed to deteriorate over time, such as the drift of gyros in the inertial navigation platforms and the length of fatigue cracks in rotating bearings [9]. Therefore, health monitoring data for these characteristics obtained from routine condition monitoring (CM) is a feasible and low-cost alternative used for the lifetime estimation task. However, perfect measurements in practical cases are impossible and the measured health monitoring data are inevitably contaminated by the uncertainty during the measurement process [10]. It is noted that, in many cases, degradation (e.g., fracture, cracks, and electronic charge trapping) cannot be directly or perfectly measured but is correlated to other measurable parameters, which may be able to reflect the degradation condition of the monitored asset [11–14]. Here the term "*degradation*" refers to the deterioration process of a certain characteristics of an asset with time. Examples can be either performance degradation (e.g. light output from an LED) or some measures of actual physical degradation (e.g. the length of a fatigue crack, the drift of a gyro, and the account of erosion), which are closely correlated with

X.-S. Si et al., *Data-Driven Remaining Useful Life Prognosis Techniques*,
Springer Series in Reliability Engineering, DOI 10.1007/978-3-662-54030-5_3

the underlying physics-of-failure of the asset. Therefore, throughout this chapter, we used the term "measured data" to represent the actually measured health monitoring data which are associated with the hidden degradation of the asset.

As a source of uncertainty, measurement error (ME) resulting from the noise, disturbance, non-ideal measurement instruments, etc., exists in almost all measurement processes [14–17]. Hence, any investigation of the lifetime estimation problem via degradation modeling must take the ME into account. From an engineering point of view, the study of ME to reveal sources contributing to its variation or the characterization of the distribution of ME has been well considered [10, 18, 19]. A variety of literature has addressed various aspects of the relationship between the ME and parameter estimation in the degradation models and the applications of lifetime estimation in maintenance, including [3, 10, 14, 20, 21], etc. Recently, Peng and Tseng in [14] investigated the effect of ME on lifetime estimation based on a Wiener process with a random drift coefficient. Si et al. in [22] presented a comprehensive survey of degradation data-based methods in the context of remaining useful life estimation.

The above-mentioned studies with respect to the effect of ME on performance characteristics of lifetime estimation could be the solution to the problem of "*the performance of lifetime estimation due to the ME*." Here we call this problem as a forward problem which focuses on the estimation from the measured data such as parameter estimation, structure identification, reliability estimation, and lifetime estimation, while the specification of ME is fixed. However, a reversed problem of "*the specification of the ME range in order to achieve a desirable lifetime estimation performance*" has not been addressed in the literature. This problem is related to the usability of the measured data for estimating the lifetime and is called an inverse problem in this chapter. It aims at specifying the ME characteristics to achieve the desired lifetime estimation performance. If the performance of ME does not satisfy certain requirements, the desirable performance of the estimated lifetime cannot be ensured. In other words, the inverse problem seeks to obtain the limitations to the ME in lifetime estimation under a given desired performance.

As a matter of fact, the inverse problem has its practical background, and reliability practitioners or users in maintenance decision and logistic scheduling are often interested in the various facets of the inverse problem. First, a practitioner in practice may make maintenance decisions based on the specified performance characteristics of lifetime estimation. He or she would like to expect accurate lifetime estimation in order to plan maintenance and other logistic support activities in a timely and cost-effective way. As discussed earlier, the higher quality the measured data has, the better performance the lifetime estimation has. Hence if the measured data with ME are used for such lifetime estimation task, he or she may wonder how to specify the allowable limits to the distribution-related parameters (e.g., bias and standard deviation) of ME in order to avoid unacceptable deterioration in the performance of lifetime estimation. Second, to ensure a desired performance for lifetime estimation based on the measured data, the ME should be controlled so that the performance of lifetime estimation can be maintained. This refers to the design of the device that takes the measurements. Third, when the lifetime estimated from the measured data with ME is used for maintenance decision, the practitioner may wonder what effect of ME will have on the final decision, because inaccurate lifetime estimation can lead to the increase of extra costs due to unscheduled maintenance. For example, Scanff et al. in [6] provided a business case study using Eurocopter's data from manufactures of two standard microelectronic subsystems in commercial helicopter to predict the lifecycle cost impact of using prognostics and health management. The results indicated that modeling failure data as a Weibull distribution was cost-effective as opposed to those whose failure data are represented by an exponential distribution. This resulted from the fact that most lifetimes of assets were not exponential. Together with these discussions, it can be concluded that the inverse problem has its practical implication for engineering practices in the fields of maintenance schedules and logistic supports requiring lifetime estimation from the measured condition monitoring data, and can be regarded as the necessity analysis of ME for the required lifetime estimation performance. However, all these very practical problems are not presented and addressed in the literature.

In this chapter, we will consider such a reverse problem and attempt to give some initial answers based on a Wiener process-based degradation model (WPDM), commonly used for modeling degradation processes where the asset operates in time-invariant environments and thus the rate of degradation can be approximated as a constant for simplicity. We note that such rate can be time-dependent and nonlinear (e.g., [9]), but it is not the focus of this chapter. WPDMs are popular degradation models which have been widely studied and applied in a variety of contexts such as LED lights, rotating bearings, and gyros drifts, and have tractable mathematical properties ([18, 23–26] and a review [22]). As indicated before, we consider that the actual degradation is unobservable, but some measured data which are related to the degradation are available. Specifically, we consider a WPDM affected by the ME for lifetime analysis and develop some expressions for permissible bias and variance of the ME under required lifetime estimation performance. In order to answer the aforementioned questions, the properties of the estimated lifetime considering the effect of ME under WPDM are derived first. Then, we define some measures to characterize the difference between lifetime estimations without/with considering the ME. Through these measures, we formulate some requirements on the ME for the sake of achieving certain required performance of lifetime estimation. Based on the obtained results, we further analyze the effect of ME on an age-based replacement decision, which is one of the most common and popular maintenance policies in maintenance scheduling [27], and often used as a benchmark model for demonstration. Finally, numerical examples and a case study are provided to illustrate the implementation procedure and usefulness of the theoretical results, where we also consider a comparison of condition-based replacement policy. The results indicate that, by specifying the ME range given a desirable lifetime estimation performance, it is possible to mitigate the conservativeness of maintenance decision so that the effectiveness of the replacement decision can be improved such as extending the operation cycle and reducing the long run average cost per unit time.

The rest of the chapter is organized as follows. In Sect. 3.2, we provide the fundamental results of WPDM for lifetime estimation. In Sect. 3.3, the properties of WPDM with ME are derived. In Sect. 3.4, given the required performance characteristics for lifetime estimation, we analyze the allowable bias and standard deviation of ME. In Sect. 3.5, we analyze the effect of ME on an age-based replacement decision. Numerical examples and a case study are provided in Sect. 3.6.

3.2 Properties of the WPDM

In this chapter, let $\{X(t), t \geq 0\}$ denote the stochastic degradation process which is correlated with the underlying physics-of-failure of the asset. A linear WPDM is typically used for modeling degradation processes, where the cumulative damage does not have a significant effect on the rate of degradation [28]. So far, this kind of degradation model has been widely adopted to characterize the degradation of a variety of assets (e.g., LED lights, rotating bearings, gyros, automotive wheels, etc.) to estimate the lifetime [22, 29, 30]. Therefore, we model the degradation process $\{X(t), t \geq 0\}$ as a Wiener process in this chapter. In general, a Wiener-process-based degradation process $\{X(t), t \geq 0\}$ can be represented by

$$X(t) = \lambda t + \sigma_B B(t), \tag{3.1}$$

where $X(t)$ is the actual degradation at time t, λ is the drift coefficient, $\sigma_B > 0$ is the diffusion coefficient, and $B(t)$ is the standard Brownian motion.

We now illustrate how to estimate the lifetime based on Eq. (3.1). From [31], we use the concept of the first hitting time (FHT) to define the lifetime. Namely, when the degradation process $\{X(t), t \geq 0\}$ reaches a pre-set critical level w, the asset is declared to be non-usable. This critical level is known as failure threshold, which is often defined by the industrial standard such as the International Standards Organization (ISO) (e.g., the ISO 2372 and ISO 10816 for defining acceptable vibration threshold levels). Therefore, it is natural to view the event of lifetime termination as the point when the degradation process $\{X(t), t \geq 0\}$ crosses threshold level $w$for the first time. From the FHT, the lifetime T can be defined as

$$T = \inf\{t : X(t) \geq w | X(0) < w\}. \tag{3.2}$$

with the probability density function (PDF) $f_T(t)$ and cumulative distribution function (CDF) $F_T(t)$.

It is well-known that the FHT of the Wiener process crossing a constant threshold follows an inverse Gaussian distribution [32]. Accordingly, we have the PDF of the lifetime T,

$$f_T(t) = \frac{w}{\sqrt{2\pi t^3 \sigma_B^2}} \exp\left(-\frac{(w - \lambda t)^2}{2\sigma_B^2 t}\right), \tag{3.3}$$

and the CDF

$$F_T(t) = 1 - \Phi\left(\frac{w - \lambda t}{\sigma_B\sqrt{t}}\right) + \exp\left(\frac{2\lambda w}{\sigma_B^2}\right)\Phi\left(\frac{-w - \lambda t}{\sigma_B\sqrt{t}}\right), \quad (3.4)$$

with the mean and variance as follows

$$E(T) = \frac{w}{\lambda}, \text{var}(T) = \frac{w\sigma_B^2}{\lambda^3}. \quad (3.5)$$

However, in practical cases, the true degradation cannot be observed directly [11–13]. Such unobservability is largely due to the noise, disturbance, non-ideal measurement instruments, etc. [15–17]. In this sense, it is better to let the true degradation process be a latent process, which is continuously fluctuating but not directly observable. For example, when measuring a temperature using a thermometer, the height of the column of mercury is the observed variable, which may be linearly related to the actual unknown temperature. For the same measured variable, different measuring devices may be used. Thus infrared-based devices may also measure the temperature, possibly with a much higher accuracy than that of the common thermometer. They therefore have different MEs, but both are driven by the latent temperature. The measurement process will, of course, in turn affect lifetime estimation. In this chapter, we consider this latent degradation process with measured health monitoring data contaminated by the ME.

3.3 Properties of the WPDM with the ME

If variable $X(t)$ representing the degradation at time t is monitored via some measurement process, then the actual observed variable $Y(t)$ is a function of $X(t)$. Following the works in statistical process control [33], we assume a linear relationship between them, namely,

$$Y(t) = A + BX(t) + \varepsilon \quad (3.6)$$

where ε is normally distributed with zero mean and standard deviation σ; A and B are parameters. Assume $X(t)$ is the true degradation characteristics to be monitored, represented by Eq. (3.1). The observed quantity $Y(t)$ is then distributed normally. It is obvious that if $A = 0$ and $B = 1$, no bias is introduced in measuring $X(t)$ through $Y(t)$, and the associated standard deviation of ME is σ. In this chapter, we only consider the case that $B = 1$ in Eq. (3.6) for simplicity. Together with Eq. (3.1), we get,

$$Y(t) = A + X(t) + \varepsilon = A + \lambda t + \sigma_B B(t) + \varepsilon, \quad (3.7)$$

where A and σ are considered as the bias and standard deviation of ME, respectively.

From the FHT concept, the lifetime T_e associated with $\{Y(t), t \geq 0\}$ can be defined as

$$T_e = \inf\{t : Y(t) \geq w | Y(0) < w\}. \quad (3.8)$$

with the PDF $f_{T_e}(t)$ and CDF $F_{T_e}(t)$.

Obviously, the estimated lifetime T_e from $Y(t)$ will be different from the lifetime T. Since $X(t)$ is not observable, and therefore in practice, only $Y(t)$ is used to define the lifetime. In the following, based on Eq. (3.7) and the definition given in Eq. (3.8), we try to obtain the distribution of T_e. The following results are presented to simplify the derivation of the distribution of T_e.

Lemma 3.1 *If* $Y \sim N(\mu_1, \sigma_1^2)$, *then* $E_Y[\Phi(Y)]$ *can be formulated as* $E_Y[\Phi(Y)] = \Phi\left(\mu_1 \Big/ \sqrt{\sigma_1^2 + 1}\right)$.

Proof Using the property of the conditional expectation, we directly have

$$E_Y[\Phi(Y)] = E\left[\, E(I_{\{Z \le Y\}} | \, Y) \big| \, Y\right] = \Pr(Z \le Y) = \Pr(Z - Y \le 0) = \Phi\left(\mu_1 \Big/ \sqrt{\sigma_1^2 + 1}\right), \tag{3.9}$$

where $\Phi(\cdot)$ denotes the standard normal CDF, $I_{\{Z \le Y\}}$ is an indicator function, and Z is a standard normal random variable and independent of Y. Then, we have $Z - Y \sim N(-\mu_1, \sigma_1^2 + 1)$. This completes the proof.

Using Lemma 3.1 and expectation manipulations, we obtain the following conclusions after some lengthy derivations.

Lemma 3.2 *If* $Z \sim N(\mu, \sigma^2)$, *and* $w, \vartheta, B, D \in \Re$, $C \in \Re^+$, *then the following hold:*

$$(1)\ \mathbb{E}_Z\left[\exp\{\vartheta Z\}\Phi(C + DZ)\right] = \exp\left\{\frac{\vartheta^2}{2}\sigma^2 + \vartheta\mu\right\} \cdot \Phi\left(\frac{C + D\mu + \vartheta D\sigma^2}{\sqrt{1 + D^2\sigma^2}}\right), \tag{3.10}$$

$$(2)\ \mathbb{E}_Z\left[(\vartheta - Z) \cdot \exp\left\{-(B - Z)^2/2C\right\}\right] = \sqrt{\frac{C}{\sigma^2 + C}}\left(\vartheta - \frac{\sigma^2 B + \mu C}{\sigma^2 + C}\right) \times \exp\left\{-\frac{(B - \mu)^2}{2\left(\sigma^2 + C\right)}\right\}. \tag{3.11}$$

Thanks to the results in Lemma 3.2, the lifetime estimation results corresponding to $\{Y(t), t \ge 0\}$ can be summarized as follows:

Theorem 3.1 *For Eq. (3.7) and the definition in Eq. (3.8), the followings hold:*

$$(1)\ f_{Te}(t) = \frac{\lambda\sigma^2 + \sigma_B^2(w - A)}{\sqrt{2\pi(\sigma^2 + \sigma_B^2 t)^3}} \exp\left\{-\frac{(w - A - \lambda t)^2}{2\left(\sigma^2 + \sigma_B^2 t\right)}\right\}, \tag{3.12}$$

$$(2)\ F_{Te}(t) = 1 - \Phi\left(\frac{w - A - \lambda t}{\sqrt{\sigma_B^2 t + \sigma^2}}\right) + \exp\left\{\frac{2\lambda(w - A)}{\sigma_B^2} + \frac{2\lambda^2\sigma^2}{\sigma_B^4}\right\} \times$$

$$\Phi\left(\frac{-w-\lambda t+A-2\lambda\sigma^2/\sigma_B^2}{\sqrt{\sigma_B^2 t+\sigma^2}}\right), \tag{3.13}$$

$$(3)\ \mathbb{E}(T_e)=\frac{w-A}{\lambda},\ var(T_e)=\frac{\lambda\sigma^2+(w-A)\sigma_B^2}{\lambda^3}. \tag{3.14}$$

Proof see Appendix A.

There are three observations regarding the results in Eqs. (3.3)–(3.5) and Theorem 3.1. The first observation is that the measurement uncertainty can propagate into the lifetime distribution in Theorem 3.1, and thus we account for the uncertainties in the stochastic degradation process and its measurement process simultaneously. The second is that the results in Theorem 3.1 can reduce to the results in Eqs. (3.3)–(3.5) by setting $A = 0$ and $\sigma^2 = 0$. Last, the most commonly used properties of T and T_e, i.e., mean and variance, can both be obtained in explicit forms. This can significantly facilitate the subsequent analysis. In the following section, we will present allowable bias and standard deviation for ME, given the specified process parameters and required performance characteristics for lifetime estimation.

3.4 Permissible ME Parameters for Lifetime Estimation

3.4.1 Performance Measures to Quantify the Difference in Lifetime Estimation with Versus Without the ME

We first introduce the following definitions to measure the difference in lifetime estimations with versus without the ME. To characterize the variability existing in a random variable, there are two frequently adopted measures, i.e., variance and coefficient of variation (CV), where the variance of a random variable is often used as a measure of spread or dispersion, but the CV of a random variable is used as a measure of scale invariant dispersion since the mean of the random variable is considered in the definition of the CV while the mean is a measure of the location of the distribution of a random variable.

As discussed before, the CV is often used to measure scale invariant dispersion existing in a random variable [34], e.g., Z. Specifically, the CV is defined as the ratio of the standard deviation to the mean of Z. It is the inverse of the signal-to-noise ratio, formulated as

$$CV(Z)=\frac{\sqrt{var(Z)}}{|E(Z)|}. \tag{3.15}$$

Since the CV can measure the relative variability existing in any random variable, it is naturally expected that the difference between the CVs of T_e and T is relatively small if the effect of ME is small. This signifies that the measurement process

$\{Y(t), t \geq 0\}$ is accurate enough. In order to quantify their difference using the CV measure, we define the relative increase ratio $R_{cv}(T_e, T)$ of the CVs between T_e and T as follows:

$$R_{cv}(T_e, T) = \frac{|CV(T_e) - CV(T)|}{CV(T)}. \tag{3.16}$$

From Eq. (3.16), we can observe that the less $R_{cv}(T_e, T)$ is, the more accurate the estimated lifetime from $\{Y(t), t \geq 0\}$ is, and vice versa. Therefore, $R_{cv}(T_e, T)$ can serve as a measure to quantify the difference between T_e and T. As such, it is expected that $R_{cv}(T_e, T)$ should be less than a given value $C_m \geq 0$ if the ME has less influence on the estimated lifetime, i.e., $0 \leq R_{cv}(T_e, T) \leq C_m$. In other words, we expect that the estimated lifetime through measurement process $\{Y(t), t \geq 0\}$ can approach the estimation from the error-free process $\{X(t), t \geq 0\}$ accurately enough.

Similarly, we can also define the relative increase ratio $R_{var}(T_e, T)$ of the variances between T_e and T as follows:

$$R_{var}(T_e, T) = \frac{\left|\sqrt{var(T_e)} - \sqrt{var(T)}\right|}{\sqrt{var(T)}}, \tag{3.17}$$

Similarly, it is expected that $R_{var}(T_e, T)$ should satisfy $v_{\min} \leq R_{var}(T_e, T) \leq v_{\max}$ if the ME has less influence on the lifetime estimation, where $v_{\max} \geq v_{\min} \geq 0$ are lower and upper limits. Note that the less $R_{var}(T_e, T)$ is, the more accurate the estimated lifetime from $\{Y(t), t \geq 0\}$ is, and vice versa. Therefore, $R_{cv}(T_e, T)$ can serve as the second measure to quantify the difference between T_e and T.

In this chapter, we utilize the above measures to analyze the requirements for ME under certain performance requirements regarding lifetime estimation. Specifically, we present the allowable bias and standard deviation for ME—A and σ, given the specified process parameters (including λ and σ_B) and the required performance characteristics for lifetime estimation. Note that, to specify the performance requirement for ME, we should first determine the required performance characteristics, i.e., C_m, $v_{\min}$, and $v_{\max}$. Generally, small values of C_m, $v_{\min}$, and $v_{\max}$ correspond to the high accuracy of the estimated lifetime, and vice versa. Therefore, these quantities can be determined according to the required accuracy for lifetime estimation and are context-specific. In the following, we investigate how to specify the performance requirement for ME given C_m, $v_{\min}$ and $v_{\max}$, and mainly focus on the case of $A \leq w$, $w \geq 0$, and $\lambda > 0$ unless otherwise specified. Other cases can be analyzed in a similar manner and thus are omitted.

3.4.2 *Permissible ME Parameters Using the Relative Increase Ratio of the CV*

According to the relative increase ratio $R_{cv}(T_e, T)$ based on the CV measure, the following theorem can be obtained to specify the requirement for ME.

Theorem 3.2 *From the CV measure, if it is required* $0 \leq R_{cv}(T_e, T) \leq C_m$*, the following conditions about* A *and* σ *should be satisfied,*

$$\begin{cases} if \lambda\sigma^2 + \sigma_B^2 A\,(w - A)\,/w \geq 0, then\, \lambda\sigma^2 + \sigma_B^2\,(w - A)\left[1 - (w - A)\,\frac{(C_m+1)^2}{w}\right] \leq 0, \\ if \lambda\sigma^2 + \sigma_B^2 A\,(w - A)\,/w < 0, then\, \lambda\sigma^2 + \sigma_B^2\,(w - A)\left[1 - (w - A)\,\frac{(C_m+1)^2}{w}\right] > 0. \end{cases} \tag{3.18}$$

Proof see Appendix.

Several observations regarding Theorem 3.2 need to be noted. First, given the performance requirement (i.e. C_m) and process parameters (including λ and σ_B), Theorem 3.2 can specify the feasible area of A and σ^2 in a two-dimension plane. In other words, if Eq. (3.18) is violated for all feasible values of A and σ^2, the required $0 \leq R_{cv}(T_e, T) \leq C_m$ cannot be achieved. Secondly, in Eq. (3.18), the ME variance is expressed in combination with the bias, and the bias can be expressed in terms of the ME variance, and vice versa. Therefore, given that $R_{cv}(T_e, T)$ should not exceed the allowable limit C_m, either A or σ^2 may arbitrarily be determined as both these sets of expressions have been derived from a single equation. Third, the expressions developed above involve two parameters, λ and σ_B, which need to be determined. We assume that these parameters are known or can be estimated from the historical degradation data. In other words, our primary focus is on how to specify the permissible ME when using degradation data-based lifetime estimation approaches.

Now we consider two special cases: one is for the case of $A = 0$ and the other is for $\sigma^2 = 0$.

Corollary 3.1 *For the case of* $A = 0$*, according to the CV measure* $0 \leq R_{cv}(T_e, T) \leq C_m$*, the following conditions about* σ^2 *should be satisfied,*

$$0 \leq \sigma^2 \leq \frac{\sigma_B^2 w\left(C_m^2 + 2C_m\right)}{\lambda}. \tag{3.19}$$

Corollary 3.2 *For the case of* $\sigma^2 = 0$*, from the CV measure* $0 \leq R_{cv}(T_e, T) \leq C_m$*, the following conditions about* A *should be satisfied,*

$$0 \leq A \leq \frac{w(2C_m + C_m^2)}{(C_m + 1)^2}. \tag{3.20}$$

Obviously, in the above two extreme cases, the required values of A and σ^2 can be expressed in a simple and explicit manner, and thus may be very easy to apply in practice.

Now suppose that we wish to determine permissible values of the bias for any desirable ME variance. Obviously, the allowable values can be derived from Eq. (3.18). Consequently, one can plot Eq. (3.18), where the horizontal axis represents the ME bias, and the vertical axis represents the corresponding permissible variance. This graph will depict the trade-off between the bias and variance of ME. Points on the curve indicate that the upper bound has been reached (i.e. Eq. (3.18)

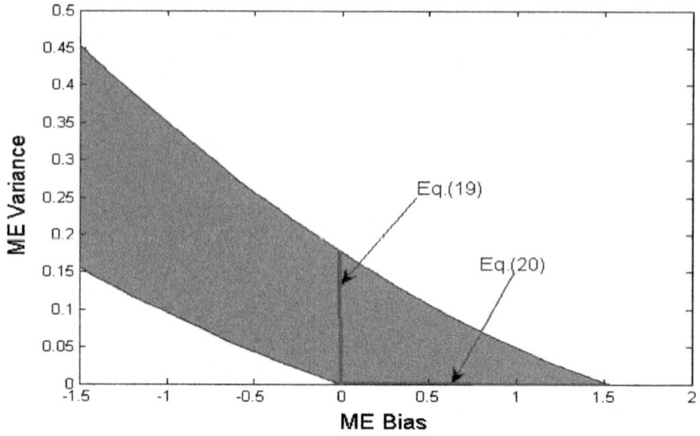

(a) $C_m = 0.2$

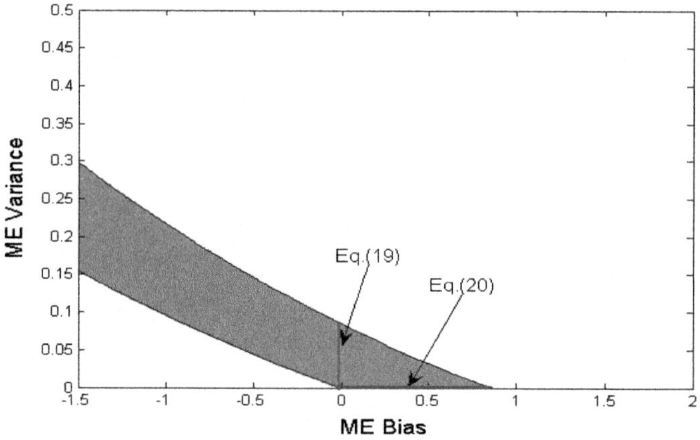

(b) $C_m = 0.1$

Fig. 3.1 Permissible ME parameters under the relative increase ratio of the CV

is maintained as equality). Conversely, points inside the area bounded by the curve represent the permissible ME bias and variance that have not reached the limiting values specified by Eq. (3.18). In other words, the deviation between the estimated lifetime for T_e and the nominal for T is smaller than the requirement C_m. Such plot is illustrated in Fig. 3.1, which is related to the numerical example in this chapter.

It is interesting to note that the permissible area of A and σ^2 becomes narrower as C_m decreases. To achieve a higher accuracy of lifetime estimation, the stricter the required performance of lifetime estimation is, the less error of the measurement process is required.

3.4.3 Permissible ME Parameters Using the Relative Increase Ratio of the Variance

From the relative increase ratio $R_{var}(T_e, T)$ of the variance defined in Eq. (3.17), the following theorem can be obtained to specify the requirement for ME.

Theorem 3.3 *From the measure using the variance, in order to satisfy* $v_{\min} \leq R_{var}(T_e, T) \leq v_{\max}$, *the following conditions about A and* σ^2 *should be satisfied,*

$$\begin{cases} if \lambda\sigma^2 - A\sigma_B^2 \geq 0, then w\sigma_B^2(2v_{\min} + v_{\min}^2) \leq \lambda\sigma^2 - A\sigma_B^2 \leq w\sigma_B^2(2v_{\max} + v_{\max}^2), \\ if \lambda\sigma^2 - A\sigma_B^2 < 0, then w\sigma_B^2(v_{\max}^2 - 2v_{\max}) \leq \lambda\sigma^2 - A\sigma_B^2 \leq w\sigma_B^2(v_{\min}^2 - 2v_{\min}). \end{cases} \quad (3.21)$$

Proof see Appendix.

Similar to Theorems 3.2, 3.3 also specifies the feasible area of parameters A and σ^2 in a two-dimension plane, given the performance requirements, $v_{\min}$ and $v_{\max}$, and process parameters λ and σ_B. From Theorem 3.3, we have the following corollaries.

Corollary 3.3 *For the case of* $A = 0$, *by the variance measure* $v_{\min} \leq R_{var}(T_e, T) \leq v_{\max}$, *the following conditions about* σ^2 *should be satisfied*

$$\frac{w\sigma_B^2}{\lambda}\left(v_{\min}^2 + 2v_{\min}\right) \leq \sigma^2 \leq \frac{w\sigma_B^2}{\lambda}\left(v_{\max}^2 + 2v_{\max}\right). \quad (3.22)$$

Corollary 3.4 *For the case of* $\sigma^2 = 0$, *by the variance measure* $v_{\min} \leq R_{var}(T_e, T) \leq v_{\max}$, *the following conditions about A should be satisfied*

$$\begin{cases} w(2v_{\min} - v_{\min}^2) \leq A \leq w(2v_{\max} - v_{\max}^2), \ A \geq 0 \\ -w(v_{\max}^2 + 2v_{\max}) \leq A \leq -w(v_{\min}^2 + 2v_{\min}), \ A < 0 \end{cases}. \quad (3.23)$$

Similar discussions such as those in Sect. 4.2 can be presented based on the analysis of Eqs. (3.21), (3.22) and (3.23) through the variance measure, $R_{var}(T_e, T)$. Also, a plot to determine the permissible ME parameters is illustrated in Fig. 3.2, which is related to the numerical example in this chapter. In addition, we also observe that the permissible area of A and σ^2 becomes narrower as $v_{\max}$ decreases in this case.

Remark 3.1 It is noted that, to use Theorems 3.2 and 3.3 in practice, A and σ^2 should be estimated from the measured data which are in turn affected by the precision of the measuring device as well as by the selected estimation method. To address this problem, it is better to use the confidence intervals of the estimated A and σ^2 to verify whether or not Theorems 3.2 and 3.3 are satisfied, rather than only using their point estimates. We will use this method in the subsequent case study.

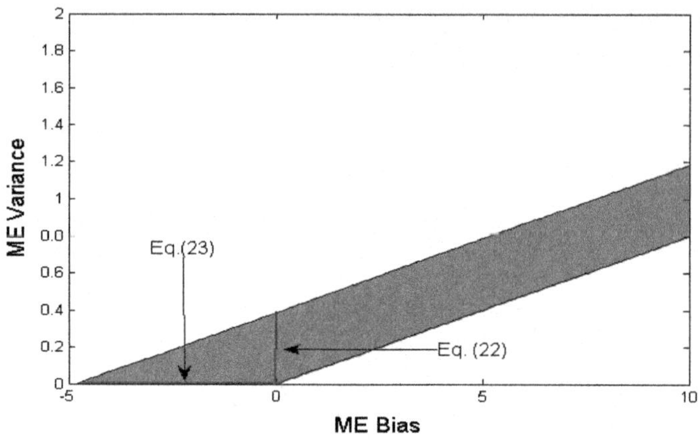

(a) $v_{\min} = 0, v_{\max} = 0.4$

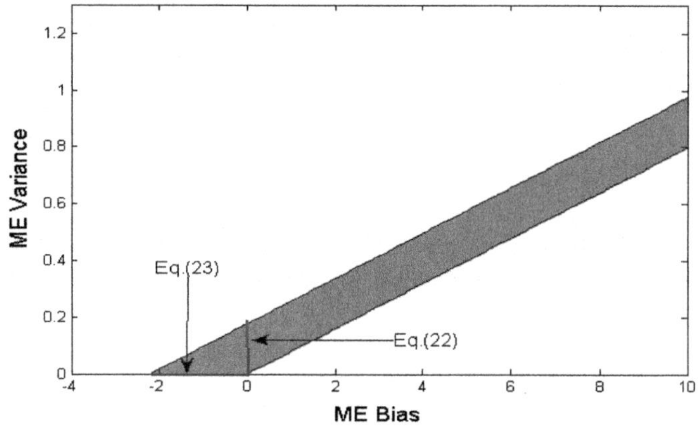

(b) $v_{\min} = 0, v_{\max} = 0.2$

Fig. 3.2 Permissible ME parameters under the relative increase ratio of the variance

3.5 Effect of Lifetime Estimation with or Without ME on an Age-Based Replacement Decision

In this section, based on the estimated lifetime distribution, we take an age-based replacement policy to analyze the effect of ME on a maintenance decision. The reason for considering an age-based replacement policy is that it is one of the most common and popular maintenance policies and often used as a benchmark model for demonstration [23, 25–27, 34, 35]. Under this policy, an asset is always replaced upon failure (the degradation hits the failure threshold for the case considered in this chapter) or upon reaching a pre-determined age τ, whichever occurs first, where τ is a decision variable. After replacement, the replacement process renews. Therefore, the replacement process is a renewal process renewed by each replacement, either scheduled or unexpected. The long run average cost per unit time under this replacement policy can thus be computed based on the theory of renewal reward processes [36].

We first consider the case without the ME and denote by T_r the corresponding time interval between two sequential replacements, also called as a replacement cycle. Then the long run average cost per unit time under the age-based replacement policy can be expressed based on the renewal reward theory as

$$CR(\tau) = \frac{E\left[C\right]}{E\left(T_r \mid w\right)} = \frac{c_p \bar{F}_T(\tau) + c_f F_T(\tau)}{\int_0^\tau \bar{F}_T(t)\mathrm{d}t}, \tag{3.24}$$

where $E\left[C\right]$ is the expected cost per cycle, $E\left(T_r \mid w\right)$ is the expected cycle length, c_p is the preventive replacement cost, and c_f is the failure replacement cost with $c_p < c_f$. In addition, $F_T(t)$ is the CDF of the lifetime obtained by Eq. (15.4) and $\bar{F}_T(t) = 1 - F_T(t)$. Therefore, the optimal replacement time τ^* is the time that minimizes Eq. (3.24).

Similarly, we further consider the case with the ME and denote by T_{er} the corresponding time interval between two sequential replacements. Then the long run average cost per unit time can be expressed as

$$CR_e(\tau) = \frac{E\left[C_e\right]}{E\left(T_{er} \mid w\right)} = \frac{c_p \bar{F}_{T_e}(\tau) + c_f F_{T_e}(\tau)}{\int_0^\tau \bar{F}_{T_e}(t)\mathrm{d}t}, \tag{3.25}$$

where $E\left[C_e\right]$ is the expected cost per cycle, $E\left(T_{er} \mid w\right)$ is the expected cycle length, and $F_{T_e}(t)$ is the CDF of the lifetime with the ME obtained by Eq. (3.13). Accordingly, the optimal replacement time τ_e^* is the time that minimizes Eq. (3.25).

In order to shed a light on the effect of ME on the replacement decision under the age based replacement policy, we present the following theorem. Its proof is given in the Appendix.

Theorem 3.4 *Consider an asset whose degradation process without the ME follows a Wiener process* $\{X(t), t \geq 0\}$ *as Eq. (3.1) or a Wiener process affected by the ME,* $\{Y(t), t \geq 0\}$*, as Eq. (3.7). The asset is replaced upon failure (the degradation hits the failure threshold* w*) or upon reaching a pre-determined age* τ*, whichever occurs first. Then the followings hold:*

(1) $E\left(T_r \mid w\right) = \tau \bar{F}_{IG}\left(\tau; \frac{w}{\lambda}, \frac{w^2}{\sigma_B^2}\right) + \frac{w}{\lambda}\bar{F}_{IG}\left(\frac{w^2}{\tau\lambda^2}; \frac{w}{\lambda}, \frac{w^2}{\sigma_B^2}\right)$*, and* $E\left(T_{er} \mid w\right) = E_{A'}\left[E\left(T_r \mid w\right)\right]$*, where* $\bar{F}_{IG}\left(t; \mu, \theta\right) = 1 - \Phi\left(\sqrt{\frac{\theta}{t}}\left(\frac{t}{\mu} - 1\right)\right) + \exp\left(\frac{2\theta}{\mu}\right)\Phi\left(-\sqrt{\frac{\theta}{t}}\left(\frac{t}{\mu} + 1\right)\right)$ *and* $A' \sim N(A, \sigma^2)$*.*

(2) If $\sigma^2 = 0$*,* T_e *is stochastically decreasing in* A*; particularly* $T \geq_{st} T_e$ *when* $A \geq 0$*.*

(3) *If* $\sigma^2 = 0$ *and* $A \geq 0$, *then* $E(T_{er}|w) \leq E(T_r|w)$, $CR_e(\tau) \geq CR(\tau)$, *and* $CR_e(\tau_e^*) \geq CR(\tau^*)$, *where* τ^* *and* τ_e^* *are the solutions of minimizing Eqs. (3.24) and (3.25), respectively.*

Theorem 3.4 shows the effect of ME on the lifetime estimation and an age based replacement decision, including the expected cycle length, and the long run average cost per unit time. Particularly, we can find that, if $\sigma^2 = 0$ and $A \geq 0$, the expected cycle length will be shortened and instead a greater long run average cost per unit time will be incurred with the ME effect, compared with the case without the ME effect. This implies that the replacement decision made under the measured data with the ME effect is conservative and thus is not economically optimal. Therefore, by specifying the ME range given a desirable lifetime estimation performance, it is possible to mitigate such conservativeness and extend the operation cycle of the asset so that the effectiveness of the replacement decision is improved. For the case $\sigma^2 \neq 0$, we show the effect of ME on the replacement decision in the next section by a real case study.

3.6 Experimental Studies

3.6.1 A Numerical Illustration

We first provide several numerical examples to demonstrate the main results about the required performance characteristics of ME for lifetime estimation. For illustrative purposes, we consider the parameters for the degradation process as $\lambda = 0.5$, $w = 5$ and $\sigma_B^2 = 0.04$. To give an intuitive impression on the effect of ME, we here consider three special cases. The first case is that we let $A = 0$ and increase σ^2 from 0 to 0.8; the second is that we let $\sigma^2 = 0$ and increase A from 0 to 1.6; and the third is that we consider the increases of both A and σ^2. The corresponding PDFs and CDFs of the lifetime for these three cases are illustrated in Figs. 3.3, 3.4 and 3.5, respectively.

It can be observed from Fig. 3.3 that, when $A = 0$, the estimated lifetime is unbiased but the variance of the estimated lifetime from the measurement process $\{Y(t), t \geq 0\}$ increases with σ^2. This is intuitively understandable since it is assumed that the ME is independent of $\{B(t), t \geq 0\}$ and thus the uncertainty within the measurement process is larger than that within the degradation process $\{X(t), t \geq 0\}$. At the same time, increasing σ^2 first leads to the decrease of the reliability function and then leads to the increase of the reliability function. Implied by Fig. 3.4, when $\sigma^2 = 0$, the estimated mean lifetime decreases and at the same time the variance of the estimated lifetime decreases as A increases (note that we limit A to be non-negative in above numerical examples). This can be well-explained since in this case increasing A makes the lifetime decrease and thus it may be not possible to have a large variance for a small lifetime. Figure 3.5 illustrates the compared results when both A and σ^2 increase.

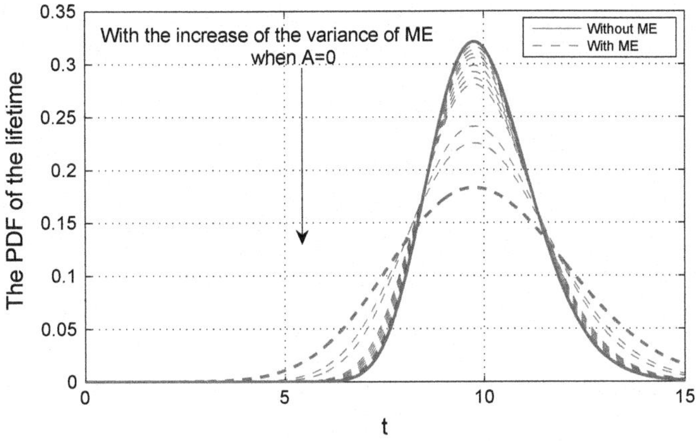

(a) Lifetime PDF

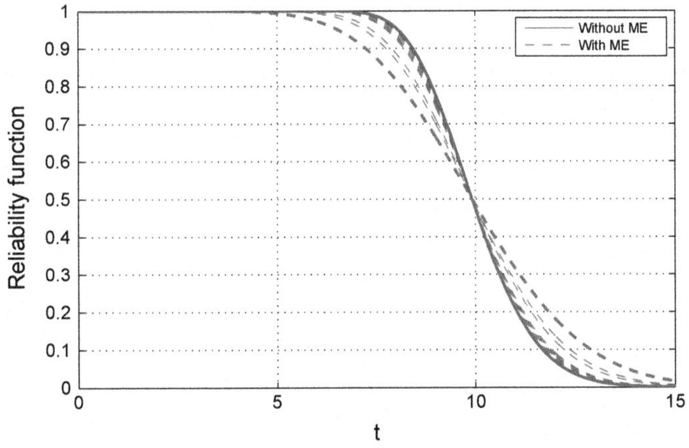

(b) Reliability function

Fig. 3.3 Comparisons between estimated lifetime distributions and reliability functions with and without ME ($A = 0$), respectively

The above results give an intuitive picture of how the ME affects the estimated lifetime. In the following, we give an example under $\lambda = 0.5$, $w = 5$ and $\sigma_B^2 = 0.04$ to illustrate how to determine the permissible ME parameters according to results in Sect. 3.4. For an illustrative purpose, we assume that the required performance characteristics for lifetime estimation are $C_m = 0.2$, $v_{\min} = 0$, and $v_{\max} = 0.2$.

First, based on the CV measure, when it is required that $C_m = 0.2$, the permissible values of A and σ^2 can be determined from Theorem 3.2, which is illustrated in Fig. 3.1. It can be found that decreasing C_m will make the permissible area of A and σ^2 narrower. Particularly, when $A = 0$, it is required that $0 \leq \sigma^2 \leq 0.176$, and when $\sigma^2 = 0$, it is required that $0 \leq A \leq 1.5278$. These results are calculated from

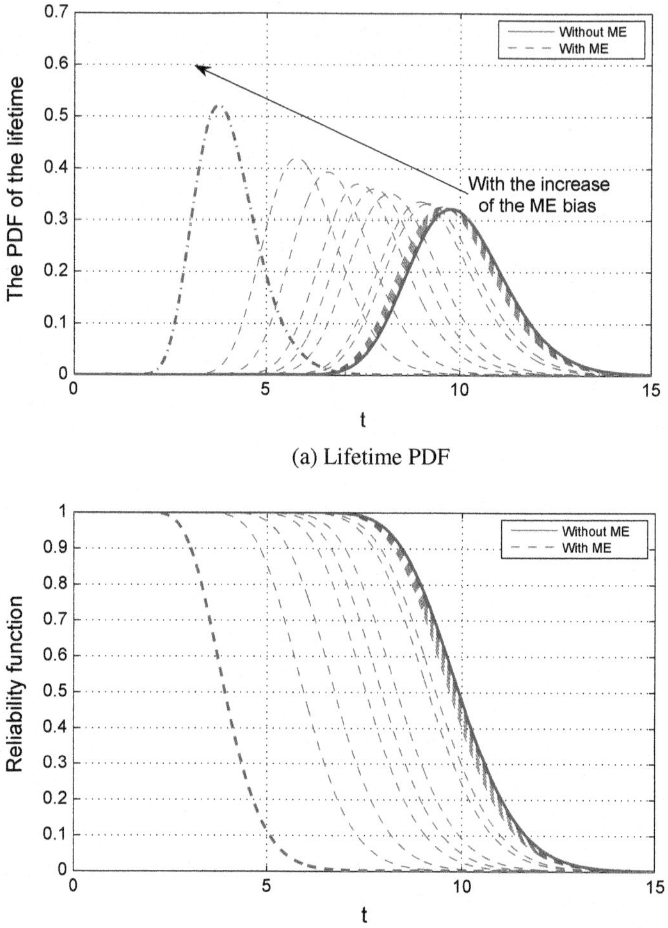

(a) Lifetime PDF

(b) Reliability function

Fig. 3.4 Comparisons between estimated lifetime distributions and reliability functions with and without ME ($\sigma^2 = 0$), respectively

Eqs. (3.19) and (3.20), respectively, and are shown in Fig. 3.1. Second, based on the variance measure, when it is required that $v_{\min} = 0$ and $v_{\max} = 0.2$, the permissible values of A and σ^2 can be determined from Theorem 3.3 as well (see Fig. 3.2). It can be found that increasing $v_{\max}$ will make the permissible area of A and σ^2 wider. Particularly, when $A = 0$, it is required that $0 \leq \sigma^2 \leq 0.176$. When $\sigma^2 = 0$, we have $-2.2 \leq A \leq 0$. These results are calculated from Eqs. (3.22) and (3.23), respectively, and are shown in Fig. 3.2. The main results of the above cases can be summarized in the following Table 3.1.

It is noted that, when $A = 0$, the permissible values of σ^2, determined by $R_{cv}(T_e, T)$ are the same as the results based on $R_{var}(T_e, T)$. Because $v_{\max} = C_m$,

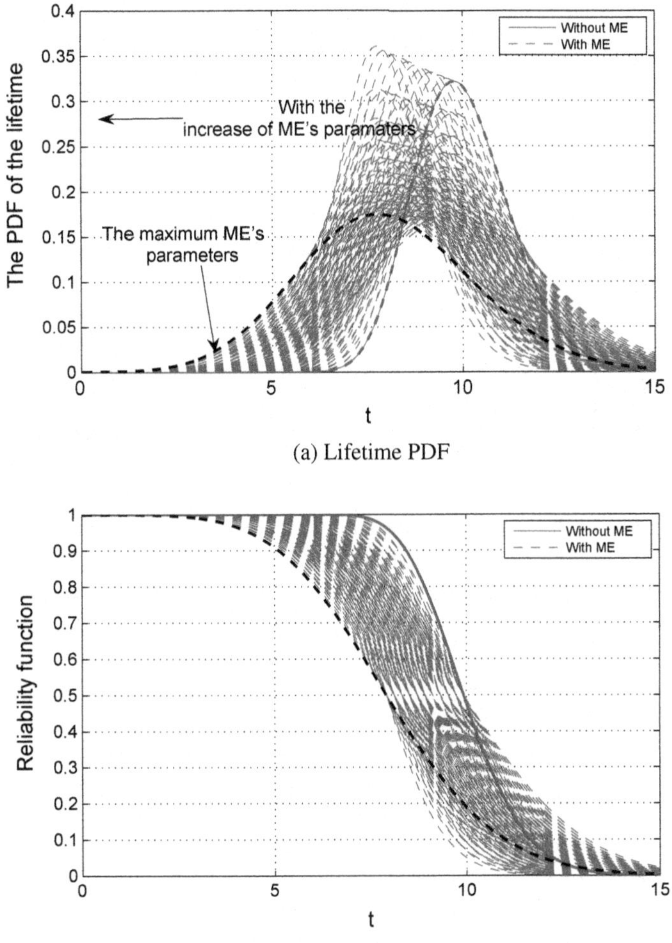

(a) Lifetime PDF

(b) Reliability function

Fig. 3.5 Comparisons between estimated lifetime distributions and reliability functions with and without ME (with the increases of both A and σ^2), respectively

then $R_{cv}(T_e, T) = R_{var}(T_e, T)$ when $A = 0$. Namely, the estimate is unbiased in this case.

3.6.2 The Case Study

In this section, we provide a practical case study for gyros in an inertial navigation system (INS) to illustrate the application of the developed approach.

Table 3.1 Permissible parameters for ME under the required performance of lifetime estimation

Measures	Required performance	$A = 0$	$\sigma^2 = 0$
$R_{cv}(T_e, T)$	$C_m = 0.2$	$0 \le \sigma^2 \le 0.176$	$0 \le A \le 1.527\,8$
$R_{var}(T_e, T)$	$v_{\min} = 0, \quad v_{\max} = 0.2$	$0 \le \sigma^2 \le 0.176$	$-2.2 \le A \le 0$
$\delta(T_e, T)$	$\delta_m = 0.9 \frac{k=40}{k=20}$	$0 \le \sigma^2 \le 0.093\,8$	$\frac{-8.881\,8e-16 \le A \le 2.519\,7}{-8.881\,8e-16 \le A \le 3.728\,8}$

As a key device of the INS in weapon systems and space equipment, an inertial platform plays an important and irreplaceable role in the INS. The sensors fixed in the inertial platform include three gyros and three accelerometers, which measure angular velocity and linear acceleration, respectively. When the inertial platform is operating, the wheels of gyros rotate at very high speeds and can lead to the rotation axis wear and finally result in gyros' drift. In our case, the gyro fixed on the inertial platform is a mechanical structure having two degrees of freedom from the driver and sense axis. As the wear is accumulated, the bearing of the gyro's electric motor will become deformed and such deformation leads to the drift of the gyro. The increasing drift finally results in the failure of gyro and then the inertial platform. Statistical analysis shows that almost 70% of the failures of inertial platforms result from gyros' drift. It can be observed that the failure of the gyro and inertial platform is largely resulted from the bearings as the case of rolling element bearings which are extensively investigated in the literature. However, the difference of our case is in that we use the drift data of gyros to estimate the lifetime rather than the vibration data as rolling element bearings. In our case, we cannot obtain the vibration data since it is not allowed to fix the vibration sensors in the inertial platform. As such, the drift of gyros is often used as a performance indicator to evaluate the health condition of the inertial platform and schedule maintenance activities. For an illustrative purpose, we provide an illustration of a deformed bearing of the gyro's electric motor (see Fig. 3.6), which is obtained by scanning electron microscopy S-3700N. It can be found that the maximum length of the metal flake is 155 um. Such deform is reflected by the drift data of gyros, which can be monitored and measured.

In this study, we assume that the CM values of gyro's drift reflect the performance of the inertial platform, and the larger the monitored drift is, the worse the performance is. For our monitored INS, we collected ten failure data of gyros in total, which are summarized in Table 3.2. For illustration, the collected monitoring data of six gyros are illustrated in Fig. 3.7 with regular CM intervals 2.5 h.

Table 3.2 Failure data of gyros

Gyros No.	1	2	3	4	5	6	7	8	9	10
Failure time (h)	145.6	175	160.3	197.5	150.5	157.5	230	137.5	192.5	267.5

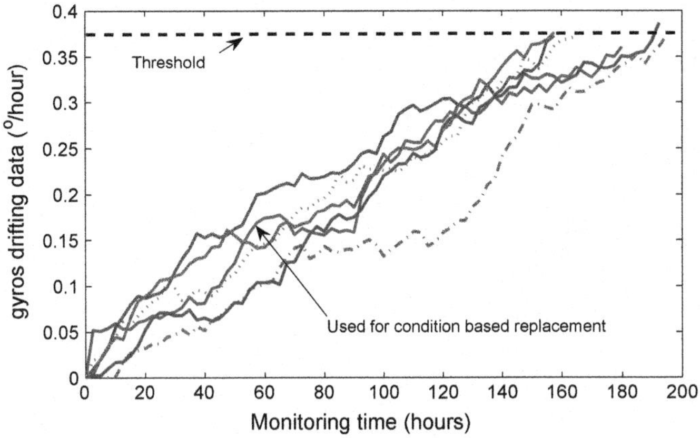

Fig. 3.6 The collected drifting data of six failed gyros

In the practice of the INS health monitoring, it is usually required that the gyro's drift measurement should not exceed 0.37 °/h, i.e., $w = 0.37$ °/h. This threshold is predetermined at the design stage and is strictly enforced in practice since an INS is a critical device used in the navigated weapon system.

Now, we use the above collected data to demonstrate theoretical results developed in this chapter. First, we use the lifetime data in Table 3.2 and lifetime distribution in Eq. (3.3) to determine λ and σ_B^2. Adopting the maximum likelihood estimation method, the estimates of λ and σ_B^2, and their standard errors (std) and confidence intervals (CI) can be obtained, as summarized in Table 3.3.

Based on the estimated λ and σ_B^2, and the collected drift data of gyros as shown in Fig. 3.7, the parameters of the ME existing in the monitoring data can be estimated using the maximum likelihood estimation method presented in [18]. Accordingly, the standard errors and confidence intervals can also be obtained, as summarized in Table 3.3. From Table 3.3, we can observe the uncertainty existing in the estimated parameters. To quantify the effect of the uncertainty in parameter estimations, the standard errors and 95% confidence intervals of the mean lifetime with the ME, the replacement cost and replacement time are also calculated as shown in Table 3.3, where the replacement cost and replacement time are obtained through minimizing Eq. (3.25) using the Nelder–Mead Simplex method in MATLAB toolbox under $c_p = 6{,}000$ RMB and $c_f = 10{,}000$ RMB, which are specified according to the purchase cost of gyro and engineering practice of gyro. From these computation results, we can find that the uncertainty in parameter estimations will have some impacts on the accuracy of the lifetime estimation and replacement decisions, and thus statistical part related to parameter estimation is an important issue. However, the issue about parameter estimation is beyond the focus of this chapter and instead the goal of this chapter is to specify the ME range to achieve a desirable lifetime estimation performance under the given model parameters. The details of the parameter estimation method can be found in [18].

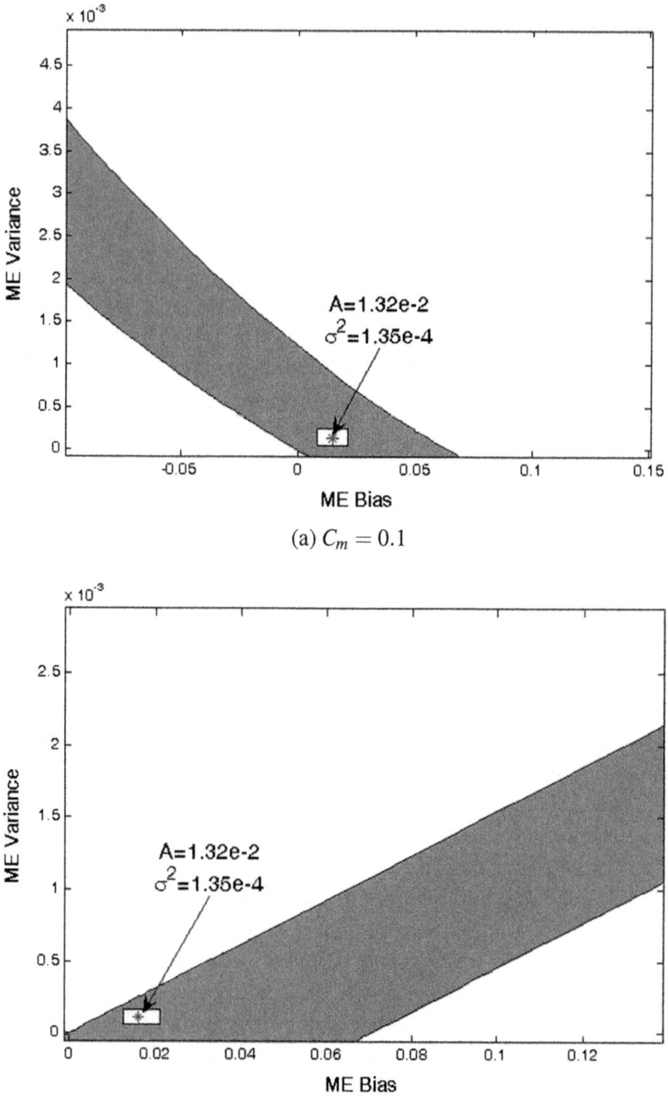

(a) $C_m = 0.1$

(b) $v_{\min} = 0$, $v_{\max} = 0.1$

Fig. 3.7 Permissible ME parameters for gyros' drift data

Considering that the INS is a critical device used in the navigated system, it is expected that the estimated lifetime from the drift data contaminated by the ME should be accurate sufficiently and hence we set the required performance characteristics for lifetime estimation as $C_m = 0.1$, $v_{\min} = 0$, and $v_{\max} = 0.1$. Therefore, the permissible values of A and σ^2 can be determined from Theorems 3.2 and 3.3, which

Table 3.3 Estimates of the parameters, mean lifetime, replacement cost, and replacement time

	MLE	std	95% CI	
			Lower limit	Upper limit
λ	2.04e-3	1.32e-4	1.96e-3	2.12e-3
σ_B^2	3.15e-5	1.41e-5	2.28e-5	4.02e-5
A	1.32e-2	2.10e-3	1.30e-2	1.34e-2
σ^2	1.35e-4	2.52e-5	1.33e-4	1.37e-4
Mean lifetime (h)	174.9	36.48	168.1	182.3
Optimal cost (RMB/h)	35.37	2.56	32.29	37.94
Optimal replacement time (h)	126.9	4.32	124.3	133.5

are illustrated in Fig. 3.7. In Fig. 3.7, the asterisk denotes the maximum likelihood estimates of A and σ^2, and the box corresponds to the 95% confidence intervals of the estimated A and σ^2. It can be observed that the 95% confidence intervals of the estimated A and σ^2 are located in the permissible ranges of A and σ^2. Specifically, it is calculated that $R_{cv}(T_e, T) = 0.0307$ and $R_{var}(T_e, T) = 0.006$ for this case study. Obviously, $R_{cv}(T_e, T) < C_m$ and $v_{\min} < R_{var}(T_e, T) < v_{\max}$. These analyses imply that the required performance for lifetime estimation can be satisfied in this case study under the given C_m, $v_{\min}$, and $v_{\max}$. In the subsequent comparisons, we use the maximum likelihood estimations of the parameters for illustration.

To have a closer look at the effect of ME on the age-based replacement decision, we compare the optimal replacement times and the corresponding long run average costs per unit time with/without the ME effect, respectively. By implementing the Nelder-Mead Simplex method in Matlab toolbox, Eqs. (3.24) and (3.25) can be minimized under the previously specified c_p and c_f. As a result, for the case without the ME effect, the optimal replacement time is $\tau^* = 132.4$ hours and the optimal long run average cost per unit time is $CR(\tau^*) = 33.72$ RMB/hour. For the case with the ME effect, it is obtained that $\tau_e^* = 126.9$ hours and $CR_e(\tau_e^*) = 35.37$ RMB/hour. From these results, it is found that the optimal replacement time τ_e^* is earlier than τ^* 4.15%, and instead the corresponding optimal long run average cost per unit time is increased by 4.89%. This shows that under the age based replacement policy using τ_e^* for replacement will render the earlier replacement and increase the operation cost. However, in this case study, the distribution-related parameters of ME, i.e. A and σ^2, are located in their permissible ranges, as explored previously. Thus the effect of ME on the replacement decision is not significant.

To further exploit the effect of ME on the replacement decision, we calculate $R_{cv}(T_e, T)$, $R_{var}(T_e, T)$, τ_e^*, and $CR_e(\tau_e^*)$ under different pairs of (A, σ^2) for illustrative purposes. The main calculation results are summarized in Table 3.4.

From Table 3.4, we can observe that the appearance of ME will shorten the operation cycle length and incur a greater long run average cost per unit time compared with the case without the ME effect (corresponding to the case with $\sigma^2 = 0$ and $A = 0$), and thus lead to unnecessary economical loss. Particularly, when $\sigma^2 = 0$,

Table 3.4 $R_{cv}(T_e, T)$, $R_{var}(T_e, T)$, τ_e^*, and $CR_e(\tau_e^*)$ under different pairs of (A, σ^2)

			σ^2				
			0	0.000 28	0.000 56	0.001 12	0.002 24
A	$A = 0$	$R_{cv}(T_e, T)$	0	0.024 2	0.047 9	0.093 6	0.179 9
		$R_{var}(T_e, T)$	0	0.024 2	0.047 9	0.093 6	0.179 9
		τ_e^*	132.4	131.9	131.5	130.8	129.9
		$CR_e(\tau_e^*)$	33.721 2	34.065 3	34.399 9	35.042 3	36.231 6
	$A = 0.01$	$R_{cv}(T_e, T)$	0.013 8	0.039 0	0.063 6	0.111 2	0.200 8
		$R_{var}(T_e, T)$	0.013 6	0.010 9	0.034 9	0.081 2	0.168 4
		τ_e^*	128.5	128.0	127.6	127.0	126.2
		$CR_e(\tau_e^*)$	34.817 7	35.186 7	35.545 1	36.232 1	37.500 6
	$A = 0.02$	$R_{cv}(T_e, T)$	0.028 2	0.054 5	0.080 1	0.129 7	0.222 8
		$R_{var}(T_e, T)$	0.027 4	0.002 5	0.021 7	0.068 6	0.156 7
		τ_e^*	124.6	124.1	123.7	123.1	122.4
		$CR_e(\tau_e^*)$	35.984 0	36.380 4	36.764 9	37.501 0	38.855 9
	$A = 0.04$	$R_{cv}(T_e, T)$	0.058 9	0.087 6	0.115 5	0.169 5	0.270 5
		$R_{var}(T_e, T)$	0.055 6	0.030 0	0.005 1	0.043 0	0.133 1
		τ_e^*	116.8	116.4	116.0	115.5	115.0
		$CR_e(\tau_e^*)$	38.553 1	39.013 4	39.458 9	40.308 5	41.861 8
	$A = 0.08$	$R_{cv}(T_e, T)$	0.129 5	0.164 3	0.198 1	0.262 9	0.383 5
		$R_{var}(T_e, T)$	0.114 7	0.087 4	0.061 0	0.010 1	0.084 4
		τ_e^*	101.4	101.0	100.8	100.6	100.5
		$CR_e(\tau_e^*)$	44.888 4	45.527 2	46.141 5	47.302 2	49.386 7

the optimal replacement time is decreasing in A but always earlier than τ^*. At the same time, the optimal long run average cost per unit time is increasing in A but greater than $CR(\tau^*)$ always. This is consistent with the results obtained in Theorem 3.4. Additionally, we observe from the results in Table 3.4 that, when $R_{cv}(T_e, T)$ increases, $CR_e(\tau_e^*)$ has an increasing trend and τ_e^* has a decreasing trend. In the worst case, corresponding to $R_{cv}(T_e, T) = 0.3835$, the optimal replacement time τ_e^* is significantly earlier than τ^* 24.09%, and the corresponding optimal long run average cost per unit time is significantly increased by 46.46%. This contrasts sharply with the case $R_{cv}(T_e, T) = 0.0307$, in which both the decrease in optimal replacement time and the increase in optimal long run average cost per unit time are less than 5%. This demonstrates the necessity of specifying the ME range. Finally, it is worth noting that, though increasing $R_{var}(T_e, T)$ may also make $CR_e(\tau_e^*)$ increase and τ_e^* decrease, there is no general trending among τ_e^*, $CR_e(\tau_e^*)$, and $R_{var}(T_e, T)$. This is due in large part to the fact that $R_{cv}(T_e, T)$ takes into account both the means and variances of T_e and T, but only the variances of T_e and T are taken into account by $R_{var}(T_e, T)$.

To have a further comparison, we here consider a condition-based replacement policy which can make replacement decision at each condition monitoring time, conditional on the condition monitoring data (see, e.g. [29, 37]). Therefore, the residual life distribution is used in such condition-based replacement instead of the lifetime distribution and the cost per sampling is also considered with $c = 30$ RMB in this case. As discussed previously, the drift of the gyro is often used as a performance degradation indicator to evaluate the health condition of the inertial platform. Thus, we use the drift data of the gyro to estimate the residual life and then determine the optimal replacement time. The specific drift data used in this case are shown in Fig. 3.7. Suppose that the degradation of the gyro is discretely monitored at time $0 < t_1 < \ldots < t_i$ and let $y_i = Y(t_i)$ denote the drift measurement at time t_i. Then, the set of the measurements up to t_i is represented by $Y_i = \{y_1, y_2, \ldots, y_i\}$ and the corresponding set of the underlying degradation states up to t_i is represented by $X_i = \{x_1, x_2, \ldots, x_i\}$, where $x_i = X(t_i)$. According to the model setting, if the ME is considered, we have $y_i = x_i + \varepsilon$; otherwise, $x_i = y_i$. Based on the condition monitoring data, the residual life can be estimated at time t_i, denoted by R_i. Further, we denote the estimated residual life distribution with considering the ME as $F_{R_i|Y_i}(r_i)$ and without considering the ME as $F_{R_i|X_i}(r_i)$. The specific formulations for $F_{R_i|Y_i}(r_i)$ and $F_{R_i|X_i}(r_i)$ are briefly summarized in Appendix E. For an illustrating purpose, at the ith monitoring time t_i, the average cost per unit time without the ME under this condition-based replacement policy can be expressed as

$$CdR(\tau) = \frac{i \times c + c_p \bar{F}_{R_i|X_i}(\tau - t_i) + c_f F_{R_i|X_i}(\tau - t_i)}{t_i + \int_0^{\tau - t_i} \bar{F}_{R_i|X_i}(u)\mathrm{d}u}, \tag{3.26}$$

where τ corresponds to the decision variable representing the replacement time and $\bar{F}_{R_i|X_i}(r_i) = 1 - F_{R_i|X_i}(r_i)$. Similarly, the average cost per unit time with the ME at t_i is

$$CdR_e(\tau) = \frac{i \times c + c_p \bar{F}_{R_i|Y_i}(\tau - t_i) + c_f F_{R_i|Y_i}(\tau - t_i)}{t_i + \int_0^{\tau - t_i} \bar{F}_{R_i|Y_i}(u)\mathrm{d}u}, \tag{3.27}$$

where $F_{R_i|Y_i}(r_i)$ is the estimated residual life distribution with considering the ME and $\bar{F}_{R_i|Y_i}(r_i) = 1 - F_{R_i|Y_i}(r_i)$. It is noted that the underlying assumption for Eqs. (3.26) and (3.27) is that the next monitoring time is sufficiently long. Thus Eqs. (3.26) and (3.27) are used to show where the minimum is located and illustrate the results regarding the replacement times and the average costs per unit time; otherwise, the result for τ more than the next monitoring time will be recalculated after obtaining new monitoring data, since the residual life distribution is updated. The specific development and clarification for Eqs. (3.26) and (3.27) are given in Appendix.

Table 3.5 Optimal long run average cost per unit time and optimal replacement time under condition based replacement

Times (h)	Without ME		With ME	
	$CdR(\tau^*)$(RMB/h)	τ^*(h)	$CdR_e(\tau_e^*)$(RMB/h)	τ_e^*(h)
$t_k = 0$	33.72	132.4	35.37	126.9
$t_k = 30$	36.52	135.2	37.15	134.5
$t_k = 60$	42.61	137.3	42.90	135.3
$t_k = 90$	35.02	141.7	37.24	139.8
$t_k = 120$	39.39	145.1	40.13	143.5
$t_k = 150$	36.85	158.6	37.86	157.3

For comparisons, we compute the optimal long run average cost per unit time and the optimal replacement time at several different monitoring times, as summarized in Table 3.5, where τ^* and τ_e^* are the optimal replacement times by minimizing Eqs. (3.26) and (3.27), respectively.

From Table 3.5, we can observe that the condition-based replacement policy will lead to the increase of the average cost per unit time, compared with the age-based replacement policy (corresponding to the results at $t_i = 0$), since the sampling cost is considered. However, the optimal replacement time is extended obviously and thus the operation time of the asset is extended, both in the case of considering the ME effect and the case not considering. Because the monitoring information is considered in the condition-based replacement policy and thus the estimated residual life is repeatedly updated to ensure that the most recently calculated residual life reflects the current reality of the asset. In addition, a common feature shared by the results in the condition-based replacement policy and age-based replacement policy is that the appearance of ME will shorten the length of the operation cycle and incur a greater long run average cost per unit time compared with the case without considering the ME effect. The above-observed phenomenon is not surprising in that the uncertainty introduced by the ME will contribute to the uncertainty of the estimated residual life and thus affect the replacement decisions.

In summary, the results in this case study imply that the replacement decision made directly using the measured data with the ME could be conservative and thus is not economically optimal. Instead, by specifying the ME range given a desirable lifetime estimation performance, it is possible to mitigate such conservativeness so that the effectiveness of the replacement decision can be improved such as extending the operation cycle of the asset and reducing the long run average cost per unit time.

Appendix

Proof of Theorem 3.1

From Eq. (3.7), we have

$$Y(t) = A + X(t) + \varepsilon = A' + X(t). \tag{3.28}$$

where $A' = A + \varepsilon \sim N(A, \sigma^2)$.

As a result, the lifetime T_e can be calculated by the time of $\{Y(t), t \geq 0\}$ hitting the threshold $w - A'$. To do so, the law of total probability is used to incorporate the randomness involved in A' due to the ME. Specifically, we calculate the PDF and CDF of T_e as follows:

$$f_{Te}(t) = \int f_{Te|A'}(t)p(A')\mathrm{d}A' = E_{A'}\left[f_{Te|A'}(t)\right], \tag{3.29}$$

$$F_{Te}(t) = \int F_{Te|A'}(t)p(A')\mathrm{d}A' = E_{A'}\left[F_{Te|A'}(t)\right], \tag{3.30}$$

where $f_{Te|A'}(t)$ and $F_{Te|A'}(t)$ can be obtained by replacing w with $w - A'$ in Eqs. (3.3) and (3.4), respectively.

The expectations in Eqs. (3.29) and (3.30) can be evaluated by Lemma 3.2 and then Eqs. 3.12 and 3.13 are proved. Finally, based on the property of the expectation operator, $E(T_e)$ and $var(T_e)$ can be obtained as follows:

$$E(T_e) = E_{A'}\left[E(T_e|\, A')\right] = E_{A'}\left[\frac{w - A'}{\lambda}\right] = \frac{w - A}{\lambda}, \tag{3.31}$$

$$\begin{aligned} \mathrm{var}(T_e) &= E(T_e^2) - [E(T_e)]^2 = E_{A'}\left[E^2(T_e|\, A') + \mathrm{var}(T_e|\, A')\right] - \left(\frac{w - A}{\lambda}\right)^2 \\ &= \frac{\lambda\sigma^2 + (w - A)\sigma_B^2}{\lambda^3}. \end{aligned} \tag{3.32}$$

Proof of Theorem 3.2

Using the CV measure defined in Eq. (3.16), we have

$$0 \leq \frac{|CV(T_e) - CV(T)|}{CV(T)} \leq C_m, \tag{3.33}$$

with

$$CV(T_e) = \frac{\sqrt{\lambda\sigma^2 + (w - A)\,\sigma_B^2}}{\sqrt{\lambda}(w - A)},\, CV(T) = \frac{\sigma_B}{\sqrt{\lambda w}}. \tag{3.34}$$

From Eqs. (3.33) and (3.34), we directly have

$$0 \le \left| \frac{\sqrt{\lambda\sigma^2 + (w - A)\,\sigma_B^2}}{\sqrt{\lambda}(w - A)} - \frac{\sigma_B}{\sqrt{\lambda w}} \right| / \frac{\sigma_B}{\sqrt{\lambda w}} \le C_m. \tag{3.35}$$

After some manipulations, we obtain

$$\begin{cases} \text{if } \lambda\sigma^2 + \sigma_B^2 A\,(w - A)\,/w \ge 0, \text{ then } \lambda\sigma^2 + \sigma_B^2\,(w - A)\left[1 - (w - A)\,\frac{(C_m^{+}1)^2}{w}\right] \le 0, \\ \text{if } \lambda\sigma^2 + \sigma_B^2 A\,(w - A)\,/w < 0, \text{ then } \lambda\sigma^2 + \sigma_B^2\,(w - A)\left[1 - (w - A)\,\frac{(C_m+1)^2}{w}\right] > 0. \end{cases} \tag{3.36}$$

This completes the proof.

Proof of Theorem 3.3

In a general case, we have

$$\begin{aligned} & v_{\min} \le R_{var}(T_e, T) = \left| \frac{\sqrt{\lambda\sigma^2 + (w-A)\sigma_B^2}}{\sqrt{\lambda^3}} - \frac{\sqrt{w}\sigma_B}{\sqrt{\lambda^3}} \right| / \frac{\sqrt{w}\sigma_B}{\sqrt{\lambda^3}} \le v_{\max} \\ \Leftrightarrow\ & v_{\min} \le R_{var}(T_e, T) = \left| \sqrt{\lambda\sigma^2 + (w - A)\,\sigma_B^2} - \sigma_B\sqrt{w} \right| / \sigma_B\sqrt{w} \le v_{\max} \\ \Leftrightarrow\ & \begin{cases} w\sigma_B^2(2v_{\min} + v_{\min}^2) \le \lambda\sigma^2 - A\sigma_B^2 \le w\sigma_B^2(2v_{\max} + v_{\max}^2), & \lambda\sigma^2 - A\sigma_B^2 \ge 0 \\ w\sigma_B^2(v_{\max}^2 - 2v_{\max}) \le \lambda\sigma^2 - A\sigma_B^2 \le w\sigma_B^2(v_{\min}^2 - 2v_{\min}), & \lambda\sigma^2 - A\sigma_B^2 < 0 \end{cases} \end{aligned} \tag{3.37}$$

This completes the proof.

Proof of Theorem 3.4

(1) Given the degradation process $\{X(t), t \ge 0\}$ as Eq. (3.1) and the threshold w, the expected cycle length can be expressed as

$$E\,(T_r|\,w) = \int_0^{\tau} \bar{F}_T(t)\mathrm{d}t = \tau \bar{F}_T(\tau) + \int_0^{\tau} t f_T(t)\mathrm{d}t. \tag{3.38}$$

Then, similar in spirit to the result in Ye et al. [26], we can complete the proof of $E\,(T_r|\,w)$. For $\{Y(t), t \ge 0\}$ with the ME effect as given in Eq. (3.7), similar to the proof of Theorem 3.2, we have

$$\begin{aligned} E\,(T_{er}|\,w) &= \int_0^{\tau} \bar{F}_{T_e}(t)\mathrm{d}t = \tau \bar{F}_{T_e}(\tau) + \int_0^{\tau} t f_{T_e}(t)\mathrm{d}t \\ &= \tau \int \bar{F}_{Te|A'}(\tau)p(A')\mathrm{d}A' + \int_0^{\tau} t \left(\int f_{Te|A'}(t)p(A')\mathrm{d}A' \right) \mathrm{d}t \\ &= E_{A'}\left[E\left(T_r|\,w - A'\right)\right]. \end{aligned}$$

This completes the proof.

(2) To complete the proof, we note that, if $\sigma^2 = 0$, $f_{T_e}(t)$ in Eq. (3.12) can be reformulated as

$$f_{Te}(t) = \frac{1}{\sqrt{2\pi\sigma_B^2 t^3}}(w - A)\exp\left(-\frac{(w - A - \lambda t)^2}{2\sigma_B^2 t}\right)$$
$$= \frac{\eta\upsilon}{\sqrt{2\pi\eta^3 t^3}}\exp\left(-\frac{(\upsilon - \eta t)^2}{2\eta t}\right),$$

where $\eta = \lambda^2/\sigma_B^2$ and $\upsilon = \lambda(w - A)/\sigma_B^2$.

In this case, $f_{Te}(t)$ is an inverse Gaussian distribution with parameters η and υ. It is known that υ is a convolution parameter of the inverse Gaussian distribution ([38], Proposition A.1, p. 454), and consequently the inverse Gaussian distribution is stochastically increasing in υ ([38], Proposition J.4, p. 262). Note that υ is decreasing in A. Therefore, we conclude that T_e is stochastically decreasing in A. Particularly, when $\sigma^2 = 0$ and $A = 0$, we have $f_T(t) = f_{Te}(t)$ and thus it follows $T \geq_{st} T_e$ for $A \geq 0$. This completes the proof.

(3) In this case, since $\sigma^2 = 0$ and $A \geq 0$, we have $T \geq_{st} T_e$ from the above result. Hence, it follows $\bar{F}_T(t) \geq \bar{F}_{Te}(t)$ from the definition of stochastic order. Considering $E\left(T_r|\, w\right) = \int_0^\tau \bar{F}_T(t)dt$ and $E\left(T_{er}|\, w\right) = \int_0^\tau \bar{F}_{T_e}(t)dt$, we directly have

$$E\left(T_{er}|\, w\right) \leq E\left(T_r|\, w\right). \tag{3.39}$$

To prove $CR_e(\tau) \geq CR(\tau)$, we first compare the expected cost per unit time as follows:

$$c_p\bar{F}_{T_e}(\tau) + c_f F_{T_e}(\tau) = c_p + (c_f - c_p)F_{T_e}(\tau) \geq c_p + (c_f - c_p)F_T(\tau)$$
$$= c_p\bar{F}_T(\tau) + c_f F_T(\tau),\ for c_f > c_p. \tag{3.40}$$

From Eqs. (3.24), (3.25), (3.39) and (3.40), we conclude that $CR_e(\tau) \geq CR(\tau)$.

Let τ^* and τ_e^* denote the solutions of minimizing Eqs. (3.24) and (3.25), respectively. According to the proved $CR_e(\tau) \geq CR(\tau)$, we obtain $CR(\tau^*) \leq CR(\tau_e^*) \leq CR_e(\tau_e^*)$. This completes the proof.

Formulations of $F_{R_i|X_i}(r_i)$ ***and*** $F_{R_i|Y_i}(r_i)$

According to the setting in Sect. 6.2, when the ME is not considered, we have $x_i = y_i$. In this case, given the condition monitoring data X_i up to t_i, we have the following results for the residual life R_i by using the Markov property of Wiener process and the results for the inverse Gaussian distribution,

$$f_{R_i|X_i}(r_i) = \frac{w - x_i}{\sqrt{2\pi\sigma_B^2 r_i^3}} \exp\left(-\frac{(w - x_i - \lambda r_i)^2}{2\sigma_B^2 r_i}\right), \tag{3.41}$$

$$F_{R_i|X_i}(r_i) = 1 - \Phi(\frac{w - x_i - \lambda r_i}{\sqrt{\sigma_B^2 r_i}}) + \exp(\frac{2\lambda(w - x_i)}{\sigma_B^2})\Phi(\frac{-w - \lambda r_i + x_i}{\sqrt{\sigma_B^2 r_i}}). \tag{3.42}$$

When the ME is considered, we have $y_i = x_i + \varepsilon$. Thus the degradation state x_i should be estimated based on Y_i. Define $\hat{x}_i = E(x_i | Y_i)$ and $P_i = var(x_i | Y_i)$ as the expectation and variance of x_i given the measurements to t_i, respectively. Then, according to the linear and Gaussian nature of the considered model, $\hat{x}_i$ and P_i can be calculated by the Kalman filter and we further have $x_i | Y_i \sim N(\hat{x}_i, P_i)$ [39]. Thus the residual life distribution R_i at t_i can be derived similarly by considering the results in Eq. (3.41) and the derivations in the above part, as follows:

$$f_{R_i|Y_i}(r_i) = \frac{\lambda P_i + \sigma_B^2(w - \hat{x}_i)}{\sqrt{2\pi(P_i + \sigma_B^2 r_i)^3}} \exp\left(-\frac{(w - \hat{x}_i - \lambda r_i)^2}{2\left(P_i + \sigma_B^2 r_i\right)}\right), \tag{3.43}$$

$$F_{R_i|Y_i}(r_i) = 1 - \Phi(\frac{w - \hat{x}_i - \lambda r_i}{\sqrt{\sigma_B^2 r_i + P_i}}) + \exp(\frac{2\lambda(w - \hat{x}_i)}{\sigma_B^2} + \frac{2\lambda^2 P_i}{\sigma_B^4})\Phi(\frac{-w - \lambda r_i + \hat{x}_i - 2\lambda P_i/\sigma_B^2}{\sqrt{\sigma_B^2 r_i + P_i}}), \tag{3.44}$$

where $\hat{x}_i$ and P_i are recursively calculated by the Kalman filtering algorithm as follows:

$$\begin{aligned}
&\hat{x}_i = \hat{x}_{i|i-1} + K(i)(y_i - A - \hat{x}_{i|i-1}), \hat{x}_{i|i-1} = \hat{x}_{i-1} + \lambda(t_i - t_{i-1}), \\
&K(i) = P_{i|i-1}(P_{i|i-1} + \sigma^2)^{-1}, P_{i|i-1} = P_{i-1} + \sigma_B^2(t_i - t_{i-1}), \\
&P_i = (1 - K(i))\, P_{i|i-1},
\end{aligned}$$

with $\hat{x}_{i|i-1} = E(x_i | Y_{i-1})$, $P_{i|i-1} = var(x_i | Y_{i-1})$, $\hat{x}_0 = 0$ and $P_0 = 0$. It can be found that the bias A and standard deviation σ for ME can propagate into the estimated residual life distribution.

Formulations of Eqs. (3.26) ***and*** (3.27)

In the considered condition-based replacement policy, because of the availability of condition monitoring information, there is no point looking further than the next monitoring point since the residual life distribution will be updated at that point, and so will the associated decision. This means that the decision should be made over a finite time horizon from now to the next monitoring check which itself could be a decision variable. For example, at the ith monitoring point, with the replacement time τ as the decision variable, we only need to consider the case that τ is less than the time till the next monitoring, which is fixed in our case study. Otherwise new CM data will be available at the next monitoring and all need updating and so the cost formula. If the optimal τ is within the time interval before the next monitoring, we perform a replacement or otherwise wait till the next monitoring. Since τ is less than the interval till the next monitoring, there will be no more monitoring cost to occur and we only need to include the past monitoring costs till the ith monitoring point. Indeed, the monitoring will go on, but the replacement variable in our case is bounded and this is the same to all age-based replacement but just in our case we limit the replacement variable to be within the interval between current and next monitoring. In the case study, for an illustrating purpose to show the replacement time and the associated cost, we calculate the average cost per unit time using a τ which is longer than the actual next monitoring point to show where the minimum is located under the assumption that the next monitoring point is sufficiently long. We use this assumption to formulate the cost function to obtain the optimal replacement time and associated cost for comparative studies.

Based on the above discussions, under the assumption that the next monitoring time is sufficiently long, the expected remaining cycle cost starting at the ith monitoring time till the end of the cycle can be calculated as $c_p \bar{F}_{R_i|X_i}(\tau - t_i) + c_f F_{R_i|X_i}(\tau - t_i)$ in the case without the ME effect, which is obtained based on the renewal reward theory as the age-based case in Eq. (3.24). At the same time, the according expected remaining cycle length is $\int_0^{\tau - t_i} \bar{F}_{R_i|X_i}(u)\mathrm{d}u$. In addition, it is noted that the past sampling cost $i \times c$ and time t_i have already occurred. Thus, the expected cost per cycle obtain at the ith monitoring point is $i \times c + c_p \bar{F}_{R_i|X_i}(\tau - t_i) + c_f F_{R_i|X_i}(\tau - t_i)$, and the corresponding cycle length is $t_i + \int_0^{\tau - t_i} \bar{F}_{R_i|X_i}(u)\mathrm{d}u$. As a result, at t_i, the average cost per unit time without the ME under this condition-based replacement policy can be expressed by Eq. (3.26) as

$$CdR(\tau) = \frac{i \times c + c_p \bar{F}_{R_i|X_i}(\tau - t_i) + c_f F_{R_i|X_i}(\tau - t_i)}{t_i + \int_0^{\tau - t_i} \bar{F}_{R_i|X_i}(u)\mathrm{d}u}.$$

The case of considering the ME effect is similar and thus omitted here.

References

1. Derman C, Lieberman GJ, Ross SM (1984) On the use of replacements to extend system life. Oper Res 32:616–627
2. Bayus BL (1998) An analysis of product lifetimes in a technologically dynamic industry. Manag Sci 44:763–775
3. Wang HZ (2002) A survey of maintenance policies of deteriorating systems. Eur J Oper Res 139(3):469–489
4. Pecht M (2008) Prognostics and health management of electronics. Wiley, New Jersey
5. Elwany A, Gebraeel NZ, Maillart L (2011) Structured replacement policies for systems with complex degradation processes and dedicated sensors. Oper Res 59(3):684–695
6. Scanff E, Feldman KL, Ghelam S, Sandborn P, Glade M, Foucher B (2007) Life cycle cost estimation of using prognostic health management (PHM) for helicopter avionics. Microelectron Reliab 47(12):1857–1864
7. Kalbfleisch JD, Prentice RL (2002) The statistical analysis of failure time data, 2nd edn. Wiley, New Jersey

8. Lawless JF (2002) Statistical models and methods for lifetime data. Wiley-Interscience, New York
9. Si XS, Wang W, Hu CH, Zhou DH, Pecht M (2012) Remaining useful life estimation based on a nonlinear diffusion degradation process. IEEE Trans Reliab 61(1):50–67
10. Meeker WQ, Escobar LA (1998) Statistical methods for reliability data. Wiley, New Jersey
11. Gu J, Barker D, Pecht M (2009) Health monitoring and prognostics of electronics subject to vibration load conditions. IEEE Sens J 9(11):1479–1485
12. Kumar S, Dolev E, Pecht M (2010) Parameter selection for health monitoring of electronic products. Microelectron Reliab 50(2):161–168
13. He W, Williard N, Osterman M, Pecht M (2011) Prognostics of lithium-ion batteries based on Dempster-Shafer theory and the Bayesian Monte Carlo method. J Power Sources 196(18):10314–10321
14. Huynh KT, Barros A, Bérengure C (2012) Maintenance decision-making for systems operating under indirect condition monitoring: value of online information and impact of measurement uncertainty. IEEE Trans Reliab 61(2):410–425
15. Kolle C, O'Leary P (1998) Low-cost, high-precision measurement system, for capacitive sensors. Meas Sci Technol 9(3):510–517
16. Oxtoby NP, Sun HB, Wiseman HM (2003) Non-ideal monitoring of a qubit state using a quantum tunnelling device. J Phys Condens Matter 15(46):8055–8064
17. Dieck RH (2006) Measurement uncertainty: methods and applications, 4th edn. The Instrument Society of America, Research Triangular Park, North Carolina
18. Whitmore GA (1995) Estimating degradation by a Wiener diffusion process subject to measurement error. Lifetime Data Anal 1:307–319
19. Peng CY, Tseng ST (2009) Mis-specification analysis of linear degradation models. IEEE Trans Reliab 58(3):444–455
20. Lu CJ, Meeker WQ (1993) Using degradation measures to estimate a time-to-failure distribution. Technometrics 35(2):161–174
21. Upadhyaya BR, Naghedolfeizi M, Raychaudhuri B (1994) Residual life estimation of plant components. P/PM Technol 7(3):22–29
22. Si XS, Wang W, Hu CH, Zhou DH (2011) Remaining useful life estimation-a review on the statistical data driven approaches. Eur J Oper Res 213:1–14
23. Ye MH (1990) Optimal replacement policy with stochastic maintenance and operation costs. Eur J Oper Res 44(1):84–94
24. Tseng ST, Tang J, Ku LH (2003) Determination of optimal burn-in parameters and residual life for highly reliable products. Nav Res Logist 50:1–14
25. Crowder M, Lawless J (2007) On a scheme for predictive maintenance. Eur J Oper Res 176:1713–1722
26. Ye ZS, Shen Y, Xie M (2012) Degradation-based burn-in with preventive maintenance. Eur J Oper Res 221(2):360–367
27. Barlow R, Hunter L (1960) Optimum preventive maintenance policies. Oper Res 8(1):90–100
28. Christer AH, Wang W, Sharp JM (1997) A state space condition monitoring model for furnace erosion prediction and replacement. Eur J Oper Res 101:1–14
29. Elwany AH, Gebraeel NZ (2008) Sensor-driven prognostic models for equipment replacement and spare parts inventory. IIE Trans 40:629–639
30. Lee MLT, Whitmore GA (2006) Threshold regression for survival analysis: modeling event times by a stochastic process reaching a boundary. Stat Sci 21(4):501–513
31. Folks JL, Chhikara RS (1978) The inverse Gaussian distribution and its statistical application–a review. J Roy Stat Soc Ser B 40(3):263–289
32. Linna KW, Woodall WH (2001) Effect of MEs on Shewhart control charts. J Qual Technol 33(2):213–222
33. Lovie P (2005) Coefficient of variation. Encyclopedia of statistics in behavioral science. Wiley, New Jersey
34. Chien YH, Sheu SH (2006) Extended optimal age-replacement policy with minimal repair of a system subject to shocks. Eur J Oper Res 174(1):169–181

35. Vagnorius Z, Rausand M, Sorby K (2010) Determining optimal replacement time for metal cutting tools. Eur J Oper Res 206(2):407–416
36. Ross SM (2007) Introduction to probability models. Academic Press, London
37. Carr MJ, Wang W (2011) An approximate algorithm for prognostic modelling using condition monitoring information. Eur J Oper Res 211(1):90–96
38. Marshall AW, Olkin I (2007) Life distributions: structure of nonparametric, semiparametric, and parametric families. Springer, New York
39. Harvey AC (1991) Forecasting, structural time series models and the Kalman filter. Cambridge University Press, Cambridge

Part II
Prognostic Techniques for Linear Degrading Systems

Chapter 4
An Adaptive Remaining Useful Life Estimation Approach with a Recursive Filter

4.1 Introduction

Enhancing safety, efficiency, availability, and effectiveness of industrial and military systems through prognostics and health management (PHM) paradigm has gained momentum over the last decade [1, 2]. PHM is a systematic approach that is used to evaluate the reliability of a system in its actual life cycle conditions, predict failure progression, and mitigate operating risks via management actions. There are two parts in PHM, namely, '*prognostics*' and '*health management*'. Prognostics is often characterized by estimating the remaining useful life (RUL) of a system using available condition monitoring (CM) information [3–7]. Once such prognosis is available, appropriate health management actions such as repair, replacement, and logistic support can be performed to achieve the required system's operational objectives [8–10]. In PHM, the term 'RUL estimation' often implies to find the probability density function (PDF) of the RUL or the mean of the RUL [11], but the emphasis is often placed more on estimating the PDF of the RUL than the mean RUL since a PDF can characterize the uncertainty associated with the RUL and is hence more informative for management decision-making.

The current RUL estimation approaches can be broadly classified as physics of failure, data-driven and fusion methods. Physics of failure approaches rely on the physics of underlying failure mechanisms. Data-driven approaches achieve RUL estimation via data fitting mainly including machine learning and statistics-based approaches. The fusion approaches are the combination of the physics of failure and data-driven approaches. However, for complex or large-scale engineering systems, it is typically difficult to obtain the physical failure mechanisms in advance or cost-expensive and time-consuming to capture the physics of failure by experiments. In contrast, data-driven approaches attempt to derive models directly from collected degradation data or life data, and thus are more appealing and have gained much attention in recent years.

Statistics-based data-driven methods for RUL estimation can be classified into the models based on indirectly observed state processes and the models based on

X.-S. Si et al., *Data-Driven Remaining Useful Life Prognosis Techniques*,
Springer Series in Reliability Engineering, DOI 10.1007/978-3-662-54030-5_4

directly observed state processes [2]. The former models considered the data partially indicating the underlying state of the system, and assumed that the available CM data were stochastically related to the underlying health state. In this case, lifetime data must be available to establish the relationship between the CM data and failure. The latter models utilized the observed degradation data directly to describe the underlying state of the system. Therefore, the RUL is defined as the time to reach the failure threshold of the monitored degradation data for the first time, namely the first hitting time (FHT) [12]. It is noted that the observed degradation data with a threshold are easier to manipulate and implement in practice when lifetime data are scarce [13, 14].

It is well recognized that degradation process is uncertain over time and thus stochastic models are frequently used to characterize the evolution of degradation process. In literature, random effect regression (RER) models and stochastic process (SP) models are two kinds of most commonly used stochastic models. Lu and Meeker in [15] first presented the RER model to characterize the degradation of a population of units, and many extensions appeared later [2, 16]. However, the degradation modeling paradigm in RER models is based on the fact that a population of 'identical' systems (or devices) has a common degradation form. However, individual systems may exhibit different degradation rates, hence the different failure times. On the other side, Pandey et al. in [17] identified that the temporal uncertainty of the degradation process was not taken into account in RER models and thus argued that SP models could remedy it well. They also showed the advantages of SP models over RER models in condition based maintenance. SP models such as Markov chain, Gamma processes, and Wiener processes have been widely used to model the degradation process [5, 18–21]. A Wiener-process-based degradation model is one of statistics-based data-driven models, which can characterize a non-monotonic degradation process, and provide a good description of system's behavior due to an increased or reduced intensity of the use [22]. This type of models has been applied to model the degradation process and to estimate the RUL of a variety of industrial assets, such as rotating element bearings [19], LED lamps [23], self-regulating heating cables [24], laser generator [25], bridge beams [26], and fatigue crack dynamics [27]. Therefore, in this chapter, we focus on RUL estimation based on a Wiener process where the degradation process can be observed directly and a failure threshold of degradation is available. It is known that the selection of the failure threshold is an important problem in practice. However, such threshold is usually set based on either engineering domain knowledge or accepted industrial standards. For example, the ISO 2372 and ISO 10816 are frequently adopted for defining acceptable vibration threshold levels. It is, therefore, an issue beyond the scope of this chapter. In this work, we assume that the failure threshold is known a priori.

A Wiener-process-based model has a drift term characterized by its drift coefficient and a noise term by Brownian motion. It has been widely used to model degradation processes which can be observed directly, and conduct lifetime analysis. Tseng et al. in [23] used a Wiener process to determine the lifetime for the light intensity of LED lamps of contact image scanners. As an extension, Tseng and Peng proposed an integrated Wiener process to model the cumulative degradation path of a product's quality characteristics [28]. Joseph and Yu used a Wiener process for degradation modeling and reliability improvement [14]. Other recent extensions in lifetime estimation can be found in [29–31]. However, the degradation modeling paradigm for lifetime analysis in these conventional Wiener-process-based models is based on an assumption that the estimated PDF of RUL depends only on the currently observed degradation data, which is a strong Markovian assumption.

To relax this assumption, Gebraeel et al. in [19] presented an exponential degradation model for rotating element bearings based on a Wiener process, but incorporated some new and important improvements for RUL estimation. Their model established a linkage between the past and current degradation data of the same system by a Bayesian mechanism. However, it is worth noting that the Brownian motion in the Wiener process was just used as an error term in their models and the availability of the explicit distribution of the FHT from the Wiener process was not utilized. Instead, they directly estimated the RUL distribution using an implicit monotonic assumption. It is well known that Wiener processes are non-monotonic. As such, the resulted RUL estimates in [19] are approximations. In addition, we note that the stochastic coefficients reflecting the individual-to-individual variability in [19] followed some prior distributions, but no elaborated method was presented to select the parameters in the prior distributions. Typically, several systems' historical degradation data of the same type are required to determine the prior parameters. However, such historical degradation data of many systems are not always available in practice, particularly for newly commissioned systems. It is shown in Sect. 4.4 that the inappropriate selection of the prior parameters can result in inaccurate estimation of the degradation and the RUL. Wang et al. in [32] recently proposed a Wiener-process-based model which used all past degradation data to date of the system for RUL estimation. Their model explicitly used the FHT from the Wiener process for RUL estimation. However, we note that the model in [32] also required the data of many same systems for parameter estimation, and the distribution of the updated drift coefficient was not considered. Our results reveal that considering such distribution can lead to uncertainty reduction in the estimated RUL. One final observation from the existing literature of Wiener-process-based RUL estimation models is that all estimated parameters are not updated in line with newly observed data.

From the above review of related researches, we observe that there are three issues remaining to be solved when applying Wiener process for RUL estimation. The first is how to estimate the model parameters from an individual system's data without the need of past data from many same systems. The second is to consider the distribution in the estimated drift coefficient, which is a critical parameter having impact on both the mean and variance of the RUL. The third is to update model parameters based on newly observed degradation data. As we know the system's life is heavily influenced by the way it is operated, maintained and the environment where it has been operating. The consideration of the above three issues will make our model tailored to an individual system through its actual monitoring data which relate to its operational and environmental characteristics.

In this chapter, we address the above issues by utilizing a Wiener-process-based model with a recursive filter algorithm for RUL estimation. We use two techniques for the updating of the RUL estimation. A state-space model is used to recursively update the drift coefficient and an expectation maximization (EM) algorithm is used to reestimate all unknown parameters at each time when new data are available. The new contributions of this chapter are summarized as follows: (1) Different from all previous works, our model estimates an individual system's RUL based on its entire monitoring information to date through a recursive filter and an EM algorithm, and does not require historical degradation data of other systems in a population; (2) Unlike the work of Wang et al. in [32] where a recursive filter was also used, the drift coefficient is treated as a random variable to incorporate its distribution in estimation; (3) Our model is also different from the approximated results in [19, 33] in that our result on the PDF of the RUL is exact in the sense of the FHT, and we also show that our result can ensure that the moments of the RUL exist, but this is not the case for the approximated results in [19, 33]; (4) We apply the proposed model to estimate the RUL of gyros in an inertial navigation system used in weapon systems as a case application.

The remainder parts are organized as follows. In Sect. 4.2, we develop a Wiener-process-based degradation model and obtain the distribution of the RUL. In Sect. 4.3, we discuss the parameter estimation algorithm in detail. Section 4.4 provides a case study to illustrate the application and usefulness of the developed model.

4.2 Wiener-Process-Based Degradation Modeling and RUL Estimation

4.2.1 An Outline of Wiener-Process-Based Degradation Model for Lifetime Analysis

In this section, we briefly present the conventional Wiener-process-based degradation model for lifetime analysis. A Wiener process is typically used for modeling degradation processes where the degradation increases linearly in time with random noise. The rate of degradation is characterized by the drift coefficient. The wear of break pads on automotive wheels is a practical example. Christer and Wang in [34] modeled the wear of break pads as a linear function of time where the thickness of the break pad decreases linearly in time with a Gaussian noise.

In general, a Wiener-process-based degradation model can be represented as,

$$X(t) = \lambda t + \sigma B(t), \tag{4.1}$$

where λ is the drift coefficient, $\sigma > 0$ is the diffusion coefficient, and $B(t)$ is the standard Brownian motion representing the stochastic dynamics of the degradation process.

In physics, a Wiener process aims at modeling the movement of small particles in fluids and air with tiny fluctuations. A characteristic feature of this process in the context of reliability is that the plant's degradation can increase or decrease gradually and accumulatively over time. The tiny increase or decrease in degradation over a small time interval behaves similarly to the random walk of small particles in fluids and air. Therefore, this type of stochastic processes has been widely used to characterize the path of degradation processes where successive fluctuations in degradation can be observed, such as the degradation observations of rotating element bearings [19], LED lamps [23], self-regulating heating cables [24], laser generator [25], bridge beams [26] and other examples in [14, 30–35] and our case of gyros' drifting. Modeling a stochastic degradation process as a Wiener process implies that the mean degradation path is a linear function of time, i.e., $E\left[X(t)\right] = \lambda t$. Therefore, the drift parameter λ is closely related with the progression of the degradation. In addition, we have the variance of the degradation process $var\left[X(t)\right] = \sigma^2 t$, which represents the uncertainty of the degradation at time t.

For in service lifetime estimation at time t_i with the obtained degradation observation x_i, we can use,

$$\begin{aligned} X(t) &= x_i + \lambda(t - t_i) + \sigma\left(B(t) - B(t_i)\right) \\ &= x_i + \lambda(t - t_i) + \sigma B(t - t_i),\ for\ t > t_i. \end{aligned} \tag{4.2}$$

At time t_i, we assume $x_i < w$, otherwise the degradation has crossed w and the system would have failed as defined. Although the above model setting is the same, there are two different ways to relate the degradation $X(t)$ to lifetime T at t_i in the literature. The first one is that the lifetime is directly defined as $T = \{t : X(t) \geq w|\, x_i\}$, and then the lifetime distribution can be represented by $F_{T|x_i}(t|\, x_i) = \Pr(X(t) \geq w|\, x_i)$, such as [19, 31, 33]. To see this, using $T = \{t : X(t) \geq w|\, x_i\}$, the PDF and the cumulative density function (CDF) of lifetime T at time t_i can be directly obtained as,

$$f_{T|x_i}(t|\, x_i) = \frac{1}{\sigma\sqrt{2\pi(t - t_i)}} \exp\left(-\frac{(w - x_i - \lambda(t - t_i))^2}{2\sigma^2(t - t_i)}\right), \tag{4.3}$$

$$F_{T|x_i}(t|\, x_i) = 1 - \Phi\left(\frac{w - x_i - \lambda(t - t_i)}{\sigma\sqrt{t - t_i}}\right), \tag{4.4}$$

where $\Phi(\cdot)$ denotes the standard normal CDF.

Another one defines the lifetime based on the concept of the FHT as $T = \inf\{t : X(t) \geq w|\, x_i\}$, such as [21, 29, 30, 32, 35]. Therefore, using the FHT, the following results can be obtained [36],

$$f_{T|x_i}(t|\, x_i) = \frac{w - x_i}{\sqrt{2\pi(t - t_i)^3\sigma^2}} \exp\left\{-\frac{(w - x_i - \lambda(t - t_i))^2}{2\sigma^2(t - t_i)}\right\}, \tag{4.5}$$

$$F_{T|x_i}(t|x_i) = 1 - \Phi\left(\frac{w - x_i - \lambda(t - t_i)}{\sigma\sqrt{t - t_i}}\right) + \exp\left\{\frac{2\lambda(w - x_i)}{\sigma^2}\right\}\Phi\left(\frac{-(w - x_i) - \lambda(t - t_i)}{\sigma\sqrt{t - t_i}}\right), \tag{4.6}$$

with the mean $E[T|x_i] = t_i + (w - x_i)/\lambda$ and variance $\mathrm{var}[T|x_i] = (w - x_i)\sigma^2/\lambda^3$. This reflects the relationship between the parameters used in the model, the current degradation measurement and the estimated future life of the plant modeled. Particularly, λ is critical for both the mean and variance of the estimated lifetime.

Clearly, the above two definitions are different from each other and also lead to different lifetime estimations. As noted by Park and Bae in [16], $T = \{t : X(t) \geq w|x_i\}$ completely ignores the possible hitting events within interval (t_i, t) and thus is only a crude approximation to $T = \inf\{t : X(t) \geq w|x_i\}$ when the degradation fluctuations are large. In addition, the CDF using the FHT as Eq. (4.6) is greater than the approximation by Eq. (4.4). For safety-critical systems, using the approximated result as Eq. (4.4) for maintenance scheduling may lead to under-maintenance because of the lower risk of failure estimated by Eq. (4.4). As such, it is necessary to consider the FHT as the lifetime, which is exact if the failure is defined as the FHT.

We note that Eq. (4.6) uses only the current degradation data, but not its history before t_i. However, as we have discussed, this is a strong Markovian assumption and ideally the future FHT should depend on the path that the degradation has involved to date. For example, in Fig. 4.1, at the same level of x_i, case (a) would be expected to fail faster than case (b), but using Eq. (4.6) will give the same prediction.

Consequently, it is desired to utilize the degradation data to date for evaluating the RUL of the degraded system. It is expected that utilizing the degradation data to date can make the RUL estimation sharper and more tailored to an individual system than only using the current data. This is our main focus in the remaining parts of this chapter.

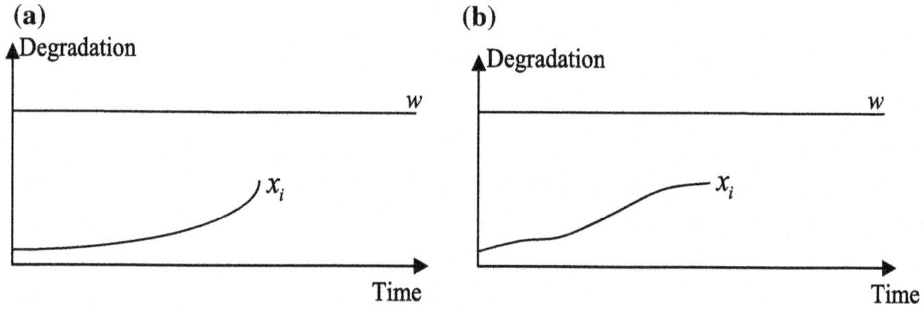

Fig. 4.1 Two exemplar sample paths with different tracks but the same x_i

4.2.2 Wiener-Process-Based Degradation Modeling

Now we address the issues discussed in the introduction. Since some variants introduced in [19] can be easily transformed into Eq. (4.1) by logarithmic transformation, we only focus on Eq. (4.1) which is used to describe the evolution of the monitored degradation variable over time in this chapter.

To incorporate the history of the observations and to maintain at the same time the nice property of the Wiener process, we consider an updating procedure for coefficient λ by a random walk model $\lambda_i = \lambda_{i-1} + \eta$ over time where $\eta \sim N(0, Q)$. Thus the drift coefficient λ evolves as a time-dependent random variable with a distribution, conditional on λ_{i-1}. In fact, the diffusion coefficient, σ, can also be made time-dependent. However the structure of Eq. (4.1) does not allow us to use the state-space model shown later, and therefore a general filter has to be used, which is computationally difficult. There is also a practical reason why we are only interested in making λ time-varying. It is known from Eq. (4.1) and the discussion in Sect. 4.2.1 that the mean degradation and the progression of the degradation are governed by the drift coefficient λ and the time while the diffusion coefficient σ controls in part the uncertainty in the degradation process. The trend in degradation is determined by the drift coefficient while the diffusion coefficient only influences the noise, which can be considered to be constant. A similar idea can also be found in statistical process control literature, in which it is frequently assumed that the process mean will change but the variance will be constant when the process shifts from 'in control' to 'out of control' [37]. Motivated by the state-space model [38], the degradation equation can be reconstructed via a linear state-space model as,

$$\lambda_i = \lambda_{i-1} + \eta, \tag{4.7}$$

$$x_i = x_{i-1} + \lambda_{i-1}(t_i - t_{i-1}) + \sigma \varepsilon_i, \tag{4.8}$$

where $t_0 = 0$, $x_0 = 0$, and $\varepsilon_i \sim N(0, t_i - t_{i-1})$. The use of $t_i - t_{i-1}$ as the variance of ε_i is required by the property of Brownian motion. Equation (4.7) is called the system equation, while Eq. (4.8) is the observation equation [38]. The reason to use a linear system equation is not only because of its simplicity. We can use a nonlinear system equation for λ_i, but we expect that the gain obtained will be minor, but at the expense of a substantially long computation time. There is also a practical problem as what form of nonlinearity we should use because λ_i is not observable. Since the system equation is to model the change of the drift coefficient over a sampling interval which is not long, we would expect that λ_i should be around λ_{i-1} adjusted by the noise term.

We assume that the initial drift coefficient λ_0 follows a normal distribution with mean a_0 and variance P_0 as required by the state space model. The drift coefficient is considered as a hidden "state" and can be estimated from the observations up to t_i, denoted by $\mathbf{X}_{0:i} = \{x_0, x_1, x_2, \ldots, x_i\}$. As such, this model establishes the linkage between the drift coefficient and the observation history up to t_i. In Eq. (4.7), λ_i

follows a distribution which can be estimated by a recursive filter once new observation x_i is available at t_i. We denote its mean by $\hat{\lambda}_i = E(\lambda_i | \mathbf{X}_{0:i})$ and its variance by $P_{i|i} = Var(\lambda_i | \mathbf{X}_{0:i})$.

In order to compute $\hat{\lambda}_i$ and $P_{i|i}$, we need to know the PDF of the λ_i given $\mathbf{X}_{0:i}$, denoted by $p(\lambda_i | \mathbf{X}_{0:i})$. Recursion solution of $p(\lambda_i | \mathbf{X}_{0:i})$ can be computed from $p(\lambda_{i-1} | \mathbf{X}_{0:i-1})$ by the well-known Bayesian rule as follows,

$$\begin{aligned} p(\lambda_i | \boldsymbol{X}_{0:i}) &= \int p(\lambda_i | \lambda_{i-1}) p(\lambda_{i-1} | \boldsymbol{X}_{0:i}) \mathrm{d}\lambda_{i-1} \\ &= \frac{\int p(\lambda_i | \lambda_{i-1}) p(x_i | \lambda_{i-1}, \boldsymbol{X}_{0:i-1}) p(\lambda_{i-1} | \boldsymbol{X}_{0:i-1}) \mathrm{d}\lambda_{i-1}}{p(x_i | \boldsymbol{X}_{0:i-1})}. \end{aligned} \tag{4.9}$$

It has been well established that if Eqs. (4.7) and (4.8) are used, Eq. (4.9) is Gaussian with mean $\hat{\lambda}_i$ and variance $P_{i|i}$ which can be computed by the Kalman filter [38]. As a result, the entire history is captured via recursively updating the estimate of λ_i, which is the advantage of the state-space model. The recursive estimations for $\hat{\lambda}_i$ and $P_{i|i}$ using Kalman filtering are summarized as

Algorithm 4.1 (*Kalman filtering algorithm*)

Step 1: Initialize $\hat{\lambda}_0 = a_0, P_0$.

Step 2: State estimation at time t_i

$$\begin{aligned} &P_{i|i-1} = P_{i-1|i-1} + Q \\ &K_i = (t_i^- t_{i-1})^2 P_{i|i-1} + \sigma^2 (t_i - t_{i-1}) \\ &\hat{\lambda}_i = \hat{\lambda}_{i-1} + P_{i|i-1}(t_i^- t_{i-1}) K_i^{-1} \left(x_i - x_{i-1} - \hat{\lambda}_{i-1}(t_i^- t_{i-1}) \right) \end{aligned}.$$

Step 3: Updating variance $P_{i|i} = P_{i|i-1} - P_{i|i-1}(t_i^- t_{i-1})^2 K_i^{-1} P_{i|i-1}$.

In literature, the Kalman filter has been successfully applied when system states (here referred the drift coefficient as a state) and observations evolve in a smooth and gradually changing way. However, degradation sometimes may have jumps or sudden changes [19]. Here, we introduce an algorithm to deal with this by a strong tracking filter (STF) [39]. STF is also Kalman filter-based, but adjusts the prediction variance $P_{i|i-1}$ so that it is sensitive to the prediction error, $x_i - x_{i-1} - \hat{\lambda}_{i-1}(t_i - t_{i-1})$, and then the filter gain K_i is sensitive to the change of the system state. The details about the STF algorithm are summarized as Algorithm 4.2.

Algorithm 4.2 (*Strong tracking filtering algorithm*)

Step 1: Initialize $\hat{\lambda}_0 = a_0, P_0, \alpha, \rho$.

Step 2: Calculating fading factor $\upsilon(t_i)$ from orthogonality principle

$$V_0(t_i) = \begin{cases} \gamma^2(t_1), \ i = 1 \\ \frac{\rho V_0(t_{i-1}) + \gamma^2(t_i)}{1+\rho}, \ i > 1 \end{cases} \quad \textit{with } \gamma(t_i) = x_i - x_{i-1} - \hat{\lambda}_{i-1}(t_i - t_{i-1})$$

$$B(t_i) = V_0(t_i) - Q(t_i - t_{i-1})^2 - \alpha\sigma^2(t_i - t_{i-1});\ C(t_i) = P_{i-1|i-1}(t_i - t_{i-1})^2;\ \upsilon_0 = B(t_i)/C(t_i).$$

$$\upsilon(t_i) = \begin{cases} \upsilon_0, \ \upsilon_0 \geq 1 \\ 1, \ \upsilon_0 < 1 \end{cases}$$

Step 3: State estimation
$P_{i|i-1} = \upsilon(t_i)P_{i-1|i-1} + Q$
$K_i = (t_i - t_{i-1})^2 P_{i|i-1} + \sigma^2(t_i - t_{i-1})$
$\hat{\lambda}_i = \hat{\lambda}_{i-1} + P_{i|i-1}(t_i - t_{i-1})K_i^{-1}\left(x_i - x_{i-1} - \hat{\lambda}_{i-1}(t_i - t_{i-1})\right)$.
Step 4: Updating variance $P_{i|i} = P_{i|i-1} - P_{i|i-1}(t_i - t_{i-1})^2 K_i^{-1} P_{i|i-1}$.

In Algorithm 4.2, $\alpha \geq 1$ and ρ denote the softening factor and the forgetting factor respectively, which can be selected heuristically. $\rho = 0.95$ has been used in general [39, 40].

Based on Eqs. (4.7), (4.8), and (4.9), the PDF of λ_i conditional on $\mathbf{X}_{0:i}$ is,

$$p(\lambda_i | \mathbf{X}_{0:i}) = \frac{1}{\sqrt{2\pi P_{i|i}}} \exp\left[-\left(\lambda_i - \hat{\lambda}_i\right)^2 \Big/ 2P_{i|i}\right], \tag{4.10}$$

where the dependence between λ_i and $\mathbf{X}_{0:i}$ is contained in $\hat{\lambda}_i$ and $P_{i|i}$. Based on this result, we derive the associated RUL distribution in the following.

4.2.3 Real-Time Updating of the RUL Distribution

Based on a predefined threshold w, the RUL modeling principle is that when degradation $X(t)$ first reaches threshold w, the system is declared to be nonoperable and its lifetime terminates. Consequently, it is natural to view the event of lifetime termination as the point that the degradation $X(t)$ exceeds threshold w for the first time. In this chapter, from the concept of the FHT, we define RUL R_i at time t_i as

$$R_i = \inf\{r_i : X(r_i + t_i) \geq w | \mathbf{X}_{0:i}\}, \tag{4.11}$$

with CDF $F_{R_i|\mathbf{X}_{0:i}}(r_i|\mathbf{X}_{0:i})$and PDF $f_{R_i|\mathbf{X}_{0:i}}(r_i|\mathbf{X}_{0:i})$.

From Eq. (4.2), it is direct to obtain the PDF and CDF of the RUL at time t_i defined in Eq. (4.11) as follows [36],

$$f_{R_i|\lambda_i,\mathbf{X}_{0:i}}(r_i|\lambda_i, \mathbf{X}_{0:i}) = \frac{w - x_i}{\sqrt{2\pi r_i^3 \sigma^2}} \exp\left(-\frac{(w - x_i - \lambda_i r_i)^2}{2\sigma^2 r_i}\right), r_i > 0. \tag{4.12}$$

$$F_{R_i|\lambda_i,\mathbf{X}_{0:i}}(r_i|\lambda_i, \mathbf{X}_{0:i}) = 1 - \Phi\left(\frac{w - x_i - \lambda_i r_i}{\sigma\sqrt{r_i}}\right) + \exp\left(\frac{2\lambda_i(w - x_i)}{\sigma^2}\right) \times \Phi\left(\frac{-(w - x_i) - \lambda_i r_i}{\sigma\sqrt{r_i}}\right), r_i > 0. \tag{4.13}$$

In Eq. (4.12) if we replace λ_i by $\hat{\lambda}_i$, then it is the RUL model used in [32]. We call this as *Wang's model* for subsequent comparisons in Sect. 4.4. However, as mentioned above, the drift coefficient evolves as a random variable in Eq. (4.7) with a distribution, $p(\lambda_i | \mathbf{X}_{0:i})$, conditional on the observed data up to time t_i as formulated by Eq. (4.10). Now we want to use $p(\lambda_i | \mathbf{X}_{0:i})$ for deriving the estimated RUL distribution. In order to achieve this aim, we first give two lemmas, which can significantly simplify the course of the derivation for the RUL distribution.

Lemma 4.1 *If* $Y \sim N(\mu_1, \sigma_1^2)$*, then* $E_Y[\Phi(Y)]$ *can be formulated as* $E_Y[\Phi(Y)] = \Phi(\mu_1 \big/ \sqrt{\sigma_1^2 + 1})$ *where* $\Phi(\cdot)$ *denotes the standard normal CDF.*

Proof From the property of the normal distribution, we have

$$\begin{aligned} E_Y[\Phi(Y)] &= E\left[\, E(I_{\{Z \le Y\}} | \, Y) | \, Y\right] = \Pr(Z \le Y) = \Pr(Z - Y \le 0) \\ &= \Phi(\mu_1 \Big/ \sqrt{\sigma_1^2 + 1}) \end{aligned} \quad . \tag{4.14}$$

In the derivation process, $I_{\{Z \le Y\}}$ is the indicator function, Z is a standard normal variable and independent of Y, and $Z - Y \sim N(-\mu_1, \sigma_1^2 + 1)$.

Lemma 4.2 *If* $\lambda \sim N(\mu_\lambda, \sigma_\lambda^2)$*, the PDF and CDF of the FHT of process* $X(t) = \lambda t + \sigma B(t)$ *to first hit threshold* w *can be formulated as*

$$f_T(t) = \frac{w}{\sqrt{2\pi t^3 \left(\sigma_\lambda^2 t + \sigma^2\right)}} \exp\left[-\frac{(w - \mu_\lambda t)^2}{2t\left(\sigma_\lambda^2 t + \sigma^2\right)}\right]. \tag{4.15}$$

$$F_T(t) = \Phi\left(\frac{\mu_\lambda t - w}{\sqrt{\sigma_\lambda^2 t^2 + \sigma^2 t}}\right) + \exp\left(\frac{2\mu_\lambda w}{\sigma^2} + \frac{2\sigma_\lambda^2 w^2}{\sigma^4}\right)\Phi\left(-\frac{2\sigma_\lambda^2 w t + \sigma^2(\mu_\lambda t + w)}{\sigma^2\sqrt{\sigma_\lambda^2 t^2 + \sigma^2 t}}\right). \tag{4.16}$$

Lemma 4.2 is similar to the results given in [41, 42]. This lemma can be obtained by some direct manipulations using Lemma 4.1 and the total law of probability [25]. From Lemma 4.2, we give the following theorem.

Theorem 4.1 *For the Wiener process defined by Eq.* (4.2) *and the state-space model as Eqs. (4.7) and (4.8), the PDF and CDF of the updated RUL at* t_i *based on the updated PDF of* λ_i *can be obtained as*

$$f_{R_i|\mathbf{X}_{0:i}}(r_i|\mathbf{X}_{0:i}) = \frac{w - x_i}{\sqrt{2\pi r_i^3 \left(P_{i|i} r_i + \sigma^2\right)}} \exp\left(-\frac{\left(w - x_i - \hat{\lambda}_i r_i\right)^2}{2r_i\left(P_{i|i} r_i + \sigma^2\right)}\right), r_i > 0. \tag{4.17}$$

$$F_{R_i|\mathbf{X}_{0:i}}(r_i|\mathbf{X}_{0:i}) = 1 - \Phi\left(\frac{w-x_i-\hat{\lambda}_i r_i}{\sqrt{P_{i|i}r_i^2+\sigma^2 r_i}}\right) + \exp\left(\frac{2\hat{\lambda}_i(w-x_i)}{\sigma^2} + \frac{2P_{i|i}(w-x_i)^2}{\sigma^4}\right)\Phi\left(-\frac{2P_{i|i}(w-x_i)r_i+\sigma^2\left(\hat{\lambda}_i r_i+w-x_i\right)}{\sigma^2\sqrt{P_{i|i}r_i^2+\sigma^2 r_i}}\right). \tag{4.18}$$

Proof Using Eqs. (4.10), (4.12), (4.13), (4.15) and (4.16) and the total law of probability, we have

$$f_{R_i|\mathbf{X}_{0:i}}(r_i|\mathbf{X}_{0:i}) = \int_{-\infty}^{+\infty} f_{R_i|\lambda_i,\mathbf{X}_{0:i}}(r_i|\lambda_i,\mathbf{X}_{0:i})p(\lambda_i|\mathbf{X}_{0:i})d\lambda_i, \tag{4.19}$$

$$F_{R_i|\mathbf{X}_{0:i}}(r_i|\mathbf{X}_{0:i}) = \int_{-\infty}^{+\infty} F_{R_i|\lambda_i,\mathbf{X}_{0:i}}(r_i|\lambda_i,\mathbf{X}_{0:i})p(\lambda_i|\mathbf{X}_{0:i})d\lambda_i. \tag{4.20}$$

Following Lemma 4.2, it is straightforward to obtain Eqs. (4.17) and (4.18).

Comparing Eq. (4.17) with Eq. (4.12), we observe that the observation history and the variance of λ_i are involved in Eqs. (4.17) and (4.18), which is also recursively updated. We call Eq. (4.17) as *our model* to distinguish other models in Sect. 4.4. Therefore, in the RUL estimation by our model, we account for both the temporal uncertainty of the degradation process and the uncertainty in drift parameter λ_i.

Remark 4.1 In [19], they directly used $\Pr(R_i \leq r_i|\mathbf{X}_{0:i}) = \Pr(X(r_i+t_i) \geq w|\mathbf{X}_{0:i})$ to calculate the RUL distribution and obtained the associated RUL distribution with similar form to Eqs. (4.3) and (4.4). However, using $\Pr(R_i \leq r_i|\mathbf{X}_{0:i}) = \Pr(X(r_i+t_i) \geq w|\mathbf{X}_{0:i})$ ignores the possible hitting events within (t_i, t_i+r_i) as discussed in Sect. 4.2.1 and thus their results are approximations. Instead, Eqs. (4.17) and (4.18) are exact in the sense of the FHT.

Remark 4.2 The moment of the RUL distribution obtained in [19, 33], does not exist since their obtained RUL distributions belong to the family of Bernstein distributions, known without moments, but this is not the case for our result. For example, the mean of RUL can be easily formulated by

$$\begin{aligned} E(R_i|\mathbf{X}_{0:i}) &= E\left[E(R_i|\lambda_i,\mathbf{X}_{0:i})|\mathbf{X}_{0:i}\right] = E\left(\frac{w-x_i}{\lambda_i}\middle|\mathbf{X}_{1:k}\right) \\ &= \frac{w-x_k}{P_{i|i}}\exp\left(-\frac{\hat{\lambda}_i^2}{2P_{i|i}}\right)\int_0^{\hat{\lambda}_i}\exp\left(\frac{u^2}{2P_{i|i}}\right)du \\ &= \frac{\sqrt{2}(w-x_i)}{\sqrt{P_{i|i}}}D\left(\frac{\hat{\lambda}_i}{\sqrt{2P_{i|i}}}\right), \end{aligned}$$

where $D(z) = \exp(-z^2)\int_0^z \exp(u^2)du$ is the Dawson integral for real z, which is known to exist. This property is desired in maintenance practice, since the expectation of the life estimation is required to be existent sometimes [43, 44].

In Eqs. (4.17) and (4.18), parameters a_0, P_0, Q and σ^2 should be estimated. As opposed to the result in [19, 32], we develop a parameter estimation algorithm in the following section for this task.

4.3 Parameter Estimation

Now we return to estimate and update a_0, P_0, Q and σ^2 in Eqs. (4.7) and (4.8). Unlike the method adopted in [32], we use only the data from one system from the time of installation and recursively update the estimate along with the observation process. We denote $\boldsymbol{\theta} = [a_0, P_0, Q, \sigma^2]^T$ as a parameter vector. We use the maximum likelihood estimation (MLE) to estimate $\boldsymbol{\theta}$ once new degradation observation x_i is available. In this case, the log-likelihood function for $\mathbf{X}_{0:i}$ can be written as

$$L_i(\boldsymbol{\theta}) = \log[p(\mathbf{X}_{0:i}|\boldsymbol{\theta})], \tag{4.21}$$

where $p(\mathbf{X}_{0:i}|\boldsymbol{\theta})$ is the joint PDF of the degradation data $\mathbf{X}_{0:i}$. Then the MLE estimate of $\boldsymbol{\theta}$, denoted by $\hat{\boldsymbol{\theta}}_i$, conditional on $\mathbf{X}_{0:i}$ can be obtained by

$$\hat{\boldsymbol{\theta}}_i = \arg\max_{\boldsymbol{\theta}} L_i(\boldsymbol{\theta}). \tag{4.22}$$

If the drift coefficient is constant then maximizing Eq. (4.21) with respect to $\boldsymbol{\theta}$ is straightforward. However, since we treat λ_i as a hidden variable which is given by Eq. (4.7), then directly maximizing Eq. (4.21) is impossible. However, the EM algorithm provides a possible framework for estimating the parameters involving hidden variables [45]. A fundamental assumption of the EM algorithm is that the hidden variables can be estimated by observed data. This is the case for our problem since $p(\lambda_i|\mathbf{X}_{0:i})$ can be obtained by Eq. (4.10).

4.3.1 EM Algorithm

The fundamental principle of the EM algorithm is to replace the hidden variables with their expectations conditional on the observed data. Then the parameter estimation can be formulated as maximizing the joint likelihood function $p(\mathbf{X}_{0:i}, \Upsilon_i|\boldsymbol{\theta})$. Specifically, by manipulating the relationship between $p(\mathbf{X}_{0:i}|\boldsymbol{\theta})$ and $p(\mathbf{X}_{0:i}, \Upsilon_i|\boldsymbol{\theta})$, $L_i(\boldsymbol{\theta})$ can be divided into two parts as

$$L_i(\boldsymbol{\theta}) = \ell_i(\boldsymbol{\theta}) - \log p(\Upsilon_i|\mathbf{X}_{0:i}, \boldsymbol{\theta}), \tag{4.23}$$

where

$$\ell_i(\boldsymbol{\theta}) = \log p(\mathbf{X}_{0:i}, \Upsilon_i|\boldsymbol{\theta}). \tag{4.24}$$

Then taking the expectation operator on both sides of Eq. (4.23) with respect to $\Upsilon_i \,|\, \mathbf{X}_{0:i}, \boldsymbol{\theta}'$, we have

$$L_i(\boldsymbol{\theta}) = \ell(\boldsymbol{\theta} \,|\, \boldsymbol{\theta}') - K(\boldsymbol{\theta} \,|\, \boldsymbol{\theta}'), \tag{4.25}$$

where

$$\ell(\boldsymbol{\theta} \,|\, \boldsymbol{\theta}') = \mathrm{E}_{\Upsilon_i|\mathbf{X}_{0:i},\boldsymbol{\theta}'} \left\{\ell_i(\boldsymbol{\theta})\right\}, \tag{4.26}$$

$$K(\boldsymbol{\theta} \,|\, \boldsymbol{\theta}') = \mathrm{E}_{\Upsilon_i|\mathbf{X}_{0:i},\boldsymbol{\theta}'} \left\{\log p(\Upsilon_i \,|\, \mathbf{X}_{0:i}, \boldsymbol{\theta})\right\}. \tag{4.27}$$

Finally, via Eq. (4.25), the following holds

$$L_i(\boldsymbol{\theta}) - L_i(\boldsymbol{\theta}') = \ell(\boldsymbol{\theta} \,|\, \boldsymbol{\theta}') - \ell(\boldsymbol{\theta}' \big| \boldsymbol{\theta}') + \underbrace{K(\boldsymbol{\theta}' \big| \boldsymbol{\theta}') - K(\boldsymbol{\theta} \,|\, \boldsymbol{\theta}')}_{\geq 0}, \tag{4.28}$$

where the positivity of the last term is implied by the Kullback–Leibler divergence metric between $p(\Upsilon_i \,|\, \mathbf{X}_{0:i}, \boldsymbol{\theta})$ and $p(\Upsilon_i \,|\, \mathbf{X}_{0:i}, \boldsymbol{\theta}')$, see [46] for reference.

Obviously, if $\ell(\boldsymbol{\theta} \,|\, \boldsymbol{\theta}') > \ell(\boldsymbol{\theta}' \big| \boldsymbol{\theta}')$, then $L_i(\boldsymbol{\theta}) - L_i(\boldsymbol{\theta}') > 0$ holds. This is achieved by the EM algorithm which takes an approximation $\hat{\boldsymbol{\theta}}_i^{(k)}$ of MLE $\hat{\boldsymbol{\theta}}_i$ given in Eq. (4.22) and updates it to a better $\hat{\boldsymbol{\theta}}_i^{(k+1)}$ according to the following two steps:

- **E-step**: Calculate

$$\ell(\boldsymbol{\theta} \,|\, \hat{\boldsymbol{\theta}}_i^{(k)}) = \mathrm{E}_{\Upsilon_i|\mathbf{X}_{0:i},\hat{\boldsymbol{\theta}}_i^{(k)}} \left\{\ell_i(\boldsymbol{\theta})\right\}, \tag{4.29}$$

where $\hat{\boldsymbol{\theta}}_i^{(k)} = [a_{0i}^{(k)}, P_{0i}^{(k)}, Q_i^{(k)}, \sigma_i^{2(k)}]^T$ denotes the estimated parameters in the kth step conditional on $\mathbf{X}_{0:i}$.

- **M-step**: Calculate

$$\hat{\boldsymbol{\theta}}_i^{(k+1)} = \arg\max_{\boldsymbol{\theta}} \left\{\mathrm{E}_{\Upsilon_i|\mathbf{X}_{0:i},\hat{\boldsymbol{\theta}}_i^{(k)}} \left\{\ell_i(\boldsymbol{\theta})\right\}\right\}. \tag{4.30}$$

Then we iterate the E-step and M-step until a criterion of convergence is satisfied. In our case, we can calculate the E-step and M-step separately but just outline the properties of the estimation algorithm. The details of the algorithm are summarized in Appendix C. Interestingly, it can be observed from Theorem 2 in Appendix C that the M-step in our approach can be solved analytically and we can obtain the unique maximum point. This implies that each iteration of the EM algorithm can be performed with a single computation, which leads to an extremely fast and simple estimation procedure. This computation advantage plus the exact RUL distribution are particularly attractive for practical applications. The convergence property of the proposed algorithm can be similarly demonstrated in [47–51]. In the next part, the specific implementation of the EM algorithm for the proposed model is provided.

4.3.2 The Implementation of EM Algorithm for the Proposed Model

Following the above-mentioned procedures for the EM algorithm, the joint log-likelihood function for our problem can be expressed as

$$\begin{aligned}\ell_i(\boldsymbol{\theta}) &= \log p(\mathbf{X}_{0:i} \mid \Upsilon_i, \boldsymbol{\theta}) + \log p(\Upsilon_i \mid \boldsymbol{\theta}) \\ &= \log p(\lambda_0 \mid \boldsymbol{\theta}) + \log \prod_{j=0}^{i} p(\lambda_j \mid \lambda_{j-1}, \boldsymbol{\theta}) + \log \prod_{j=0}^{i} p(x_j \mid \lambda_{j-1}, \boldsymbol{\theta}).\end{aligned} \tag{4.31}$$

From Eqs. (4.7) and (4.8), we directly have

$$\begin{aligned}&\lambda_j \mid \lambda_{j-1} \sim N(\lambda_{j-1}, Q), \\ &x_j \mid \lambda_{j-1} \sim N\left(x_{j-1} + \lambda_{j-1}(t_j - t_{j-1}), \sigma^2(t_j - t_{j-1})\right), \\ &\lambda_0 \sim N(a_0, P_0).\end{aligned}$$

Using Eq. (4.31) and ignoring the constant terms, the joint log-likelihood function can be formulated as

$$\begin{aligned}2\ell_i(\boldsymbol{\theta}) = &-\log P_0 - (\lambda_0 - a_0)^2/P_0 - \sum_{j=1}^{i} \left(\log Q + (\lambda_j - \lambda_{j-1})^2/Q\right) \\ &- \sum_{j=1}^{i} \left(\log \sigma^2 + \left(x_j - x_{j-1} - \lambda_{j-1}(t_j - t_{j-1})\right)^2 \Big/ \sigma^2(t_j - t_{j-1})\right).\end{aligned} \tag{4.32}$$

To calculate the conditional expectation $\ell(\boldsymbol{\theta} \mid \hat{\boldsymbol{\theta}}_i^{(k)})$ defined in Eq. (4.29), we have

$$\begin{aligned}&2\ell(\boldsymbol{\theta} \mid \hat{\boldsymbol{\theta}}_i^{(k)}) = E_{\Upsilon_i \mid \mathbf{X}_{1:i}, \hat{\boldsymbol{\theta}}_i^{(k)}}[2\ell_i(\boldsymbol{\theta})] \\ &= E_{\Upsilon_i \mid \mathbf{X}_{0:i}, \hat{\boldsymbol{\theta}}_i^{(k)}}\Big[-\log P_0 - (\lambda_0 - a_0)^2/P_0 - \sum_{j=1}^{i} \left(\log Q + (\lambda_j - \lambda_{j-1})^2/Q\right) \\ &- \sum_{j=1}^{i} \left(\log \sigma^2 + \left(x_j - x_{j-1} - \lambda_{j-1}(t_j - t_{j-1})\right)^2 \Big/ \left(\sigma^2(t_j - t_{j-1})\right)\right)\Big].\end{aligned} \tag{4.33}$$

Clearly, to calculate the expectation of this expression requires to obtain $E_{\Upsilon_i \mid \mathbf{X}_{0:i}, \hat{\boldsymbol{\theta}}_i^{(k)}}(\lambda_j)$, $E_{\Upsilon_i \mid \mathbf{X}_{0:i}, \hat{\boldsymbol{\theta}}_i^{(k)}}(\lambda_j^2)$ and $E_{\Upsilon_i \mid \mathbf{X}_{0:i}, \hat{\boldsymbol{\theta}}_i^{(k)}}(\lambda_j \lambda_{j-1})$, which are the conditional expectations with respect to Υ_i, given the observed history $\mathbf{X}_{0:i}$. In this chapter, we use the Rauch–Tung–Striebel (RTS) smoother to provide an optimal estimation of $E_{\Upsilon_i \mid \mathbf{X}_{0:i}, \hat{\boldsymbol{\theta}}_i^{(k)}}(\lambda_j)$, $E_{\Upsilon_i \mid \mathbf{X}_{0:i}, \hat{\boldsymbol{\theta}}_i^{(k)}}(\lambda_j^2)$ and $E_{\Upsilon_i \mid \mathbf{X}_{0:i}, \hat{\boldsymbol{\theta}}_i^{(k)}}(\lambda_j \lambda_{j-1})$, summarized as Algorithm 4.3, [48, 52]. In Algorithm 4.3, we define $M_{j|i} = Cov(\lambda_j, \lambda_{j-1} \mid \mathbf{X}_{0:i})$.

Algorithm 4.3 (*RTS smoothing algorithm*)

Step 1: Forwards iteration by Algorithm 4.1 or Algorithm 4.2

Step 2: Backwards iteration

$S_j = P_{j|j} P_{j+1|j}^{-1}$

$\hat{\lambda}_{j|i} = \hat{\lambda}_j + S_j(\hat{\lambda}_{j+1|i} - \hat{\lambda}_{j+1|j}) = \hat{\lambda}_j + S_j(\hat{\lambda}_{j+1|i} - \hat{\lambda}_j)$

$P_{j|i} = P_{j|j} + S_j^2(P_{j+1|i} - P_{j+1|j})$

Step 3: Initialize

$M_{i|i} = (1 - K_i(t_i - t_{i-1}))\, P_{i-1|i-1}$

Step 4: Backwards iteration for smoothing covariance

$M_{j|i} = P_{j|j} S_{j-1} + S_j(M_{j+1|i} - P_{j|j}) S_{j-1}.$

From the RTS smoothing algorithm, we can obtain the conditional expectations of $E_{\Upsilon_i|\mathbf{X}_{0:i},\hat{\boldsymbol{\theta}}_i^{(k)}}(\lambda_j)$, $E_{\Upsilon_i|\mathbf{X}_{0:i},\hat{\boldsymbol{\theta}}_i^{(k)}}(\lambda_j^2)$ and $E_{\Upsilon_i|\mathbf{X}_{0:i},\hat{\boldsymbol{\theta}}_i^{(k)}}(\lambda_j\lambda_{j-1})$ in the following lemma.

Lemma 4.3 *Conditional on current estimated parameter $\hat{\boldsymbol{\theta}}_i^{(k)}$ and observations history $\mathbf{X}_{0:i}$, the values of $E_{\Upsilon_i|\mathbf{X}_{0:i},\hat{\boldsymbol{\theta}}_i^{(k)}}(\lambda_j)$, $E_{\Upsilon_i|\mathbf{X}_{0:i},\hat{\boldsymbol{\theta}}_i^{(k)}}(\lambda_j^2)$ and $E_{\Upsilon_i|\mathbf{X}_{0:i},\hat{\boldsymbol{\theta}}_i^{(k)}}(\lambda_j\lambda_{j-1})$ are given by*

$$
\begin{aligned}
&E_{\Upsilon_i|\mathbf{X}_{0:i},\hat{\boldsymbol{\theta}}_i^{(k)}}(\lambda_j) = \hat{\lambda}_{j|i}\,,\\
&E_{\Upsilon_i|\mathbf{X}_{0:i},\hat{\boldsymbol{\theta}}_i^{(k)}}(\lambda_j^2) = \hat{\lambda}_{j|i}^2 + P_{j|i}\,,\\
&E_{\Upsilon_i|\mathbf{X}_{0:i},\hat{\boldsymbol{\theta}}_i^{(k)}}(\lambda_j\lambda_{j-1}) = \hat{\lambda}_{j|i}\hat{\lambda}_{j-1|i} + M_{j|i} = P_{j|j}S_{j-1} + S_j(M_{j+1|i} - P_{j|j})S_{j-1} + \hat{\lambda}_{j|i}\hat{\lambda}_{j-1|i}\,.
\end{aligned}
\tag{4.34}
$$

Proof These equations are the direct results of applying the properties of variance–covariance and RTS smoothing algorithm and the proof is hence omitted.

From Eq. (4.34) and Lemma 4.3, $\ell(\boldsymbol{\theta}\,|\,\hat{\boldsymbol{\theta}}_i^{(k)})$ can be written as

$$
\begin{aligned}
&2\ell(\boldsymbol{\theta}\,|\,\hat{\boldsymbol{\theta}}_i^{(k)})\\
&=\mathbb{E}_{\Upsilon_i|X_{0:i},\hat{\boldsymbol{\theta}}_i^{(k)}}\left[2\ell_i(\boldsymbol{\theta})\right]\\
&=\mathbb{E}_{\Upsilon_i|X_{0:i},\hat{\boldsymbol{\theta}}_i^{(k)}}\Bigg[-\log P_0 - \frac{(\lambda_0 - a_0)^2}{P_0} - \sum\nolimits_{j=1}^{i}\left(\log Q + \frac{(\lambda_j - \lambda_{j-1})^2}{Q}\right) -\\
&\quad \sum\nolimits_{j=1}^{i}\left(\log\sigma^2 + (x_j - x_{j-1} - \lambda_{j-1}(t_j - t_{j-1}))^2\Big/\left(\sigma^2(t_j - t_{j-1})\right)\right)\Bigg]\\
&= -\log P_0 - \frac{(C_{0|i} - 2a_{0|i}\,a_0 + a_0^2)}{P_0} - \sum\nolimits_{j=1}^{i}\left(\log Q + \frac{(C_{j|i} - 2C_{j,j-1|i} + C_{j-1|i})}{Q}\right) -\\
&\quad \sum\nolimits_{j=1}^{i}\left(\log\sigma^2 + \frac{(x_j - x_{j-1})^2 - 2\hat{\lambda}_{j-1|i}(x_j - x_{j-1})(t_j - t_{j-1}) + (t_j - t_{j-1})^2 C_{j-1|i}}{\sigma^2(t_j - t_{j-1})}\right).
\end{aligned}
\tag{4.35}
$$

This completes the E-step and in the following we handle the M-step.

After obtaining $\ell(\boldsymbol{\theta}\,|\,\hat{\boldsymbol{\theta}}_i^{(k)})$, the results of estimated parameter $\hat{\boldsymbol{\theta}}_i^{(k+1)}$ in the $(k+1)$th step can be summarized in the following theorem.

Theorem 4.2 *$\hat{\boldsymbol{\theta}}_i^{(k+1)}$, by maximizing $\ell(\boldsymbol{\theta}\,|\,\hat{\boldsymbol{\theta}}_i^{(k)})$, is given by*

$$\begin{aligned}
a_{0i}^{(k+1)} &= a_{0|i}\,,\\
P_{0i}^{(k+1)} &= C_{0|i} - a_{0|i}^2 = P_{0|i}\,,\\
Q_i^{(k+1)} &= \tfrac{1}{i}\sum_{j=1}^{i}\left(C_{j|i} - 2C_{j,j-1|i} + C_{j-1|i}\right),\\
\left(\sigma^2\right)_i^{(k+1)} &= \tfrac{1}{i}\sum_{j=1}^{i}\left(\frac{\left(x_j-x_{j-1}\right)^2 - 2\hat{\lambda}_{j-1|i}\left(x_j-x_{j-1}\right)\left(t_j-t_{j-1}\right)+\left(t_j-t_{j-1}\right)^2C_{j-1|i}}{t_j-t_{j-1}}\right).
\end{aligned}$$

with $C_{j|i} = E_{\Upsilon_i|\mathbf{X}_{0:i},\hat{\boldsymbol{\theta}}_i^{(k)}}(\lambda_j^2)$, $a_{0|i} = \hat{\lambda}_{0|i}$, $C_{j,j-1|i} = E_{\Upsilon_i|\mathbf{X}_{0:i},\hat{\boldsymbol{\theta}}_i^{(k)}}(\lambda_j\lambda_{j-1})$ and $\hat{\boldsymbol{\theta}}_i^{(k+1)}$ is uniquely determined and located at the maximum.

Proof The unknown parameters $\hat{\boldsymbol{\theta}}_i^{(k+1)}$ can be obtained by maximizing the $\ell(\boldsymbol{\theta}\,|\,\hat{\boldsymbol{\theta}}_i^{(k)})$ with $\boldsymbol{\theta}$. Therefore, the central goal in this step is to find the maximum of function $\ell(\boldsymbol{\theta}\,|\,\hat{\boldsymbol{\theta}}_i^{(k)})$ on $\boldsymbol{\theta}$, i.e.,

$$\hat{\boldsymbol{\theta}}_i^{(k+1)} = \arg\max_{\boldsymbol{\theta}}\left\{\mathrm{E}_{\Upsilon_i|\mathbf{X}_{0:i},\hat{\boldsymbol{\theta}}_i^{(k)}}\left\{\ell_i(\boldsymbol{\theta})\right\}\right\} = \arg\max_{\boldsymbol{\theta}}\ell(\boldsymbol{\theta}\,|\,\hat{\boldsymbol{\theta}}_i^{(k)}). \tag{4.36}$$

From Eq. (4.35), taking $\partial\ell(\boldsymbol{\theta}\,|\,\hat{\boldsymbol{\theta}}_i^{(k)})\big/\partial\boldsymbol{\theta}$, we obtain the solution by $\partial\ell(\boldsymbol{\theta}\,|\,\hat{\boldsymbol{\theta}}_i^{(k)})\big/\partial\boldsymbol{\theta} = \mathbf{0}$, which leads to the maximum, and taking $\partial^2\ell(\boldsymbol{\theta}\,|\,\hat{\boldsymbol{\theta}}_i^{(k)})\big/\partial\boldsymbol{\theta}\partial\boldsymbol{\theta}^T$, the following is obtained,

$$\frac{\partial^2\ell(\boldsymbol{\theta}\,|\,\hat{\boldsymbol{\theta}}_i^{(k)})}{\partial\boldsymbol{\theta}\partial\boldsymbol{\theta}^T} = \frac{1}{2}\begin{bmatrix} -\frac{2}{P_0} & \frac{2(a_0-a_{0|i})}{P_0^2} & 0 & 0\\ \frac{2(a_0-a_{0|i})}{P_0^2} & \frac{1}{P_0^2} - \frac{2(C_{0|i}-2a_{0|i}a_0+a_0^2)}{P_0^3} & 0 & 0\\ 0 & 0 & \frac{k}{Q^2}-\frac{2\psi}{Q^3} & 0\\ 0 & 0 & 0 & \frac{i}{\sigma^4}-\frac{2\varphi}{\sigma^6}\end{bmatrix}, \tag{4.37}$$

with

$$\begin{aligned}
\psi &= \sum_{j=1}^{i}\left(C_{j|i} - 2C_{j,j-1|i} + C_{j-1|i}\right),\\
\varphi &= \sum_{j=1}^{i}\left(\frac{\left(x_j-x_{j-1}\right)^2 - 2\hat{\lambda}_{j-1|i}\left(x_j-x_{j-1}\right)\left(t_j-t_{j-1}\right)+\left(t_j-t_{j-1}\right)^2C_{j-1|i}}{t_j-t_{j-1}}\right).
\end{aligned} \tag{4.38}$$

We show that the matrix in (4.37) is negative definite at $\boldsymbol{\theta} = \hat{\boldsymbol{\theta}}_i^{(k+1)}$, by calculating the order principal minor determinant as follows,

$$\Delta_1 = -\frac{1}{P_0},\ \Delta_2 = -\frac{1}{2P_0}\left(\frac{1}{P_0^2} - \frac{2(C_{0|i} - 2a_{0|i}\,a_0 + a_0^2)}{P_0^3}\right) - \frac{(a_0 - a_{0|i})^2}{P_0^4},$$
$$\Delta_3 = \frac{1}{2}\left(\frac{i}{Q^2} - \frac{2\psi}{Q^3}\right)\Delta_2,\ \Delta_4 = \frac{1}{2}\left(\frac{i}{\sigma^4} - \frac{2\varphi}{\sigma^6}\right)\Delta_3.$$

Then, at $\boldsymbol{\theta} = \hat{\boldsymbol{\theta}}_i^{(k+1)}$, the followings are obtained,

$$\Delta_1\big|_{\boldsymbol{\theta}=\hat{\boldsymbol{\theta}}_i^{(k+1)}} = -\frac{1}{P_{0|i}} < 0,$$
$$\Delta_2\big|_{\boldsymbol{\theta}=\hat{\boldsymbol{\theta}}_i^{(k+1)}} = -\frac{1}{2P_{0|i}}\left(\frac{1}{P_{0|i}^2} - \frac{2(C_{0|i} - 2a_{0|i}\,a_{0|i} + a_{0|i}^2)}{P_{0|i}^3}\right) - \frac{(a_{0|i} - a_{0|i})^2}{P_{0|i}^4}$$
$$= -\frac{1}{2P_{0|i}}\left(\frac{1}{P_{0|i}^2} - \frac{2}{P_{0|i}^2}\right) = \frac{1}{2P_{0|k}^3} > 0,$$
$$\Delta_3\big|_{\boldsymbol{\theta}=\hat{\boldsymbol{\theta}}_i^{(k+1)}} = \frac{1}{2}\left(\frac{i^3}{\psi^2} - \frac{2i^3\psi}{\psi^3}\right)\Delta_2\big|_{\boldsymbol{\theta}=\hat{\boldsymbol{\theta}}_i^{(k+1)}} = -\frac{i^3}{2\psi^2}\Delta_2\big|_{\boldsymbol{\theta}=\hat{\boldsymbol{\theta}}_i^{(k+1)}} < 0,$$
$$\Delta_4\big|_{\boldsymbol{\theta}=\hat{\boldsymbol{\theta}}_i^{(k+1)}} = \frac{1}{2}\left(\frac{i}{\sigma^4} - \frac{2\varphi}{\sigma^6}\right)\Delta_3\big|_{\boldsymbol{\theta}=\hat{\boldsymbol{\theta}}_i^{(k+1)}} = \frac{1}{2}\left(\frac{i^3}{\varphi^2} - \frac{2i^3\varphi}{\varphi^3}\right)\Delta_3\big|_{\boldsymbol{\theta}=\hat{\boldsymbol{\theta}}_i^{(k+1)}}$$
$$= -\frac{i^3}{2\varphi^2}\Delta_3\big|_{\boldsymbol{\theta}=\hat{\boldsymbol{\theta}}_i^{(k+1)}} > 0.$$

This completes the proof that the matrix in (4.37) is negative definite at $\boldsymbol{\theta} = \hat{\boldsymbol{\theta}}_i^{(k+1)}$, verifying that $\hat{\boldsymbol{\theta}}_i^{(k+1)}$ is located at a maximum. In addition, $\hat{\boldsymbol{\theta}}_i^{(k+1)}$ is unique since $\hat{\boldsymbol{\theta}}_i^{(k+1)}$ is the only solution satisfying $\partial\ell(\boldsymbol{\theta}\,|\,\hat{\boldsymbol{\theta}}_i^{(k)})\big/\partial\boldsymbol{\theta} = \mathbf{0}$.

The preceding derivations are summarized as Algorithm 4.4 via a complete specification of the EM-based algorithm for estimating the parameters in $\boldsymbol{\theta}$ when new observation x_i is available.

Algorithm 4.4 (*EM algorithm for parameter estimation*)

(1) **Initialization**: Initialize the initial parameters in $\hat{\boldsymbol{\theta}}_i^{(0)}$.

(2) **E-Step**: Calculate the expectation quantities defined in Eq. (4.34) using Algorithm 4.3, with state-space model of Eqs. (4.7) and (4.8) parameterized by $\hat{\boldsymbol{\theta}}_i^{(k)}$.

(3) **M-Step**: Maximize $\ell(\boldsymbol{\theta}\,|\,\hat{\boldsymbol{\theta}}_i^{(k)})$ to obtain the updated parameter estimates by Theorem 4.2.

(4) **Test convergence**

Test the convergence of the algorithm. If converged, then stop. Otherwise set $k = k + 1$, go to *Step 2* and repeat.

4.3.3 Convergence Analysis of Adaptive Model Parameter Estimation Algorithm

Regarding the convergence of Algorithm 4.4, we first the following result.

Theorem 4.3 *Let $\hat{\boldsymbol{\theta}}_i^{(k+1)}$ be generated from $\hat{\boldsymbol{\theta}}_i^{(k)}$ by an iteration algorithm of Algorithm 4.4. Then, $L(\hat{\boldsymbol{\theta}}_i^{(k+1)}) \geq L(\hat{\boldsymbol{\theta}}_i^{(k)})$, with equality held if and only if both $\hat{\boldsymbol{\theta}}_i^{(k)} = \hat{\boldsymbol{\theta}}_i^{(k+1)}$.*

Proof Using the above-defined notation and $p(\boldsymbol{X}_{0:i}|\boldsymbol{\theta})p(\Upsilon_i|\boldsymbol{X}_{0:i},\boldsymbol{\theta}) = p(\boldsymbol{X}_{0:i},\Upsilon_i|\boldsymbol{\theta})$, there is

$$\log p(\boldsymbol{X}_{0:i}|\boldsymbol{\theta}) = \log p(\boldsymbol{X}_{0:i},\Upsilon_i|\boldsymbol{\theta}) - \log p(\Upsilon_i|\boldsymbol{X}_{0:i},\boldsymbol{\theta}),$$

or

$$L(\boldsymbol{\theta}) = \ell(\boldsymbol{\theta}) - \log p(\Upsilon_k|\boldsymbol{X}_{0:i},\boldsymbol{\theta}). \tag{4.39}$$

In this case, taking the expectation operator on both sides of (4.39), with respect to the distribution $p(\Upsilon_i|\boldsymbol{X}_{0:i},\hat{\boldsymbol{\theta}}_i^{(k)})$, we obtain the following result due to the fact$\mathbb{E}(g(Y)|Y) = g(Y)$,

$$L(\boldsymbol{\theta}) = \ell(\boldsymbol{\theta}|\hat{\boldsymbol{\theta}}_i^{(k)}) - \mathbb{E}_{\Upsilon_i|\boldsymbol{X}_{0:i},\hat{\boldsymbol{\theta}}_i^{(k)}}\{\log p(\Upsilon_i|\boldsymbol{X}_{0:i},\boldsymbol{\theta})\}. \tag{4.40}$$

The EM algorithm proceeds by maximizing $\ell(\boldsymbol{\theta}|\hat{\boldsymbol{\theta}}_i^{(k)})$ on $\boldsymbol{\theta}$ in the hope of delivering a new estimate $\hat{\boldsymbol{\theta}}_i^{(k+1)}$ which is an improvement relative to $\hat{\boldsymbol{\theta}}_i^{(k)}$, i.e., $\ell(\hat{\boldsymbol{\theta}}_i^{(k+1)}\big|\hat{\boldsymbol{\theta}}_i^{(k)}) \geq \ell(\hat{\boldsymbol{\theta}}_i^{(k)}\big|\hat{\boldsymbol{\theta}}_i^{(k)})$. The difference between the likelihoods can be written as

$$\begin{aligned}
L(\hat{\boldsymbol{\theta}}_i^{(k+1)}) - L(\hat{\boldsymbol{\theta}}_i^{(k)}) =& \ell(\hat{\boldsymbol{\theta}}_i^{(k+1)}\big|\hat{\boldsymbol{\theta}}_i^{(k)}) - \ell(\hat{\boldsymbol{\theta}}_i^{(k)}\big|\hat{\boldsymbol{\theta}}_i^{(k)}) + \mathbb{E}_{\Upsilon_i|\boldsymbol{X}_{0:i},\hat{\boldsymbol{\theta}}_i^{(k)}}\left\{\log p(\Upsilon_i|\boldsymbol{X}_{0:i},\hat{\boldsymbol{\theta}}_i^{(k)})\right\} \\
&- \mathbb{E}_{\Upsilon_i|\boldsymbol{X}_{0:i},\hat{\boldsymbol{\theta}}_i^{(k)}}\{\log p(\Upsilon_i|\boldsymbol{X}_{0:i},\boldsymbol{\theta})\} \\
=& \ell(\boldsymbol{\theta}_i^{(\hat{k+1})}\big|\hat{\boldsymbol{\theta}}_i^{(k)}) - \ell(\hat{\boldsymbol{\theta}}_i^{(k)}\big|\hat{\boldsymbol{\theta}}_i^{(k)}) + \\
&\int \log\frac{p(\Upsilon_i|\boldsymbol{X}_{0:i},\hat{\boldsymbol{\theta}}_i^{(k)})}{p(\Upsilon_i|\boldsymbol{X}_{0:i},\hat{\boldsymbol{\theta}}_i^{(k+1)})}p(\Upsilon_i|\boldsymbol{X}_{0:i},\hat{\boldsymbol{\theta}}_i^{(k)})\mathrm{d}\Upsilon_i \\
\geq& \int \log\frac{p(\Upsilon_i|\boldsymbol{X}_{0:i},\hat{\boldsymbol{\theta}}_i^{(k)})}{p(\Upsilon_i|\boldsymbol{X}_{0:i},\hat{\boldsymbol{\theta}}_i^{(k+1)})}p(\Upsilon_i|\boldsymbol{X}_{0:i},\hat{\boldsymbol{\theta}}_i^{(k)})\mathrm{d}\Upsilon_i \\
\geq& -\log\mathbb{E}_{\Upsilon_i|\boldsymbol{X}_{0:i},\hat{\boldsymbol{\theta}}_i^{(k)}}\left[\frac{p(\Upsilon_i|\boldsymbol{X}_{0:i},\hat{\boldsymbol{\theta}}_i^{(k+1)})}{p(\Upsilon_i|\boldsymbol{X}_{0:i},\hat{\boldsymbol{\theta}}_i^{(k)})}\right] \\
=& -\log\int p(\Upsilon_i|\boldsymbol{X}_{0:i},\hat{\boldsymbol{\theta}}_i^{(k+1)})\mathrm{d}\Upsilon_i \\
=& 0.
\end{aligned} \tag{4.41}$$

According to (4.41), $L(\hat{\boldsymbol{\theta}}_i^{(k+1)})$ is an increasing function with k, i.e., $L(\hat{\boldsymbol{\theta}}_i^{(k+1)}) \geq L(\hat{\boldsymbol{\theta}}_i^{(k)})$. If $\hat{\boldsymbol{\theta}}_i^{(k)} = \hat{\boldsymbol{\theta}}_i^{(k+1)}$, then $L(\hat{\boldsymbol{\theta}}_i^{(k+1)}) = L(\hat{\boldsymbol{\theta}}_i^{(k)})$. As for the equality part in $L(\hat{\boldsymbol{\theta}}_i^{(k+1)}) \geq L(\hat{\boldsymbol{\theta}}_i^{(k)})$, the following result can be established.

According to the result in [49], to prove the necessary condition, it is only required to prove that, if $\ell(\hat{\boldsymbol{\theta}}_i^{(k+1)} \,|\, \hat{\boldsymbol{\theta}}_i^{(k)}) = \ell(\hat{\boldsymbol{\theta}}_i^{(k)} \,|\, \hat{\boldsymbol{\theta}}_i^{(k)})$ and $p(\Upsilon_i | \boldsymbol{X}_{0:i}, \hat{\boldsymbol{\theta}}_i^{(k+1)}) = p(\Upsilon_i | \boldsymbol{X}_{0:i}, \hat{\boldsymbol{\theta}}_i^{(k)})$, then $\hat{\boldsymbol{\theta}}_i^{(k)} = \hat{\boldsymbol{\theta}}_i^{(k+1)}$. It is noted that, if $\ell(\hat{\boldsymbol{\theta}}_i^{(k+1)} \,|\, \hat{\boldsymbol{\theta}}_i^{(k)}) = \ell(\hat{\boldsymbol{\theta}}_i^{(k)} \,|\, \hat{\boldsymbol{\theta}}_i^{(k)})$, then $\hat{\boldsymbol{\theta}}_i^{(k+1)}$ and $\hat{\boldsymbol{\theta}}_i^{(k)}$ must be the maximal points of $\ell(\boldsymbol{\theta} | \hat{\boldsymbol{\theta}}_i^{(k)})$, according to the underlying definition of the EM algorithm. Further, Theorem 4.3 shows that the maximal point of $\ell(\boldsymbol{\theta} | \hat{\boldsymbol{\theta}}_i^{(k)})$ is unique, and thus $\hat{\boldsymbol{\theta}}_i^{(k)} = \hat{\boldsymbol{\theta}}_i^{(k+1)}$. As for the sufficient condition of the equality in $L(\hat{\boldsymbol{\theta}}_i^{(k+1)}) \geq L(\hat{\boldsymbol{\theta}}_i^{(k)})$, the conclusion is trivial.

This completes the proof.

In addition, we have the following result regarding the relationship among $\boldsymbol{\theta}_i^*$, $L(\boldsymbol{\theta})$ and $\ell(\boldsymbol{\theta} | \boldsymbol{\theta}_i^*)$.

Theorem 4.4 *If the EM-based algorithm defined in Algorithm 4.4 terminates at* $\boldsymbol{\theta}_i^*$, $\boldsymbol{\theta}_i^*$ *is both stationary point of* $L(\boldsymbol{\theta})$ *and* $\ell(\boldsymbol{\theta} | \boldsymbol{\theta}_i^*)$, *i.e.,*

$$\left.\frac{\partial L(\boldsymbol{\theta})}{\partial \boldsymbol{\theta}}\right|_{\boldsymbol{\theta}=\boldsymbol{\theta}_i^*} = \left.\frac{\partial \ell(\boldsymbol{\theta} | \boldsymbol{\theta}_i^*)}{\partial \boldsymbol{\theta}}\right|_{\boldsymbol{\theta}=\boldsymbol{\theta}_i^*} = \mathbf{0}. \tag{4.42}$$

Proof According to (4.40), it is obtained that

$$L(\boldsymbol{\theta}) = \ell(\boldsymbol{\theta} | \boldsymbol{\theta}_i^*) - \mathbb{E}_{\Upsilon_i | \boldsymbol{X}_{0:i}, \boldsymbol{\theta}_i^*} \left\{ \log p(\Upsilon_i | \boldsymbol{X}_{0:i}, \boldsymbol{\theta}) \right\}. \tag{4.43}$$

Then, there is

$$\frac{\partial L(\boldsymbol{\theta})}{\partial \boldsymbol{\theta}} = \frac{\partial \ell(\boldsymbol{\theta} | \boldsymbol{\theta}_i^*)}{\partial \boldsymbol{\theta}} - \frac{\partial}{\partial \boldsymbol{\theta}} \mathbb{E}_{\Upsilon_i | \boldsymbol{X}_{0:i}, \boldsymbol{\theta}_i^*} \left\{ \log p(\Upsilon_i | \boldsymbol{X}_{0:i}, \boldsymbol{\theta}) \right\}. \tag{4.44}$$

In the deriving process of Theorem 4.4, we know that, for any $\boldsymbol{\theta}$, the following holds

$$\mathbb{E}_{\Upsilon_i | \boldsymbol{X}_{0:i}, \boldsymbol{\theta}_i^*} \left\{ \log p(\Upsilon_i | \boldsymbol{X}_{0:i}, \boldsymbol{\theta}_i^*) \right\} - \mathbb{E}_{\Upsilon_i | \boldsymbol{X}_{0:i}, \boldsymbol{\theta}_i^*} \left\{ \log p(\Upsilon_i | \boldsymbol{X}_{0:i}, \boldsymbol{\theta}) \right\} \geq \mathbf{0}. \tag{4.45}$$

That is to say, $\boldsymbol{\theta}_i^*$ is the maximum of $\mathbb{E}_{\Upsilon_i | \boldsymbol{X}_{0:i}, \boldsymbol{\theta}_i^*} \left\{ \log p(\Upsilon_i | \boldsymbol{X}_{0:i}, \boldsymbol{\theta}) \right\}$ on $\boldsymbol{\theta}$. Then we directly obtain

$$\left.\frac{\partial}{\partial \boldsymbol{\theta}} \mathbb{E}_{\Upsilon_i | \boldsymbol{X}_{0:i}, \boldsymbol{\theta}_i^*} \left\{ \log p(\Upsilon_i | \boldsymbol{X}_{0:i}, \boldsymbol{\theta}) \right\}\right|_{\boldsymbol{\theta}=\boldsymbol{\theta}_i^*} = 0. \tag{4.46}$$

As a result of (4.44), we have

$$\left.\frac{\partial L(\boldsymbol{\theta})}{\partial \boldsymbol{\theta}}\right|_{\boldsymbol{\theta}=\boldsymbol{\theta}_i^*} = \left.\frac{\partial \ell(\boldsymbol{\theta} \mid \boldsymbol{\theta}_i^*)}{\partial \boldsymbol{\theta}}\right|_{\boldsymbol{\theta}=\boldsymbol{\theta}_i^*}. \tag{4.47}$$

From Theorem 4.3 and the assumption that the EM-based algorithm in Algorithm 4.4 terminates at point $\boldsymbol{\theta}_i^*$, we easily obtain

$$\left.\frac{\partial \ell(\boldsymbol{\theta} \mid \boldsymbol{\theta}_i^*)}{\partial \boldsymbol{\theta}}\right|_{\boldsymbol{\theta}=\boldsymbol{\theta}_i^*} = \mathbf{0}. \tag{4.48}$$

This completes the proof.

Theorem 4.5 *Let* $\{\hat{\boldsymbol{\theta}}_i^{(k)}\}$ *be a sequence of the estimates at* t_i *by the EM algorithm in Algorithm 4.4. Then, a limiting point* $\boldsymbol{\theta}_i^*$ *of* $\{\hat{\boldsymbol{\theta}}_i^{(k)}\}$ *is a stationary point of* $L(\boldsymbol{\theta})$ *and* $\left\{L(\hat{\boldsymbol{\theta}}_i^{(k)})\right\}$ *converges monotonically to* $L(\boldsymbol{\theta}_i^*)$.

Proof The first part can be easily proved by Theorem 4.3. From Theorem 4.4, [47, 49], the second part can also be easily proved.

4.4 A Practical Case Study

In this section, we provide a practical case study for gyros in an inertial navigation system (INS) to illustrate the application of our model and compare the performance of our model with the models presented in [19, 32]. In our study, it is found that STF can generate superior results to Kalman filter to a certain extent. As such, in the following, we only use STF in the filtering step for illustration. Of course, in practice which one to use needs to be tested against the model fit. However, due to limited space, we do not discuss this issue in this chapter. However, the detailed comparisons between STF and Kalman filter with sudden state changing can be found in [40, 53].

4.4.1 Problem Description

As a key device of the INS in weapon systems and space equipment, an inertial platform plays an important and irreplaceable role in the INS. Its operating state has a direct influence on navigation precision. The sensors fixed in an inertial platform include three gyros and three accelerometers, which measure angular velocity and linear acceleration, respectively. The gyro fixed on an inertial platform is a mechanical structure having two degrees of freedom from the driver and sense axis (see [54] for a general description of inertial navigation platforms and gyros). When the inertial platform is operating, the wheels of the gyros rotate at very high speeds and can lead to rotation axis wear. As the wear is accumulated, the bearings on the gyros' electric motor will become deformed and such deformation can lead to the drift of the gyros. The increasing drift finally results in the failure of gyros and then the inertial platform. Past data show that almost 70% of the failures of inertial platforms result from gyroscopic drift and such drift is largely resulted from the wear of bearings as the case of rolling element bearings which are extensively investigated in the literature. However, the difference of our case is that we use the drift data of gyros to estimate the RUL rather than the vibration data as rolling element bearings since we cannot obtain the vibration data as it is not allowed to fix the vibration sensors in the inertial platform. As such, the drift of gyros is often used as a performance indicator to evaluate the health condition of an inertial platform and to schedule maintenance activities.

In this study, we assume that CM values of drift coefficients reflect the performance of the inertial platform, and the larger the drift coefficients monitored are, the worse the performance is. Therefore, according to the CM data and technical index of the inertial platform, failure prediction can be implemented by modeling the drift coefficients. The drift coefficients of an inertial platform mainly include K_{OX}, K_{OY}, K_{OZ}, K_{SX}, K_{SY}, K_{IZ}, in which K_{OX}, K_{OY}, K_{OZ} denote constant drift coefficients, and K_{SX}, K_{SY}, K_{IZ} are stochastic drift coefficients, where K_{SX}, K_{SY} denote the coefficients related to the first moment of specific force along the sense axis, and K_{IZ} denotes the coefficient related to the first moment of specific force along the input axis. Generally, the drift degradation measurement along the sense axis, K_{SX}, plays a dominant role in the assessment of gyro degradation. In our study, we take the CM data of K_{SX} as the degradation signals and use them for RUL estimation of the INS. For our monitored INS in certain weapon system with the terminated life 180.5 h, 73 points of drift coefficients data were collected with regular CM intervals 2.5 h in field condition. The collected data are illustrated in Fig. 4.2.

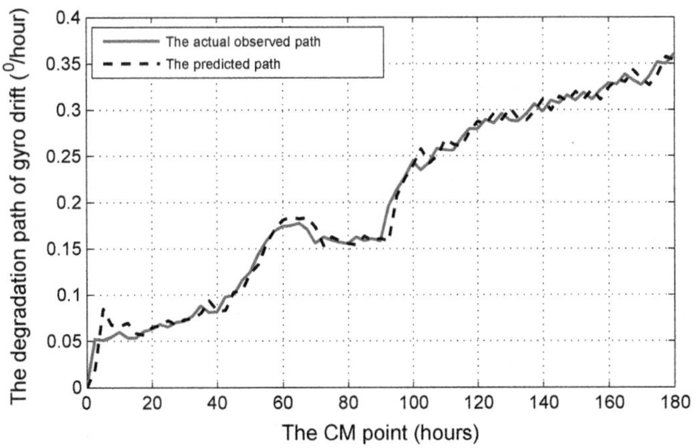

Fig. 4.2 He actual gyro's drift data and the predictions of our model

In the practice of the INS health monitoring, it is usually required that the drift measurement along the sense axis should not exceed 0.37 (°/hour). This threshold is predetermined at the design stage and is strictly enforced in practice since an INS is a critical device used in a navigated weapon system.

4.4.2 The Implementation of Our Model for RUL Estimation of the INS

Using our model, the predictions of the gyro's drift and the distribution of the RUL can be obtained at each CM point. Specifically, using our approach initialized by parameter vector $\boldsymbol{\theta}_0 = [0.002, 0.001, 0.01, 0.01]^T$, the one step predicted drifting path by $\hat{x}_{i+1} = x_i + \hat{\lambda}_i(t_{i+1} - t_i)$ is illustrated in Fig. 4.4 to show the fitness of our model to the gyro's drift degradation data. Clearly, the predicted results match with the actual data well and the mean squared error (MSE) of the predictions is 1.1962E-4 which is small. This demonstrates that our developed model can model the gyro's drift degradation data effectively. Correspondingly, the evolving path of the estimated parameter vector $\hat{\boldsymbol{\theta}}$, consisting of a_0, P_0, Q and σ^2, is illustrated in Fig. 4.3, which are estimated by Algorithm 4.4.

Figure 4.3 shows that the updated parameters converge quickly as the observed degradation data are accumulated. In this case, once the parameters converge, further

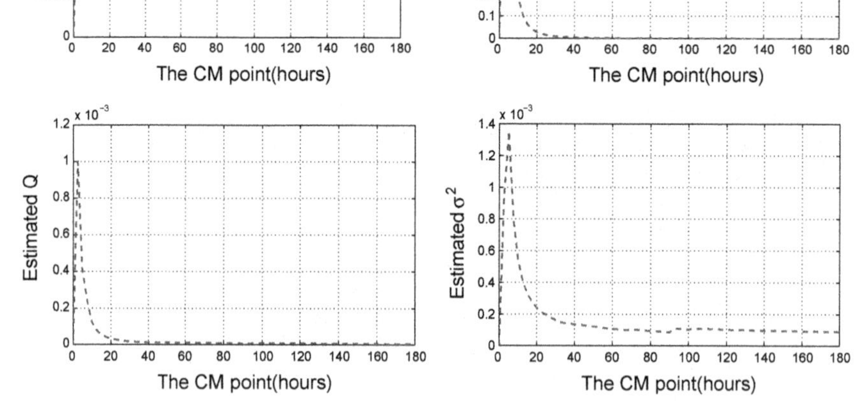

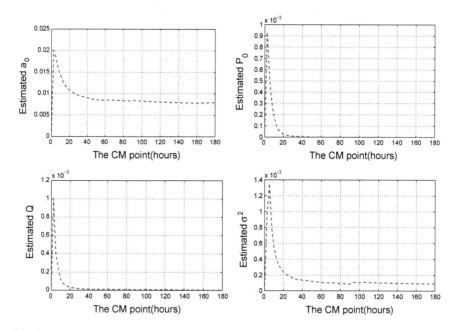

Fig. 4.3 The updated parameters at monitoring points $t_0, t_1, \ldots, t_{73}$

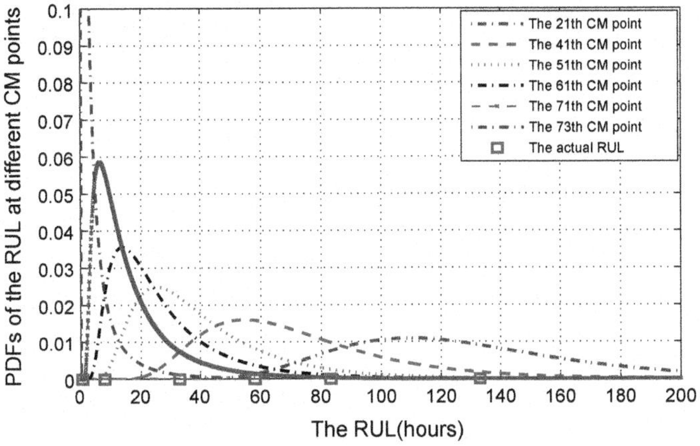

Fig. 4.4 Illustration of the PDF of the RUL at six different CM points

updating may be unnecessary. However, such updating is needed for the case that the degradation process may be subject to unusual changes in its progression. Once the parameters in the model are updated, the PDF of the estimated RUL can be calculated at each CM point. It is noted that the RUL estimation at the CM point is not achieved by successively predicting the degradation state and instead achieved by calculating the FHT of the degradation process. For example, at the current CM point t_i, after parameters updating we use Eq. (4.2) to describe the degradation process and estimate the RUL by calculating the FHT distribution of this process to the failure threshold according to Eqs. (4.11) and (4.17). In the RUL estimation, both the temporal uncertainty of the degradation process and the uncertainty in drift parameter are taken into account. Figure 4.4 illustrates the RUL distributions at six different CM points.

It is noted that the first drift reading is zero according to the model setting. In our used practical data, the last drift reading is 0.3566 (°/hour) at the monitoring time $t_{73} = 180$ h. It can be found that this drift reading is very close to the failure threshold ($w = 0.37$ (°/hour)). Therefore, we can consider that the data captures full life cycle history (useful life) of the machine component. In other words, the actual lifetime of the gyros in an INS is approximated to be 180.5 h and thus the actual RUL at each CM point is known from the full life cycle data. As shown in Fig. 4.4, the actual RUL (denoted by square) falls within the range of the estimated PDF of the RUL at each CM point from our model and further the estimated PDF of the RUL becomes sharper as the degradation data are accumulated. This implies that the uncertainty of the estimated RUL is reduced since more data are utilized during estimating the model parameters. When we use our predictive model for RUL estimation at a given CM point, we use the CM data up to that CM point. In other words, if we estimate the RUL at t_i, the data $\mathbf{X}_{0:i}$ are used to update the model and for RUL estimation. Therefore, at the last CM time t_{73}, all the available CM data are used to estimate the

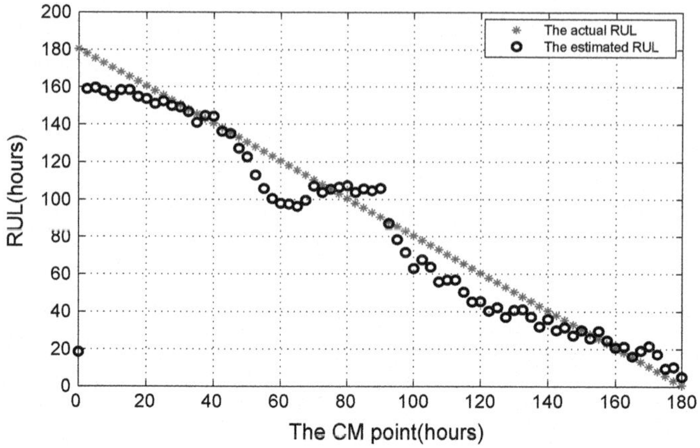

Fig. 4.5 The estimated mean RUL and the actual RUL at monitoring points $t_0, t_1, \ldots, t_{73}$

RUL. Figure 4.5 illustrates the performance of our predictive model against the full life cycle data.

It can be observed that the estimated mean RUL and the actual RUL match each other well. For example, the relative error between the actual life and the estimated mean life at t_{73} is 0.16%. This reflects that the estimated result of our developed model can match the actual result from the full life cycle data closely.

4.4.3 Comparative Studies

In this part, we conduct some comparative studies with the models presented in [19, 32]. We first compare our model with Wang's model about their performance of the RUL estimation for the INS. The detailed implementation process of Wang's model can be found in [32].

In order to compare our model with Wang's model, a loss function is employed to enable a direct comparison of the distributions for the RUL estimation between two models. The loss function is the MSE about the actual RUL obtained at each observation point [4], defined as

$$MSE_i = \int_0^\infty (r_i - \tilde{r}_i)^2 f_{R_i|\mathbf{X}_{0:i}}(r_i|\mathbf{X}_{0:i})\, dr_i, \tag{4.49}$$

where $\tilde{r}_i$ is the actual RUL obtained at t_i and $f_{R_i|\mathbf{X}_{0:i}}(r_i|\mathbf{X}_{0:i})$ is the estimated PDF of the RUL.

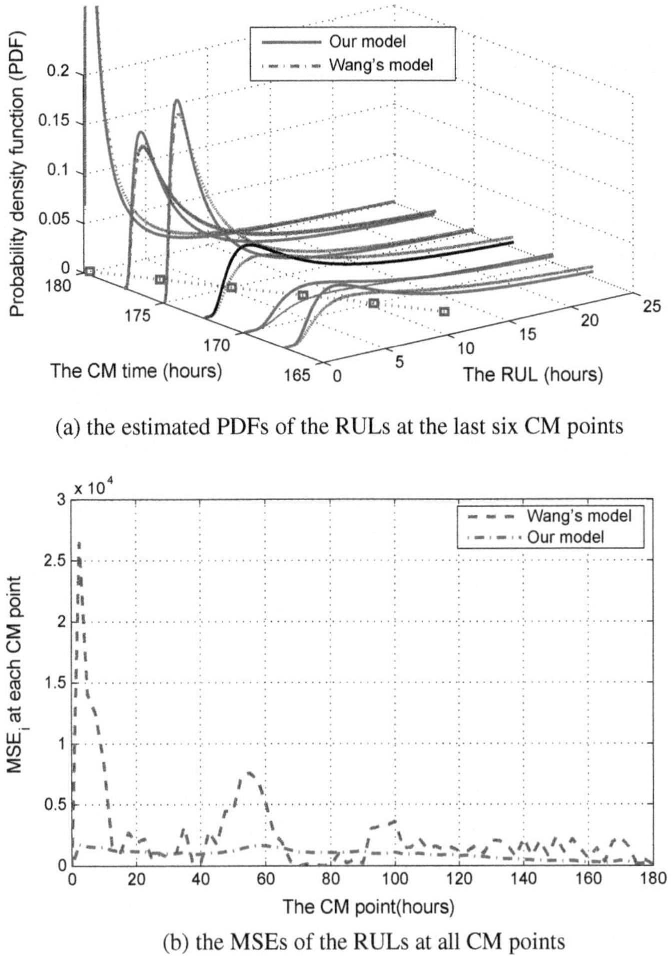

Fig. 4.6 Comparative results of our model with Wang's model (□–actual RUL)

Figure 4.6a compares the estimated RUL distributions from our model with Wang's model. In both cases, the unknown parameters, such as a_0, P_0, Q and σ^2, are obtained by the EM algorithm, but the difference is that our model recursively updates all parameter estimates whenever a new piece of information is available, whereas the model in [32] uses the data of past systems to estimate the model parameters and once estimated they are fixed. Only $\hat{\lambda}_i$ is updated with the new data in Wang' model. Clearly, the PDFs of the RULs from both models can cover the actual RULs well as the obtained data are accumulated. However, it is observed from Fig. 4.6a that the PDFs of the estimated RULs are typically more dispersed when using Wang's model. Figure 4.6b shows the calculated MSE between the actual RUL and the estimated RUL at each CM point. We can observe that the MSEs using Wang' model change irregularly at different sampling points with relatively large fluctuations, particularly in the earlier stage of estimation. This implies that the estimated RUL PDFs by Wang's model are sensitive to small changes in observations shown in Fig. 4.4. This may also lead to completely different health management decisions at two consecutive CM points where the actual degradation observations may only have changed a little as seen from Fig. 4.2. The reason for such behavior seen from Fig. 4.6b stems from the factor that the distribution $p(\lambda_i \mid \mathbf{X}_{0:i})$ of the updated drift coefficient is not considered and the estimated parameters are not recursively updated in [32]. From Fig. 4.6b, we observe that our model can make the MSEs about the actual RUL at each CM point less sensitive to small changes. But most importantly our model can improve the accuracy of the estimated PDFs of the RULs with reduced variances, see Fig. 4.6. This reduction in variance is due to the use of the complete distribution of λ_i via the Bayesian rule shown earlier. These comparisons reflect the superiority of our model to Wang's model in the RUL estimation for the INS.

Now, we further compare our model with Gebraeel's model about their performance in the RUL estimation. The detailed implementation process of Gebraeel's model can be found in [19]. It is noted that the model in [19] needs prior parameters as well our model. In order to compare the goodness of fit, the prior parameters of our model in the following comparison are selected at random at time zero, but we consider two cases for the prior parameters of Gebraeel's model. One uses the

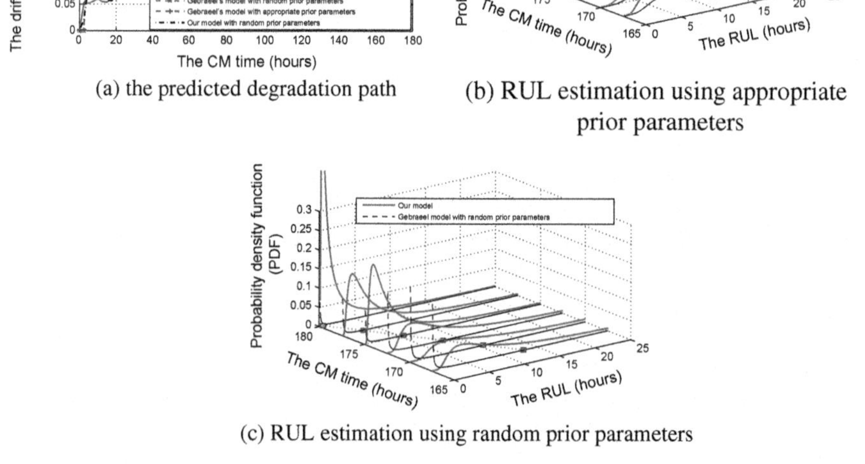

(a) the predicted degradation path

(b) RUL estimation using appropriate prior parameters

(c) RUL estimation using random prior parameters

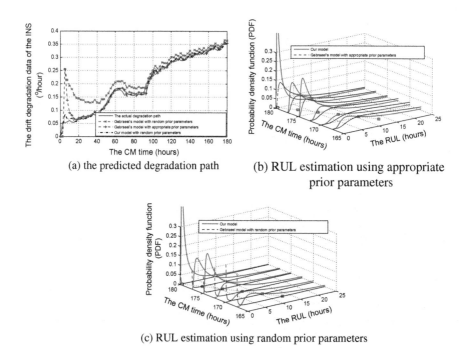

Fig. 4.7 Comparative results of our model with Gebraeel's model (□–actual RUL)

appropriately selected parameters as the prior parameters and the other uses random prior parameters. Figure 4.7 shows the comparative results.

Figure 4.7a shows that the predicted degradation path. The MSEs between the actual data and the corresponding predictions of the degradation path for our model with random prior parameters, Gebraeel's model with random prior parameters, and Gebraeel's model with appropriate prior parameters are 1.179E-4, 0.0026, and 7.9864E-4, respectively. Thus, our model has the best fitting for the degradation data. We also find that our model is robust with the choice of the prior parameters, but Gebraeel's model is not. Figure 4.7b illustrates several estimated PDFs of the RULs over time using Gebraeel's model with appropriate prior parameters. It is clear that the range of PDFs of RULs covers the actual RULs, but the uncertainty in the estimated RUL of our model is less than that from Gebraeel's model, particular for the last few sampling points, since our estimated PDFs of the RULs are sharper than the results of Gebraeel's model. However, if the prior parameters in Gebraeel's model are selected at random, the estimated RUL may be incorrect as illustrated in Fig. 4.7c (the observed RULs are outside of the predicted RUL PDF ranges). In comparison, our model using the random selected prior parameters can produce reasonable estimates. This demonstrates the merit of our model for RUL estimation of a particular system. Therefore, if there are no sufficient historical degradation data at the beginning for selecting the prior parameters, our model may be more appropriate for RUL estimation.

As noted by Remark 4.2, the moments of the RUL distribution obtained by Gebraeel's model do not exist. Therefore, we cannot calculate the MSE about the actual RUL defined in Eq. (4.49). But, to have a further look at the difference of the estimated RUL distributions between our exact result and Gebraeel's approximate result, it is desired to compare the estimated reliability from both models. Here, we use $\bar{F}_{R_i|\mathbf{X}_{0:i}}(r_i|\mathbf{X}_{0:i}) = 1 - F_{R_i|\mathbf{X}_{0:i}}(r_i|\mathbf{X}_{0:i})$ as the conditional reliability obtained

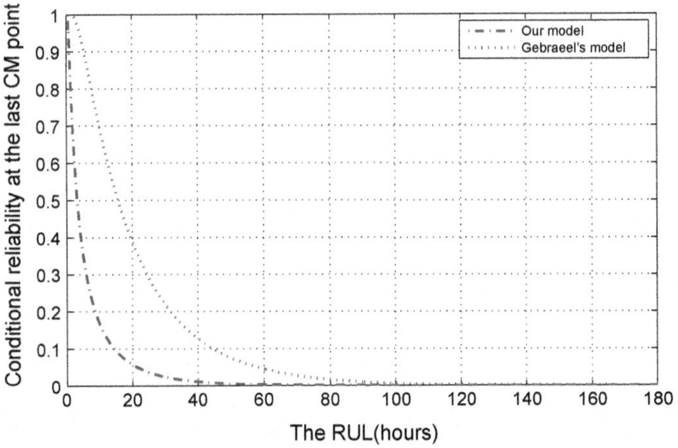

Fig. 4.8 Conditional reliability at t_{73}

from our exact approach at t_i, while $\bar{F}_{R_i'|\mathbf{X}_{0:i}}(r_i|\mathbf{X}_{0:i})$ as the conditional reliability obtained from Gebraeel's model by $\Pr(R_i' \le r_i | \mathbf{X}_{0:i}) = \Pr(X(r_i + t_i) \ge w | \mathbf{X}_{0:i})$ at t_i. The difference between them at the last CM point is illustrated in Fig. 4.8.

It can be found that the difference between both results is significant. This is due in part to the many fluctuations existing in the degradation path of gyro's drift so that using $\Pr(L_k' \le l_k | \mathbf{X}_{1:k}) = \Pr(X(l_k + t_k) \ge w | \mathbf{X}_{1:k})$ to approximate the RUL will lead to overestimating of the reliability, as discussed in Remark 4.1. Consequently, when the approximated result is applied to reliability-centered maintenance, the obtained decision may be far from the reality.

In sum, this practical case study demonstrates that our developed model can work well and efficiently. On the other hand, we verify that incorporating the observation history to date can improve the accuracy of the RUL estimation indeed.

References

1. Pecht M (2008) Prognostics and health management of electronics. Wiley, New Jersey
2. Si XS, Wang W, Hu CH, Zhou DH (2011) Remaining useful life estimation-A review on the statistical data driven approaches. Eur J Oper Res 213:1–14
3. Camci F, Chinnam RB (2010) Health state estimation and prognostics in machining processes. IEEE Trans Autom Sci Eng 7:581–597
4. Carr MJ, Wang W (2010) Modeling failure modes for residual life prediction using stochastic filtering theory. IEEE Trans Reliab 59:346–355
5. Dong M, He D (2007) Hidden semi-Markov model-based methodology for multi-sensor equipment health diagnosis and prognosis. Eur J Oper Res 178:858–878
6. Peng Y, Dong M (2011) A prognosis method using age-dependent hidden semi-Markov model for equipment health prediction. Mech Syst Signal Process 25:237–252
7. Sikorska J, Hodkiewicz M, Ma L (2011) Prognostic modelling options for remaining useful life estimation by industry. Mech Syst Signal Process 25:1803–1836
8. Mazhar MI, Kara S, Kaebernick H (2007) Remaining life estimation of used components in consumer products Life cycle data analysis by Weibull and artificial neural networks. J Oper Manag 25:1184–1193
9. Wang W (2007) A two-stage prognosis model in condition based maintenance. Eur J Oper Res 182:1177–1187
10. You MY, Li L, Meng G, Ni J (2010) Statistically planned and individual improved predictive maintenance management for continuously monitored degrading systems. IEEE Trans Reliab 59:744–753
11. Jardine AKS, Lin D, Banjevic D (2006) A review on machinery diagnostics and prognostics implementing condition-based maintenance. Mech Syst Signal Process 20:1483–1510
12. Lee M-LT, Whitmore GA (2006) Threshold regression for survival analysis: modeling event times by a stochastic process reaching a boundary. Stat Sci 21:501–513
13. Escobar LA, Meeker WQ (2006) A review of accelerated test models. Stat Sci 21:552–577
14. Joseph VR, Yu IT (2006) Reliability improvement experiments with degradation data. IEEE Trans Reliab 55:149–157
15. Lu CJ, Meeker WQ (1993) Using degradation measures to estimate a time-to-failure distribution. Technometrics 35:161–174
16. Park JI, Bae SJ (2010) Direct prediction methods on lifetime distribution of organic light-emitting diodes from accelerated degradation tests. IEEE Trans Reliab 59:74–90
17. Pandey MD, Yuan X-X, van Noortwijk JM (2009) The influence of temporal uncertainty of deterioration on life-cycle management of structures. Struct Infrastruct Eng 5:145–156

18. Bloch-Mecier S (2002) A preventive maintenance policy with sequential checking procedure for a Markov deteriorating system. Eur J Oper Res 147:548–576
19. Gebraeel N, Lawley MA, Li R, Ryan JK (2005) Residual-life distributions from component degradation signals: a Bayesian approach. IIE Trans 37:543–557
20. Delia MC, Rafael PO (2008) A maintenance model with failures and inspection following Markovian arrival processes and two repair modes. Eur J Oper Res 186:694–707
21. Liao HT, Elsayed EA (2006) Reliability inference for field conditions from accelerated degradation testing. Naval Res Logist 53:576–587
22. Barker CT, Newby MJ (2009) Optimal non-periodic inspection for a multivariate degradation model. Reliab Eng Syst Safety 94:33–43
23. Tseng ST, Tang J, Ku LH (2003) Determination of optimal burn-in parameters and residual life for highly reliable products. Naval Res Logist 50:1–14
24. Whitmore GA, Schenkelberg F (1997) Modelling accelerated degradation data using Wiener diffusion with a time scale transformation. Lifetime Data Anal 3:27–45
25. Peng CY, Tseng ST (2009) Mis-specification analysis of linear degradation models. IEEE Trans Reliab 58:444–455
26. Wang X (2010) Wiener processes with random effects for degradation data. J Multivar Anal 101:340–351
27. Ray A, Tangirala S (1996) Stochastic modeling of fatigue crack dynamics for on-line failure prognostics. IEEE Trans Control Syst Technol 4:443–450
28. Tseng ST, Peng CY (2004) Optimal burn-in policy by using an integrated Wiener process. IIE Trans 36:1161–1170
29. Crowder M, Lawless J (2007) On a scheme for predictive maintenance. Eur J Oper Res 176:1713–1722
30. Tang J, Su TS (2008) Estimating failure time distribution and its parameters based on intermediate data from a Wiener degradation model. Naval Res Logist 55:265–276
31. Xu ZG, Ji YD, Zhou DH (2008) Real-time reliability prediction for a dynamic system based on the hidden degradation process identification. IEEE Trans Reliab 57:230–242
32. Wang W, Carr M, Xu W, Kobbacy AKH (2011) A model for residual life prediction based on Brownian motion with an adaptive drift. Microeletron Reliab 51:285–293
33. Elwany AH, Gebraeel NZ (2009) Real-time estimation of mean remaining life using sensor-based degradation models. J Manuf Sci Eng 131:1–9(051005)
34. Christer AH, Wang W (1992) A model of condition monitoring of a production plant. Int J Prod Res 30:2199–2211
35. Li R, Ryan JK (2011) A Bayesian inventory model using real-time condition monitoring information. Prod Oper Manag 20:754–771
36. Cox DR, Miller HD (1965) The theory of stochastic processes. Methuen and Company, London
37. Montgomery DC (2008) Introduction to statistical quality control, 6th edn. Wiley, New York
38. Harvey AC (1989) Forecasting, structural time series models and the Kalman filter. Cambradge University Press, Cambridge
39. Zhou DH, Frank PM (1996) Strong tracking filtering of nonlinear time-varying stochastic systems with colored noise: application to parameter estimation and empirical robustness analysis. Int J Control 65:295–307
40. Zhou DH, Frank PM (1998) Fault diagnostics and fault tolerant control. IEEE Trans Aerosp Electron Syst 34:420–427
41. Whitmore GA (1986) Normal-Gamma mixture of inverse Gaussian distributions. Scand J Stat 13:211–220
42. Aalen OO (1994) Effects of frailty in survival analysis. Stat Methods Med Res 3:227–243
43. Derman C, Lieberman GJ, Ross SM (1984) On the use of replacements to extend system life. Oper Res 32:616–627
44. Shechter SM, Bailey MD, Schaefer AJ (2008) Replacing nonidentical vital components to extend system life. Naval Res Logist 55:700–703
45. Dempster AP, Laird NM, Rubin DB (1977) Maximum likelihood from incomplete data via the EM algorithm. J R Stat Soc Ser B 39:1–38

46. Kailath T, Sayed A, Hassabi B (2000) Linear estimation. Prentice-Hall, Upper Saddle River
47. Wu CFJ (1983) On the convergence property of the EM algorithm. Ann Stat 11:95–103
48. Dewar M, Scerri K, Kadirkamanathan V (2009) Data-driven spatio-temporal modeling using the integro-difference equation. IEEE Trans Signal Process 57:83–91
49. Gibson S, Wills A, Ninness B (2005) Maximum-likelihood parameter estimation of bilinear systems. IEEE Trans Autom Control 50:1581–1596
50. Wills A, Ninness B, Gibson S (2009) Maximum likelihood estimation of state space models from frequency domain data. IEEE Trans Autom Control 54:19–33
51. Schön TB, Wills A, Ninness B (2011) System identification of nonlinear state-space models. Automatica 47:39–49
52. Rauch HE, Tung F, Striebel CT (1965) Maximum Likelihood estimates of linear dynamic systems. AIAA J 3:1145–1450
53. Jwo DJ, Wang SH (2007) Adaptive fuzzy strong tracking extended Kalman filtering for GPS navigation. IEEE Sensors J 7:778–789
54. Woodman OJ (2007) An introduction to inertial navigation, Technical report, published by the University of Cambridge Computer Laboratory. http://www.cl.cam.ac.uk/techreports/UCAM-CL-TR-696.html (available via Internet)

Chapter 5
An Exact and Closed-Form Solution to Degradation Path-Dependent RUL Estimation

5.1 Introduction

Prognostics and health management (PHM) is an efficient and systematic approach for evaluating the reliability of a system in its actual operating conditions, predicting failure progression, and mitigating operating risks via management actions [1]. In PHM, prognostics can yield an advance warning of impending failure in a system, thereby helping in making maintenance decisions and executing preventive actions. The past decade has witnessed a constant research interest on various aspects of PHM due, primarily, to the fact that PHM has been extensively applied in a variety of fields including electronics, smart grid, nuclear plant, power industry, aerospace and military application, fleet-industrial maintenance, and public health management [2–7].

In each of these applications and documents, one critical quantity during prognostics for a system is the prognostic distance within which management decisions and repair actions can be planned effectively prior to failure occurrence to extend system life [8, 9]. This prognostic distance is closely associated with the definition of the remaining useful life (RUL) which is the length of the time from the present to the end of useful life. In fact, RUL estimation is always a key part in any PHM program and management can make use of RUL information in condition-based maintenance (CBM) to produce economic benefits in engineering, maintenance, logistics, and operations. Therefore, over the past few decades, significant advances have been made in developing RUL estimation approaches [9].

Stochasticity is one of the main characteristics in system operations that contributes to the uncertainty in estimating the RUL of the system. Therefore, one fundamental issue in RUL estimation is to find the probability density function (PDF) of the RUL. However this also leads to the main difficulty of RUL estimation since how to make full use of condition monitoring (CM) information to infer a RUL distribution is a not-well-solved problem. So far, RUL estimation has been regarded as one of the most central components in PHM [1, 10]. Thus, our primary interest

X.-S. Si et al., *Data-Driven Remaining Useful Life Prognosis Techniques*,
Springer Series in Reliability Engineering, DOI 10.1007/978-3-662-54030-5_5

of this chapter is to utilize CM information to adaptively estimate the RUL of the system and then apply it to support decision-related applications.

The current RUL estimation approaches can be generally classified as physics of failure, data driven and fusion. Physics of failure approaches rely on the physics of underlying failure mechanisms. Data-driven approaches achieve RUL estimation via data fitting mainly including machine learning and statistics based approaches. The fusion approaches are the combination of the physics of failure and data-driven approaches. However, for complex or large-scale engineering systems, it is typically difficult to obtain the physical failure mechanisms in advance or cost-expensive and time-consuming to capture the physics of failure. In contrast, data-driven approaches attempt to derive models directly from collected CM and life data, and thus are more appealing and have gained much attention in recent years [9].

In conventional data-based approaches, estimating the RUL is achieved by evaluating the conditional lifetime distribution given that a system has survived up to a specific time, e.g., $T - t| T > t$, where T denotes the lifetime [11, 12]. The obtained RUL distributions from these approaches are generally based on the life characteristics of a population of identical systems and lifetime data are required. However, such data are scarce in reality or even nonexistent at all for systems which are costly or time-consuming to collect the life data [13]. With the advances in CM technologies, degradation data can be obtained from routine CM as feasible and low-cost alternatives to estimate the RUL. These data are usually correlated with the underlying physical degradation process. If they are properly modeled, degradation data can be used to predict unexpected failures and accurately estimate the lifetime of gradually degraded systems [14, 15]. In general, degradation-data-based methods for RUL estimation can be classified into the models based on indirectly observed degradation processes and the models based on directly observed degradation processes [9]. The former models considered that the degradation state was hidden and assumed that the available CM data were stochastically related to the underlying degradation state. In this case, lifetime data must be available to establish the relationship between the CM data and failure. The latter models utilized the observed degradation data directly to describe the underlying degradation state of the system. In this chapter, we mainly focus on the directly observed degradation processes.

One common definition of RUL in the directly observed case is related to the concept of the first passage time (FPT) of the degradation process crossing a predefined threshold level. The use of the FPT concept as the definition of failure or a terminating event has a long history of application in diverse fields, including medicine, environmental science, engineering, business, economics, and sociology [16–19]. It is also acknowledged as a mainstream definition of failure in reliability literature based on degradation data [20–25]. Thus, in this chapter, we pay particular attention to a type of degradation-data-based models and derive the RUL distribution based on the concept of the FPT. Since degradation data are part of CM data, throughout this chapter, we use terms ‘CM information’ and ‘degradation data’ interchangeably.

In most of degradation-data-based models for RUL estimation, an exact and closed-form of the RUL distribution in the FPT sense is only available for some special cases. Frequently, a stepwise approximation or numerical simulation has to be used for finding an approximated RUL [2, 24, 26, 27]. In addition, most of these models either do not use the in situ degradation data during lifetime inference or only use information contained at the current observation point. However, the degradation data over the path collected up to date could contain more useful information to make the RUL estimation more accurate.

The type of models we specifically consider in this work follows the idea in [28] where two exponential-like degradation models were proposed. In their models, stochastic parameters were updated via a Bayesian approach to incorporate real-time CM information. Following [28], many variants and applications have been reported in prognostics, maintenance, and inventory management [25, 29–32]. However, in these chapters, they estimated the RUL distribution as the distribution of the time that it takes the trajectory of the degradation signal to cross the failure threshold based on an approximated method. In reality, this is not the FPT since the signal may have already crossed the failure threshold, signifying failure, prior to predicting the RUL. In extreme cases where the degradation fluctuations are large, this approximation could be significantly crude from the FPT concept. Even when the Brownian motion (BM) used as an error term, the availability of the explicit distribution of the FPT from the BM with a drift, i.e., the inverse Gaussian distribution, was not utilized in their models. Elwany and Gebraeel in [32] used the FPT to approximate the mean RUL but the distribution of the RUL was still evaluated by their approximate approach. As such, the results of RUL estimation in [28] and the followed works in applications are approximations as opposed to the FPT concept. Furthermore, in above works, the obtained RUL distributions belong to a family of Bernstein distributions. Consequently, the moments of the RUL do not exist. But in maintenance practice, the expectation of the RUL is required to be existent sometimes [7, 33]. Also, the stochastic coefficients in [28] and other following works had some prior distributions but no elaborated method is presented to select the hyperparameters of the prior distributions. Typically, several systems' historical degradation data of the same type are required to determine the deterministic coefficient and the unknown parameters in the prior distributions of the stochastic coefficients. But the scarcity of such historical degradation data of multiple systems is a commonly encountered case in practice, particularly for newly armed systems. As shown in Sect. 5.5 of this chapter, an inappropriate selection of these parameters can result in an incorrect estimate of the RUL.

Driven by the above survey over the related works, the purpose of this chapter is to develop a degradation path-dependent approach for RUL estimation that allows the estimated RUL distribution to be dependent on a system's degradation data history and to be adaptively updated, at the moment that a newly observed data is available. In particular, our goal is to shed light on three fundamental issues: (i) RUL estimation for an individual fielded device without the need of offline data of other similar systems, (ii) parameter estimation/updating of the degradation model from the observed degradation data, and (iii) an exact yet closed-form expression of the RUL distribution given (i) and (ii).

In response to the above issues, the dependency of RUL estimation with a system's past degradation path is presented through the combination of Bayesian updating and expectation maximization (EM) algorithm. This is a novel contribution of the chapter and is not fully explored in the conventional RUL modeling paradigms. As such, the deterministic coefficient and the unknown hyperparameters in the prior distributions of the stochastic coefficients can be updated when the new degradation observation is available. Two specific cases of our general approach, a linear model and an exponential-based model which were considered by [28], are used to illustrate the implementation of our presented approach. A major contribution of this chapter under these two special cases is that our approach can obtain an exact yet closed-form RUL distribution respectively, and we show that the moment of the obtained RUL distribution from our approach exists. This contrasts sharply with the approximated results obtained in the literature for the same cases. To our knowledge, the RUL estimation approach presented in this chapter for the two special cases is *the only one* that can provide an exact but closed-form RUL distribution utilizing the monitoring history.

Furthermore, in our approach the parameter updates in each iteration of the EM algorithm have explicit formulas. This implies that each iteration of the EM algorithm can be performed with a single computation, which leads to an extremely fast and simple estimation procedure. This computation advantage plus the exact yet closed-form RUL distribution are particularly attractive for practical applications.

We have performed extensive numerical studies to substantiate the superiority of the proposed approach in comparison with previously reported models. Since the presented method allows real-time updating the RUL distribution as new observations from CM are available, such updating mechanism enables the estimates less sensitive to the selection of parameters in the prior distributions, i.e., our estimation method is robust with prior or initial parameters, as revealed by the experimental results. This is another important character since it can make the engineering implementation rather reliable. We also provide a practical case study to test the performance of the developed approach in condition-based replacement decision-making. The use of our estimated RUL in CBM decision-making allows us to generate new insights on the effect of the estimated RUL from CM data upon decision-making and to explore in more detail of how estimated parameters influence the RUL estimation and further the replacement decision.

The remainder parts are organized as follows. Section 5.2 first constructs a general stochastic process based degradation model and then presents a degradation path-dependent approach for adaptive RUL estimation via real-time CM data. Sections 5.3 and 5.4 consider a linear model and an exponential model for illustrating the working mechanism of the proposed approach, respectively. Section 5.5 provides several simulations and a case study to illustrate the application and usefulness of the developed approach.

5.2 A Degradation Path-Dependent Approach for Adaptive RUL Estimation

In the following, a general parametric degradation model is developed first and then we present a degradation path-dependent approach that utilizes online CM sensory information to adaptively compute RUL distribution.

5.2.1 *A General Description of Stochastic Process Based Degradation Models*

As discussed previously, a degradation process is stochastic in nature due to inherent randomness in manufacturing and operations. Therefore, it is natural to model a degradation process as a stochastic process [20, 21, 23]. In this chapter, the degradation model is represented as a stochastic process $\{X(t), t \geq 0\}$ where $X(t)$ is the degradation signal at t. As mentioned earlier, a degradation signal is a characteristic pattern from the sensory information that captures the physical transitions associated with the degradation process. Some examples of degradation signals have been extensively illustrated in the literature, such as [34, 35].

Usually, a degradation model consists of deterministic and stochastic parts. The deterministic part represents a constant physical phenomenon common to all systems of a given population. While the stochastic part captures the variation of the degradation process of an individual system, particularly represented by a probability distribution. The stochastic part of the noise and random effects associated with the degradation signals are usually represented by a random term $\varepsilon(t)$, which is modeled as a stochastic process in this chapter.

With the above considerations in place, and without loss of generality, we assume that the degradation $X(t)$ at time t can be represented by the following general expression,

$$X(t) = h(t, \varepsilon(t); \boldsymbol{\theta}, \boldsymbol{\phi}), \tag{5.1}$$

where $X(t)$ is driven by a function $h(\cdot)$ with stochastic process $\varepsilon(t)$, characterizing the dynamics/uncertainty of the degradation process with $\boldsymbol{\theta}$ and $\boldsymbol{\phi}$ as the parameters. The functional form $h(\cdot)$ depends on the type of the system under consideration and represents a relationship between the operating time and the degradation signal. This functional form may follow a linear, polynomial, exponential, or any other trend. Considering that each system possibly experiences different sources of variations during its operation, for a degradation model to be realistic, we treat $\boldsymbol{\theta}$ as a random-effect vector representing unit-to-unit variability, and $\boldsymbol{\phi}$ as a fixed effect vector that is common to all systems. For simplicity, we assume that $\boldsymbol{\theta}$ and $\varepsilon(t)$ are s-independent. The ideas of random effects and the independent assumption between $\boldsymbol{\theta}$ and $\varepsilon(t)$ have been widely used in degradation modeling literature [22, 24, 28, 34].

5.2.2 A Degradation Path-Dependent Approach for Adaptive RUL Estimation via Real-Time CM Data

We have established the degradation model using a general stochastic process. We now illustrate how to estimate the RUL based on the established model. Define $\boldsymbol{X}_{1:k} = \{x_1, x_2, \ldots, x_k\}$ as the observed degradation at CM times $t_1, t_2, \ldots, t_k$, which could be irregularly spaced. It is noted that the degradation modeling paradigm for RUL estimation in most of conventional models is based on one assumption that the estimated PDF of RUL depends only on the currently observed degradation data, x_k. As such, it is highly desired to construct a model which can be conditional on all the data up to t_k, that is, $\boldsymbol{X}_{1:k}$. Consequently, using the FPT of the degradation process $\{X(t), t \geq 0\}$ crossing the threshold w and conditional on the observation history $\boldsymbol{X}_{1:k}$, we define RUL L_k at time t_k as

$$L_k = \inf\{l_k : X(l_k + t_k) \geq w | \boldsymbol{X}_{1:k}\}. \tag{5.2}$$

with PDF $f_{L_k|\boldsymbol{X}_{1:k}}(l_k|\boldsymbol{X}_{1:k})$ and cumulative distribution function (CDF) $F_{L_k|\boldsymbol{X}_{1:k}}(l_k|\boldsymbol{X}_{1:k})$. It can be observed from Eq. (5.2) that the defined RUL is degradation path-dependent.

Now we need to focus on how to estimate $f_{L_k|\boldsymbol{X}_{1:k}}(l_k|\boldsymbol{X}_{1:k})$ in an adaptive way, namely, when the newly observed data is available, the PDF of the RUL can be updated in order to make the estimated RUL depend on $\boldsymbol{X}_{1:k}$. In order to compute $f_{L_k|\boldsymbol{X}_{1:k}}(l_k|\boldsymbol{X}_{1:k})$, considering the stochastic nature of $\boldsymbol{\theta}$, we can formulate $f_{L_k|\boldsymbol{X}_{1:k}}(l_k|\boldsymbol{X}_{1:k})$ by the law of total probability as follows,

$$f_{L_k|\boldsymbol{X}_{1:k}}(l_k|\boldsymbol{X}_{1:k}) = \int f_{L_k|\boldsymbol{\theta},\boldsymbol{X}_{1:k}}(l_k|\boldsymbol{\theta}, \boldsymbol{X}_{1:k}) p(\boldsymbol{\theta}|\boldsymbol{X}_{1:k}) d\boldsymbol{\theta}, \tag{5.3}$$

From Eq. (5.3), $f_{L_k|\boldsymbol{\theta},\boldsymbol{X}_{1:k}}(l_k|\boldsymbol{\theta}, \boldsymbol{X}_{1:k})$ and $p(\boldsymbol{\theta}|\boldsymbol{X}_{1:k})$ must be known, and the unknown parameter $\boldsymbol{\phi}$ is needed to be estimated from $\boldsymbol{X}_{1:k}$, for the sake of calculating $f_{L_k|\boldsymbol{X}_{1:k}}(l_k|\boldsymbol{X}_{1:k})$. There are four steps which are shown below to accomplish the task.

***Step 1*: Determine prior information for θ**

As for the stochastic parameter vector, $\boldsymbol{\theta}$, it can be specified as a prior distribution as $p(\boldsymbol{\theta})$ in a Bayesian framework. The prior distribution of $p(\boldsymbol{\theta})$ contains hyperparameter $\boldsymbol{a}$. Once the observed data and the sampling distribution $p(\boldsymbol{X}_{1:k}|\boldsymbol{\theta})$ are available, the posterior distribution of $\boldsymbol{\theta}$, $p(\boldsymbol{\theta}|\boldsymbol{X}_{1:k})$, can be computed by the Bayesian rule. During the course of selecting the prior distribution, $p(\boldsymbol{\theta})$, one convenient way is to make $p(\boldsymbol{\theta})$ belong to the conjugate family of the sampling distributions, $p(\boldsymbol{X}_{1:k}|\boldsymbol{\theta})$, which can lead to a tractable posterior distribution of $\boldsymbol{\theta}$. In our illustration cases, we use such method to make the posterior distribution of $\boldsymbol{\theta}$ tractable.

Step 2*: *Update the posterior distribution for θ

Once a new observation at t_k is available, the posterior distribution of $\boldsymbol{\theta}$ can be updated via the Bayesian rule as follows,

$$p(\boldsymbol{\theta}|\,\boldsymbol{X}_{1:k}) = \frac{p(\boldsymbol{X}_{1:k}|\,\boldsymbol{\theta}) \cdot p(\boldsymbol{\theta})}{p(\boldsymbol{X}_{1:k})} \propto p(\boldsymbol{X}_{1:k}|\,\boldsymbol{\theta}) \cdot p(\boldsymbol{\theta}). \tag{5.4}$$

If analytical $p(\boldsymbol{\theta}|\,\boldsymbol{X}_{1:k})$is not available, Gibbs sampling or Metropolis–Hastings algorithm can be used to simulate all distributions [36, 37]. In this chapter, we do not consider the general case involving such as Gibbs sampling, but construct a conjugate prior distribution for the special cases considered in this chapter instead.

From Eq. (5.4) and using the chain rule in probability, we have

$$\begin{aligned}
p(\boldsymbol{\theta}|\,\boldsymbol{X}_{0:k}) &= \frac{p(\boldsymbol{X}_{0:k}, \boldsymbol{\theta})}{p(\boldsymbol{X}_{0:k})} = \frac{p(x_k|\,\boldsymbol{X}_{0:k-1}, \boldsymbol{\theta}) \cdot p(\boldsymbol{\theta}|\,\boldsymbol{X}_{0:k-1}) p(\boldsymbol{X}_{0:k-1})}{p(\boldsymbol{X}_{0:k})} \\
&= \frac{p(x_k|\,\boldsymbol{X}_{0:k-1}, \boldsymbol{\theta}) \cdot p(\boldsymbol{\theta}|\,\boldsymbol{X}_{0:k-1})}{p(x_k|\,\boldsymbol{X}_{0:k-1})} \\
&\propto p(x_k|\,\boldsymbol{X}_{0:k-1}, \boldsymbol{\theta}) \cdot p(\boldsymbol{\theta}|\,\boldsymbol{X}_{0:k-1}).
\end{aligned} \tag{5.5}$$

Equation (5.5) shows the recursive relationship between the prior at t_{k-1} and the newly observed information at t_k.

Step 3*: *Estimate the unknown parameters via the EM algorithm

Now we return to estimate unknown parameters $\boldsymbol{\phi}$ in Eq. (5.1) and parameters $\boldsymbol{a}$ in prior distribution $p(\boldsymbol{\theta})$. For simplicity, we denote unknown parameter vector consisting of $\boldsymbol{\phi}$ and $\boldsymbol{a}$ as $\boldsymbol{\Theta} = [\boldsymbol{\phi}, \boldsymbol{a}]$. In order to estimate $\boldsymbol{\Theta}$, we calculate the maximum likelihood estimation (MLE) of $\boldsymbol{\Theta}$ once new observation x_k is available. In this case, the log-likelihood function for $\boldsymbol{X}_{1:k}$ can be written as

$$\ell_k(\boldsymbol{\Theta}) = \log[p(\boldsymbol{X}_{1:k}|\boldsymbol{\Theta})], \tag{5.6}$$

where $p(\boldsymbol{X}_{1:k}|\boldsymbol{\Theta})$ is the joint PDF of the degradation data $\boldsymbol{X}_{1:k}$.

Then the MLE $\hat{\boldsymbol{\Theta}}_k$ of $\boldsymbol{\Theta}$ conditional on $\boldsymbol{X}_{1:k}$ can be obtained by

$$\hat{\boldsymbol{\Theta}}_k = \arg\max_{\boldsymbol{\Theta}} \ell_k(\boldsymbol{\Theta}). \tag{5.7}$$

Due to the random effect and unobservability of $\boldsymbol{\theta}$, Eq. (5.7) will be too difficult to maximize with respect to $\boldsymbol{\Theta}$. However, the EM algorithm [38] provides a possible way for resolving this difficulty. The essential idea in the EM algorithm is to manipulate the relationship between $p(\boldsymbol{X}_{1:k}|\,\boldsymbol{\Theta})$ and $p(\boldsymbol{X}_{1:k}, \boldsymbol{\theta}|\,\boldsymbol{\Theta})$ via the Bayesian rule so that estimating $\boldsymbol{\Theta}$ can be achieved by two steps: E-step and M-step.

- ***E-step***: Calculate

$$\ell(\boldsymbol{\Theta}|\,\hat{\boldsymbol{\Theta}}_k^{(i)}) = \mathbb{E}_{\boldsymbol{\theta}|\boldsymbol{X}_{0:k}, \hat{\boldsymbol{\Theta}}_k^{(i)}} \left\{\log p(\boldsymbol{X}_{0:k}, \boldsymbol{\theta}|\,\boldsymbol{\Theta})\right\}, \tag{5.8}$$

where $\hat{\boldsymbol{\Theta}}_k^{(i)}$ denotes the estimated parameters in the ith step conditional on $\boldsymbol{X}_{1:k}$.

- ***M-step***: Calculate

$$\hat{\boldsymbol{\Theta}}_k^{(i+1)} = \arg\max_{\boldsymbol{\Theta}} \ell(\boldsymbol{\Theta} \mid \hat{\boldsymbol{\Theta}}_k^{(i)}). \tag{5.9}$$

The above steps are iterated multiple times to produce a sequence $\{\hat{\boldsymbol{\Theta}}_k^{(0)}, \hat{\boldsymbol{\Theta}}_k^{(1)}, \hat{\boldsymbol{\Theta}}_k^{(2)}, \ldots\}$ of increasingly good approximations $\hat{\boldsymbol{\Theta}}_k^{*}$ to $\hat{\boldsymbol{\Theta}}_k$. The iterations are usually terminated using a standard criterion such as the difference between $\hat{\boldsymbol{\Theta}}_k^{(i)}$ and $\hat{\boldsymbol{\Theta}}_k^{(i+1)}$ falling below a predefined threshold. The properties of the convergence of the log-likelihood function and parameter estimates are discussed in [38, 39].

Step 4*: *Update the RUL distribution conditional on the observed information

After obtaining the required $f_{L_k|\boldsymbol{\theta},\boldsymbol{X}_{1:k}}(l_k|\boldsymbol{\theta}, \boldsymbol{X}_{1:k})$, $p(\boldsymbol{\theta}|\boldsymbol{X}_{1:k})$, and estimated $\boldsymbol{\Theta}$ conditional on $\boldsymbol{X}_{1:k}$ in previous three steps in place, the updated PDF and CDF of the RUL at time t_k can be formulated via the law of total probability as follows,

$$f_{L_k|\boldsymbol{X}_{1:k}}(l_k|\boldsymbol{X}_{1:k}) = \int_{-\infty}^{+\infty} f_{L_k|\boldsymbol{\theta},\boldsymbol{X}_{1:k}}(l_k|\boldsymbol{\theta}, \boldsymbol{X}_{1:k}) p(\boldsymbol{\theta}|\boldsymbol{X}_{1:k}) d\boldsymbol{\theta} = E_{\boldsymbol{\theta}|\boldsymbol{X}_{1:k}}\left[f_{L_k|\boldsymbol{\theta},\boldsymbol{X}_{1:k}}(l_k|\boldsymbol{\theta}, \boldsymbol{X}_{1:k})\right], \tag{5.10}$$

$$F_{L_k|\boldsymbol{X}_{1:k}}(l_k|\boldsymbol{X}_{1:k}) = \int_{-\infty}^{+\infty} F_{L_k|\boldsymbol{\theta},\boldsymbol{X}_{1:k}}(l_k|\boldsymbol{\theta}, \boldsymbol{X}_{1:k}) p(\boldsymbol{\theta}|\boldsymbol{X}_{1:k}) d\boldsymbol{\theta} = E_{\boldsymbol{\theta}|\boldsymbol{X}_{1:k}}\left[F_{L_k|\boldsymbol{\theta},\boldsymbol{X}_{1:k}}(l_k|\boldsymbol{\theta}, \boldsymbol{X}_{1:k})\right]. \tag{5.11}$$

Clearly, $f_{L_k|\boldsymbol{X}_{1:k}}(l_k|\boldsymbol{X}_{1:k})$ contains the whole history of observation to t_k, which is introduced by two updating procedures, updating $\boldsymbol{\theta}$ via the Bayesian rule and then updating $\boldsymbol{\Theta}$ via the EM algorithm. In the subsequent sections, we will give two specific models based on above-presented framework to illustrate the implementation process of our presented degradation path-dependent approach. Let us first consider a linear degradation model for RUL estimation.

5.3 Linear Model

The linear degradation model is typically used for modeling degradation processes where the degradation rate is approximately a constant, see [28, 31, 32, 40]. In this chapter, we consider a linear degradation model based on a Wiener process as follows,

$$X(t) = \phi + \theta t + \sigma B(t), \tag{5.12}$$

where ϕ is the initial degradation, θ and σ are the drift and diffusion parameters, and $B(t)$ denotes the standard BM, which represents the stochastic dynamics of the degradation process, as denoted by $\varepsilon(t)$ in Eq. (5.1). In this model, we assume that θ is the stochastic coefficient while ϕ and σ are deterministic. Without loss of generality, we further assume $t_0 = 0$ and $x_0 = 0$ and thus $\phi = 0$ in this case. Now, we illustrate the step-by-step implementation of our approach presented in Sect. 5.2.

***Step 1*: Determine the prior distribution**

In Eq. (5.12), stochastic parameter θ is generally assumed to follow a prior distribution, $p(\theta)$. Here we assume that θ is normally distributed with mean μ_0 and variance σ_0^2. Then according to the properties of the standard BM, for given θ, the sampling distribution of $\boldsymbol{X}_{1:k} = \{x_1, x_2, \ldots, x_k\}$ is multivariable normal, distributed by the following expression,

$$p(\boldsymbol{X}_{1:k}|\theta) = \frac{1}{\prod_{j=1}^{k}\sqrt{2\pi\sigma^2(t_j - t_{j-1})}}\exp\left[-\sum\nolimits_{j=1}^{k}\frac{\left(x_j - x_{j-1} - \theta(t_j - t_{j-1})\right)^2}{2\sigma^2(t_j - t_{j-1})}\right]. \tag{5.13}$$

In Bayesian framework, in order to calculate posterior $p(\theta|\boldsymbol{X}_{1:k})$, it is assumed that the prior distribution of θ follows $N(\mu_0, \sigma_0^2)$. Note that such prior distribution actually falls into the conjugate family of sampling distribution $p(\boldsymbol{X}_{1:k}|\theta)$. Consequently, the posterior estimate of θ conditional on $\boldsymbol{X}_{1:k}$ is still normal, that is, $\theta|\boldsymbol{X}_{1:k} \sim N(\mu_{\theta,k}, \sigma_{\theta,k}^2)$. Other prior distributions can also be used, but evaluating the posterior may involve numerical techniques such as Gibbs sampler.

***Step 2*: Posterior estimate of random parameter**

Given $\theta \sim N(\mu_0, \sigma_0^2)$, $p(\theta|\mathbf{X}_{1:k})$ can be calculated from Eq. (5.4) as

$$\begin{aligned} p(\theta|\boldsymbol{X}_{0:k}) &\propto p(\boldsymbol{X}_{0:k}|\theta)\cdot p(\theta) \\ &\propto \exp\left\{-\sum\nolimits_{j=1}^{k}\frac{\left(x_j - x_{j-1} - \theta(t_j - t_{j-1})\right)^2}{2\sigma^2(t_j - t_{j-1})}\right\}\cdot\exp\left\{-\frac{(\theta-\mu_0)^2}{2\sigma_0^2}\right\} \\ &\propto \exp\left\{-\frac{(\theta-\mu_{\theta,k})^2}{2\sigma_{\theta,k}^2}\right\}. \end{aligned} \tag{5.14}$$

Due to the property of the normal distribution of $\theta|\mathbf{X}_{1:k}$, we can obtain,

$$p(\theta|\boldsymbol{X}_{1:k}) = \frac{1}{\sigma_{\theta,k}\sqrt{2\pi}}\exp\left[-\frac{(\theta-\mu_{\theta,k})^2}{2\sigma_{\theta,k}^2}\right], \tag{5.15}$$

with

$$\begin{aligned} \mu_{\theta,k} &= (\mu_0\sigma^2 + x_k\sigma_0^2)/(t_k\sigma_0^2 + \sigma^2) \\ \sigma_{\theta,k}^2 &= \sigma^2\sigma_0^2/(t_k\sigma_0^2 + \sigma^2) \end{aligned}, \tag{5.16}$$

where we can learn that the posterior estimate of θ can be easily updated once new observation is available.

Remark 5.1 It is noted that, if we write the mean drift as $\overline{\mu_{\theta,k}} = x_k / t_k$, the posterior estimate of θ in Eq. (5.16) can be rewritten as $\mu_{\theta,k} = w_1\overline{\mu_{\theta,k}} + w_2\mu_0$, where $w_1 =$

$t_k\sigma_0^2/(t_k\sigma_0^2+\sigma^2)$ and $w_2=\sigma^2/(t_k\sigma_0^2+\sigma^2)$, and $\sigma_{\theta,k}^2=\sigma^2\sigma_0^2/(t_k\sigma_0^2+\sigma^2)<\sigma_0^2$. It is easily verified that $\sigma_{\theta,k}^2$ decreases monotonically approaching σ^2/t_k as $t_k\to\infty$, namely, the uncertainty about the true value of θ decreases. $\mu_{\theta,k}$ is a linearly weighted combination of $\overline{\mu_{\theta,k}}$ and μ_0 with weighted coefficients w_1 and w_2. Thus $\mu_{\theta,k}$ always lies somewhere between $\overline{\mu_{\theta,k}}$ and μ_0. It approaches $\overline{\mu_{\theta,k}}$ as $t_k\to\infty$. Therefore, if we have the adaptive adjustment mechanism for the prior parameters μ_0, σ_0^2, and fixed parameter σ^2, the initial guess of the true θ will be hopefully improved so as to make it closer to the new mean drift with less uncertainty. This motivates us to develop such adjustment mechanism in Step 3.

Now let us first focus on how to calculate $f_{L_k|\theta,\boldsymbol{X}_{1:k}}(l_k|\theta,\boldsymbol{X}_{1:k})$ at t_k and then go the further details about the parameters updating for μ_0, σ_0^2, and σ^2. Note that we consider the case of the directly observed degradation process (e.g., at the current CM point t_k, the current degradation state x_k is observed). Therefore, for in service RUL estimation at t_k, given θ and x_k, we can translate the original degradation process as,

$$X(t)=\ x_k+\theta(t-t_k)+\sigma\left(B(t)-B(t_k)\right),t\geq t_k. \tag{5.17}$$

Further translating this model with time scale over residual time l_k, i.e., RUL, as,

$$X(l_k+t_k)=\ x_k+\theta l_k+\sigma\left(B(l_k+t_k)-B(t_k)\right). \tag{5.18}$$

In order to calculate $f_{L_k|\theta,\boldsymbol{X}_{1:k}}(l_k|\theta,\boldsymbol{X}_{1:k})$, we first show that the following holds in general.

Theorem 5.1 *Given t_k, for any $t\geq 0$, the stochastic process, $\{W(t),t\geq 0\}$, with $W(t)=B(t+t_k)-B(t_k)$ is still a standard BM, where $\{B(t),t\geq 0\}$ is a standard BM.*

Proof We only need to check that the following properties hold for $\{W(t),t\geq 0\}$: (i) $\{W(t),t\geq 0\}$ is a Gaussian process with continuous path; (ii) $E\left[W(t)\right]=0$; and (iii) $E\left[W(t)W(s)\right]=\min\{s,t\}$.The first two are easy to check via the properties of the standard BM $\{B(t),t\geq 0\}$. Here, we only need to show that the last property is true for $\{W(t),t\geq 0\}$. We have

$$\begin{aligned}&\mathbb{E}\left[W(t)W(s)\right]=\mathbb{E}\left\{\left[B(t+t_k)-B(t_k)\right]\left[B(s+t_k)-B(t_k)\right]\right\}\\&=\mathbb{E}\left[B(t+t_k)B(s+t_k)\right]-\mathbb{E}\left[B(t_k)B(s+t_k)\right]-\mathbb{E}\left[B(t_k)B(t+t_k)\right]-\mathbb{E}\left[B(t_k)B(t_k)\right]\\&=\min\{t+t_k,s+t_k\}-t_k=\min\{s,t\}.\end{aligned} \tag{5.19}$$

This completes the proof.

Based on Theorem 5.1 and Eq. (5.18), the estimated RUL at t_k can be calculated as the FPT of the following process $\left\{X'(l_k),\,l_k\geq 0\right\}$ crossing threshold w,

$$X'(l_k)=x_k+\theta l_k+\sigma\left(B(l_k+t_k)-B(t_k)\right)=x_k+\theta l_k+\sigma W(l_k),\,l_k\geq 0, \tag{5.20}$$

with

$$W(l_k) = B(l_k + t_k) - B(t_k). \tag{5.21}$$

Therefore, $\{X'(l_k), l_k \geq 0\}$ is still a BM with a drift part θl_k and initial value $X'(0) = x_k$.

We further show that the following holds.

Theorem 5.2 *Once $\boldsymbol{X}_{1:k}$ is available at t_k, the followings hold,*

$$\begin{aligned} F_{L_k|\theta,\boldsymbol{X}_{0:k}}(l_k|\theta, \boldsymbol{X}_{0:k}) &= F_{L_k|\theta,X(t_k)=x_k}(l_k|\theta, x_k), f_{L_k|\theta,\boldsymbol{X}_{0:k}}(l_k|\theta, \boldsymbol{X}_{0:k}) \\ &= f_{L_k|\theta,X(t_k)=x_k}(l_k|\theta, x_k). \end{aligned} \tag{5.22}$$

Proof By the Markov property of the Wiener process,

$$\begin{aligned} F_{L_k|\theta,\boldsymbol{X}_{0:k}}(l_k|\theta, \boldsymbol{X}_{0:k}) &= \Pr(L_k \leq l_k|\theta, \boldsymbol{X}_{0:k}) \\ &= \Pr(\inf\{l_k : X(l_k + t_k) \geq w|\,\boldsymbol{X}_{0:k}\} \leq l_k|\theta, \boldsymbol{X}_{0:k}) \\ &= \Pr(\inf\{l_k : X(l_k + t_k) \geq w|\,X(t_k) = x_k\} \leq l_k) \\ &= F_{L_k|\theta,X(t_k)=x_k}(l_k|\theta, x_k). \end{aligned} \tag{5.23}$$

The second term of Theorem 5.2 follows immediately. This completes the proof.

Note the FPT of BM with a drift follows an inverse Gaussian distribution. From Theorem 5.2 and Eq. (5.21), it is direct to obtain the PDF and CDF of the RUL at t_k, associated with Eq. (5.19), as follows,

$$f_{L_k|\theta,\boldsymbol{X}_{0:k}}(l_k|\theta, \boldsymbol{X}_{0:k}) = f_{L_k|\theta,X(t_k)=x_k}(l_k|\theta, x_k) = \frac{w - x_k}{\sqrt{2\pi l_k^3 \sigma^2}} \exp\left\{-\frac{(w - x_k - \theta l_k)^2}{2\sigma^2 l_k}\right\}, \tag{5.24}$$

$$\begin{aligned} F_{L_k|\theta,\boldsymbol{X}_{0:k}}(l_k|\theta, \boldsymbol{X}_{0:k}) =& 1 - \Phi\left(\frac{w - x_k - \theta l_k}{\sigma\sqrt{l_k}}\right) + \\ & \exp\left\{\frac{2\theta(w - x_k)}{\sigma^2}\right\} \Phi\left(\frac{-(w - x_k) - \theta l_k}{\sigma\sqrt{l_k}}\right). \end{aligned} \tag{5.25}$$

In the next step, we illustrate how to estimate the unknown parameters $\boldsymbol{\Theta} = [\sigma^2, \boldsymbol{a}] = [\sigma^2, \mu_0, \sigma_0^2]$. In order to incorporate the updating nature of $\boldsymbol{\Theta}$, we use $\boldsymbol{\Theta}_k = [\sigma_k^2, \mu_{0,k}, \sigma_{0,k}^2]$ to denote the parameter needed to be estimated based on $\boldsymbol{X}_{1:k}$ and the estimated parameters are denoted by $\hat{\boldsymbol{\Theta}}_k = [\hat{\sigma}_k^2, \hat{\mu}_{0,k}, \hat{\sigma}_{0,k}^2]$.

***Step 3*: Estimate deterministic parameters based on EM algorithm**

In order to estimate $\boldsymbol{\Theta}_k$, from Eq. (5.8), we first evaluate the complete log-likelihood function $\ln p(\boldsymbol{X}_{1:k}, \theta|\,\boldsymbol{\Theta}_k)$, which is

$$\begin{aligned}\ln p(\boldsymbol{X}_{0:k}, \theta | \boldsymbol{\Theta}_k) &= \ln p(\boldsymbol{X}_{0:k} | \theta, \boldsymbol{\Theta}_k) + \ln p(\theta | \boldsymbol{\Theta}_k) \\ &= -\frac{k+1}{2}\ln 2\pi - \frac{1}{2}\sum_{j=1}^{k}\ln(t_j - t_{j-1}) - \frac{k}{2}\ln \sigma_k^2 - \\ &\sum_{j=1}^{k}\frac{\left(x_j - x_{j-1} - \theta(t_j - t_{j-1})\right)^2}{2\sigma_k^2(t_j - t_{j-1})} - \frac{1}{2}\ln\sigma_{0,k}^2 - \frac{(\theta - \mu_{0,k})^2}{2\sigma_{0,k}^2}.\end{aligned} \tag{5.26}$$

Given $\hat{\boldsymbol{\Theta}}_k^{(i)} = [\hat{\sigma_k^2}^{(i)}, \hat{\mu}_{0,k}^{(i)}, \hat{\sigma_{0,k}^2}^{(i)}]$ as the estimate in the ith step based on $\boldsymbol{X}_{1:k}$, the expectation $\ell(\boldsymbol{\Theta}_k | \hat{\boldsymbol{\Theta}}_k^{(i)})$ of $\ln p(\boldsymbol{X}_{1:k}, \boldsymbol{\theta} | \boldsymbol{\Theta}_k)$, can be computed as follows.

$$\begin{aligned}\ell(\boldsymbol{\Theta}_k | \hat{\boldsymbol{\Theta}}_k^{(i)}) &= \mathbb{E}_{\theta | \boldsymbol{X}_{0:k}, \hat{\boldsymbol{\Theta}}_k^{(i)}}\{\ln p(\boldsymbol{X}_{0:k}, \theta | \boldsymbol{\Theta}_k)\} \\ &= -\frac{k+1}{2}\ln 2\pi - \frac{1}{2}\sum_{j=1}^{k}\ln(t_j - t_{j-1}) - \frac{k}{2}\ln\sigma_k^2 - \\ &\sum_{j=1}^{k}\frac{(x_j - x_{j-1})^2 - 2\mu_{\theta,k}(t_j - t_{j-1})(x_j - x_{j-1}) + (t_j - t_{j-1})^2(\mu_{\theta,k}^2 + \sigma_{\theta,k}^2)}{2\sigma_k^2(t_j - t_{j-1})} - \\ &\frac{1}{2}\ln\sigma_{0,k}^2 - \frac{\mu_{\theta,k}^2 + \sigma_{\theta,k}^2 - 2\mu_{\theta,k}\mu_{0,k} + \mu_{0,k}^2}{2\sigma_{0,k}^2}.\end{aligned} \tag{5.27}$$

Let $\frac{\partial \ell(\boldsymbol{\Theta}_k | \hat{\boldsymbol{\Theta}}_k^{(i)})}{\partial \boldsymbol{\Theta}_k} = 0$, we obtain $\hat{\boldsymbol{\Theta}}_k^{(i+1)}$ as follows,

$$\hat{\sigma_k^2}^{(i+1)} = \frac{1}{k}\sum_{j=1}^{k}\frac{(x_j - x_{j-1})^2 - 2\mu_{\theta,k}(t_j - t_{j-1})(x_j - x_{j-1}) + (t_j - t_{j-1})^2(\mu_{\theta,k}^2 + \sigma_{\theta,k}^2)}{(t_j - t_{j-1})}, \tag{5.28}$$

$$\begin{aligned}\hat{\mu}_{0,k}^{(i+1)} &= \mu_{\theta,k} \\ \hat{\sigma_{0,k}^2}^{(i+1)} &= \sigma_{\theta,k}^2\end{aligned}. \tag{5.29}$$

Theorem 5.3 $\hat{\boldsymbol{\Theta}}_k^{(i+1)}$ *obtained by Eqs. (5.28) and (5.29) is uniquely determined and located at the maximum of* $\ell(\boldsymbol{\Theta}_k | \hat{\boldsymbol{\Theta}}_k^{(i)})$.

Proof From Eq. (5.25), we can learn that $\hat{\boldsymbol{\Theta}}_k^{(i+1)}$ obtained by Eqs. (5.26) and (5.27) is the only solution satisfying $\partial \ell(\boldsymbol{\Theta}_k | \hat{\boldsymbol{\Theta}}_k^{(i)})/\partial \boldsymbol{\Theta}_k = 0$. Consequently, taking $\partial^2 \ell(\boldsymbol{\Theta}_k | \hat{\boldsymbol{\Theta}}_k^{(i)})/\partial \boldsymbol{\Theta}_k \partial \boldsymbol{\Theta}_k^T$, the following is obtained,

$$\frac{\partial^2 \ell(\boldsymbol{\Theta}_k | \hat{\boldsymbol{\Theta}}_k^{(i)})}{\partial \boldsymbol{\Theta}_k \partial \boldsymbol{\Theta}_k^T} = \begin{bmatrix} \frac{k}{2\sigma_k^4} - \frac{\phi}{\sigma_k^6} & 0 & 0 \\ 0 & -\frac{1}{\sigma_{0,k}^2} & \frac{\mu_{\theta,k} - \mu_{0,k}}{\sigma_{0,k}^4} \\ 0 & \frac{\mu_{\theta,k} - \mu_{0,k}}{\sigma_{0,k}^4} & \frac{1}{2\sigma_{0,k}^4} - \frac{\psi}{\sigma_{0,k}^6} \end{bmatrix}, \tag{5.30}$$

with

$$
\psi = \mu_{\theta,k}^2 + \sigma_{\theta,k}^2 - 2\mu_{\theta,k}\mu_{0,k} + \mu_{0,k}^2,
$$
$$
\phi = \sum_{j=1}^{k} \frac{(x_j - x_{j-1})^2 - \mu_{\theta,k}(t_j - t_{j-1})(x_j - x_{j-1}) + (t_j - t_{j-1})^2(\mu_{\theta,k}^2 + \sigma_{\theta,k}^2)}{(t_j - t_{j-1})}. \tag{5.31}
$$

We show that the matrix in (5.31) is negative definite at $\boldsymbol{\Theta}_k = \hat{\boldsymbol{\Theta}}_k^{(i+1)}$, by calculating the order principal minor determinant as follows,

$$
\Delta_1 = \frac{k}{2\sigma_k^4} - \frac{\phi}{\sigma_k^6},\ \Delta_2 = -\frac{1}{\sigma_{0,k}^2}\Delta_1,\ \Delta_3 = \left(\frac{1}{2\sigma_{0,k}^4} - \frac{\psi}{\sigma_{0,k}^6} - \frac{(\mu_{\theta,k} - \mu_{0,k})^2}{\sigma_{0,k}^6}\right)\Delta_2. \tag{5.32}
$$

Then, at $\boldsymbol{\Theta}_k = \hat{\boldsymbol{\Theta}}_k^{(i+1)}$, the followings are obtained,

$$
\Delta_1|_{\boldsymbol{\Theta}=\hat{\boldsymbol{\Theta}}_k^{(i+1)}} = -\frac{k^3}{2\phi^2} < 0,
$$
$$
\Delta_2|_{\boldsymbol{\Theta}=\hat{\boldsymbol{\Theta}}_k^{(i+1)}} = -\frac{1}{\sigma_{\theta,k}^2}\Delta_1|_{\boldsymbol{\Theta}=\hat{\boldsymbol{\Theta}}_k^{(i+1)}} > 0,
$$
$$
\Delta_3|_{\boldsymbol{\Theta}=\hat{\boldsymbol{\Theta}}_k^{(i+1)}} = \Delta_2|_{\boldsymbol{\Theta}=\hat{\boldsymbol{\Theta}}_k^{(i+1)}} = -\frac{1}{2\sigma_{\theta,k}^4}\Delta_2|_{\boldsymbol{\Theta}=\hat{\boldsymbol{\Theta}}_k^{(i+1)}} < 0.
$$

This proves that the matrix in (5.30) is negative definite at $\boldsymbol{\Theta}_k = \hat{\boldsymbol{\Theta}}_k^{(i+1)}$. This conclusion plus the result that $\hat{\boldsymbol{\Theta}}_k^{(i+1)}$ is the only solution satisfying $\partial\ell(\boldsymbol{\Theta}_k|\hat{\boldsymbol{\Theta}}_k^{(i)})/\partial\boldsymbol{\Theta}_k = 0$ verifies Theorem 5.3.

Remark 5.2 It is observed from Theorem 5.3 that the M-step in our approach can be solved analytically and obtains the unique maximum point. In other words, parameter updates in each iteration of the EM algorithm have explicit formulas. This implies that each iteration of the EM algorithm can be performed with a single computation, which leads to an extremely fast and simple estimation procedure. This computation advantage is particularly attractive for practical applications.

***Step 4*: Exact and closed-form solution to degradation path-dependent RUL estimation**

We note that the estimated RUL by Eqs. (5.24) and (5.25) only uses the current degradation data, but not the system's degradation history before t_k. As discussed previously, ideally the future FPT depends on the path that the degradation has involved to date. In this step, we attempt to achieve such desired feature, i.e., to obtain $f_{L_k|\boldsymbol{X}_{1:k}}(l_k|\boldsymbol{X}_{1:k})$.

In order to calculate $f_{L_k|\mathbf{X}_{1:k}}(l_k|\mathbf{X}_{1:k})$, we first present the following two results.

Lemma 5.1 *If* $Y \sim N(0, 1)$ *and* $\lambda, \gamma, \in \boldsymbol{R}$*, then* $E_Y\left[\Phi(\lambda + \gamma Y)\right]$ *can be formulated as*

$$E_Y\left[\Phi(\lambda + \gamma Y)\right] = \Phi\left(\lambda\Big/\sqrt{\gamma^2 + 1}\right).$$

Proof We have

$$\begin{aligned} E_Y\left[\Phi(\lambda + \gamma Y)\right] &= E\left[\,E(I_{\{Z \le \lambda + \gamma Y\}} \big| \,Y) \big|\, Y\right] = \Pr(Z \le \lambda + \gamma Y) \\ &= \Pr(Z - \lambda - \gamma Y \le 0) = \Phi\left(\lambda\Big/\sqrt{\gamma^2 + 1}\right). \end{aligned}$$

In the derivation process, Φ denotes the standard normal CDF, $I_{\{Z \le \lambda + \gamma Y\}}$ is the indicator function, Z is standard normal and independent of Y and $Z - Y \sim N(-\lambda, \gamma^2 + 1)$. This completes the proof.

Theorem 5.4 *If* $Z \sim N(\mu, \sigma^2)$*, and* $w, A, B, D \in \boldsymbol{R}$*,* $C \in \boldsymbol{R}^+$*, then the following holds:*

$$E_Z\left[\exp(AZ)\Phi(C + DZ)\right] = \exp\left(A\mu + \frac{A^2}{2}\sigma^2\right) \cdot \Phi\left(\frac{C + D\mu + AD\sigma^2}{\sqrt{1 + D^2\sigma^2}}\right), \tag{5.33}$$

$$E_Z\left[(A - Z) \cdot \exp\left(-(B - Z)^2/2C\right)\right] = \sqrt{\frac{C}{\sigma^2 + C}}\left(A - \frac{\sigma^2 B + \mu C}{\sigma^2 + C}\right)\exp\left(-\frac{(B - \mu)^2}{2\left(\sigma^2 + C\right)}\right). \tag{5.34}$$

Proof (1) Through some algebraic manipulations, using Lemma, we have

$$\begin{aligned} E_Z\left[\exp(AZ)\Phi(C + DZ)\right] &= \frac{1}{\sqrt{2\pi\sigma^2}}\int \exp\left(-\frac{(z - \mu)^2}{2\sigma^2}\right)\exp(Az)\Phi(C + Dz)dz \\ &= \frac{1}{\sqrt{2\pi\sigma^2}}\int \exp\left(-\frac{z^2 - 2(\mu + A\sigma^2)z + \mu^2}{2\sigma^2}\right)\Phi(C + Dz)dz \\ &= \frac{1}{\sqrt{2\pi\sigma^2}} \cdot \exp\left(A\mu + \frac{A^2}{2}\sigma^2\right) \cdot \int \exp\left(-\frac{\left(z - (\mu + A\sigma^2)\right)^2}{2\sigma^2}\right)\Phi(C + Dz)dz \\ &= \frac{1}{\sigma}\exp\left(A\mu + \frac{A^2}{2}\sigma^2\right)\int \varphi\left(\frac{z - (\mu + A\sigma^2)}{\sigma}\right)\Phi(C + Dz)dz \\ &= \exp\left(A\mu + \frac{A^2}{2}\sigma^2\right)\int \varphi(u)\Phi(C + D\mu + AD\sigma^2 + D\sigma u)du \\ &= \exp\left(A\mu + \frac{A^2}{2}\sigma^2\right) \cdot \Phi\left(\frac{C + D\mu + AD\sigma^2}{\sqrt{1 + D^2\sigma^2}}\right). \end{aligned}$$

This completes the proof of the first equation in Theorem 5.4.

(2) Due to the limited space, we only summarize the main results below:

$$E_Z\left[(A-Z)\cdot\exp\left(-\frac{(B-Z)^2}{2C}\right)\right]=AI_1-I_2,$$

where I_1 and I_2 can be formulated separately as follows,

$$\begin{aligned}
I_1&=E_Z\left[\exp\left(-\frac{(B-Z)^2}{2C}\right)\right]=\frac{1}{\sqrt{2\pi\sigma^2}}\exp\left(-\frac{\sigma^2B^2+\mu C}{2\sigma^2C}\right)\int_{-\infty}^{\infty}\exp\left(-\frac{z^2-2\phi z}{\psi}\right)dz\\
&=\frac{1}{\sqrt{2\pi\sigma^2}}\exp\left(-\frac{\sigma^2B^2+\mu^2C}{2\sigma^2C}\right)\cdot\exp\left(\frac{\phi^2}{\psi}\right)\int_{-\infty}^{\infty}\exp\left(-\frac{(z-\phi)^2}{\psi}\right)dz\\
&=\frac{\sqrt{\psi\pi}}{\sqrt{2\pi\sigma^2}}\exp\left(-\frac{\sigma^2B^2+\mu^2C}{2\sigma^2C}\right)\cdot\exp\left(\frac{\phi^2}{\psi}\right)\\
&=\sqrt{\frac{C}{\sigma^2+C}}\exp\left(-\frac{(B-\mu)^2}{2\left(\sigma^2+C\right)}\right).
\end{aligned}$$

with $\phi=\left(\sigma^2B+\mu C\right)/\left(\sigma^2+C\right)$, ψ=$2\sigma^2C/\left(\sigma^2+C\right)$. In a similar way, I_2 can be written as

$$\begin{aligned}
I_2&=E_Z\left[Z\exp\left(-\frac{(B-Z)^2}{2C}\right)\right]\\
&=\frac{1}{\sqrt{2\pi\sigma^2}}\exp\left(-\frac{\sigma^2w^2+\mu C}{2\sigma^2C}+\frac{\phi^2}{\psi}\right)\cdot\int_{-\infty}^{\infty}z\exp\left(-\frac{(z-\phi)^2}{\psi}\right)dz\\
&=\frac{\sqrt{\psi}}{\sqrt{2\pi\sigma^2}}\exp\left(-\frac{\sigma^2B^2+\mu^2C}{2\sigma^2C}\right)\cdot\exp\left(\frac{\phi^2}{\psi}\right)\cdot\int_{-\infty}^{\infty}\left(\phi+u\sqrt{\psi}\right)\exp\left(-u^2\right)du\\
&=\frac{\sqrt{\psi}}{\sqrt{2\pi\sigma^2}}\exp\left(-\frac{\sigma^2B^2+\mu^2C}{2\sigma^2C}\right)\cdot\exp\left(\frac{\phi^2}{\psi}\right)\cdot\phi\sqrt{\pi}=\phi I_1.
\end{aligned}$$

Then, the final result can be written directly

$$\begin{aligned}
&E_Z\left[(A-Z)\cdot\exp\left(-(B-Z)^2/2C\right)\right]=AI_1-I_2=(A-\phi)I_1\\
&=\sqrt{\frac{C}{\sigma^2+C}}\left(A-\frac{\sigma^2B+\mu C}{\sigma^2+C}\right)\exp\left(-\frac{(B-\mu)^2}{2\left(\sigma^2+C\right)}\right).
\end{aligned}$$

This completes the proof of the second equation in Theorem 5.4.

The updated RUL distribution at time t_k can be summarized in the following theorem by using Lemma 5.1 and Theorem 5.5.

Theorem 5.5 *The PDF and CDF of the RUL conditional on the observations up to* t_k *can be written as*

$$f_{L_k|\boldsymbol{X}_{0:k}}(l_k|\boldsymbol{X}_{0:k}) = \frac{w-x_k}{\sqrt{2\pi l_k^3\left(\sigma_{\theta,k}^2 l_k+\sigma^2\right)}}\exp\left\{-\frac{\left(w-x_k-\mu_{\theta,k}l_k\right)^2}{2l_k\left(\sigma_{\theta,k}^2 l_k+\sigma^2\right)}\right\}, l_k>0, \tag{5.35}$$

$$F_{L_k|\boldsymbol{X}_{0:k}}(l_k|\boldsymbol{X}_{0:k}) = 1-\Phi\left(\frac{w-x_k-\mu_{\theta,k}l_k}{\sqrt{\sigma_{\theta,k}^2 l_k^2+\sigma^2 l_k}}\right)+$$
$$\exp\left\{\frac{2\mu_{\theta,k}(w-x_k)}{\sigma^2}+\frac{2\sigma_{\theta,k}^2(w-x_k)^2}{\sigma^4}\right\}\Phi\left(-\frac{2\sigma_{\theta,k}^2(w-x_k)l_k+\sigma^2\left(\mu_{\theta,k}l_k+w-x_k\right)}{\sigma^2\sqrt{\sigma_{\theta,k}^2 l_k^2+\sigma^2 l_k}}\right), \tag{5.36}$$

where the degradation history is introduced by parameters updating.

Proof Using Eqs. (5.15), (5.16), (5.22), (5.23) and the total law of probability, we have

$$\begin{aligned}
f_{L_k|\boldsymbol{X}_{0:k}}(l_k|\boldsymbol{X}_{0:k}) &= \int_{-\infty}^{\infty} f_{L_k|\theta,\boldsymbol{X}_{0:k}}(l_k|\theta,\boldsymbol{X}_{0:k})p(\theta|\,\boldsymbol{X}_{0:k})\mathrm{d}\theta \\
&= \mathbb{E}_{\theta|\boldsymbol{X}_{0:k}}\left[f_{L_k|\theta,\boldsymbol{X}_{0:k}}(l_k|\theta,\boldsymbol{X}_{0:k})\right] \\
&= \frac{w-x_k}{\sqrt{2\pi l_k^3\sigma^2}}\mathbb{E}_{\theta|\boldsymbol{X}_{0:k}}\left[\exp\left\{-\frac{\left(w-x_k-\theta l_k\right)^2}{2\sigma^2 l_k}\right\}\right] \\
&= \frac{w-x_k}{\sqrt{2\pi l_k^3\sigma^2}}\sqrt{\frac{\sigma^2 l_k}{\sigma^2 l_k+\sigma_{\theta,k}^2 l_k^2}}\exp\left\{-\frac{(w-x_k-\mu_{\theta,k}l_k)^2}{2l_k\left(\sigma^2+\sigma_{\theta,k}^2 l_k\right)}\right\}.
\end{aligned} \tag{5.37}$$

The last equation is implied by the second result of Theorem 5.4.

$$\begin{aligned}
F_{L_k|\boldsymbol{X}_{0:k}}(l_k|\boldsymbol{X}_{0:k}) &= \int_{-\infty}^{\infty} F_{L_k|\theta,\boldsymbol{X}_{0:k}}(l_k|\theta,\boldsymbol{X}_{0:k})p(\theta|\,\boldsymbol{X}_{0:k})\mathrm{d}\theta \\
&=\mathbb{E}_{\theta|\boldsymbol{X}_{0:k}}\left[F_{L_k|\theta,\boldsymbol{X}_{0:k}}(l_k|\theta,\boldsymbol{X}_{0:k})\right] \\
&=1-\mathbb{E}_{\theta|\boldsymbol{X}_{0:k}}\left[\Phi\left(\frac{w-x_k-\theta l_k}{\sigma\sqrt{l_k}}\right)\right]+ \\
&\quad \mathbb{E}_{\theta|\boldsymbol{X}_{0:k}}\left[\exp\left\{\frac{2\theta(w-x_k)}{\sigma^2}\right\}\Phi\left(\frac{-(w-x_k)-\theta l_k}{\sigma\sqrt{l_k}}\right)\right].
\end{aligned} \tag{5.38}$$

Following the first result in Theorem 5.4, it is straightforward to obtain Eq. (5.32). This completes the proof.

So far, we have accomplished our presented degradation path-dependent approach for RUL estimation in the linear case. In the following, we give some remarks through comparing our approach with previously reported approaches.

Remark 5.3 The results in [28] and the related works directly used $\Pr(L_k \leq l_k | \boldsymbol{X}_{1:k}) = \Pr\left(X(l_k + t_k) \geq w | \boldsymbol{X}_{1:k}\right)$ to calculate the RUL distribution, which ignored the possible hitting events within $(t_k, t_k + l_k)$. This implies that their results are approximate in the sense of the FPT. However, our obtained results in Eqs. (5.35) and (5.36) are exact but with explicit form.

In order to further compare the obtained results with the results by [28], we first introduce the concept of stochastic comparison between two random variables.

Definition 5.1 ([41]) Given random variables ξ and ζ, ξ is stochastically greater than ζ if $\Pr(\xi > v) \geq \Pr(\zeta > v)$, for all real v, denoted by $\xi \geq_{st} \zeta$.

Denote the estimated RUL by Gebraeel's approach as L'_k, then we have the following conclusion.

Theorem 5.6 *Conditional on the degradation history to t_k, i.e., $\boldsymbol{X}_{1:k}$, and adopting the same parameter estimation procedure, we have $L'_k \geq_{st} L_k$.*

Proof As mentioned in Remark 5.2, in Gebraeel et al. (2005), they directly used $\Pr(L'_k \leq l_k \big| \boldsymbol{X}_{1:k}) = \Pr\left(X(l_k + t_k) \geq w | \boldsymbol{X}_{1:k}\right)$ to estimate the RUL distribution. As a result, the CDF of the estimated RUL can be written as

$$F_{L'_k|\boldsymbol{X}_{0:k}}(l_k|\boldsymbol{X}_{0:k}) = 1 - \Phi\left(\frac{w - x_k - \mu_{\theta,k} l_k}{\sqrt{\sigma^2_{\theta,k} l_k^2 + \sigma^2 l_k}}\right). \tag{5.39}$$

Compared with Eq. (5.32), it is obviously observed that the following holds,

$$F_{L'_k|\boldsymbol{X}_{0:k}}(l_k|\boldsymbol{X}_{0:k}) \leq F_{L_k|\boldsymbol{X}_{0:k}}(l_k|\boldsymbol{X}_{0:k}), \tag{5.40}$$

Namely, we have

$$\Pr(L'_k \geq l_k|\boldsymbol{X}_{0:k}) \geq \Pr(L_k \geq l_k|\boldsymbol{X}_{0:k}). \tag{5.41}$$

Then according the given definition by Eq. (5.33), the proof is completed.

From Theorem 5.6, the result following the approach developed by [28] overestimates the RUL and then can lead to under-maintenance or delayed-maintenance.

Remark 5.4 The moment of the RUL distribution obtained by [28, 32], does not exist since their obtained RUL distributions belong to the family of Bernstein distributions, known without moments, but this is not the case for our result. For example, the mean of RUL can be easily formulated by

$$\begin{aligned}\mathbb{E}(L_k|\boldsymbol{X}_{0:k}) &= \mathbb{E}\left[\mathbb{E}(L_k|\theta,\boldsymbol{X}_{0:k})|\boldsymbol{X}_{0:k}\right] = \mathbb{E}\left(\frac{w-x_k}{\theta}\middle|\boldsymbol{X}_{0:k}\right)\\ &= \frac{w-x_k}{\sigma_{\theta,k}^2}\exp\left\{-\frac{\mu_{\theta,k}^2}{2\sigma_{\theta,k}^2}\right\}\int_0^{\mu_{\theta,k}}\exp\left\{\frac{u^2}{2\sigma_{\theta,k}^2}\right\}du\\ &= \frac{\sqrt{2}\,(w-x_k)}{\sigma_{\theta,k}}D\left(\frac{\mu_{\theta,k}}{\sqrt{2}\sigma_{\theta,k}}\right),\end{aligned} \tag{5.42}$$

where $D(z)=\exp(-z^2)\int_0^z\exp(u^2)du$ is the Dawson integral, which is known to exist. Particularly, if we assume $\Pr(\theta<0)=0$, which implies $\mu_{\theta,k}\gg\sigma_{\theta,k}$. Then using the approximation property of Dawson integral for large z, $D(z)\approx 1/2z$, then $E(L_k|\boldsymbol{X}_{1:k})=(w-x_k)/\mu_{\theta,k}$.

The PDF and CDF in Eqs. (5.35) and (5.36) enable the construction of a replacement decision model that incorporates the probability of failure before a particular instant conditioned on the degradation history to date. At each monitoring point throughout the life of a system, an optimal replacement time can be scheduled using the renewal-reward theory and the long run expected cost per unit time. When the RUL distribution is used in condition-based replacement, the following is usually minimized to decide the optimal replacement time [2],

$$C(T_{R,k})=\frac{c_p+(c_f-c_p)\Pr(L_k<T_{R,k}-t_k|\boldsymbol{X}_{1:k})}{t_k+(T_{R,k}-t_k)\left(1-\Pr(L_k<T_{R,k}-t_k|\boldsymbol{X}_{1:k})\right)+\int\limits_{l_k=0}^{T_{R,k}-t_k} l_k f_{L_k|\mathbf{X}_{1:k}}(l_k|\mathbf{X}_{1:k})dl_k}, \tag{5.43}$$

where $T_{R,k}$ is the decision variable representing the planned replacement time determined at the kth CM point, c_p is the cost of a preventive replacement, and c_f is the replacement cost with the failure. It is well known that if $c_p\ge c_f$, no preventive replacement is optimal. In this case, $T_{R,k}$ will approach positive infinity, i.e., $T_{R,k}\to+\infty$. Then the last term in the denominator of Eq. (5.43) will be $E(L_k|\boldsymbol{X}_{1:k})$. Therefore, the nonexistence of $E(L_k|\boldsymbol{X}_{1:k})$ may lead to the nonexistence of Eq. (5.43). However, our result can avoid this problem and makes Eq. (5.43) hold in general.

Additionally, it can be proved that the above cost function equals the cost function used in [31].

Theorem 5.7 *Let* $F_{L_k|\mathbf{X}_{1:k}}(T_{R,k}-t_k|\boldsymbol{X}_{1:k})=\Pr(L_k<T_{R,k}-t_k|\boldsymbol{X}_{1:k})$ *and* $\bar{F}_{L_k|\mathbf{X}_{1:k}}(l_k|\boldsymbol{X}_{1:k})=1-F_{L_k|\mathbf{X}_{1:k}}(l_k|\boldsymbol{X}_{1:k})$. *Then*

$$C(T_{R,k})=\frac{c_p+(c_f-c_p)F_{L_k|\mathbf{X}_{1:k}}(T_{R,k}-t_k|\boldsymbol{X}_{1:k})}{t_k+\int\limits_{l_k=0}^{T_{R,k}-t_k}\bar{F}_{L_k|\mathbf{X}_{1:k}}(l_k|\boldsymbol{X}_{1:k})dl_k}.$$

Proof From Eq. (5.35), we directly have

$$C(T_{R,k}) = \frac{c_p + (c_f - c_p)F_{L_k|\mathbf{X}_{0:k}}(T_{R,k} - t_k \big| \mathbf{X}_{0:k})}{t_k + (T_{R,k} - t_k)\bar{F}_{L_k|\mathbf{X}_{0:k}}(T_{R,k} - t_k \big| \mathbf{X}_{0:k}) + \int\limits_{l_k=0}^{T_{R,k}-t_k} l_k f_{L_k|\mathbf{X}_{0:k}}(l_k|\mathbf{X}_{0:k})\mathrm{d}l_k}. \tag{5.44}$$

For the third term of the denominator in the above equation, we get

$$\begin{aligned}
\int\limits_{l_k=0}^{T_{R,k}-t_k} l_k f_{L_k|\mathbf{X}_{0:k}}(l_k|\mathbf{X}_{0:k})\mathrm{d}l_k &= \int\limits_{l_k=0}^{T_{R,k}-t_k} l_k \mathrm{d}F_{L_k|\mathbf{X}_{0:k}}(l_k|\mathbf{X}_{0:k}) \\
&= l_k F_{L_k|\mathbf{X}_{0:k}}(l_k|\mathbf{X}_{0:k})\Big|_{l_k=0}^{l_k=T_{R,k}-t_k} - \int\limits_{l_k=0}^{T_{R,k}-t_k} F_{L_k|\mathbf{X}_{0:k}}(l_k|\mathbf{X}_{0:k})\mathrm{d}l_k \\
&= (T_{R,k} - t_k)F_{L_k|\mathbf{X}_{0:k}}(T_{R,k} - t_k \big| \mathbf{X}_{0:k}) + \int\limits_{l_k=0}^{T_{R,k}-t_k} \bar{F}_{L_k|\mathbf{X}_{0:k}}(l_k|\mathbf{X}_{0:k})\mathrm{d}l_k - (T_{R,k} - t_k) \\
&= \int\limits_{l_k=0}^{T_{R,k}-t_k} \bar{F}_{L_k|\mathbf{X}_{0:k}}(l_k|\mathbf{X}_{0:k})\mathrm{d}l_k - (T_{R,k} - t_k)\bar{F}_{L_k|\mathbf{X}_{0:k}}(T_{R,k} - t_k \big| \mathbf{X}_{0:k}).
\end{aligned} \tag{5.45}$$

Replacing the third term in the denominator of Eq. (5.44) with above equation will complete the proof.

Based on above results, we have the following conclusion for the cost functions when using the approximated RUL distribution by Gebraeel's approach and our exact RUL distribution. Denote the cost function using the RUL distribution obtained from Gebraeel's approach as $C'(T_{R,k})$.

Theorem 5.8 *Conditional on the degradation history to t_k, i.e., $\mathbf{X}_{1:k}$, and adopting the same parameter estimation procedure, if $c_f \geq c_p$, then $C(T_{R,k}) \geq C'(T_{R,k})$.*

Proof Based on Theorem 5.7, we have the following formulas of $C(T_{R,k})$ and $C'(T_{R,k})$,

$$C(T_{R,k}) = \frac{c_p + (c_f - c_p)F_{L_k|\mathbf{X}_{0:k}}(T_{R,k} - t_k \big| \mathbf{X}_{0:k})}{t_k + \int\limits_{l_k=0}^{T_{R,k}-t_k} \bar{F}_{L_k|\mathbf{X}_{0:k}}(l_k \big| \mathbf{X}_{0:k})\mathrm{d}l_k}, \tag{5.46}$$

and

$$C'(T_{R,k}) = \frac{c_p + (c_f - c_p)F_{L'_k|\mathbf{X}_{0:k}}(T_{R,k} - t_k \big| \mathbf{X}_{0:k})}{t_k + \int\limits_{l_k=0}^{T_{R,k}-t_k} \bar{F}_{L'_k|\mathbf{X}_{0:k}}(l_k \big| \mathbf{X}_{0:k})\mathrm{d}l_k}. \tag{5.47}$$

From Theorem 5.6, we have $F_{L_k|X_{1:k}}(l_k|\mathbf{X}_{1:k}) \geq F_{L_k'|X_{1:k}}(l_k|\mathbf{X}_{1:k})$. Then we have

$$c_p + (c_f - c_p)F_{L_k|X_{0:k}}(T_{R,k} - t_k \big| X_{0:k}) \geq c_p + (c_f - c_p)F_{L_k'|X_{0:k}}(T_{R,k} - t_k \big| X_{0:k}), \tag{5.48}$$

and

$$t_k + \int_{l_k=0}^{T_{R,k}-t_k} \bar{F}_{L_k|X_{0:k}}(l_k|X_{0:k})\mathrm{d}l_k \leq t_k + \int_{l_k=0}^{T_{R,k}-t_k} \bar{F}_{L_k'|X_{0:k}}(l_k|X_{0:k})\mathrm{d}l_k. \tag{5.49}$$

As a result, the proof is completed.

Theorem 5.8 implies that, when using approximated RUL distribution in decision-making, the operating risk represented by the expected cost per unit time will be underestimated and then the maintenance action may be delayed. This further confirms the statement implied by Theorem 5.6.

Remark 5.5 Note that variance parameter σ^2 and parameters μ_0, σ_0^2 in prior distribution $p(\theta)$ of [28] and the related works are prior determined from the offline degradation data of multiple other systems. However, once these parameters are determined and they are then fixed even if real-time CM data are available. This makes the RUL estimation non-robust over these parameters. Particularly, if these parameters are not determined accurately enough, then the estimated RUL may be hardly accurate. In contrast, our approach can adaptively adjust $\boldsymbol{\Theta} = [\sigma^2, \mu_0, \sigma_0^2]$ via the EM algorithm in line with real-time data. In this sense, our approach relies less on prior information.

To facilitate the implementation of the developed approach, the main steps are summarized in the following algorithm.

Algorithm 5.1 (*RUL estimation and associated replacement decision algorithm*)

Step 1: At the initial time t_0, select $\sigma^2 = \hat{\sigma}_0^2$ and the initial prior parameters $\mu_0 = \mu_{\theta,0}, \sigma_0^2 = \sigma_{\theta,0}^2$.

Step 2: Once obtaining the degradation observation x_k at time t_k for $k \geq 1$, let $\sigma^2 = \hat{\sigma}_{k-1}^2$ and the prior parameters $\mu_0 = \mu_{\theta,k-1}, \sigma_0^2 = \sigma_{\theta,k-1}^2$. Then Eq. (5.16) is used to calculate $\mu_{\theta,k}, \sigma_{\theta,k}^2$.

Step 3: Based on $\mu_{\theta,k}, \sigma_{\theta,k}^2$, the parameters $\sigma^2 = \hat{\sigma}_k^2$ and $\mu_0 = \mu_{\theta,k}, \sigma_0^2 = \sigma_{\theta,k}^2$ are obtained by Eqs. (5.28) and (5.29), respectively.

Step 4: Based on $\mu_{\theta,k}, \sigma_{\theta,k}^2$ from Step 2 and σ^2 from Step 3, estimate the RUL by Eq. (5.35) and make the replacement decision by minimizing Eq. (5.43).

Step 5: Once obtaining new degradation observation x_{k+1} at time t_{k+1}, go to Step 2 and repeat the above steps.

5.4 Exponential Model

The exponential-like degradation model is another typical model representing a degradation process where the cumulative damage has a particular effect on the rate of degradation, but the degradation path can be linearized by log-transformation. So far, it has long been thought to be a good approximation for nonlinear degradation processes such as corrosion, bearing degradation, deterioration of LED lighting, see [28, 42–44]. In this section, we borrow this kind of model but obtain some novel results which were not reported before. In general, an exponential degradation model can be represented as,

$$X(t) = \phi + \theta' \exp\left(\beta' t + \sigma B(t) - \frac{\sigma^2}{2} t\right), \tag{5.50}$$

where ϕ is a known constant, σ is a constant representing the deterministic parameter, θ' and β' are random variables characterizing the unit-to-unit variability, and $B(t)$ is also a standard BM.

For an exponential-like model, it is more convenient to work with its logged format. Thus, we define $S(t)$ at t as follows,

$$S(t) = \ln\left[X(t) - \phi\right] = \ln\theta' + \left(\beta' - \frac{\sigma^2}{2}\right) t + \sigma B(t) = \theta + \beta t + \sigma B(t), \tag{5.51}$$

where $\theta = \ln\theta'$ and $\beta = \beta' - \sigma^2/2$, and we denote $\boldsymbol{\theta} = [\theta, \beta]$.

As well as the case study part in [28, 35, 42], it is assumed that $\phi = 0$ to simplify the analysis. In general, however, ϕ could be any known constant. In the following, we mainly focus on the transformed degradation observations from $S(t)$ and illustrate the implementation process of our approach for RUL estimation in this exponential case.

***Steps 1–2*: Bayesian estimate of random parameters**

As the same assumption used for the prior distributions of θ' and β' in [28], we let $\ln\theta'$ and β' follow $N(\mu_0, \sigma_0^2)$ and $N(\mu_1', \sigma_1^2)$, respectively, where $\mu_1' = \mu_1 + \sigma^2/2$, and we further assume that θ', β' and $B(t)$ are mutually independent. Therefore, we have $\theta \sim N(\mu_0, \sigma_0^2)$ and $\beta \sim N(\mu_1, \sigma_1^2)$. As a result, once the new CM data is available, the posterior estimates of θ and β can be evaluated by the Bayesian rule.

Let $\boldsymbol{S}_{1:k} = \{s_1, s_2, \ldots, s_k\}$, where $s_k = \ln x_k$. Then, given θ, β, the sampling distribution of $\boldsymbol{S}_{1:k}$ is multivariable normal as

$$p(\boldsymbol{S}_{1:k} | \theta, \beta) = \frac{1}{\prod_{j=1}^{k} \sqrt{2\pi\sigma^2(t_j - t_{j-1})}} \times \exp\left\{ -\frac{(s_1 - \theta - \beta t_1)^2}{2\sigma^2 t_1} - \sum_{j=2}^{k} \frac{\left(s_j - s_{j-1} - \beta(t_j - t_{j-1})\right)^2}{2\sigma^2(t_j - t_{j-1})} \right\}. \tag{5.52}$$

Then the joint posterior estimate of θ and β conditional on $\boldsymbol{S}_{1:k}$ is still normal resulted from the fact of the normal distribution assumption of θ and β. In other words, $\theta, \beta | \boldsymbol{S}_{1:k} \sim N(\mu_{\theta,k}, \sigma_{\theta,k}^2, \mu_{\beta,k}, \sigma_{\beta,k}^2, \rho_k)$. To be more precise, we have

$$\begin{aligned} p(\theta, \beta | \boldsymbol{S}_{1:k}) & \propto p(\boldsymbol{S}_{1:k} | \theta, \beta) \cdot p(\theta, \beta) \\ & \propto \exp\left\{ \frac{(s_1 - \theta - \beta t_1)^2}{2\upsilon^2 t_1} - \sum_{j=2}^{k} \frac{(s_j - s_{j-1} - \beta(t_j - t_{j-1}))^2}{2\upsilon^2(t_j - t_{j-1})} \right\} \times \\ & \quad \exp\left\{ -\frac{(\theta - \mu_0)^2}{2\sigma_0^2} \right\} \cdot \exp\left\{ -\frac{(\beta - \mu_1)^2}{2\sigma_1^2} \right\} \\ & \propto \frac{1}{2\pi\sigma_{\theta,k}\sigma_{\beta,k}\sqrt{1-\rho_k^2}} \times \\ & \quad \exp\left\{ -\frac{1}{2(1-\rho_k^2)} \left(\frac{(\theta - \mu_{\theta,k})^2}{\sigma_{\theta,k}^2} - 2\rho_k \frac{(\theta - \mu_{\theta,k})(\beta - \mu_{\beta,k})}{\sigma_{\theta,k}\sigma_{\beta,k}} + \frac{(\beta - \mu_{\beta,k})^2}{\sigma_{\beta,k}^2} \right) \right\}, \end{aligned} \tag{5.53}$$

with

$$\begin{aligned} \mu_{\theta,k} &= \frac{(s_1\sigma_0^2 + \mu_0\sigma^2 t_1)(\sigma^2 + \sigma_1^2 t_k) - \sigma_0^2 t_1(\sigma_1^2 s_k + \mu_1\sigma^2 - 0.5\sigma^4)}{(\sigma_0^2 + \sigma^2 t_1)(\sigma_1^2 t_k + \sigma^2) - \sigma_0^2\sigma_1^2 t_1} \\ \sigma_{\theta,k}^2 &= \frac{\sigma_0^2\sigma^2 t_1(\sigma^2 + \sigma_1^2 t_k)}{(\sigma_0^2 + \sigma^2 t_1)(\sigma_1^2 t_k + \sigma^2) - \sigma_0^2\sigma_1^2 t_1} \\ \mu_{\beta,k} &= \frac{(s_k\sigma_1^2 + \mu_1\sigma^2 - 0.5\sigma^4)(\sigma_0^2 + \sigma^2 t_1) - \sigma_1^2(\sigma_0^2 s_1 + \mu_0\sigma^2 t_1)}{(\sigma_0^2 + \sigma^2 t_1)(\sigma_1^2 t_k + \sigma^2) - \sigma_0^2\sigma_1^2 t_1} \\ \sigma_{\beta,k}^2 &= \frac{\sigma_1^2\sigma^2 t_1(\sigma_0^2 + \sigma^2 t_1)}{(\sigma_0^2 + \sigma^2 t_1)(\sigma_1^2 t_k + \sigma^2) - \sigma_0^2\sigma_1^2 t_1} \\ \rho_k &= \frac{-\sigma_0\sigma_1\sqrt{t_1}}{\sqrt{(\sigma_0^2 + \sigma^2 t_1)(\sigma_1^2 t_k + \sigma^2)}} \end{aligned}. \tag{5.54}$$

As we have discussed that $\theta, \beta | \boldsymbol{S}_{1:k} \sim N(\mu_{\theta,k}, \sigma_{\theta,k}^2, \mu_{\beta,k}, \sigma_{\beta,k}^2, \rho_k)$, according to the properties of the bivariate normal distribution we then have

$$\begin{aligned} & \theta | \boldsymbol{S}_{1:k} \sim N(\mu_{\theta,k}, \sigma_{\theta,k}^2),\ \beta | \boldsymbol{S}_{1:k} \sim N(\mu_{\beta,k}, \sigma_{\beta,k}^2) \\ & \beta | \theta, \boldsymbol{S}_{1:k} \sim N\left(\mu_{\beta|\theta,k}, \sigma_{\beta|\theta,k}^2\right) \end{aligned}, \tag{5.55}$$

with

$$\begin{aligned} \mu_{\beta|\theta,k} &= \mu_{\beta,k} + \rho_k\sigma_{\beta,k}(\theta - \mu_{\theta,k})/\sigma_{\theta,k} \\ \sigma_{\beta|\theta,k}^2 &= \sigma_{\beta,k}^2(1 - \rho_k^2) \end{aligned}. \tag{5.56}$$

Additionally, the following equations hold

$$\begin{aligned} & E(\theta\beta | \boldsymbol{S}_{1:k}) = \rho_k\sigma_{\theta,k}\sigma_{\beta,k} + \mu_{\theta,k}\mu_{\beta,k},\ E(\theta | \boldsymbol{S}_{1:k}) = \mu_{\theta,k},\ E(\beta | \boldsymbol{S}_{1:k}) = \mu_{\beta,k}, \\ & E(\theta^2 | \boldsymbol{S}_{1:k}) = \mu_{\theta,k}^2 + \sigma_{\theta,k}^2,\ E(\beta^2 | \boldsymbol{S}_{1:k}) = \mu_{\beta,k}^2 + \sigma_{\beta,k}^2. \end{aligned} \tag{5.57}$$

Now let us focus on calculating $f_{L_k|\boldsymbol{\theta},\boldsymbol{X}_{1:k}}(l_k|\boldsymbol{\theta},\boldsymbol{S}_{1:k})$ at t_k. For in service RUL estimation at t_k, given θ and β and the current observation $s_k = \ln x_k$, we can translate the original degradation as

$$S(t) = s_k + \beta(t - t_k) + \sigma\left(B(t) - B(t_k)\right), t \geq t_k. \tag{5.58}$$

Based on Theorem 5.1 and Eq. (5.44), given θ and β, the estimated RUL at t_k can be calculated from the FPT of the following process crossing threshold w,

$$S'(l_k) = s_k + \beta l_k + \sigma W(l_k), l_k \geq 0, \tag{5.59}$$

with

$$W(l_k) = B(l_k + t_k) - B(t_k). \tag{5.60}$$

Theorem 5.1 tells us that $\{W(l_k), l_k \geq 0\}$ is still a standard BM. Therefore, $\left\{S'(l_k), l_k \geq 0\right\}$ is still a BM with a drift part βl_k and initial value $S'(0) = s_k$. Therefore, given θ and β, from Theorem 5.2, it is direct to obtain the PDF and CDF of the RUL at t_k as follows,

$$f_{L_k|\theta,\beta,\boldsymbol{S}_{1:k}}(l_k|\theta,\beta,\boldsymbol{S}_{1:k}) = \frac{w - s_k}{\sqrt{2\pi l_k^3 \sigma^2}} \exp\left\{-\frac{(w - s_k - \beta l_k)^2}{2\sigma^2 l_k}\right\}, l_k > 0, \tag{5.61}$$

$$\begin{aligned} F_{L_k|\theta,\beta,\boldsymbol{S}_{1:k}}(l_k|\theta,\beta,\boldsymbol{S}_{1:k}) =& 1 - \Phi\left(\frac{w - s_k - \beta l_k}{\sigma\sqrt{l_k}}\right) + \\ & \exp\left\{\frac{2\beta(w - s_k)}{\sigma^2}\right\}\Phi\left(\frac{-(w - s_k) - \beta l_k}{\sigma\sqrt{l_k}}\right). \end{aligned} \tag{5.62}$$

In the next step, we illustrate how to estimate unknown parameters $\boldsymbol{\Theta} = [\sigma^2, \boldsymbol{a}] = [\sigma^2, \mu_0, \mu_1, \sigma_0^2, \sigma_1^2]$. In the following, we use $\boldsymbol{\Theta}_k = [\sigma_k^2, \mu_{0,k}, \mu_{1,k}, \sigma_{0,k}^2, \sigma_{1,k}^2]$ to denote the parameter needed to be estimated based on $\boldsymbol{S}_{1:k}$.

***Step 3*: Estimate deterministic parameters based on EM algorithm**

Similar to the linear case, the ith estimate is represented as $\hat{\boldsymbol{\Theta}}_k^{(i)} = [\hat{\sigma}_k^{2(i)}, \hat{\mu}_{0,k}^{(i)}, \hat{\mu}_{1,k}^{(i)}, \sigma_{0,k}^{2(i)}, \sigma_{1,k}^{2(i)}]$. Then the complete log-likelihood function can be written as

$$\begin{aligned} \ln p(\boldsymbol{S}_{1:k}, \theta, \beta|\boldsymbol{\Theta}_k) =& \ln p(\boldsymbol{S}_{1:k}|\theta, \beta, \boldsymbol{\Theta}_k) + \ln p(\theta, \beta|\boldsymbol{\Theta}_k) \\ =& -\frac{1}{2}\sum_{j=1}^k \ln(t_j - t_{j-1}) - \frac{k+2}{2}\ln 2\pi - \frac{k}{2}\ln\sigma_k^2 - \frac{(s_1 - \theta - \beta t_1)^2}{2\sigma_k^2 t_1} - \\ & \sum_{j=2}^k \frac{\left(s_j - s_{j-1} - \beta(t_j - t_{j-1})\right)^2}{2\sigma_k^2(t_j - t_{j-1})} - \frac{1}{2}\ln\sigma_{0,k}^2 - \frac{1}{2}\ln\sigma_{1,k}^2 - \\ & \frac{(\theta - \mu_{0,k})^2}{2\sigma_{0,k}^2} - \frac{(\beta - \mu_{1,k})^2}{2\sigma_{1,k}^2}. \end{aligned} \tag{5.63}$$

Then, from Eq. (5.8), $\ell(\boldsymbol{\Theta}_k|\hat{\boldsymbol{\Theta}}_k^{(i)})$ can be computed as

$$\begin{aligned}
&\ell(\boldsymbol{\Theta}_k|\hat{\boldsymbol{\Theta}}_k^{(i)})\\
&=\mathbb{E}_{\theta,\beta|\boldsymbol{S}_{1:k},\hat{\boldsymbol{\Theta}}_k^{(i)}}\{\ln p(\boldsymbol{S}_{1:k},\theta,\beta|\boldsymbol{\Theta}_k)\}\\
&=\frac{k+2}{2}\ln 2\pi-\frac{1}{2}\sum_{j=1}^{k}\ln(t_j-t_{j-1})-\frac{k}{2}\ln\sigma_k^2-\\
&\frac{s_1^2-2s_1(\mu_{\theta,k}+\mu_{\beta,k}t_1)+\mu_{\theta,k}^2+\sigma_{\theta,k}^2+2t_1\left(\rho_k\sigma_{\theta,k}\sigma_{\beta,k}+\mu_{\theta,k}\mu_{\beta,k}\right)+t_1^2(\mu_{\beta,k}^2+\sigma_{\beta,k}^2)}{2\sigma_k^2t_1}-\\
&\sum_{j=2}^{k}\frac{(s_j-s_{j-1})^2-2(s_j-s_{j-1})(t_j-t_{j-1})\mu_{\beta,k}+(t_j-t_{j-1})^2(\mu_{\beta,k}^2+\sigma_{\beta,k}^2)}{2\sigma_k^2(t_j-t_{j-1})}-\\
&\frac{1}{2}\ln\sigma_{0,k}^2-\frac{1}{2}\ln\sigma_{1,k}^2-\frac{\mu_{\theta,k}^2+\sigma_{\theta,k}^2-2\mu_{\theta,k}\mu_{0,k}+\mu_{0,k}^2}{2\sigma_{0,k}^2}-\frac{\mu_{\beta,k}^2+\sigma_{\beta,k}^2-2\mu_{\beta,k}\mu_{1,k}+\mu_{1,k}^2}{2\sigma_{1,k}^2}.
\end{aligned}\tag{5.64}$$

Let $\frac{\partial\ell(\boldsymbol{\Theta}_k|\hat{\boldsymbol{\Theta}}_k^{(i)})}{\partial\boldsymbol{\Theta}_k}=0$, we obtain $\hat{\boldsymbol{\Theta}}_k^{(i+1)}$ as follows,

$$\begin{aligned}
&\hat{\sigma}_k^{2\,(i+1)}\\
&=\frac{1}{k}\left(\frac{s_1^2-2s_1(\mu_{\theta,k}+\mu_{\beta,k}t_1)+\sigma_{\theta,k}^2+\sigma_{\beta,k}^2+2t_1(\rho_k\sigma_{\theta,k}\sigma_{\beta,k}+\mu_{\theta,k}\mu_{\beta,k})+t_1^2(\sigma_{\beta,k}^2+\mu_{\beta,k}^2)}{t_1}+\right.\\
&\left.\sum_{j=2}^{k}\frac{(s_j-s_{j-1})^2-\mu_{\beta,k}(t_j-t_{j-1})(s_j-s_{j-1})+(t_j-t_{j-1})^2(\mu_{\beta,k}^2+\sigma_{\beta,k}^2)}{(t_j-t_{j-1})}\right).
\end{aligned}\tag{5.65}$$

$$\begin{aligned}
\mu_{0,k}^{(i+1)}&=\mu_{\theta,k},\ \sigma_{0,k}^{2\ (i+1)}=\sigma_{\theta,k}^2\\
\mu_{1,k}^{(i+1)}&=\mu_{\beta,k},\ \sigma_{1,k}^{2\ (i+1)}=\sigma_{\beta,k}^2.
\end{aligned}\tag{5.66}$$

Theorem 5.9 $\hat{\boldsymbol{\Theta}}_k^{(i+1)}$ *obtained by Eqs. (5.65) and (5.66) is uniquely determined and located at the maximum of* $\ell(\boldsymbol{\Theta}_k|\hat{\boldsymbol{\Theta}}_k^{(i)})$.

Proof: the proof is similar to the proof of Theorem 5.3.

The above theorem guarantees that each iteration of the EM algorithm can be performed with a single computation, which leads to an extremely fast and simple estimation procedure.

***Step 4*: Exact and closed-form solution to degradation path-dependent RUL estimation**

Note that the RUL estimation by Eqs. (5.61) and (5.62) only uses the current degradation data. In this step, we attempt to make the estimated RUL depend on the CM history through the updating of parameters, i.e., $f_{L_k|\boldsymbol{S}_{1:k}}(l_k|\boldsymbol{S}_{1:k})$. The updated RUL distribution at t_k can be summarized in the following theorem.

Theorem 5.10 *For the exponential model, the PDF and CDF of the RUL conditional on* $\boldsymbol{S}_{1:k}$ *can be respectively expressed by*

$$f_{L_k|\boldsymbol{S}_{1:k}}(l_k|\boldsymbol{S}_{1:k}) = \frac{w-s_k}{\sqrt{2\pi l_k^3\left(\sigma_{\beta,k}^2 l_k+\sigma^2\right)}}\exp\left\{-\frac{\left(w-s_k-\mu_{\beta,k}l_k\right)^2}{2l_k\left(\sigma_{\beta,k}^2 l_k+\sigma^2\right)}\right\},\ l_k>0, \tag{5.67}$$

$$F_{L_k|\boldsymbol{S}_{1:k}}(l_k|\boldsymbol{S}_{1:k}) = 1-\Phi\left(\frac{w-s_k-\mu_{\beta,k}l_k}{\sqrt{\sigma_{\beta,k}^2 l_k^2+\sigma^2 l_k}}\right)+$$
$$\exp\left\{\frac{2\mu_{\beta,k}(w-s_k)}{\sigma^2}+\frac{2\sigma_{\beta,k}^2(w-s_k)^2}{\sigma^4}\right\}\Phi\left(-\frac{2\sigma_{\beta,k}^2(w-s_k)l_k+\sigma^2\left(\mu_{\beta,k}l_k+w-s_k\right)}{\sigma^2\sqrt{\sigma_{\beta,k}^2 l_k^2+\sigma^2 l_k}}\right). \tag{5.68}$$

Proof (1) For $f_{L_k|\boldsymbol{S}_{1:k}}(l_k|\boldsymbol{S}_{1:k})$, we have

$$\begin{aligned}
f_{L_k|\boldsymbol{S}_{1:k}}(l_k|\boldsymbol{S}_{1:k}) &= \iint f_{L_k|\theta,\beta,\boldsymbol{S}_{1:k}}(l_k|\theta,\beta,\boldsymbol{S}_{1:k})p(\theta,\beta|\,\boldsymbol{S}_{1:k})\mathrm{d}\theta\mathrm{d}\beta \\
&= \int f_{L_k|\theta,\beta,\boldsymbol{S}_{1:k}}(l_k|\theta,\beta,\boldsymbol{S}_{1:k})p(\beta|\,\theta,\boldsymbol{S}_{1:k})p(\theta|\,\boldsymbol{S}_{1:k})\mathrm{d}\theta\mathrm{d}\beta \\
&= \int p(\theta|\,\boldsymbol{S}_{1:k})\left(\int f_{L_k|\theta,\beta,\boldsymbol{S}_{1:k}}(l_k|\theta,\beta,\boldsymbol{S}_{1:k})p(\beta|\,\theta,\boldsymbol{S}_{1:k})\mathrm{d}\beta\right)\mathrm{d}\theta \\
&= \mathbb{E}_{\theta|\boldsymbol{S}_{1:k}}\left[\mathbb{E}_{\beta|\theta,\boldsymbol{S}_{1:k}}\left[f_{L_k|\theta,\beta,\boldsymbol{S}_{1:k}}(l_k|\theta,\beta,\boldsymbol{S}_{1:k})\right]\right].
\end{aligned} \tag{5.69}$$

Therefore, we first have

$$\begin{aligned}
&\mathbb{E}_{\beta|\theta,\boldsymbol{S}_{1:k}}\left[f_{L_k|\theta,\beta,\boldsymbol{S}_{1:k}}(l_k|\theta,\beta,\boldsymbol{S}_{1:k})\right] \\
&\quad= \frac{w-s_k}{\sqrt{2\pi l_k^3\sigma^2}}\mathbb{E}_{\beta|\theta,\boldsymbol{S}_{1:k}}\left[\exp\left\{-\frac{(w-s_k-\beta l_k)^2}{2\sigma^2 l_k}\right\}\right] \\
&\quad= \frac{w-s_k}{\sqrt{2\pi l_k^3(\sigma^2+\sigma_{\beta|\theta,k}^2 l_k)}}\exp\left\{-\frac{(w-s_k-\mu_{\beta|\theta,k}l_k)^2}{2l_k\left(\sigma^2+\sigma_{\beta|\theta,k}^2 l_k\right)}\right\} \\
&\quad= \frac{w-s_k}{\sqrt{2\pi l_k^3(\sigma^2+\sigma_{\beta|\theta,k}^2 l_k)}}\exp\left\{-\frac{(w-s_k-\mu_{\beta,k}l_k+\varphi\mu_{\theta,k}l_k-\varphi\theta l_k)^2}{2l_k\left(\sigma^2+\sigma_{\beta|\theta,k}^2 l_k\right)}\right\},
\end{aligned} \tag{5.70}$$

with $\varphi=\rho_k\sigma_{\beta,k}/\sigma_{\theta,k}$.

From Eq. (5.69), we have

$$
\begin{aligned}
&\mathbb{E}_{\theta|\mathbf{S}_{1:k}}\left[\mathbb{E}_{\beta|\theta,\mathbf{S}_{1:k}}\left[f_{L_k|\theta,\beta,\mathbf{S}_{1:k}}(l_k|\theta,\beta,\mathbf{S}_{1:k})\right]\right] \\
&=\frac{w-s_k}{\sqrt{2\pi l_k^3(\sigma^2+\sigma^2_{\beta|\theta,k}l_k^)}}\mathbb{E}_{\theta|\mathbf{S}_{1:k}}\left[\exp\left\{-\frac{(w-s_k-\mu_{\beta,k}l_k+\varphi\mu_{\theta,k}l_k-\varphi\theta l_k)^2}{2l_k\left(\sigma^2+\sigma^2_{\beta|\theta,k}l_k\right)}\right\}\right] \\
&=\frac{w-s_k}{\sqrt{2\pi l_k^3(\sigma^2+\sigma^2_{\beta|\theta,k}l_k^)}}\mathbb{E}_{\theta|\mathbf{S}_{1:k}}\left[\exp\left\{-\frac{(\frac{w-s_k-\mu_{\beta,k}l_k+\varphi\mu_{\theta,k}l_k}{\varphi l_k}-\theta)^2}{2l_k\left(\sigma^2+\sigma^2_{\beta|\theta,k}l_k\right)/(\varphi l_k)^2}\right\}\right] \\
&=A\mathbb{E}_{\theta|\mathbf{S}_{1:k}}\left[\exp\left\{-\frac{(B-\theta)^2}{2C}\right\}\right],
\end{aligned}
\tag{5.71}
$$

with

$$
\begin{aligned}
A&=\frac{w-s_k}{\sqrt{2\pi l_k^3(\sigma^2+\sigma^2_{\beta|\theta,k}l_k^)}},\ B=\frac{w-s_k-\mu_{\beta,k}l_k+\varphi\mu_{\theta,k}l_k}{\varphi l_k}, \\
C&=l_k\left(\sigma^2+\sigma^2_{\beta|\theta,k}l_k\right)/(\varphi l_k)^2,\ \varphi=\rho_k\frac{\sigma_{\beta,k}}{\sigma_{\theta,k}}.
\end{aligned}
\tag{5.72}
$$

Then using the second result in Theorem 5.4 and simplifying the expression can complete the proof of the first equation in Theorem 5.10.

(2) For $F_{L_k|\mathbf{S}_{1:k}}(l_k|\mathbf{S}_{1:k})$, we have,

$$
\begin{aligned}
F_{L_k|\mathbf{S}_{1:k}}(l_k|\mathbf{S}_{1:k})&=\iint F_{L_k|\theta,\beta,\mathbf{S}_{1:k}}(l_k|\theta,\beta,\mathbf{S}_{1:k})p(\theta,\beta|\mathbf{S}_{1:k})\mathrm{d}\theta\mathrm{d}\beta \\
&=\mathbb{E}_{\theta,\beta|\mathbf{S}_{1:k}}\left[F_{L_k|\theta,\beta,\mathbf{S}_{1:k}}(l_k|\theta,\beta,\mathbf{S}_{1:k})\right] \\
&=\mathbb{E}_{\theta|\mathbf{S}_{1:k}}\left[\mathbb{E}_{\beta|\theta,\mathbf{S}_{1:k}}\left[F_{L_k|\theta,\beta,\mathbf{S}_{1:k}}(l_k|\theta,\beta,\mathbf{S}_{1:k})\right]\right].
\end{aligned}
\tag{5.73}
$$

Therefore, we first calculate $E_{\beta|\theta,\mathbf{S}_{1:k}}\left[F_{L_k|\theta,\beta,\mathbf{S}_{1:k}}(l_k|\theta,\beta,\mathbf{S}_{1:k})\right]$ as

$$
\begin{aligned}
&\mathbb{E}_{\beta|\theta,\mathbf{S}_{1:k}}\left[F_{L_k|\theta,\beta,\mathbf{S}_{1:k}}(l_k|\theta,\beta,\mathbf{S}_{1:k})\right] \\
&=1-\mathbb{E}_{\beta|\theta,\mathbf{S}_{1:k}}\left[\Phi\left(\frac{w-s_k-\beta l_k}{\sigma\sqrt{l_k}}\right)\right]+ \\
&\quad\mathbb{E}_{\beta|\theta,\mathbf{S}_{1:k}}\left[\exp\left\{\frac{2\beta(w-s_k)}{\sigma^2}\right\}\Phi\left(\frac{-(w-s_k)-\beta l_k}{\sigma\sqrt{l_k}}\right)\right] \\
&=1-\Phi\left(\frac{w-s_k-\mu_{\beta|\theta,k}l_k}{\sqrt{\sigma^2_{\beta|\theta,k}l_k^2+\sigma^2 l_k}}\right)+\exp\left\{\frac{2\mu_{\beta|\theta,k}(w-s_k)}{\sigma^2}+\right. \\
&\quad\left.\frac{2\sigma^2_{\beta|\theta,k}(w-s_k)^2}{\sigma^4}\right\}\Phi\left(-\frac{2\sigma^2_{\beta|\theta,k}(w-s_k)l_k+\sigma^2(\mu_{\beta|\theta,k}l_k+w-s_k)}{\sigma^2\sqrt{\sigma^2_{\beta|\theta,k}l_k^2+\sigma^2 l_k}}\right)
\end{aligned}
$$

$$
\begin{aligned}
&=1-\Phi\left(\frac{w-s_k-\mu_{\beta,k}l_k+\varphi\mu_{\theta,k}l_k-\varphi\theta l_k}{\sqrt{\sigma^2_{\beta|\theta,k}l_k^2+\sigma^2 l_k}}\right)+\\
&\exp\left\{\frac{2(\mu_{\beta,k}-\varphi\mu_{\theta,k}+\varphi\theta)(w-s_k)}{\sigma^2}+\frac{2\sigma^2_{\beta|\theta,k}(w-s_k)^2}{\sigma^4}\right\}\times\\
&\Phi\left(-\frac{2\sigma^2_{\beta|\theta,k}(w-s_k)l_k+\sigma^2(\mu_{\beta,k}l_k-\varphi\mu_{\theta,k}l_k+\varphi\theta l_k+w-s_k)}{\sigma^2\sqrt{\sigma^2_{\beta|\theta,k}l_k^2+\sigma^2 l_k}}\right)\\
&=1-\Phi\left(a+b\theta\right)+E\exp\{A'\theta\}\Phi(C'+D'\theta),
\end{aligned}
\tag{5.74}
$$

where a, b, E, A', C', D' are defined as follows

$$
\begin{aligned}
a&=\frac{w-s_k-\mu_{\beta,k}l_k+\varphi\mu_{\theta,k}l_k}{\sqrt{\sigma^2_{\beta|\theta,k}l_k^2+\sigma^2 l_k}},\; b=\frac{-\varphi l_k}{\sqrt{\sigma^2_{\beta|\theta,k}l_k^2+\sigma^2 l_k}},\\
E&=\exp\left\{\frac{2(\mu_{\beta,k}-\varphi\mu_{\theta,k})(w-s_k)}{\sigma^2}+\frac{2\sigma^2_{\beta|\theta,k}(w-s_k)^2}{\sigma^4}\right\},\; D'=-\frac{\sigma^2\varphi l_k}{\sigma^2\sqrt{\sigma^2_{\beta|\theta,k}l_k^2+\sigma^2 l_k}}\\
A'&=\frac{2\varphi(w-s_k)}{\sigma^2},\; C'=-\frac{2\sigma^2_{\beta|\theta,k}(w-s_k)l_k+\sigma^2(\mu_{\beta,k}l_k-\varphi\mu_{\theta,k}l_k+w-s_k)}{\sigma^2\sqrt{\sigma^2_{\beta|\theta,k}l_k^2+\sigma^2 l_k}}.
\end{aligned}
$$

Then, using Theorem 5.4, it is straightforward to show,

$$
\begin{aligned}
F_{L_k|\boldsymbol{S}_{1:k}}(l_k|\boldsymbol{S}_{1:k}) &=\mathbb{E}_{\theta|\boldsymbol{S}_{1:k}}\left[\mathbb{E}_{\beta|\theta,\boldsymbol{S}_{1:k}}\left[F_{L_k|\theta,\beta,\boldsymbol{S}_{1:k}}(l_k|\theta,\beta,\boldsymbol{S}_{1:k})\right]\right]\\
&=\mathbb{E}_{\theta|\boldsymbol{S}_{1:k}}\left[1-\Phi\left(a+b\theta\right)+E\exp\{A\theta\}\Phi(C+D\theta)\right]\\
&=1-\Phi\left(\frac{a+b\mu_{\theta,k}}{\sqrt{1+b^2\sigma^2_{\theta,k}}}\right)+E\cdot\exp\left\{A\mu_{\theta,k}+\frac{A^2}{2}\sigma^2_{\theta,k}\right\}\times\\
&\quad\Phi\left(\frac{C+D\mu_{\theta,k}+AD\sigma^2_{\theta,k}}{\sqrt{1+D^2\sigma^2_{\theta,k}}}\right).
\end{aligned}
\tag{5.75}
$$

Simplifying above expression completes the proof of the second equation in Theorem 5.10.

It can be found that the form of the estimated RUL distribution in this exponential case is much similar to the case of the linear model. This is largely due to the fact that exponential model can be easily linearized by log-transformation. Therefore, based on the transformed data $\boldsymbol{S}_{1:k}$, it is not strange to see that both cases have a similar style of the RUL distribution.

If we denote the estimated RUL by [28] as L_k', then the following conclusion holds in general.

Theorem 5.11 *Conditional on the degradation history to* t_k*, i.e.,* $\boldsymbol{S}_{1:k}$*, and using the same parameter estimation procedure, we have* $L'_k \geq_{st} L_k$.

Proof As [28], following the procedure $\Pr(L'_k \leq l_k | \boldsymbol{S}_{1:k}) = \Pr(S(l_k + t_k) \geq w | \boldsymbol{S}_{1:k})$, we have

$$S(l_k + t_k) | \boldsymbol{S}_{1:k} \sim N(s_k + \mu_{\theta,k} l_k, \sigma^2_{\theta,k} l_k^2 + \sigma^2 l_k), \tag{5.76}$$

As a result, the CDF of the estimated RUL can be written as

$$F_{L'_k|\boldsymbol{S}_{1:k}}(l_k|\boldsymbol{S}_{1:k}) = 1 - \Phi\left(\frac{w - s_k - \mu_{\beta,k} l_k}{\sqrt{\sigma^2_{\beta,k} l_k^2 + \sigma^2 l_k}}\right). \tag{5.77}$$

Compared with Eq. (5.54), we have,

$$F_{L'_k|\boldsymbol{S}_{1:k}}(l_k|\boldsymbol{S}_{1:k}) \leq F_{L_k|\boldsymbol{S}_{1:k}}(l_k|\boldsymbol{S}_{1:k}), \tag{5.78}$$

Namely,

$$\Pr(L'_k \geq l_k|\boldsymbol{S}_{1:k}) \geq \Pr(L_k \geq l_k|\boldsymbol{S}_{1:k}). \tag{5.79}$$

Then according to the given definition by Eq. (5.33), the proof is completed.

Resulted from a similar style of the RUL distribution to the linear case, other remarks and conclusions in Sect. 5.3 are also applied to this exponential case. In addition, the RUL estimation and associated replacement decision algorithm for the exponential case is also similar to the linear case.

5.5 Experimental Studies

In this section, we provide several numerical simulations to compare the performance of the presented approach with some known works in the literature for the same case. Then a practical case study for the condition-based replacement of gyros in the inertial navigation system (INS) is illustrated to demonstrate the application of our approach. For an illustrative purpose, we only consider the case of the linear degradation model. From our previous theoretical derivations, we can observe that the implementation of the exponential case is much similar to the linear case.

5.5.1 Numerical Example

In this experiment, we first show that the parameters of the degradation model and the PDFs of RULs obtained from our approach adapt to the data as they accumulate. Some comparisons with the approximated results by [28] are performed subsequently.

In order to illustrate and validate our developed approach in estimating the RUL, we use a nonlinear process $X(t) = \lambda t^b + \sigma_B B(t)$ to generate the simulated degradation data. This is to show that our linear model has an adaptive nature so it can also be used for modeling nonlinear degradation. By *Euler* approximation, we can simulate the degradation data as $X((k+1)\Delta t) = X(k\Delta t) + \lambda b (k\Delta t)^{b-1} \Delta t + \sigma_B Y \sqrt{\Delta t}$ where $Y \sim N(0, 1)$ and Δt is the discretization step. In the simulation, we let $\lambda = 0.05$, $b = 2$, $\Delta t = 0.1$, $\sigma_B = 0.05$. One particular simulated sample path is shown in Fig. 5.1 with 136 sampling points, terminated at $X(13.6) = 10.1759$. Given threshold $w = 10.176$, and the FHT for this path is approximately to be 13.7.

In following experiments, we assume that the parameters in prior distribution $p(\theta)$ are $\mu_0 = 7, \sigma_0^2 = 0.05$, and the variance parameter $\sigma^2 = 0.5$ for the linear model represented by Eq. (5.12). Then according to our developed approach for the linear case, the estimated degradation and the updated parameters are illustrated with sequential sampling as Figs. 5.2 and 5.3, respectively.

Figure 5.2 shows that our model has a quick and good predictive ability and the degradation paths of both actual and predicted almost overlap, where the predicted path is the estimated expectation of the degradation path, i.e., $E(X(t) - X(0)) = \mu_{\theta,k} t$. The updates of the parameters $\mu_0, \sigma_0^2, \sigma^2$ are illustrated by Fig. 5.3. In comparison, Gebraeel's method cannot trace the degradation path as well as our approach, since once $\mu_0, \sigma_0^2, \sigma^2$ are selected, they are fixed. Regarding the parameter updat-

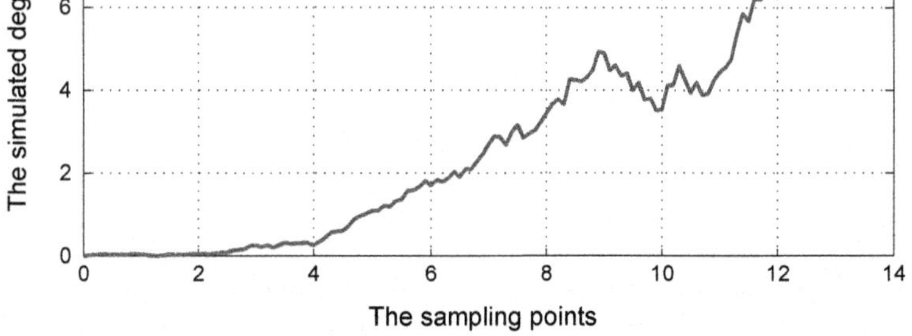

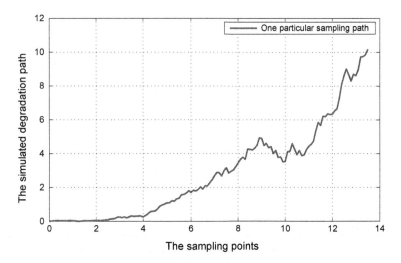

Fig. 5.1 The simulated degradation path

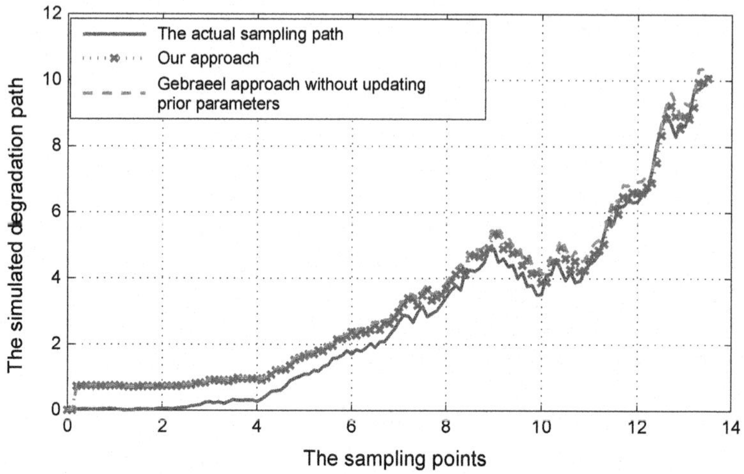

Fig. 5.2 The predicted degradation path

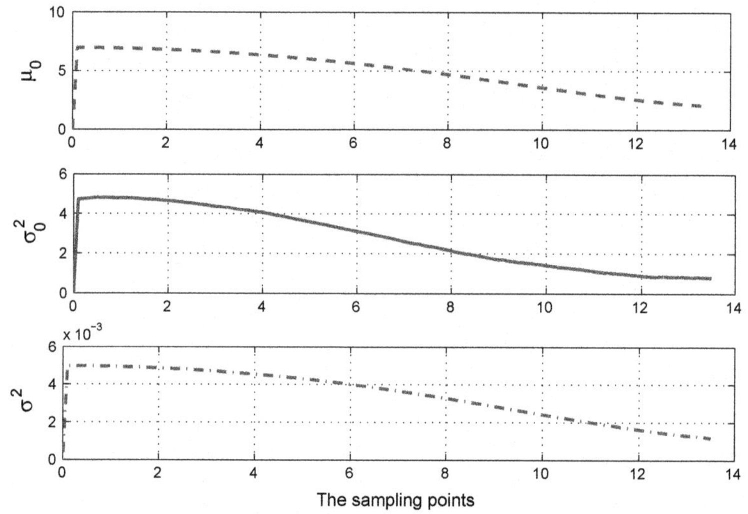

Fig. 5.3 The updates of parameters $\mu_0, \sigma_0^2, \sigma^2$

ing, it is interesting to note that, however, the early predictions of both methods are not accurate enough. Namely, the confidence in the true value of $\mu_{\theta,k}$ is not high at the early stage of the updating. This is resulted from the inappropriate selection of the initial parameters, particularly μ_0, which is selected as a relatively large value $\mu_0 = 7$ to verify the adaptive ability of our updating approach. Notwithstanding, it can be observed that the predictive ability of our method is improved as the data are accumulated. Another observation is that $\mu_{0,k}$ is not approaching $\overline{\mu_{\theta,k}} = x_k / t_k$

exactly in the used sampling data (e.g., $\mu_{0,k} = 1.8201$ but $\overline{\mu_{\theta,k}} = 0.7538$ at the last sampling point), since at the last sampling point, $t_k = 13.5$ which is not large enough to make $\mu_{0,k}$ approach $\overline{\mu_{\theta,k}}$ according to Remark 5.1. Therefore, inappropriate selection of the initial parameters will require more data to make the parameter estimation approach $\overline{\mu_{\theta,k}}$.

To be more precise, the mean squared errors (MSEs) of our approach and Gebraeel's approach for degradation predictions are calculated over time for further comparisons, where the MSE at time t_k is defined as $MSE_k = \sum_{i=1}^{k}[x_k - E(X(t_k))]^2 \big/ k = \sum_{i=1}^{k}[x_k - \mu_{\theta,k}t_k]^2 \big/ k$. Figure 5.4 shows the results of the MSE_k over time, which also reflect the superiority of our approach.

In addition, at the final sampling point with the actual value 10.1759, the MSEs of our approach and Gebraeel's approach are 0.2683 and 0.3895 with the predicted degradation values 10.0945 and 10.4828, respectively. At this particular point, the absolute error of our approach and Gebraeel's approach are 0.0814 and 0.3069, respectively. This further verifies that the updating mechanism adopted in this chapter can effectively improve the predictive ability of the degradation modeling.

Figure 5.5 compares the estimated RUL distributions from our approach with Gebraeel's approach at the last six sampling points, in which the prior parameters $\mu_0, \sigma_0^2, \sigma^2$ selected at random are not updated. In this case, the difference between our results and Gebraeel's results is significant. It is shown that if the prior parameters in Gebraeel's approach are selected inappropriately, the estimated RUL may be incorrect (the observed RULs are outside of the predicted RUL PDF ranges). However our approach can produce reasonable estimates, and the estimated RUL distributions cover the actual remaining time well. This further demonstrates the merit of our approach. In fact, other cases of inappropriate selection of $\mu_0, \sigma_0^2, \sigma^2$

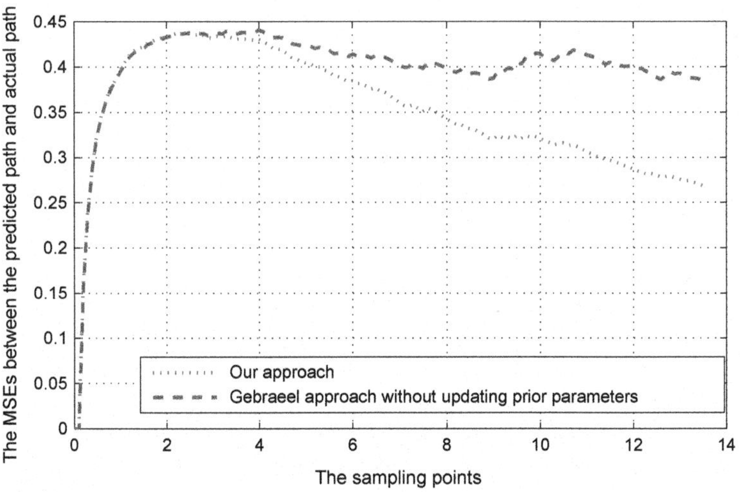

Fig. 5.4 The MSEs between the predicted path and actual path

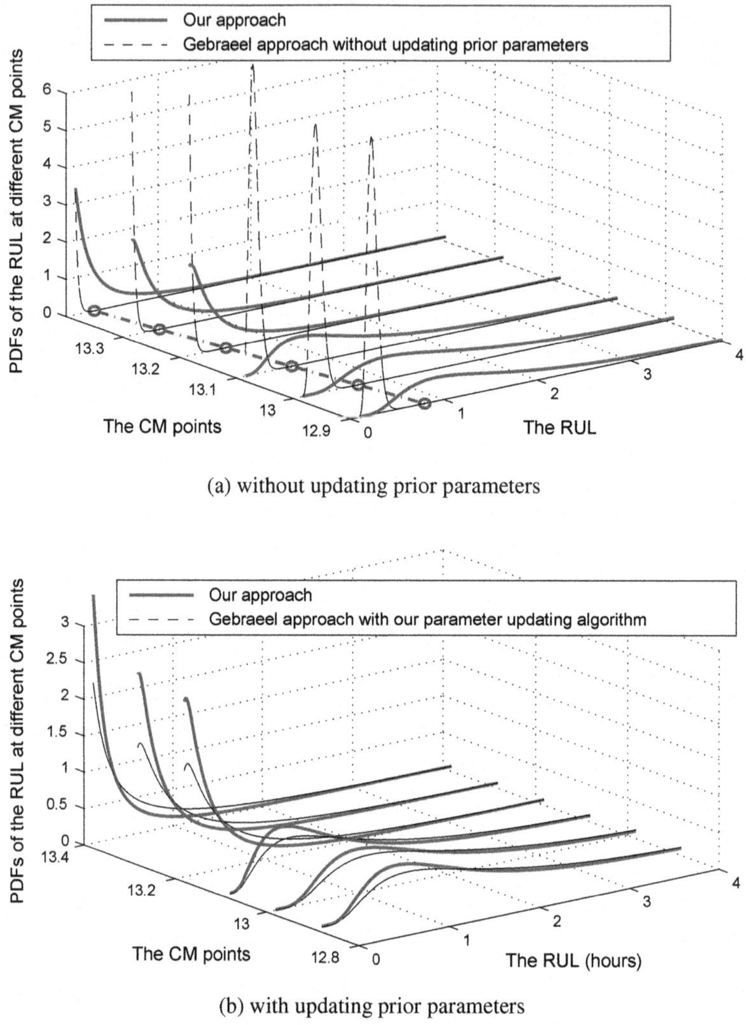

Fig. 5.5 The estimated PDFs of the RUL by our approach and Gebraeel's approach (○–end of life)

can also be tested and the results are similar to above conclusion. This shows the robustness of our approach over the selection of these prior parameters.

Due to the sensitivity of Gebraeel's approach to the prior parameters, we adopt our developed algorithm to update $\mu_0, \sigma_0^2, \sigma^2$ in Gebraeel's approach in the following to achieve a fair comparison. Accordingly, the estimated RUL distributions of our approach and Gebraeel's approach are shown in the right-hand side of Fig. 5.5. It can be found that using our updating approach of parameters can improve Gebraeel's results, in contrast with the results shown in the left-hand side of Fig. 5.5. However, there are some differences among their RUL distributions and our obtained results can reduce the uncertainty of the estimated RUL distributions since the PDF is higher than Gebraeel's results. This is largely due to the approximated nature of the RUL distribution of Gebraeel's approach. In order to shed a light to the effect of RUL approximation on condition-based replacement decision, we minimize Eq. (5.43) to determine the replacement time. Let $c_p = 4000$, $c_f = 9000$, then the expected cost per unit time at the 135th sampling time ($t_{135} = 13.5$) based both approaches is illustrated by the left-hand side of Fig. 5.6. It is noted that in this case the fail-

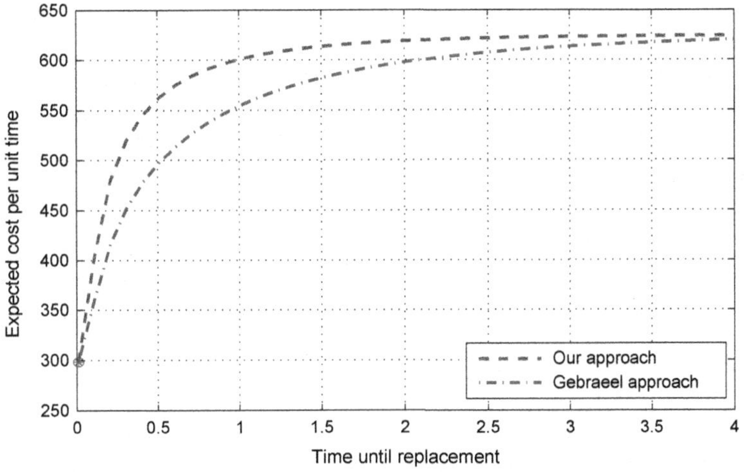

(a) $c_p = 4000$

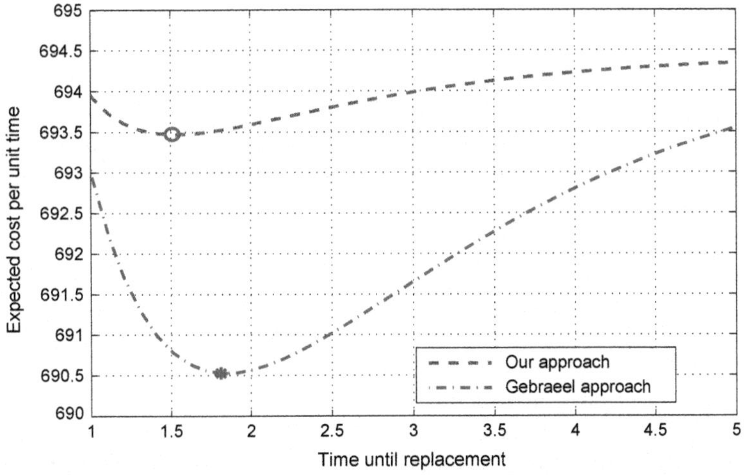

(b) $c_p = 8900$

Fig. 5.6 Illustration of the condition-based replacement decision at $t_{135} = 13.5$

ure replacement cost is much larger than the preventive replacement cost so that an immediate replacement is recommended from both approaches. However, if we change the preventive replacement cost as $c_p = 8900$, then the according expected cost per unit time is illustrated by the right-hand side of Fig. 5.6.

Two observations can be drawn from the results illustrated by Fig. 5.6. First, the expected cost per unit time using the RUL distribution of our exact approach is always larger than that of Gebraeel's approximated approach. This result is ensured by Theorem 5.8. Second, Gebraeel's approximated approach will lead to the delay of condition-based replacement time. Specifically, at $t_{135} = 13.5$, the optimal time until replacement of Gebraeel's approximated approach is 1.9 with $C'(13.5 + 1.9) = 690.5252$. The according result of our exact approach is 1.5 with $C(13.5 + 1.5) = 693.4734$. These two observations are also applied to other sampling points.

The above example shows the advantages of the presented approach and the effects of approximated RUL distributions on decision-making, compared with our exact result. In order to demonstrate the potential of the proposed approaches in engineering practice, a practical case study is provided in the next subsection using the real-world CM data of gyros in INS.

5.5.2 A Practical Case Study of the Developed Approach in Condition-Based Replacement

As a vital device in the INS, the inertial platform plays an important role in the INS. Its condition monitoring parameters have a direct influence on navigation precision. In engineering practice, health condition of an inertial platform is evaluated by its precision testing. Furthermore, the precision of the inertial platform depends on the drift coefficients of the gyro. As such, for simplicity, the drift coefficients are selected as the characteristic parameters for weighing the health condition of an inertial platform. In this study, we assume that CM values of drift coefficients reflect the performance of the inertial platform, and the larger the drift coefficients monitored are, the worse the performance is. Therefore, according to the CM data and technical index of the inertial platform, failure prediction can be implemented by modeling the drift coefficients. The drift coefficients of an inertial platform mainly include $K_{0X}, K_{0Y}, K_{0Z}, K_{SX}, K_{SY}, K_{IZ}$, in which K_{0X}, K_{0Y}, K_{0Z} denote constant drift, and K_{SX}, K_{SY}, K_{IZ} are stochastic drift, where K_{SX}, K_{SY} denote the coefficients related to the first moment of specific force along the sense axis, and K_{IZ} denotes the coefficient related to the first moment of specific force along the input axis. 109 points of drift coefficients data with CM intervals 2.5 h were collected in test conditions. The collected data are illustrated in Fig. 5.7. In practice, the rotating part of inertial platform with a high speed can lead to rotation axis wear. With the accumulation of wear, the drift coefficients increase and system performance suffers degradation and finally failure occurs. Generally, the drift degradation measurement along the sense axis, K_{SX}, plays a dominant role in the assessment of gyro degradation. In our

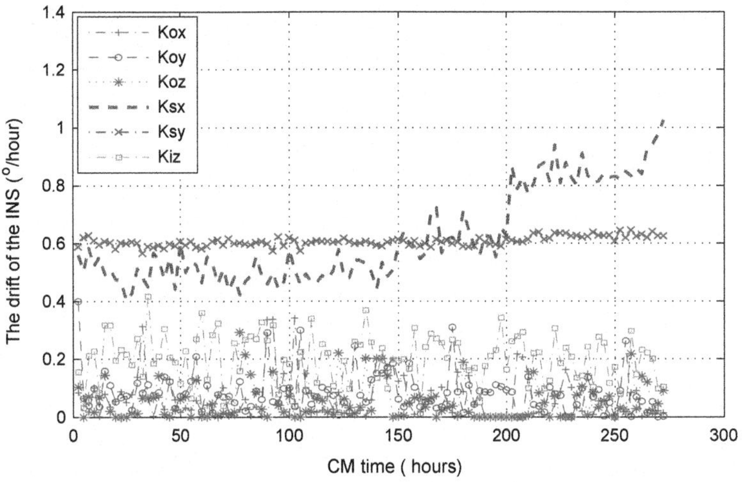

Fig. 5.7 Drift coefficients of the INS

study, we take the CM data of K_{SX} as the degradation signals and use them for RUL estimation of the tested INSs.

In practice of the INS health monitoring of this type, it is usually required that the drift measurement along the sense axis should not exceed 1.2(°/h). This threshold is determined at the design stage and is strictly enforced in practice since an INS is a critical device used in weapon systems. As for this specific application case, using the proposed method, the predictions of the gyro's drift and the distribution of the RUL can be obtained at each CM point. As mentioned previously, we adopt our developed algorithm to update $\mu_0, \sigma_0^2, \sigma^2$ in Gebraeel's approach as our approach to achieve a fair comparison. The predicted drifting path is shown in Fig. 5.8. Figure 5.9 illustrates the estimated RUL distributions at some CM points with our exact approach and Gebraeel's approximated approach.

It can also be observed from Fig. 5.8 that our approach can quickly adjust and update the possibly inappropriate prior parameter in $p(\theta)$ and thus the predicted path matches the actual path well after some updates. For the updated RUL distribution, the difference between the results of our exact approach and Gebraeel's approximated approach is significant in this case, as shown in Fig. 5.9. In contrast, our obtained results can reduce the uncertainty of the RUL estimation obviously.

To give some insights to the influence of these differences on replacement decision-making, in this case study, we first set $c_f = 10000$ RMB and $c_p = 4000$ RMB and use Eq. (5.43) to obtain the optimal replacement time. Figure 5.10 illustrates the expected costs per unit time against the associated time until replacement at the last CM point by applying our exact RUL distribution and Gebraeel's result.

Obviously seen from the left-hand side of Fig. 5.10, the expected cost per unit time using the RUL distribution of our exact approach is always larger than that of

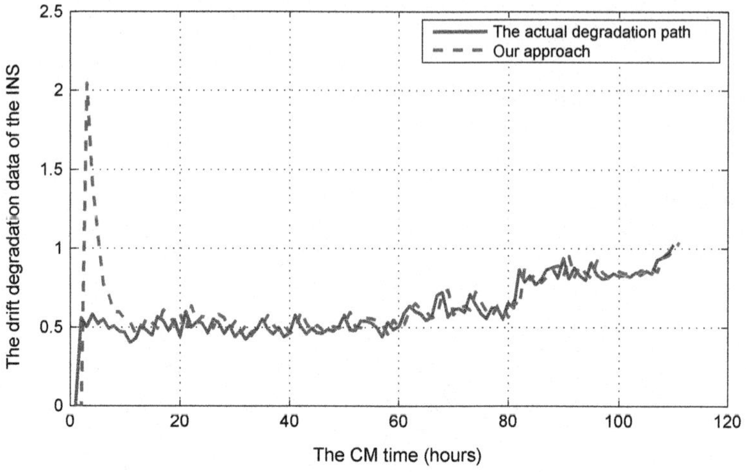

Fig. 5.8 The degradation path and fitted path of INS's drift

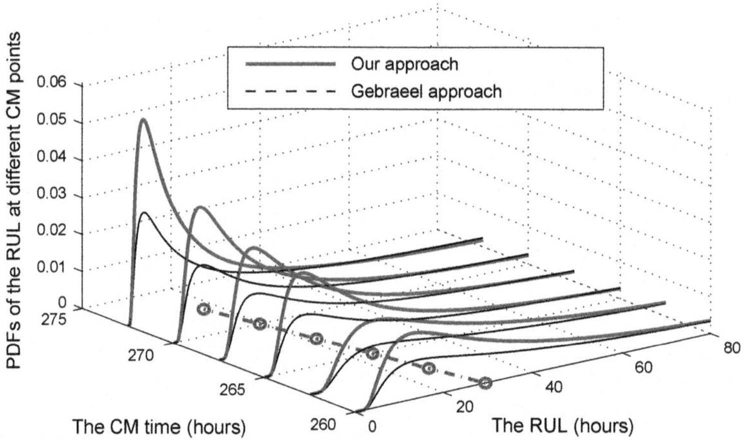

Fig. 5.9 Illustration of the RUL distributions at the last six CM points (○–actual RUL)

Gebraeel's approximated approach, ensured by Theorem 5.8. Therefore, Gebraeel's approach underestimates the expected cost and so the operating risk of the INS. Similar to the simulation, Gebraeel's approximated approach will lead to the delay of the recommended replacement time. Specifically, at the last CM point, i.e., $t_{109} = 272.5$ (h), the optimal time until replacement using Gebraeel's approximated approach is 1 h with $C'(272.5 + 1) = 14.6390$ (RMB/h). The according result of our exact approach is 0.8 h with $C(272.5 + 0.8) = 14.6436$ (RMB/h). However, when we fix the replacement cost with failure c_f, but increase the preventive replacement cost to $c_p = 9500$ RMB, the associated expected costs per unit time of our exact RUL distribution and Gebraeel's result are illustrated by the right-hand side of Fig. 5.10.

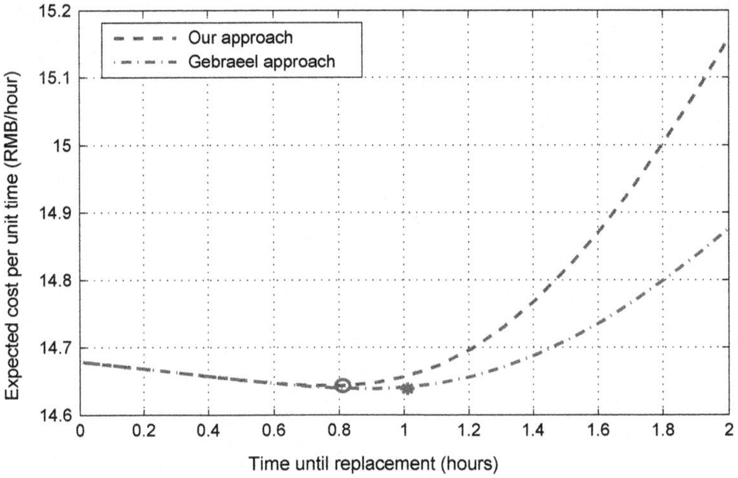

(a) $c_p = 4000$

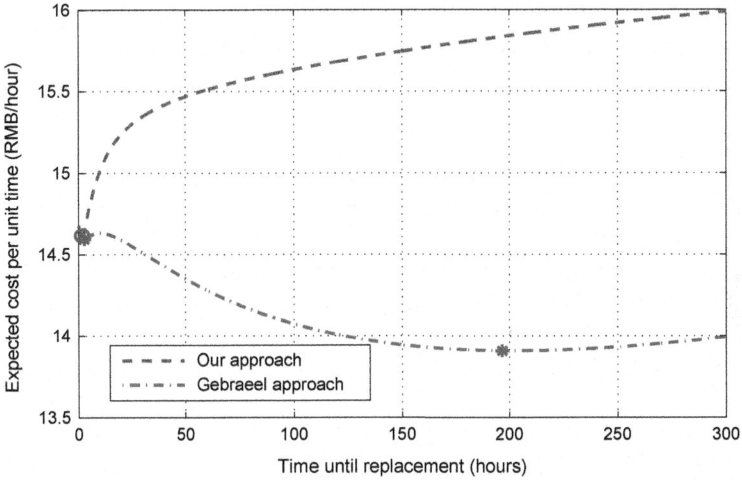

(b) $c_p = 9500$

Fig. 5.10 Illustration of the condition-based replacement decision at $t_{110} = 272.5$ (h)

It is interesting to note that in this case the curve of expected cost per unit time of Gebraeel's result has two locally minimum points. This is against the nature of the cost function of Eq. (5.43) since it should only have one minimum point. The result is due to the effect of the approximation in the Gebraeel's approach. However, our approach has only one minimum point. This shows that the approximated result can lead to a wrong recommendation if the minimum at 200 h is chosen since the cost at

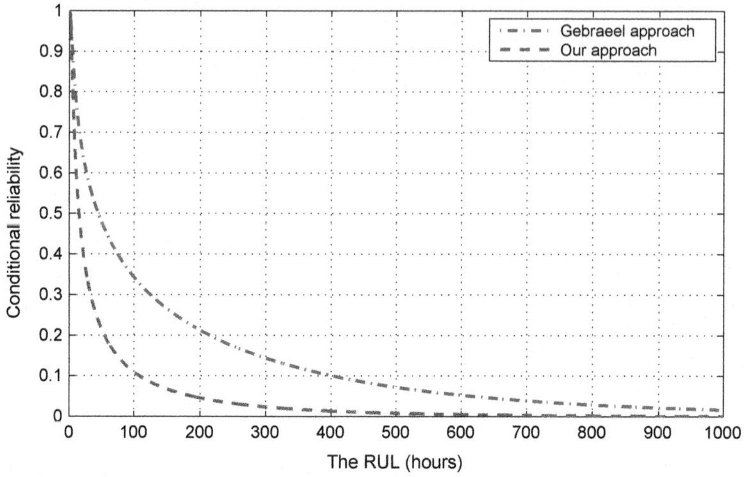

Fig. 5.11 Conditional reliability at $t_{110} = 272.5$ (h)

it is lower than that of the other minimum, but it is obviously wrong since it can be seen from Fig. 5.9 that the remaining life of the gyro is much less than 200 h.

On the other hand, many reliability-centered maintenance policies have been investigated in the literature and applied in practice. Therefore, it is useful to have an intuition of the reliability associated with the approximated result and exact result. Here, we use $\bar{F}_{L_k|\mathbf{X}_{1:k}}(l_k|\boldsymbol{X}_{1:k})$ as the conditional reliability obtained from our exact approach, while $\bar{F}_{L'_k|X_{1:k}}(l_k|\boldsymbol{X}_{1:k})$ as the conditional reliability obtained from Gebraeel's approach. The difference is illustrated at the last CM point in Fig. 5.11. It can be found that the difference is significant and the maximum difference between $\bar{F}_{L_k|X_{1:k}}(l_k|\boldsymbol{X}_{1:k})$ and $\bar{F}_{L'_k|X_{1:k}}(l_k|\boldsymbol{X}_{1:k})$ is 0.265. This is due in part to the many fluctuations existing in the degradation path of the drift. As a result, using $\Pr(L'_k \leq l_k \big| \boldsymbol{X}_{1:k}) = \Pr\left(X(l_k + t_k) \geq w \middle| \boldsymbol{X}_{1:k}\right)$ to approximate the RUL will lead to overestimating of the reliability, as discussed in Remark 5.3 and Theorem 5.6. Consequently, when the approximated result is applied to reliability-centered maintenance, the obtained decision may be far from the reality.

Overall, these simulations and the case study imply that the approximated result has significant effects on RUL estimation and further on the replacement decision-making. In extreme case, a unique and optimal replacement decision cannot be achieved and a much delayed repair action may be incurred when it is applied to decision-making and optimization. In contrast, our exact approach can effectively overcome the disadvantages of the approximated result and lead to a timely replacement, which is important, particularly when the monitored system is vital or expensive, such as the case in most military or medical applications.

References

1. Pecht M (2008) Prognostics and health management of electronics. Wiley, New Jersey
2. Smith JB, Navarro S, Haldeman D (1997) Development of a prognostics and health management capability for the joint strike fighter. In: IEEE autotestcon proceedings, 22–25 September, pp 676–682
3. Wang W, Zhang W (2005) A model to predict the residual life of aircraft engines based upon oil analysis data. Naval Res Logist 52(3):276–284
4. Nikhil MV, Pecht M (2006) Prognostics and health management of electronics. IEEE Trans Components Pack Technol 29(1):222–229
5. Lall P, Islam N, Rahim K, Suhling J, Gale S (2006) Prognostics and health management of electronic packaging. IEEE Trans Components Pack Technol 29(3):666–677
6. Mazhar MI, Kara S, Kaebernick H (2007) Remaining life estimation of used components in consumer products life cycle data analysis by Weibull and artificial neural networks. J Oper Manag 25(6):1184–1193
7. Tsui KL, Wong SY, Jiang W, Lin CJ (2011) Recent research and developments in temporal and spatiotemporal surveillance for public health. IEEE Trans Reliab 60(1):49–58
8. Derman C, Lieberman GJ, Ross SM (1984) On the use of replacements to extend system life. Oper Res 32(3):616–627
9. Si XS, Wang W, Hu CH, Zhou DH (2011) Remaining useful life estimation-A review on the statistical data driven approaches. Eur J Oper Res 213(1):1–14
10. Camci F, Chinnam RB (2010) Health state estimation and prognostics in machining processes. IEEE Trans Autom Sci Eng 7(3):581–597
11. Maguluri G, Zhang CH (1994) Estimation in the mean residual life regression model. J R Stat Soc Ser B 56(3):477–489
12. Alam MM, Suzuki K (2009) Lifetime estimation using only failure information from warranty database. IEEE Trans Reliab 58(4):573–582
13. Ma ZM, Krings AW (2011) Dynamic hybrid fault modeling and extended evolutionary game theory for reliability, survivability and fault tolerance analyses. IEEE Trans Reliab 60(1):180–196
14. Escobar LA, Meeker WQ (2006) A review of accelerated test models. Stat Sci 21(4):552–577
15. Joseph VR, Yu IT (2006) Reliability improvement experiments with degradation data. IEEE Trans Reliab 55(1):149–157
16. Whitmore GA (1986) First-passage-time models for duration data regression structures and competing risks. J R Stat Soc Ser D 35(2):207–219
17. Lee M-LT, Whitmore GA, Laden F, Hart JE, Garshick E (2004) Assessing lung cancer risk in railroad workers using a first hitting time regression model. Environmetrics 15(5):501–512
18. Lee M-LT, Whitmore GA (2006) Threshold regression for survival analysis: modeling event times by a stochastic process reaching a boundary. Stat Sci 21(4):501–513
19. Balka J, Desmond AF, McNicholas PD (2009) Review and implementation of cure models based on first hitting times for Wiener processes. Lifetime Data Anal 15(2):147–176
20. Singpurwalla ND (1995) Survival in dynamic environments. Stat Sci 10(1):86–103
21. Aalen OO, Borgan O, Gjessing HK (2008) Survival and event history analysis: a process point of view. Springer, New York
22. Peng CY, Tseng ST (2009) Mis-specification analysis of linear degradation models. IEEE Trans Reliab 58(3):444–455
23. Pandey MD, Yuan X-X, van Noortwijk JM (2009) The influence of temporal uncertainty of deterioration on life-cycle management of structures. Struct Infrastruct Eng 5(2):145–156
24. Park JI, Bae SJ (2010) Direct prediction methods on lifetime distribution of organic light-emitting diodes from accelerated degradation tests. IEEE Trans Reliab 59(1):74–90
25. Li R, Ryan JK (2011) A Bayesian inventory model using real-time condition monitoring information. Prod Oper Manag 20(5):754–771
26. Yuan X-X, Pandey MD (2009) A nonlinear mixed-effects model for degradation data obtained from in-service inspections. Reliab Eng Syst Safety 94(2):509–519

27. Wang W, Zhang W (2008) An asset residual life prediction model based on expert judgments. Eur J Oper Res 188(2):496–505
28. Gebraeel N, Lawley MA, Li R, Ryan JK (2005) Residual-life distributions from component degradation signals: a Bayesian approach. IIE Trans 37(6):543–557
29. You MY, Li L, Meng G, Ni J (2010) Statistically planned and individual improved predictive maintenance management for continuously monitored degrading systems. IEEE Trans Reliab 59(4):744–753
30. Gebraeel NZ (2006) Sensory-updated residual life distributions for components with exponential degradation patterns. IEEE Trans Autom Sci Eng 3(4):382–393
31. Elwany AH, Gebraeel NZ (2008) Sensor-driven prognostic models for equipment replacement and spare parts inventory. IIE Trans 40(7):629–639
32. Elwany A, Gebraeel NZ (2009) Real-time estimation of mean remaining life using sensor-based degradation models. J Manuf Sci Eng 131(5):051005(1–9)
33. Shechter SM, Bailey MD, Schaefer AJ (2008) Replacing nonidentical vital components to extend system life. Naval Res Logist 55(1):700–703
34. Meeker WQ, Escobar LA (1998) Statistical methods for reliability data. Wiley, New York
35. Elwany A, Gebraeel NZ, Maillart L (2011) Structured replacement policies for systems with complex degradation processes and dedicated sensors. Oper Res 59(3):684–695
36. Gelfand AE, Simth AFM (1990) Sampling based approaches to calculating marginal densities. J Am Stat Assoc 85(410):398–409
37. Tierney L (1994) Markov chains for exploring posterior distributions. Ann Stat 22(1):1701–1728
38. Dempster AP, Laird NM, Rubin DB (1977) Maximum likelihood from incomplete data via the EM algorithm. J R Stat Soc Ser B 39(1):1–38
39. Wu CFJ (1983) On the convergence property of the EM algorithm. Ann Stat 11(1):95–103
40. Christer AH, Wang W (1992) A model of condition monitoring of a production plant. Int J Prod Res 30(9):2199–2211
41. Ross SM (2007) Introduction to probability models. Academic, London
42. Chen N, Tsui KL (2012) Condition monitoring and remaining useful life prediction using degradation signals: revisited. IIE Trans 45(9):939–952
43. Park C, Padgett WJ (2006) Stochastic degradation models with several accelerating variables. IEEE Trans Reliab 55(2):379–390
44. Tseng ST, Tang J, Ku LH (2003) Determination of optimal burn-in parameters and residual life for highly reliable products. Naval Res Logist 50(1):1–14

Chapter 6
Estimating RUL with Three-Source Variability in Degradation Modeling

6.1 Introduction

Prognostics and health management (PHM) can make full use of condition monitoring (CM) data from a functioning system to assess the reliability of the system in its actual life-cycle conditions, to determine the advent of failure, and to mitigate system risk through managerial activities [1–4]. A requirement of a PHM-enabled system is the ability to estimate the remaining useful life (RUL), which can provide the decision-maker with enough lead time to perform the necessary maintenance actions prior to failure. This RUL estimation is a fundamental prerequisite for proactive maintenance because utilizing information about the system's RUL in maintenance decisions can improve the system's availability, extend the system's life, and further reduce life-cycle costs. So far, the estimated RUL, conditional on the CM data, has been considered as one of the most central components in PHM and has been attached great importance in practice. Therefore, the purpose of this chapter is to investigate how to estimate the RUL from the CM data, and emphasis is placed on handling various sources of variability in stochastic degradation modeling.

6.1.1 *Motivation*

With advances in information and sensing technologies, degradation signals of the system can be obtained relatively easily through CM techniques, and the past decade has witnessed an increasingly growing research interest on various aspects of PHM of systems based on measured degradation signals (see [5–9], and the references therein). However, it is quite common in practice that the degradation occurs in a stochastic way for a number of engineering systems such as bearings, gyros, and battery systems. As a result, the RUL is also a random variable, resulting in the difficulty to estimate the RUL with certainty; see [8] for an overview of this topic.

X.-S. Si et al., *Data-Driven Remaining Useful Life Prognosis Techniques*,
Springer Series in Reliability Engineering, DOI 10.1007/978-3-662-54030-5_6

To characterize the uncertainty of the RUL, attention is usually paid to the estimation of the probability distribution of the RUL by modeling the degradation process. Generally, the degradation processes of technological systems are usually affected by numerous sources of variability contributing to the uncertainty of the estimated RUL: (1) temporal variability, (2) unit-to-unit variability, (3) measurement variability, and (4) model error; see [6, 10–16] for some examples. Recent advances in managing uncertainties associated with prognostics can be found in [17–19]. Particularly, Baraldi et al. in [17] investigated the capabilities of different prognostic approaches to deal with various sources of uncertainty in RUL estimation, and formulated the effect of model error (e.g., resulted from model assumptions and simplifications made on the form and structure of the degradation model) on RUL estimation. In this chapter, we mainly focus on the first three sources of variability.

First, temporal variability is referred to as the inherent uncertainty associated with the progression of the degradation over time [12, 20–22]. Due to the stochastic nature of a degradation process, it is appropriate to model such degradation by a stochastic process. Second, the unit-to-unit variability determines heterogeneity among the degradation paths of different units. Taking gyros in an inertial navigation system for example, as a gyro degrades, the drift tends to increase. When the drift reaches a threshold (which is typically determined by an associated industrial standard), the gyro is considered to have failed. In this case, the degradation rate of gyros differs from each other. Such difference in degradation rates can usually be modeled by introducing unit-specific random effects through some model parameters [11, 23, 24]. Last, perfect measurements for degradation are practically impossible to achieve, and the measured data are inevitably contaminated by measurement error resulting from disturbance, non-ideal measurement instruments, and other noise [6, 13, 16, 25–27]. In this case, the observed degradation signals can only partially reflect the underlying degradation state.

From the above introduction, it has been recognized that temporal variability, unit-to-unit variability, and measurement variability are three important factors having to be taken into account simultaneously when investigating the RUL estimation in the framework of stochastic modeling. Most published works on RUL estimation using stochastic models focused on models that only considered one source or two sources of variability [10, 13, 23, 28–37]. In contrast, the research on RUL estimation using degradation models with three-source variability is very limited.

On the other hand, even for the few models that considered three-source variability, the research is limited to the case of modeling the degradation process only or estimating the lifetime for a population of statistically identical systems, ignoring the impact of imperfect measurements on the lifetime distribution [13, 14, 16, 29]. In addition, updating the estimated RUL for a particular system in service using real-time data is increasingly important and desirable so that the most recently calculated RUL value can accurately reflect the current reality of the system. This updating mechanism has been considered by others (see for example [23, 32, 35, 38]), but not in the context with three-source variability in degradation modeling. Therefore, an important, practical problem of how to achieve real-time RUL estimation with three-source variability remains unsolved. This gap leads to the primary motivation of this chapter.

Throughout this chapter, the term three-source variability means considering temporal variability, unit-to-unit variability, and measurement variability simultaneously.

6.1.2 Related Works

There is a great need to develop stochastic models to model the degradation process for RUL estimation, and a significant volume of research has been published, such as random coefficient regression models [28, 31, 39], Gamma processes [12, 30, 37], inverse Gaussian processes [40–42], and Wiener processes [10, 13, 29, 30, 32, 34–36, 43]. A recent review on these models can be found in [8].

In these research works, degradation processes described by Wiener processes are particularly attractive because they not only have some favorable mathematical properties but also can model non-monotonic degradation signals frequently encountered in practice. Therefore, this type of stochastic process has been widely used to characterize the path of the degradation process where successive fluctuations in degradation are observed, such as in the examples in [10, 23, 32–34, 43]. It is noted however that the above works considered mostly the temporal variability or unit-to-unit variability, or both. For example, in [35], the authors considered a Wiener process for RUL estimation using a deterministic drift parameter, and adaptively updated this parameter by a state-space model and Kalman filtering technique. However, in the estimated RUL distribution, only the point estimation of the drift parameter is considered, and the issue regarding the presence of the unit-to-unit variability and measurement variability is missing. Recently, an extension in [32] considered the effect of the estimation uncertainty of the drift parameter on the estimated RUL, but omitted the measurement variability as well.

There are reported works considering measurement variability using state-space models in [35–38, 44]. However, the other two sources of variability were not considered simultaneously together in these works. For example, in [38], a state-space based prognostic model was proposed to characterize the nonlinearity and measurement variability by applying the extended Kalman filter. However, only the approximate probability density function (PDF) of the RUL is obtained, and the cumulative distribution function (CDF), mean, and variance of the estimated RUL cannot be derived analytically because of the involved nonlinearity. In addition, all the model parameters in the work of [38] are deterministic. Therefore, the method in [38] ignores the unit-to-unit variability. There are some exceptions such as the works by [13, 14, 16, 29], where three-source variability was taken into account at the same time. However, all these works only consider the measurement variability in parameter estimation, and ignore the effect of the uncertainty in the estimated degradation state on the RUL distribution due to measurement errors. They also did not consider the updating mechanism for the RUL in line with the newly measured degradation signals unique to an individual system in service. In addition, when three-source variability is involved in the stochastic degradation model, it is inevitable to consider the identifiability problem of the parameters in the degradation model. However, all the existing works do not address this problem.

6.1.3 Main Works of This Chapter

In this chapter, we present a Wiener process-based degradation modeling framework for RUL estimation with three-source variability. The main novelty of the presented model lies in the concerned problem of how to enable us to consider three-source variability simultaneously to estimate and update the RUL distribution using the degradation data across the population, and the real-time observed measurements of an individual system in service. This novelty in considering three-source variability distinguishes our model from the existing results such as [10, 13, 14, 16, 23, 29, 32, 33, 35, 36, 38] in several major aspects.

1. Three sources of variability are simultaneously considered; and by constructing a state-space model, the correlated posterior distributions of the underlying degradation state and random effect parameter are estimated by employing the Kalman filtering technique. This approach differs from the cases that the parameters of the degradation model are deterministic.
2. On the basis of the posterior distributions of the underlying degradation state, and the random effect parameter, the analytical forms of the PDF, CDF, mean, and variance of the estimated RUL are derived. It is found that three kinds of variability can propagate into the estimated results; thus, we account for the effect of three-source variability on the estimated RUL simultaneously.
3. The results with three-source variability can be updated with the newly measured degradation data unique to the system in service, and the RUL estimation results only considering one- or two-source variability are obviously special cases of our results.
4. The identifiability of the parameters in the presented model is investigated. Then, a MLE method is given to specify the initial parameters of the state-space model based on the historical degradation observations of multiple units. Thus the linkage between the past and real-time data is established.

Finally, we provide a practical case study for gyros in an inertial navigation platform to illustrate the application of the developed approach. With realistic data, we compare the estimated RUL results considering three-source variability with the results only considering one- or two-source variability based on several measures: Akaike information criterion (AIC), mean squared error (MSE), and relative error (RE). The results verify that considering three-source variability can improve the model fit, and the accuracy of the RUL estimation.

Section 6.2 gives the description of the degradation modeling framework with three-source variability. In Sect. 6.3, we estimate the RUL under three sources of variability. Section 6.4 investigates the issues regarding the identifiability and parameter estimation of the presented model. Section 6.5 provides a case study for demonstration.

6.2 Description of Degradation Modeling with Three-Source Variability for RUL Estimation

Let $\{X(t), t \geq 0\}$ denote the stochastic process describing the underlying degradation progression over operating time t, which is modeled by a Wiener process. In general, a Wiener process based degradation model can be represented as

$$X(t) = X(0) + \theta t + \sigma B(t), \tag{6.1}$$

where $\sigma B(t) \sim N(0, \sigma^2 t)$ for $t > 0$, representing the stochastic dynamics of the degradation process, and $X(0) = x_0$ is a known initial degradation state. Without loss of generality, we assume $X(0) = x_0 = 0$ in the following.

Equation (6.1) can characterize the temporal variability due to the dynamics of $\{B(t), t \geq 0\}$. However, each system possibly experiences different operating conditions, and thus the degradation paths of different systems exhibit different degradation rates to increase. Therefore, for a degradation model to be realistic, it is more appropriate to incorporate unit-to-unit variability in the degradation process. As such, we treat the parameter θ to be a random effect representing between-unit variation, and σ to be a fixed parameter representing the degradation feature common to all systems in the population. For simplicity, we assume that θ follows $\theta \sim N(\mu_\theta, \sigma_\theta^2)$, and is statistically independent of $\{B(t), t \geq 0\}$. The ideas of random effects and Gaussian assumptions are widely used in degradation modeling literature [8, 11].

Additionally, perfect measurement of the underlying degradation state is often impossible or costly. Instead, the obtained measurements are inevitably subject to measurement variability, resulting from noise, disturbance, non-ideal instruments, etc. In this case, the observed measurements are imperfect, and can partially reflect the underlying degradation state. To characterize the effect of the measurement variability, the measurement process $\{Y(t), t \geq 0\}$, which describes the relationship between the observable but uncertain measurements and the underlying degradation state at time t, is formulated as

$$Y(t) = X(t) + \varepsilon, \tag{6.2}$$

where ε is the random measurement error, assumed to be statistically independent and identically distributed (*i.i.d.*) with $\varepsilon \sim N(0, \gamma^2)$ at any time point t. It is further assumed that ε, θ, and $B(t)$ are mutually statistically independent. All these assumptions are widely adopted in the practice of degradation modeling and RUL estimation.

Before addressing the RUL estimation issues based on the above model setting, we have the following remark regarding the linkage of the above model, which we refer to as our model, with the existing results.

Remark 6.1 The choice of the parameters σ^2, γ^2, σ_θ^2 is made to obtain a model that includes, as special cases, other models that have already found practical applications in the literature. In fact, it is easy to verify the following points.

1. When $\sigma^2 = 0$, our model with three-source variability reduces to the linear random effect regression model in [11, 24, 31], where only the measurement uncertainty and unit-to-unit variability are considered.
2. when $\gamma^2 = 0$, our model with three-source variability reduces to the linear Wiener process model with random drift in [5, 7, 10, 32, 33], where only the temporal uncertainty and unit-to-unit variability are considered.
3. When $\sigma_\theta^2 = 0$, our model with three-source variability reduces to the linear but hidden degradation model in [36, 44], where only the temporal uncertainty and measurement uncertainty are considered.
4. When $\gamma^2 = 0, \sigma_\theta^2 = 0$, our model with three-source variability reduces to the traditional linear Wiener process model in [15, 34, 35], where only the temporal uncertainty is considered. ■

Remark 6.2 From the above work, see that the underlying assumptions of the proposed model include: (1) the system operates in time-invariant environments, and thus the rate of degradation can be approximated as a constant for simplicity, i.e., the degradation process is linear or can be treated as locally linear; and (2) the measurement noise is Gaussian. These assumptions have been widely used in degradation modeling. ■

As a result of Remarks 6.1 and 6.2, our model can include many existing models as special cases, and thus is more general and flexible. As in other degradation modeling works [8, 45], we adopt the concept of the first hitting time (FHT) to define the lifetime, and then proceed to deducing the RUL. In other words, once the degradation process $\{X(t), t \geq 0\}$ is equal to or beyond a predefined threshold level, the system will be considered to be failed, and therefore, the lifetime can be interpreted as the FHT of the degradation process to the threshold level. According to the concept of FHT, the lifetime T of a system is defined as

$$T = \inf\{t : X(t) \geq \omega | X(0) < \omega\}, \tag{6.3}$$

where ω is the predefined threshold level.

The above formulation is mainly focused on a population of the system. The main objective here is to estimate and update the RUL distribution of an individual system in service based on the real-time observations of the degradation process. Suppose the degradation process is discretely monitored at time $0 = t_0 < t_1 < \cdots < t_k$, and let

$y_k = Y(t_k)$ denote the degradation observation at time t_k. The set of the degradation measurements up to t_k is represented by $\boldsymbol{Y}_{1:k} = \{y_1, y_2, \ldots, y_k\}$, and the corresponding set of the degradation states up to t_k is represented by $\boldsymbol{X}_{1:k} = \{x_1, x_2, \ldots, x_k\}$, where $x_k = X(t_k)$. As a result, we further express discrete measurement at time t_k as $y_k = x_k + \varepsilon_k$, where the measurement errors ε_k are assumed to be *i.i.d.* realizations of ε.

Therefore, using the concept of the FHT, we define the RUL L_k of a system at time t_k as

$$L_k = \inf\left\{l_k > 0 : X(l_k + t_k) \geq \omega\right\}. \tag{6.4}$$

with conditional PDF $f_{L_k|\boldsymbol{Y}_{1:k}}(l_k|\,\boldsymbol{Y}_{1:k})$, and conditional CDF

$$F_{L_k|\boldsymbol{Y}_{1:k}}(l_k|\,\boldsymbol{Y}_{1:k}) = \Pr(L_k \leq l_k|\boldsymbol{Y}_{1:k}) = \Pr\left(\sup_{l_k>0} X(t_k + l_k) \geq \omega \,\middle|\, \boldsymbol{Y}_{1:k}\right), \tag{6.5}$$

where $\boldsymbol{Y}_{1:k}$ is the observed measurements available up to t_k.

In the remaining sections, our primary goal is to derive $f_{L_k|\boldsymbol{Y}_{1:k}}(l_k|\,\boldsymbol{Y}_{1:k})$ and $F_{L_k|\boldsymbol{Y}_{1:k}}(l_k|\,\boldsymbol{Y}_{1:k})$ based on $\boldsymbol{Y}_{1:k}$ with three-source variability.

6.3 RUL Estimation with Three-Source Variability

To investigate RUL estimation with three-source variability, we consider three cases in this section: (1) temporal variability and unit-to-unit variability together, (2) temporal variability and measurement variability together, and (3) three-source variability simultaneously.

6.3.1 RUL Estimation with Temporal Variability and Unit-to-Unit Variability

Initially, we only consider the temporal variability in the degradation process $\{X(t), t \geq 0\}$ by making θ be deterministic. It is well-known that the FHT of the Wiener process crossing a fixed threshold follows an inverse Gaussian distribution [46]. In the following, we summarize the main results and properties of the lifetime T in this case only considering the temporal variability.

Lemma 6.1 *For the Wiener process as in (6.1) and (6.3), given θ, the following equations hold.*

$$f_{T|\theta}(t|\theta) = \frac{\omega}{\sqrt{2\pi t^3 \sigma^2}} \exp\left(-\frac{(\omega - \theta t)^2}{2\sigma^2 t}\right), \tag{6.6}$$

$$F_{T|\theta}(t|\theta) = 1 - \Phi\left(\frac{\omega - \theta t}{\sigma\sqrt{t}}\right) + \exp\left(\frac{2\theta\omega}{\sigma^2}\right)\Phi\left(\frac{-\omega - \theta t}{\sigma\sqrt{t}}\right), \tag{6.7}$$

$$E(T|\theta) = \frac{\omega}{\theta}, \tag{6.8}$$

$$var(T|\theta) = \frac{\omega\sigma^2}{\theta^3}, \tag{6.9}$$

$\Phi(\cdot)$ is the CDF of standard normal distribution.

When we consider the random effect of θ representing unit-to-unit variability, the PDF and CDF of the lifetime can be respectively computed by the law of total probability as

$$f_T(t) = \int_{-\infty}^{+\infty} f_{T|\theta}(t|\theta)p(\theta)\mathrm{d}\theta = E_\theta\left[f_{T|\theta}(t|\theta)\right], \tag{6.10}$$

and

$$F_T(t) = \int_{-\infty}^{+\infty} F_{T|\theta}(t|\theta)p(\theta)\mathrm{d}\theta = E_\theta\left[F_{T|\theta}(t|\theta)\right], \tag{6.11}$$

where $p(\theta)$ is the PDF of θ, and $E_\theta[\cdot]$ is the expectation operator with respect to θ.

To facilitate the derivation and explicitly calculate the integrals such as those in (6.10) and (6.11), we first give the following lemmas, which can be proved directly through some lengthy algebraic manipulations; thus the proofs are omitted here, and the interested readers can refer to [47].

Lemma 6.2 *If $Z \sim N(\mu, \sigma^2)$, and $w_1, w_2, A, B \in \mathbf{R}$, $C \in \mathbf{R}^+$, then*

$$E_Z\left[(w_1 - AZ)\exp\left(-\frac{(w_2 - BZ)^2}{2C}\right)\right] = \sqrt{\frac{C}{B^2\sigma^2 + C}}\left(w_1 - A\frac{Bw_2\sigma^2 + \mu C}{B^2\sigma^2 + C}\right)\cdot\exp\left(-\frac{(w_2 - B\mu)^2}{2(B^2\sigma^2 + C)}\right). \tag{6.12}$$

Lemma 6.3 *If $Z \sim N(\mu, \sigma^2)$, and $w, A, B, C, D \in \boldsymbol{R}$, $1 - 2B\sigma^2 > 0$, then*

$$E_Z\left[\exp(AZ + BZ^2)\Phi(C + DZ)\right] = \frac{1}{\sqrt{1 - 2B\sigma^2}}\exp\left(\frac{2A\mu + A^2\sigma^2 + 2B\mu^2}{2(1 - 2B\sigma^2)}\right) \times\Phi\left(\frac{C + D\mu + AD\sigma^2 - 2BC\sigma^2}{\sqrt{(1 - 2B\sigma^2)^2 + D^2\sigma^2(1 - 2B\sigma^2)}}\right). \tag{6.13}$$

Based on Lemmas 6.1–6.3, we can calculate (6.10) and (6.11) explicitly. The main results are summarized in the following theorem:

Theorem 6.1 *For the Wiener process in (6.1) and (6.3), taking into account the random effect of θ with $\theta \sim N(\mu_\theta, \sigma_\theta^2)$, the following results hold.*

$$f_T(t) = \frac{\omega}{\sqrt{2\pi t^3(\sigma_\theta^2 t + \sigma^2)}} \exp\left(-\frac{(\omega - \mu_\theta t)^2}{2t(\sigma_\theta^2 t + \sigma^2)}\right), \tag{6.14}$$

$$F_T(t) = 1 - \Phi\left(\frac{\omega - \mu_\theta t}{\sqrt{\sigma_\theta^2 t^2 + \sigma^2 t}}\right) + \exp\left(\frac{2\mu_\theta \omega}{\sigma^2} + \frac{2\sigma_\theta^2 \omega^2}{\sigma^4}\right) \Phi\left(-\frac{2\sigma_\theta^2 \omega t + \sigma^2(\omega + \mu_\theta t)}{\sigma^2\sqrt{\sigma_\theta^2 t^2 + \sigma^2 t}}\right), \tag{6.15}$$

$$E(T) = \frac{\sqrt{2}\omega}{\sigma_\theta} D\left(\frac{\mu_\theta}{\sqrt{2\sigma_\theta^2}}\right), \tag{6.16}$$

$$var(T) = \frac{E(T)}{2\sigma_\theta^2}\left[\frac{\sigma^2\mu_\theta^2}{\sigma_\theta^2} + 2\omega\mu_\theta - \sigma^2 - 2\sigma_\theta^2 E(T)\right] - \frac{\omega}{2\sigma_\theta^2}\left(\frac{\sigma^2\mu_\theta}{\sigma_\theta^2} + 2\omega\right), \tag{6.17}$$

where $D(z) = \exp(-z^2)\int_0^z \exp(u^2)du$ is the Dawson integral for real z.

Proof The proofs for the PDF and CDF can be derived following the method in [13]. In the following, we derive the formulations of the mean and variance of the lifetimes. For facilitating the subsequent derivations of the variance of the lifetime or the RUL, we first give the following facts. Assume a random variable following $\theta \sim N(\mu_\theta, \sigma_\theta^2)$. Then the following facts hold.

$$E_\theta\left[\frac{1}{\theta}\right] = \frac{1}{\sigma_\theta^2}\exp\left(-\frac{\mu_\theta^2}{2\sigma_\theta^2}\right)\int_0^{\mu_\theta} \exp\left(\frac{u^2}{2\sigma_\theta^2}\right)du = \frac{\sqrt{2}}{\sigma_\theta} D\left(\frac{\mu_\theta}{\sqrt{2}\sigma_\theta}\right), \tag{6.18}$$

$$E_\theta\left[\frac{1}{\theta^2}\right] = \frac{\mu_\theta}{\sigma_\theta^2} E_\theta\left[\frac{1}{\theta}\right] - \frac{1}{\sigma_\theta^2},\ E_\theta\left[\frac{1}{\theta^3}\right] = \frac{1}{2\sigma_\theta^2}\left[\mu_\theta E_\theta\left[\frac{1}{\theta^2}\right] - E_\theta\left[\frac{1}{\theta}\right]\right], \tag{6.19}$$

where $D(z) = \exp(-z^2)\int_0^z \exp(u^2)du$ is the Dawson integral for real z.

Based on the property of the expectation operator, and (73) and (74), we can compute $E(T)$, and var(T) as

$$E(T) = E_\theta\left[E(T\,|\,\theta)\right] = E_\theta\left[\frac{\omega}{\theta}\right] = \frac{\sqrt{2}\omega}{\sigma_\theta} D\left(\frac{\mu_\theta}{\sqrt{2}\sigma_\theta}\right), \tag{6.20}$$

and

$$\begin{aligned}
\text{var}(T) &= E_\theta\left[\text{var}(T\,|\,\theta)\right] + \text{var}_\theta\left[E(T\,|\,\theta)\right] = E_\theta\left[\frac{\omega\sigma^2}{\theta^3}\right] + \text{var}_\theta\left[\frac{\omega}{\theta}\right] \\
&= \omega\sigma^2 E_\theta\left[\frac{1}{\theta^3}\right] + E_\theta\left[\frac{\omega^2}{\theta^2}\right] - \left(E_\theta\left[\frac{\omega}{\theta}\right]\right)^2 = \omega\sigma^2 E_\theta\left[\frac{1}{\theta^3}\right] + \omega^2 E_\theta\left[\frac{1}{\theta^2}\right] - E^2(T) \\
&= \frac{E(T)}{2\sigma_\theta^2}\left[\frac{\sigma^2\mu_\theta^2}{\sigma_\theta^2} + 2\omega\mu_\theta - \sigma^2 - 2\sigma_\theta^2 E(T)\right] - \frac{\omega}{2\sigma_\theta^2}\left(\frac{\sigma^2\mu_\theta}{\sigma_\theta^2} + 2\omega\right)
\end{aligned} \tag{6.21}$$

This completes the proof of Theorem 6.1.

The above formulations in (6.14) and (6.15) are consistent with the results in [13], which are mainly focused on a population of the systems. Namely, in (14), the estimated PDF of the FHT does not consider the real-time degradation state of the monitored system.

If the degradation state $X(t_k) = x_k$ at current time t_k can be observed directly, and there is $x_k < \omega$, we assume that the system has been functioning before t_k. In this chapter, we use this assumption to update the knowledge of the degradation process. Note that this assumption is often adopted in the literature including in [7, 23, 32, 33]. This approach differs from that of [38], where the made assumption is used to approximate the PDF of the lifetime T. Thus, the time t in the assumption of [38] corresponds to a future time because the current time there is $t_0 = 0$. As a result, using the assumption in [38] to estimate the RUL L_k at time t_k, the time in the assumption is transformed to $t_k + l_k$, which is still a future time. However, for our used assumption, if $x_k < \omega$ at current time t_k, the system has been functioning before t_k. Unlike [38], our treatment does not impose additional constraints and ignore the probability of the event occurring for calculating the PDF of the RUL.

As a result, for $t \geq t_k$, given θ and x_k ($x_k < \omega$), we revise the degradation process over time since t_k according to the Markov property of the Wiener process as

$$X(t) = x_k + \theta(t - t_k) + \sigma\left(B(t) - B(t_k)\right), t \geq t_k. \tag{6.22}$$

In such a case, the residual $t - t_k$ corresponds to the realization of the RUL at time t_k if t is the FHT of $\{X(t), t \geq t_k\}$, according to the definition of the RUL in (6.4). Therefore, we first take the transformation $l_k = t - t_k$ for (16), where $l_k > 0$. Then the process $\{X(t), t \geq t_k\}$ can be represented with the residual time l_k as

$$X(l_k + t_k) = x_k + \theta l_k + \sigma\left(B(l_k + t_k) - B(t_k)\right), l_k > 0. \tag{6.23}$$

As a result, the RUL at time t_k is equal to the FHT of the process $\left\{\tilde{X}(l_k), l_k \geq 0\right\}$ crossing threshold $\omega_k = \omega - x_k$, where $\tilde{X}(l_i) = X(l_k + t_k) - x_k$, and $\tilde{X}(0) = 0$. That is to say, given t_k,

$$\tilde{X}(l_k) = \theta l_k + \sigma W(l_k), W(l_k) = B(l_k + t_k) - B(t_k). \tag{6.24}$$

To derive $f_{L_k|x_k}(l_k|\, x_k)$, it is necessary to prove that the stochastic process $\{W(l_k), l_k \geq 0\}$, with $W(l_k) = B(l_k + t_k) - B(t_k)$, is still a BM. This result is guaranteed by the following lemma [33].

Lemma 6.4 *Given t_k, the stochastic process $\{W(t), t \geq 0\}$ with $W(t) = B(t + t_k) - B(t_k)$ for any $t \geq 0$ is still a standard BM, where $\{B(t), t \geq 0\}$ is a standard BM.*

Therefore, the estimated PDF of the RUL, $f_{L_k|\theta,x_k}(l_k|\,\theta,x_k)$, conditional on the current degradation state $x_k(x_k<\omega)$ and θ, can be obtained by the following theorem.

Theorem 6.2 *For the Wiener process defined in (6.1), and the definition of the RUL in (6.4), given θ and the current degradation state $x_k(x_k<\omega)$, the following equations for RUL estimation at time t_k hold.*

$$f_{L_k|\theta,x_k}(l_k|\,\theta,x_k)=\frac{\omega-x_k}{\sqrt{2\pi l_k^3\sigma^2}}\exp\left(-\frac{(\omega-x_k-\theta l_k)^2}{2\sigma^2 l_k}\right),\tag{6.25}$$

$$F_{L_k|\theta,x_k}(l_k|\,\theta,x_k)=1-\Phi\left(\frac{\omega-x_k-\theta l_k}{\sigma\sqrt{l_k}}\right)+\exp\left(\frac{2\theta(\omega-x_k)}{\sigma^2}\right)\Phi\left(\frac{-\omega+x_k-\theta l_k}{\sigma\sqrt{l_k}}\right),\tag{6.26}$$

$$E(L_k|\,\theta,x_k)=\frac{\omega-x_k}{\theta},\tag{6.27}$$

$$var(L_k|\,\theta,x_k)=\frac{(\omega-x_k)\sigma^2}{\theta^3}.\tag{6.28}$$

In general, as the age of the system grows, the system's degradation gradually increases. Therefore, prognostics is more commonly applied to aged systems (corresponding to large x_k) rather than new systems (corresponding to small x_k). Actually, this interest can be reflected by the above obtained results. Take (21) and (22) for example. If $x_k\to\omega$, then the estimated PDF of the RUL $f_{L_k|\theta,x_k}(l_k|\,\theta,x_k)\to 0$, and the CDF $F_{L_k|\theta,x_k}(l_k|\,\theta,x_k)\to 1$. This result reflects the aging character with x_k increasing.

When the random effect of θ is considered, according to Theorems 6.1 and 6.2, we have the following results for $f_{L_k|x_k}(l_k|\,x_k)$, and $F_{L_k|x_k}(l_k|\,x_k)$.

Theorem 6.3 *For the Wiener process in (6.1), and the definition of the RUL in (6.4), given the current degradation state x_k and $\theta\sim N(\mu_\theta,\sigma_\theta^2)$, the following equations for RUL estimation at time t_k hold.*

$$f_{L_k|x_k}(l_k|\,x_k)=\frac{\omega-x_k}{\sqrt{2\pi l_k^3(\sigma_\theta^2 l_k+\sigma^2)}}\exp\left(-\frac{(\omega-x_k-\mu_\theta l_k)^2}{2l_k(\sigma_\theta^2 l_k+\sigma^2)}\right),\tag{6.29}$$

$$F_{L_k|x_k}(l_k|\,x_k)=1-\Phi\left(\frac{\omega_k-\mu_\theta l_k}{\sqrt{\sigma_\theta^2 l_k^2+\sigma^2 l_k}}\right)+\exp\left(\frac{2\mu_\theta\omega_k}{\sigma^2}+\frac{2\sigma_\theta^2\omega_k^2}{\sigma^4}\right)\Phi\left(-\frac{2\sigma_\theta^2\omega_k l_k+\sigma^2(\omega_k+\mu_\theta l_k)}{\sigma^2\sqrt{\sigma_\theta^2 l_k^2+\sigma^2 l_k}}\right),\tag{6.30}$$

$$E(L_k|x_k)=\frac{\sqrt{2}(\omega-x_k)}{\sigma_\theta}D\left(\frac{\mu_\theta}{\sqrt{2\sigma_\theta^2}}\right),\tag{6.31}$$

$$var(L_k|x_k)=\frac{E(L_k|x_k)}{2\sigma_\theta^2}\left[\frac{\sigma^2\mu_\theta^2}{\sigma_\theta^2}+2\omega_k\mu_\theta-\sigma^2-2\sigma_\theta^2 E(L_k|x_k)\right]-\frac{\omega_k}{2\sigma_\theta^2}\left(\frac{\sigma^2\mu_\theta}{\sigma_\theta^2}+2\omega_k\right)\tag{6.32}$$

where $\omega_k=\omega-x_k$, and $D(z)=\exp(-z^2)\int_0^z\exp(u^2)du$ is the Dawson integral for real z.

The above results are based on the condition that the current degradation state x_k can be observed directly and exactly. However, due to the measurement variability, the true degradation state x_k at time t_k is unobservable, and its accurate value is impossible to be known as discussed previously. In this case, it is better to let the true degradation process be a latent process, which is continuously fluctuating but not directly observable. Thus the degradation state is needed to be estimated from the measurements. This approach is the focus in the next part.

6.3.2 RUL Estimation with Temporal Variability and Uncertain Measurements

Here, given θ, we derive the RUL estimation results taking into account the temporal variability and measurement variability. In this case, only uncertain measurements $\boldsymbol{Y}_{1:k}$ up to the current time t_k are available, and the degradation state x_k cannot be directly used, so we have to estimate the distribution of x_k at time t_k to account for the effect of the measurement variability on the RUL estimation.

To identify the degradation state, the state and measurement equations should be converted into the discrete time equations to facilitate the state estimation once the new observations are available at the CM point. Then, we can obtain the transformed dynamic system equations at the discrete time point t_k, $k = 1, 2, \ldots$

$$\begin{cases} x_k = x_{k-1} + \theta(t_k - t_{k-1}) + v_k \\ y_k = x_k + \varepsilon_k \end{cases}, \tag{6.33}$$

where $v_k = \sigma\left[B(t_k) - B(t_{k-1})\right]$, and ε_k is the realization of ε at t_k. $\{v_k\}_{k\geq 1}$ and $\{\varepsilon_k\}_{k\geq 1}$ are *i.i.d.* noise sequences. According to the model setting in Sect. 6.2, we further have $v_k \sim N\left(0, \sigma^2(t_k - t_{k-1})\right)$, and $\varepsilon_k \sim N(0, \gamma^2)$.

According to the established model (6.33), we utilize a Kalman filter to estimate the underlying degradation state. First, we define $\hat{x}_{k|k} = E(x_k | \boldsymbol{Y}_{1:k}, \theta)$, and $P_{k|k} = \text{var}(x_k | \boldsymbol{Y}_{1:k}, \theta)$ as the expectation, and variance of x_k that are conditional on the measurement history $\boldsymbol{Y}_{1:k}$, respectively. We also define $\hat{x}_{k|k-1} = E(x_k | \boldsymbol{Y}_{1:k-1}, \theta)$, and $P_{k|k-1} = \text{var}(x_k | \boldsymbol{Y}_{1:k-1}, \theta)$ as the one-step-ahead predicted expectation, and variance, respectively. Therefore, at time t_k, the Kalman filter for state estimation can be summarized as follows.

State estimation,

$$\begin{aligned}
&\hat{x}_{k|k-1} = \hat{x}_{k-1|k-1} + \theta(t_k - t_{k-1}) \\
&\hat{x}_{k|k} = \hat{x}_{k|k-1} + K(k)(y_k - \hat{x}_{k|k-1}) \\
&K(k) = P_{k|k-1}(P_{k|k-1} + \gamma^2)^{-1} \\
&P_{k|k-1} = P_{k-1|k-1} + \sigma^2(t_k - t_{k-1})
\end{aligned}$$

Variance update,

$$P_{k|k} = (1 - K(k))\, P_{k|k-1},$$

The initial values of the degradation state are specified as $\hat{x}_{0|0} = 0$ and $P_{0|0} = 0$ because of the model setting $x_0 = 0$ in Sect. 6.2.

Applying the above Kalman filtering algorithm, the posterior of x_k conditional on the measurement sequence $\boldsymbol{Y}_{1:k}$ up to t_k is Gaussian and analytically tractable, i.e., $x_k|\,\theta, \boldsymbol{Y}_{1:k} \sim N(\hat{x}_{k|k}, P_{k|k})$. In this case, due to measurement variability, the degradation state is estimated from the measurements, and thus estimation uncertainty is involved. To account for estimation uncertainty, the RUL estimation in this case is derived by

$$\begin{aligned} f_{L_k|\theta,\boldsymbol{Y}_{1:k}}(l_k|\,\theta, \boldsymbol{Y}_{1:k}) &= \int_{-\infty}^{+\infty} f_{L_k|\theta,x_k,\boldsymbol{Y}_{1:k}}(l_k|\,\theta, x_k, \boldsymbol{Y}_{1:k})p(x_k|\,\theta, \boldsymbol{Y}_{1:k})\mathrm{d}x_k \\ &= E_{x_k|\theta,\boldsymbol{Y}_{1:k}}\left[f_{L_k|\theta,x_k,\boldsymbol{Y}_{1:k}}(l_k|\,\theta, x_k, \boldsymbol{Y}_{1:k})\right], \end{aligned} \tag{6.34}$$

and

$$\begin{aligned} F_{L_k|\theta,\boldsymbol{Y}_{1:k}}(l_k|\,\theta, \boldsymbol{Y}_{1:k}) &= \int_{-\infty}^{+\infty} F_{L_k|\theta,x_k,\boldsymbol{Y}_{1:k}}(l_k|\,\theta, x_k, \boldsymbol{Y}_{1:k})p(x_k|\,\theta, \boldsymbol{Y}_{1:k})\mathrm{d}x_k \\ &= E_{x_k|\theta,\boldsymbol{Y}_{1:k}}\left[F_{L_k|\theta,x_k,\boldsymbol{Y}_{1:k}}(l_k|\,\theta, x_k, \boldsymbol{Y}_{1:k})\right], \end{aligned} \tag{6.35}$$

where $p(x_k|\,\theta, \boldsymbol{Y}_{1:k})$ is the conditional PDF of $x_k|\,\theta, \boldsymbol{Y}_{1:k}$, with mean $\hat{x}_{k|k}$, and variance $P_{k|k}$.

In addition, we present the following lemma to help evaluate the integrals in (6.34) and (6.35).

Lemma 6.5 *Given the current degradation state x_k, θ, and $\boldsymbol{Y}_{1:k}$ at time t_k, we have*

$$f_{L_k|\theta,x_k,\boldsymbol{Y}_{1:k}}(l_k|\,\theta, x_k, \boldsymbol{Y}_{1:k}) = f_{L_k|\theta,x_k}(l_k|\,\theta, x_k), \tag{6.36}$$

$$F_{L_k|\theta,x_k,\boldsymbol{Y}_{1:k}}(l_k|\,\theta, x_k, \boldsymbol{Y}_{1:k}) = F_{L_k|\theta,x_k}(l_k|\,\theta, x_k). \tag{6.37}$$

Proof According to the definition in (6.5), utilizing the Markov property of the Wiener process, we have

$$\begin{aligned} F_{L_k|\theta,x_k,\boldsymbol{Y}_{1:k}}(l_k|\,\theta, x_k, \boldsymbol{Y}_{1:k}) &= \Pr(L_k \le l_k|\theta, x_k, \boldsymbol{Y}_{1:k}) = \Pr(\sup_{l_k>0} X(t_k + l_k) \ge \omega|\theta, x_k, \boldsymbol{Y}_{1:k}) \\ &= \Pr(\sup_{l_k>0} X(t_k + l_k) \ge \omega|\theta, x_k) = F_{L_k|\theta,x_k}(l_k|\,\theta, x_k). \end{aligned} \tag{6.38}$$

This completes the proof.

Based on Lemmas 6.2, 6.3, and 6.5, we have the following results for RUL estimation in the case of taking temporal variability and measurement variability into account.

Theorem 6.4 *For the Wiener process (6.1) and the definition of the RUL in (6.4), given* θ*, and uncertain measurements* $\boldsymbol{Y}_{1:k}$ *up to current time* t_k*, the following results for RUL estimation at time* t_k *hold.*

$$f_{L_k|\theta,\boldsymbol{Y}_{1:k}}(l_k|\theta,\boldsymbol{Y}_{1:k}) = \frac{(\omega-\hat{x}_{k|k})\sigma^2 + P_{k|k}\theta}{\sqrt{2\pi(P_{k|k}+\sigma^2 l_k)^3}}\exp\left(-\frac{(\omega-\hat{x}_{k|k}-\theta l_k)^2}{2\left(P_{k|k}+\sigma^2 l_k\right)}\right), \tag{6.39}$$

$$F_{L_k|\theta,\boldsymbol{Y}_{1:k}}(l_k|\theta,\boldsymbol{Y}_{1:k}) = 1-\Phi\left(\frac{\omega-\hat{x}_{k|k}-\theta l_k}{\sqrt{P_{k|k}+\sigma^2 l_k}}\right) + \exp\left(\frac{2\theta(\omega-\hat{x}_{k|k})}{\sigma^2} + \frac{2\theta^2 P_{k|k}}{\sigma^4}\right) \times \Phi\left(\frac{-\omega+\hat{x}_{k|k}-\theta l_k - 2\theta\frac{P_{k|k}}{\sigma^2}}{\sqrt{P_{k|k}+\sigma^2 l_k}}\right), \tag{6.40}$$

$$E(L_k|\theta,\boldsymbol{Y}_{1:k}) = \frac{\omega-\hat{x}_{k|k}}{\theta}, \tag{6.41}$$

$$var(L_k|\theta,\boldsymbol{Y}_{1:k}) = \frac{(\omega-\hat{x}_{k|k})\sigma^2+\theta P_{k|k}}{\theta^3}. \tag{6.42}$$

Proof From (30) and (31), we have

$$f_{L_k|\theta,\boldsymbol{Y}_{1:k}}(l_k|\theta,\boldsymbol{Y}_{1:k}) = E_{x_k|\theta,\boldsymbol{Y}_{1:k}}\left[f_{L_k|\theta,x_k,\boldsymbol{Y}_{1:k}}(l_k|\theta,x_k,\boldsymbol{Y}_{1:k})\right], \tag{6.43}$$

and

$$F_{L_k|\theta,\boldsymbol{Y}_{1:k}}(l_k|\theta,\boldsymbol{Y}_{1:k}) = E_{x_k|\theta,\boldsymbol{Y}_{1:k}}\left[F_{L_k|\theta,x_k,\boldsymbol{Y}_{1:k}}(l_k|\theta,x_k,\boldsymbol{Y}_{1:k})\right], \tag{6.44}$$

where $p(x_k|\theta,\boldsymbol{Y}_{1:k})$ is the conditional PDF of $x_k|\theta,\boldsymbol{Y}_{1:k}$.

Due to the linear nature of the state-space model and Kalman filter, it is known that $x_k|\theta,\boldsymbol{Y}_{1:k} \sim N(\hat{x}_{k|k}, P_{k|k})$. Furthermore, due to Lemma 6.5 and Theorem 6.2, we have

$$f_{L_k|\theta,\boldsymbol{Y}_{1:k}}(l_k|\theta,\boldsymbol{Y}_{1:k}) = E_{x_k|\theta,\boldsymbol{Y}_{1:k}}\left[\frac{\omega-x_k}{\sqrt{2\pi l_k^3\sigma^2}}\exp\left(-\frac{(\omega-x_k-\theta l_k)^2}{2\sigma^2 l_k}\right)\right]. \tag{6.45}$$

Using Lemma 6.2 by setting $w_1=\omega/\sqrt{2\pi l_k^3\sigma^2}$, $w_2=\omega-\theta l_k$, $A=1/\sqrt{2\pi l_k^3\sigma^2}$, $B=1$, $C=\sigma^2 l_k$, and taking the expectation with respect to x_k, we have

$$f_{L_k|\theta,\boldsymbol{Y}_{1:k}}(l_k|\theta,\boldsymbol{Y}_{1:k}) = \frac{(\omega-\hat{x}_{k|k})\sigma^2 + P_{k|k}\theta}{\sqrt{2\pi(P_{k|k}+\sigma^2 l_k)^3}}\exp\left(-\frac{(\omega-\hat{x}_{k|k}-\theta l_k)^2}{2\left(P_{k|k}+\sigma^2 l_k\right)}\right). \tag{6.46}$$

Similarly, for $F_{L_k|\theta,\boldsymbol{Y}_{1:k}}(l_k|\theta,\boldsymbol{Y}_{1:k})$, we have

$$F_{L_k|\theta,\boldsymbol{Y}_{1:k}}(l_k|\theta,\boldsymbol{Y}_{1:k}) = 1 - E_{x_k|\theta,\boldsymbol{Y}_{1:k}}\left[\Phi\left(\frac{\omega - x_k - \theta l_k}{\sigma\sqrt{l_k}}\right)\right] + \exp\left(\frac{2\theta\omega}{\sigma^2}\right)E_{x_k|\theta,\boldsymbol{Y}_{1:k}}\left[\exp\left(-\frac{2\theta x_k}{\sigma^2}\right)\Phi\left(\frac{-\omega + x_k - \theta l_k}{\sigma\sqrt{l_k}}\right)\right]. \tag{6.47}$$

In (81), for the first expectation term, using Lemma 6.3 by setting $A = 0$, $B = 0$, $C = (\omega - \theta l_k)/(\sigma\sqrt{l_k})$, $D = -1/(\sigma\sqrt{l_k})$, and then taking the expectation with respect to x_k, we have

$$E_{x_k|\theta,\boldsymbol{Y}_{1:k}}\left[\Phi\left(\frac{\omega - x_k - \theta l_k}{\sigma\sqrt{l_k}}\right)\right] = \Phi\left(\frac{\omega - \hat{x}_{k|k} - \theta l_k}{\sqrt{P_{k|k} + \sigma^2 l_k}}\right). \tag{6.48}$$

For the second expectation term in (81), using Lemma 6.3 by setting $A = -2\theta/\sigma^2$, $B = 0, C = -(\omega + \theta l_k)/(\sigma\sqrt{l_k})$, $D = 1/(\sigma\sqrt{l_k})$, and then taking the expectation with respect to θ, we have

$$\begin{aligned} E_{x_k|\theta,\boldsymbol{Y}_{1:k}}&\left[\exp\left(-\frac{2\theta x_k}{\sigma^2}\right)\Phi\left(\frac{-\omega + x_k - \theta l_k}{\sigma\sqrt{l_k}}\right)\right] \\ &= \exp\left(-\frac{2\theta\hat{x}_{k|k}}{\sigma^2} + \frac{2\theta^2 P_{k|k}}{\sigma^4}\right)\Phi\left(\frac{-\omega + \hat{x}_{k|k} - \theta l_k - \frac{2\theta P_{k|k}}{\sigma^2}}{\sqrt{P_{k|k} + \sigma^2 l_k}}\right). \end{aligned} \tag{6.49}$$

As a result, we conclude

$$\begin{aligned} F_{L_k|\theta,\boldsymbol{Y}_{1:k}}(l_k|\theta,\boldsymbol{Y}_{1:k}) =& 1 - \Phi\left(\frac{\omega - \hat{x}_{k|k} - \theta l_k}{\sqrt{P_{k|k} + \sigma^2 l_k}}\right) \\ &+ \exp\left(\frac{2\theta(\omega - \hat{x}_{k|k})}{\sigma^2} + \frac{2\theta^2 P_{k|k}}{\sigma^4}\right)\Phi\left(\frac{-\omega + \hat{x}_{k|k} - \theta l_k - \frac{2\theta P_{k|k}}{\sigma^2}}{\sqrt{P_{k|k} + \sigma^2 l_k}}\right). \end{aligned} \tag{6.50}$$

As for the mean RUL $E(L_k|\theta,\boldsymbol{Y}_{1:k})$ and variance $\mathrm{var}(L_k|\theta,\boldsymbol{Y}_{1:k})$, based on the property of the expectation operator and Theorem 6.2, we obtain

$$\begin{aligned} E(L_k|\theta,\boldsymbol{Y}_{1:k}) &= E_{x_k|\theta,\boldsymbol{Y}_{1:k}}[E(L_k|\theta,x_k,\boldsymbol{Y}_{1:k})] = E_{x_k|\theta,\boldsymbol{Y}_{1:k}}[E(L_k|\theta,x_k)] \\ &= E_{x_k|\theta,\boldsymbol{Y}_{1:k}}\left[\frac{\omega - x_k}{\theta}\right] = \frac{\omega - \hat{x}_{k|k}}{\theta}, \end{aligned} \tag{6.51}$$

and

$$\begin{aligned}\mathrm{var}(L_k|\theta, \boldsymbol{Y}_{1:k}) &= E_{x_k|\theta,\boldsymbol{Y}_{1:k}}[\mathrm{var}(L_k|\theta, x_k, \boldsymbol{Y}_{1:k})] + \mathrm{var}_{x_k|\theta,\boldsymbol{Y}_{1:k}}[E(L_k|\theta, x_k, \boldsymbol{Y}_{1:k})] \\ &= E_{x_k|\theta,\boldsymbol{Y}_{1:k}}[\mathrm{var}(L_k|\theta, x_k)] + \mathrm{var}_{x_k|\theta,\boldsymbol{Y}_{1:k}}[E(L_k|\theta, x_k)] \\ &= E_{x_k|\theta,\boldsymbol{Y}_{1:k}}\left[\frac{(\omega - x_k)\sigma^2}{\theta^3}\right] + \mathrm{var}_{x_k|\theta,\boldsymbol{Y}_{1:k}}\left[\frac{\omega - x_k}{\theta}\right] \\ &= \frac{(\omega - \hat{x}_{k|k})\sigma^2}{\theta^3} + \frac{P_{k|k}}{\theta^2} = \frac{(\omega - \hat{x}_{k|k})\sigma^2 + \theta P_{k|k}}{\theta^3}. \end{aligned} \tag{6.52}$$

This completes the proof.

In the above case, the estimated RUL by Theorem 6.4 accounts for the temporal variability, and the estimation uncertainty due to the measurement variability. Unlike the existing work in the similar framework as [38], which only derives an approximate PDF of the estimated RUL, we give the analytical PDF, CDF, mean, and variance of the estimated RUL in (35) through (38). These analytical results will facilitate their computations and applications in practice.

However, the parameter θ is assumed to be given, and the random effect from θ is not considered in this subsection. In addition, the parameter θ is not updated by the newly obtained measurements $\boldsymbol{Y}_{1:k}$. In the next part, we will investigate how to incorporate the random effect and the updating mechanism for θ into the estimated RUL distribution.

6.3.3 RUL Estimation with Three-Source Variability

To incorporate three-source variability in the degradation process jointly, we consider an updating procedure for the parameter θ by $\theta_k = \theta_{k-1}$ over time, where $\theta_0 = \theta \sim N(\mu_\theta, \sigma_\theta^2)$ is the initial distribution. The posterior distribution of θ is obtained by utilizing the measurements up to t_k. To do so, based on (6.33), the degradation equation taking three-source variability into account can be reconstructed within a state-space modeling framework as

$$\begin{cases} x_k = x_{k-1} + \theta_{k-1}(t_k - t_{k-1}) + v_k \\ \theta_k = \theta_{k-1} \\ y_k = x_k + \varepsilon_k \end{cases}. \tag{6.53}$$

where $\{v_k\}_{k\geq 1}$, and $\{\varepsilon_k\}_{k\geq 1}$ are i.i.d. noise sequences following $v_k \sim N\left(0, \sigma^2(t_k - t_{k-1})\right)$, and $\varepsilon_k \sim N(0, \gamma^2)$, respectively.

As such, the underlying degradation state and random parameter θ are considered to be hidden states, and can only be estimated from the uncertain measurements to date, $\boldsymbol{Y}_{1:k}$. On the basis of the established state-space model (6.53), the Kalman filter is used to estimate the underlying degradation state and random parameter. To apply a Kalman filter in this case, we further reorganize the state-space model (6.53) as

$$\begin{cases} z_k = A_k z_{k-1} + \eta_k \\ y_k = C z_k + \varepsilon_k \end{cases}, \tag{6.54}$$

where $z_k \in R^{2\times 1}$, $\eta_k \in R^{2\times 1}$, $A_k \in R^{2\times 2}$, $C \in R^{1\times 2}$, and $\eta_k \sim N(0,\ Q_k)$ with

$$z_k = \begin{bmatrix} x_k \\ \theta_k \end{bmatrix}, \eta_k = \begin{bmatrix} v_k \\ 0 \end{bmatrix}, A_k = \begin{bmatrix} 1 & t_k - t_{k-1} \\ 0 & 1 \end{bmatrix}, C = \begin{bmatrix} 1 \\ 0 \end{bmatrix}^T, Q_k = \begin{bmatrix} \sigma^2(t_k - t_{k-1}) & 0 \\ 0 & 0 \end{bmatrix},$$

respectively.

Similarly, we here define the expectation and variance of z_k that are conditional on the measurement history available until the current point as

$$\hat{z}_{k|k} = \begin{bmatrix} \hat{x}_{k|k} \\ \hat{\theta}_{k|k} \end{bmatrix} = E(z_k | Y_{1:k}), \tag{6.55}$$

$$P_{k|k} = \begin{bmatrix} \kappa^2_{x,k} & \kappa^2_{x\theta,k} \\ \kappa^2_{x\theta,k} & \kappa^2_{\theta,k} \end{bmatrix} = \text{cov}(z_k | Y_{1:k}), \tag{6.56}$$

where $\hat{x}_{k|k} = E(x_k | Y_{1:k})$, $\hat{\theta}_{k|k} = E(\theta_k | Y_{1:k})$, $\kappa^2_{x,k} = \text{var}(x_k | Y_{1:k})$, $\kappa^2_{\theta,k} = \text{var}(\theta_k | Y_{1:k})$, $\kappa^2_{x\theta,k} = \text{cov}(x_k \theta_k | Y_{1:k})$.

Accordingly, the one-step-ahead predicted expectation and variance are defined respectively as

$$\hat{z}_{k|k-1} = \begin{bmatrix} \hat{x}_{k|k-1} \\ \hat{\theta}_{k|k-1} \end{bmatrix} = E(z_k | Y_{1:k-1}), \tag{6.57}$$

$$P_{k|k-1} = \begin{bmatrix} \kappa^2_{x,k|k-1} & \kappa^2_{x\theta,k|k-1} \\ \kappa^2_{x\theta,k|k-1} & \kappa^2_{\theta,k|k-1} \end{bmatrix} = \text{cov}(z_k | Y_{1:k-1}). \tag{6.58}$$

Based on the above settings and definitions, the Kalman filtering algorithm can be used to jointly estimate the underlying degradation state and random parameter, i.e., z_k. The estimating process is summarized as follows.

State estimation:

$$\begin{aligned} &\hat{z}_{k|k-1} = A_k \hat{z}_{k-1|k-1}, \\ &\hat{z}_{k|k} = \hat{z}_{k|k-1} + K(k)(y_k - C\hat{z}_{k|k-1}), \\ &K(k) = P_{k|k-1} C^T [C P_{k|k-1} C^T + \gamma^2]^{-1}, \\ &P_{k|k-1} = A_k P_{k-1|k-1} A_k^T + Q_k, \end{aligned}$$

Variance update:

$$P_{k|k} = P_{k|k-1} - K(k) C P_{k|k-1},$$

where the initial values of the states in the filtering are specified as

$$\hat{z}_{0|0} = \begin{bmatrix} 0 \\ \mu_\theta \end{bmatrix}, \text{ and } P_{0|0} = \begin{bmatrix} 0 & 0 \\ 0 & \sigma_\theta^2 \end{bmatrix}.$$

Based on (6.54), and the Gaussian nature of the Kalman filter, the PDF of z_k conditional on $Y_{1:k}$ is bivariate Gaussian with $z_k \sim N(\hat{z}_{k|k}, P_{k|k})$. Namely, the posterior distributions of the underlying degradation state x_k and random effect parameter θ are correlated at time t_k, which contrast with the cases for the deterministic parameters in [35, 38]. According to the properties of the bivariate normal distribution, we have

$$\theta_k | Y_{1:k} \sim N(\hat{\theta}_{k|k}, \kappa_{\theta,k}^2), \tag{6.59}$$

$$x_k | Y_{1:k} \sim N(\hat{x}_{k|k}, \kappa_{x,k}^2), \tag{6.60}$$

$$x_k | \theta_k, Y_{1:k} \sim N\left(\mu_{x_k|\theta,k}, \sigma_{x_k|\theta,k}^2\right), \tag{6.61}$$

with

$$\mu_{x_k|\theta,k} = \hat{x}_{k|k} + \rho_k \frac{\kappa_{x,k}}{\kappa_{\theta,k}} (\theta_k - \hat{\theta}_{k|k}) \tag{6.62}$$

$$\sigma_{x_k|\theta,k}^2 = \kappa_{x,k}^2 (1 - \rho_k^2). \tag{6.63}$$

where $\rho_k = \kappa_{x\theta,k}^2 / \kappa_{x,k}\kappa_{\theta,k}$.

Now let us focus on calculating $f_{L_k|Y_{1:k}}(l_k | Y_{1:k})$ at t_k once the three-source variability is taken into account. By the law of total probability, we have

$$\begin{aligned} f_{L_k|Y_{1:k}}(l_k | Y_{1:k}) &= \int_{-\infty}^{+\infty} f_{L_k|z_k,Y_{1:k}}(l_k | z_k, Y_{1:k}) p(z_k | Y_{1:k}) \mathrm{d}z_k \\ &= \int_{-\infty}^{+\infty} [p(\theta_k | Y_{1:k}) \int_{-\infty}^{+\infty} f_{L_k|\theta_k,x_k,Y_{1:k}}(l_k | \theta_k, x_k, Y_{1:k}) p(x_k | \theta_k, Y_{1:k}) \mathrm{d}x_k] \mathrm{d}\theta_k \\ &= E_{\theta_k|Y_{1:k}} \left[E_{x_k|\theta_k,Y_{1:k}} [f_{L_k|\theta_k,x_k,Y_{1:k}}(l_k | \theta_k, x_k, Y_{1:k})] \right] \end{aligned} \tag{6.64}$$

Similar manipulations also hold for $F_{L_k|Y_{1:k}}(l_k | Y_{1:k})$, and thus we directly have

$$\begin{aligned} F_{L_k|Y_{1:k}}(l_k | Y_{1:k}) &= \int_{-\infty}^{+\infty} F_{L_k|z_k,Y_{1:k}}(l_k | z_k, Y_{1:k}) p(z_k | Y_{1:k}) \mathrm{d}z_k \\ &= E_{\theta_k|Y_{1:k}} \left[E_{x_k|\theta_k,Y_{1:k}} [F_{L_k|\theta_k,x_k,Y_{1:k}}(l_k | \theta_k, x_k, Y_{1:k})] \right]. \end{aligned} \tag{6.65}$$

Based on Lemmas 6.1–6.3, and Theorem 6.4, we have the following results for the RUL estimation based on the degradation model with three-source variability.

Theorem 6.5 *For the Wiener process as (6.1), and the measurement process (6.4), given the uncertain measurements $Y_{1:k}$ up to current time t_k, the following results for RUL L_k at time t_k hold.*

Table 6.1 Quantities in Theorem 6.5

$w_{1,k} = (\omega - \hat{x}_{k\mid k} + \rho_k \frac{\kappa_{x,k}}{\kappa_{\theta,k}} \hat{\theta}_{k\mid k})\sigma^2$, $w_{2,k} = \frac{w_{1,k}}{\sigma^2}$	$A_k = \rho_k \frac{\kappa_{x,k}}{\kappa_{\theta,k}} \sigma^2 - \sigma^2_{x_k\mid\theta,k}$, $A_{2,k} = 2\frac{w_{2,k}}{\sigma^2}$
$B_k = \rho_k \frac{\kappa_{x,k}}{\kappa_{\theta,k}} + l_k$, $B_{1,k} = \rho_k \frac{\kappa_{x,k}}{\kappa_{\theta,k}} - l_k$, $B_{2,k} = -2\frac{A_k}{\sigma^4}$	$C_k = \sigma^2_{x_k\mid\theta,k} + \sigma^2 l_k$, $C_{2,k} = -\frac{w_{2,k}}{\sqrt{C_k}}$
$D_k = \frac{\sqrt{2}}{\kappa_{\theta,k}} D(\frac{\hat{\theta}_{k\mid k}}{\sqrt{2\kappa^2_{\theta,k}}})$, $D_{2,k} = \frac{1}{\sqrt{C_k}}(B_{1,k} - 2\frac{\sigma^2_{x_k\mid\theta,k}}{\sigma^2})$	$H_k = \rho_k \frac{\kappa_{x,k}}{\kappa_{\theta,k}} (\hat{x}_{k\mid k} - \rho_k \frac{\kappa_{x,k}}{\kappa_{\theta,k}} \hat{\theta}_{k\mid k}) - \omega\rho_k \frac{\kappa_{x,k}}{\kappa_{\theta,k}}$
$J_k = \frac{w_{1,k}\hat{\theta}^2_{k\mid k}}{\kappa^4_{\theta,k}} - \frac{w_{1,k}}{2\kappa^2_{\theta,k}} - \rho_k \frac{\kappa_{x,k}}{\kappa^3_{\theta,k}} \sigma^2$	$Q_k = \rho_k \frac{\kappa_{x,k}}{\kappa^3_{\theta,k}} \sigma^2 - \frac{w_{1,k}\hat{\theta}_{k\mid k}}{2\kappa^4_{\theta,k}} + \rho^2_k \frac{\kappa^2_{x,k}}{\kappa^2_{\theta,k}} - \frac{G_k}{\kappa^2_{\theta,k}}$
$G_k = \omega^2 - 2\omega(\hat{x}_{k\mid k} - \rho_k \frac{\kappa_{x,k}}{\kappa_{\theta,k}} \hat{\theta}_{k\mid k}) + \sigma^2_{x_k\mid\theta,k} + (\hat{x}_{k\mid k} - \rho_k \frac{\kappa_{x,k}}{\kappa_{\theta,k}} \hat{\theta}_{k\mid k})^2$	$\sigma^2_{x_k\mid\theta,k} = \kappa^2_{x,k}(1 - \rho^2_k)$

$$f_{L_k\mid \boldsymbol{Y}_{1:k}}(l_k \mid \boldsymbol{Y}_{1:k}) = \frac{w_{1,k}(B^2_k\kappa^2_{\theta,k} + C_k) - A_k B_k w_{2,k}\kappa^2_{\theta,k} - A_k C_k \hat{\theta}_{k\mid k}}{C_k\sqrt{2\pi(B^2_k\kappa^2_{\theta,k} + C_k)^3}} \exp\left(-\frac{(\omega - \hat{x}_{k\mid k} - \hat{\theta}_{k\mid k} l_k)^2}{2\left(B^2_k\kappa^2_{\theta,k} + C_k\right)}\right), \tag{6.66}$$

$$F_{L_k\mid \boldsymbol{Y}_{1:k}}(l_k \mid \boldsymbol{Y}_{1:k}) = 1 - \Phi\left(\frac{\omega - \hat{x}_{k\mid k} - \hat{\theta}_{k\mid k} l_k}{\sqrt{B^2_k\kappa^2_{\theta,k} + C_k}}\right) + \frac{1}{\sqrt{1 - 2B_{2,k}\kappa^2_{\theta,k}}} \exp\left(\frac{2A_{2,k}\hat{\theta}_{k\mid k} + A^2_{2,k}\kappa^2_{\theta,k} + 2B_{2,k}\hat{\theta}^2_{k\mid k}}{2(1 - 2B_{2,k}\kappa^2_{\theta,k})}\right)$$
$$\times \Phi\left(\frac{C_{2,k} + D_{2,k}\hat{\theta}_{k\mid k} + A_{2,k} D_{2,k}\kappa^2_{\theta,k} - 2B_{2,k} C_{2,k}\kappa^2_{\theta,k}}{\sqrt{(1 - 2B_{2,k}\kappa^2_{\theta,k})^2 + D^2_{2,k}\kappa^2_{\theta,k}(1 - 2B_{2,k}\kappa^2_{\theta,k})}}\right), \tag{6.67}$$

$$E(L_k \mid \boldsymbol{Y}_{1:k}) = w_{2,k} D_k - \rho_k \frac{\kappa_{x,k}}{\kappa_{\theta,k}}, \tag{6.68}$$

$$var(L_k \mid \boldsymbol{Y}_{1:k}) = Q_k + (J_k + 2H_k + \frac{G_k}{\kappa^2_{\theta,k}} \hat{\theta}_{k\mid k}) \cdot D_k - [E(L_k \mid \boldsymbol{Y}_{1:k})]^2, \tag{6.69}$$

where $w_{1,k}$, $w_{2,k}$, A_k, $A_{2,k}$, B_k, $B_{2,k}$, C_k, $C_{2,k}$, D_k, $D_{2,k}$, H_k, J_k, Q_k, *and* G_k *are specified in Table 6.1.*

Proof To derive the result for $f_{L_k\mid \boldsymbol{Y}_{1:k}}(l_k \mid \boldsymbol{Y}_{1:k})$, we first have the following result according to Theorem 6.2, and Lemma 6.5.

$$f_{L_k\mid\theta_k, x_k, \boldsymbol{Y}_{1:k}}(l_k \mid \theta_k, x_k, \boldsymbol{Y}_{1:k}) = f_{L_k\mid\theta_k, x_k}(l_k \mid \theta_k, x_k) = \frac{\omega - x_k}{\sqrt{2\pi l^3_k \sigma^2}} \exp\left(-\frac{(\omega - x_k - \theta_k l_k)^2}{2\sigma^2 l_k}\right). \tag{6.70}$$

Furthermore, because $x_k \mid \theta_k, \boldsymbol{Y}_{1:k} \sim N\left(\mu_{x_k\mid\theta,k}, \sigma^2_{x_k\mid\theta,k}\right)$, we get the following from Theorem 6.4.

$$E_{x_k\mid\theta_k, \boldsymbol{Y}_{1:k}}[f_{L_k\mid\theta_k, x_k, \boldsymbol{Y}_{1:k}}(l_k \mid \theta_k, x_k, \boldsymbol{Y}_{1:k})] = \frac{(\omega - \mu_{x_k\mid\theta,k})\sigma^2 + \sigma^2_{x_k\mid\theta,k}\theta_k}{\sqrt{2\pi(\sigma^2_{x_k\mid\theta,k} + \sigma^2 l_k)^3}} \exp\left(-\frac{(\omega - \mu_{x_k\mid\theta,k} - \theta_k l_k)^2}{2\left(\sigma^2_{x_k\mid\theta,k} + \sigma^2 l_k\right)}\right), \tag{6.71}$$

where $\mu_{x_k|\theta,k}$, and $\sigma^2_{x_k|\theta,k}$ are specified in (48) and (49); and $\mu_{x_k|\theta,k}$ is a function of θ_k.

From (48) through (50), and $\theta_k|\,\boldsymbol{Y}_{1:k} \sim N(\hat{\theta}_{k|k}, \kappa^2_{\theta,k})$, we know

$$
\begin{aligned}
f_{L_k|\boldsymbol{Y}_{1:k}}(l_k|\,\boldsymbol{Y}_{1:k}) &= E_{\theta_k|\boldsymbol{Y}_{1:k}}\left[E_{x_k|\theta_k,\boldsymbol{Y}_{1:k}}[f_{L_k|\theta_k,x_k,\boldsymbol{Y}_{1:k}}(l_k|\,\theta_k, x_k, \boldsymbol{Y}_{1:k})]\right] \\
&= E_{\theta_k|\boldsymbol{Y}_{1:k}}\left[\frac{(\omega-\mu_{x_k|\theta,k})\sigma^2+\sigma^2_{x_k|\theta,k}\theta_k}{\sqrt{2\pi(\sigma^2_{x_k|\theta,k}+\sigma^2 l_k)^3}}\exp\left(-\frac{(\omega-\mu_{x_k|\theta,k}-\theta_k l_k)^2}{2\left(\sigma^2_{x_k|\theta,k}+\sigma^2 l_k\right)}\right)\right] \\
&= E_{\theta_k|\boldsymbol{Y}_{1:k}}\left[\frac{(\omega-\hat{x}_{k|k}-\rho_k\frac{\kappa_{x,k}}{\kappa_{\theta,k}}(\theta_k-\hat{\theta}_{k|k}))\sigma^2+\sigma^2_{x_k|\theta,k}\theta_k}{\sqrt{2\pi(\sigma^2_{x_k|\theta,k}+\sigma^2 l_k)^3}}\exp\left(-\frac{(\omega-\hat{x}_{k|k}-\rho_k\frac{\kappa_{x,k}}{\kappa_{\theta,k}}(\theta_k-\hat{\theta}_{k|k})-\theta_k l_k)^2}{2\left(\sigma^2_{x_k|\theta,k}+\sigma^2 l_k\right)}\right)\right] \\
&= \frac{1}{\sqrt{2\pi(\sigma^2_{x_k|\theta,k}+\sigma^2 l_k)^3}} \\
&\quad\times E_{\theta_k|\boldsymbol{Y}_{1:k}}\left[\left[(\omega-\hat{x}_{k|k}-\rho_k\frac{\kappa_{x,k}}{\kappa_{\theta,k}}(\theta_k-\hat{\theta}_{k|k}))\sigma^2+\sigma^2_{x_k|\theta,k}\theta_k\right]\exp\left(-\frac{(\omega-\hat{x}_{k|k}-\rho_k\frac{\kappa_{x,k}}{\kappa_{\theta,k}}(\theta_k-\hat{\theta}_{k|k})-\theta_k l_k)^2}{2\left(\sigma^2_{x_k|\theta,k}+\sigma^2 l_k\right)}\right)\right] \\
&= \frac{1}{\sqrt{2\pi(\sigma^2_{x_k|\theta,k}+\sigma^2 l_k)^3}}E_{\theta_k|\boldsymbol{Y}_{1:k}}\left[(w_{1,k}-A_k\theta_k)\exp\left(-\frac{(w_{2,k}-B_k\theta_k)^2}{2C_k}\right)\right] \\
&= \frac{w_{1,k}(B_k^2\kappa^2_{\theta,k}+C_k)-A_kB_kw_{2,k}\kappa^2_{\theta,k}-A_kC_k\hat{\theta}_{k|k}}{C_k\sqrt{2\pi(B_k^2\kappa^2_{\theta,k}+C_k)^3}}\exp\left(-\frac{(\omega-\hat{x}_{k|k}-\hat{\theta}_{k|k}l_k)^2}{2\left(B_k^2\kappa^2_{\theta,k}+C_k\right)}\right)
\end{aligned}
\tag{6.72}
$$

where the last equality is implied by Lemma 6.2, and $w_{1,k}$, $w_{2,k}$, A_k, B_k, C_k are specified as follows.

$$
w_{1,k} = (\omega-\hat{x}_{k|k}+\rho_k\frac{\kappa_{x,k}}{\kappa_{\theta,k}}\hat{\theta}_{k|k})\sigma^2,\ w_{2,k} = \frac{w_{1,k}}{\sigma^2},\quad A_k = \rho_k\frac{\kappa_{x,k}}{\kappa_{\theta,k}}\sigma^2-\sigma^2_{x_k|\theta,k},
$$
$$
B_k = \rho_k\frac{\kappa_{x,k}}{\kappa_{\theta,k}}+l_k,\quad C_k = \sigma^2_{x_k|\theta,k}+\sigma^2 l_k.
$$

This completes the proof part for $f_{L_k|\boldsymbol{Y}_{1:k}}(l_k|\,\boldsymbol{Y}_{1:k})$.

Similarly, for $F_{L_k|\boldsymbol{Y}_{1:k}}(l_k|\,\boldsymbol{Y}_{1:k})$ with $x_k|\,\theta_k, \boldsymbol{Y}_{1:k} \sim N\left(\mu_{x_k|\theta,k}, \sigma^2_{x_k|\theta,k}\right)$, the following result is obtained according to Theorem 6.4.

$$
\begin{aligned}
E_{x_k|\theta_k,\boldsymbol{Y}_{1:k}}[F_{L_k|\theta_k,x_k,\boldsymbol{Y}_{1:k}}(l_k|\,\theta_k, x_k, \boldsymbol{Y}_{1:k})] &= 1-\Phi\left(\frac{\omega-\mu_{x_k|\theta,k}-\theta_k l_k}{\sqrt{\sigma^2_{x_k|\theta,k}+\sigma^2 l_k}}\right) \\
&+\exp\left(\frac{2\theta_k(\omega-\mu_{x_k|\theta,k})}{\sigma^2}+\frac{2\theta_k^2\sigma^2_{x_k|\theta,k}}{\sigma^4}\right)\Phi\left(\frac{-\omega+\mu_{x_k|\theta,k}-\theta_k l_k-\frac{2\theta_k\sigma^2_{x_k|\theta,k}}{\sigma^2}}{\sqrt{\sigma^2_{x_k|\theta,k}+\sigma^2 l_k}}\right)
\end{aligned}
\tag{6.73}
$$

where $\mu_{x_k|\theta,k}$ and $\sigma^2_{x_k|\theta,k}$ are obtained in (48) and (49). and $\mu_{x_k|\theta,k}$ is a function of θ_k.

From (51) and $\theta_k | \boldsymbol{Y}_{1:k} \sim N(\hat{\theta}_{k|k}, \kappa_{\theta,k}^2)$, we further have

$$\begin{aligned} F_{L_k|\boldsymbol{Y}_{1:k}}(l_k|\boldsymbol{Y}_{1:k}) &= E_{\theta_k|\boldsymbol{Y}_{1:k}}\left[E_{x_k|\theta_k,\boldsymbol{Y}_{1:k}}[F_{L_k|\theta_k,x_k,\boldsymbol{Y}_{1:k}}(l_k|\theta_k,x_k,\boldsymbol{Y}_{1:k})]\right] \\ &= 1 - E_{\theta_k|\boldsymbol{Y}_{1:k}}\left[\Phi\left(\frac{\omega-\mu_{x_k|\theta,k}-\theta_k l_k}{\sqrt{\sigma^2_{x_k|\theta,k}+\sigma^2 l_k}}\right)\right] \\ &\quad + E_{\theta_k|\boldsymbol{Y}_{1:k}}\left[\exp\left(\frac{2\theta_k(\omega-\mu_{x_k|\theta,k})}{\sigma^2}+\frac{2\theta_k^2\sigma^2_{x_k|\theta,k}}{\sigma^4}\right)\Phi\right. \\ &\quad \left.\left(\frac{-\omega+\mu_{x_k|\theta,k}-\theta_k l_k-\frac{2\theta_k\sigma^2_{x_k|\theta,k}}{\sigma^2}}{\sqrt{\sigma^2_{x_k|\theta,k}+\sigma^2 l_k}}\right)\right]. \end{aligned} \tag{6.74}$$

For the first expectation part in (6.91), the following result is derived from Lemma 6.3.

$$\begin{aligned} &E_{\theta_k|\boldsymbol{Y}_{1:k}}\left[\Phi\left(\frac{\omega-\mu_{x_k|\theta,k}-\theta_k l_k}{\sqrt{\sigma^2_{x_k|\theta,k}+\sigma^2 l_k}}\right)\right] \\ &= E_{\theta_k|\boldsymbol{Y}_{1:k}}\left[\Phi\left(\frac{\omega-\hat{x}_{k|k}-\rho_k\frac{\kappa_{x,k}}{\kappa_{\theta,k}}(\theta_k-\hat{\theta}_{k|k})-\theta_k l_k}{\sqrt{\sigma^2_{x_k|\theta,k}+\sigma^2 l_k}}\right)\right] \\ &= E_{\theta_k|\boldsymbol{Y}_{1:k}}\left[\Phi(C_{1,k}+D_{1,k}\theta_k)\right] \\ &= \Phi\left(\frac{C_{1,k}+D_{1,k}\hat{\theta}_{k|k}}{\sqrt{1+D^2_{1,k}\kappa^2_{\theta,k}}}\right) = \Phi\left(\frac{\omega-\hat{x}_{k|k}-\hat{\theta}_{k|k}l_k}{\sqrt{B_k^2\kappa^2_{\theta,k}+C_k}}\right) \end{aligned} \tag{6.75}$$

where $C_{1,k} = w_{2,k}/\sqrt{C_k}$, $D_{1,k} = -B_k/\sqrt{C_k}$.

For the second expectation part in (91), we can obtain the following through some lengthy algebraic manipulations.

$$\begin{aligned} &E_{\theta_k|\boldsymbol{Y}_{1:k}}\left[\exp\left(\frac{2\theta_k(\omega-\mu_{x_k|\theta,k})}{\sigma^2}+\frac{2\theta_k^2\sigma^2_{x_k|\theta,k}}{\sigma^4}\right)\Phi\left(\frac{-\omega+\mu_{x_k|\theta,k}-\theta_k l_k-\frac{2\theta_k\sigma^2_{x_k|\theta,k}}{\sigma^2}}{\sqrt{\sigma^2_{x_k|\theta,k}+\sigma^2 l_k}}\right)\right] \\ &= E_{\theta_k|\boldsymbol{Y}_{1:k}}\left[\exp(A_{2,k}\theta_k+B_{2,k}\theta_k^2)\Phi(C_{2,k}+D_{2,k}\theta_k)\right] \\ &= \frac{1}{\sqrt{1-2B_{2,k}\kappa^2_{\theta,k}}}\exp\left(\frac{2A_{2,k}\hat{\theta}_{k|k}+A^2_{2,k}\kappa^2_{\theta,k}+2B_{2,k}\hat{\theta}^2_{k|k}}{2(1-2B_{2,k}\kappa^2_{\theta,k})}\right)\Phi \\ &\quad\times\left(\frac{C_{2,k}+D_{2,k}\hat{\theta}_{k|k}+A_{2,k}D_{2,k}\kappa^2_{\theta,k}-2B_{2,k}C_{2,k}\kappa^2_{\theta,k}}{\sqrt{(1-2B_{2,k}\kappa^2_{\theta,k})^2+D^2_{2,k}\kappa^2_{\theta,k}(1-2B_{2,k}\kappa^2_{\theta,k})}}\right) \end{aligned} \tag{6.76}$$

with

$$A_{2,k} = 2\frac{w_{2,k}}{\sigma^2},\ B_{1,k} = \rho_k \frac{\kappa_{x,k}}{\kappa_{\theta,k}} - l_k,\ B_{2,k} = -2\frac{A_k}{\sigma^4},$$

$$C_{2,k} = -\frac{w_{2,k}}{\sqrt{C_k}},\ D_{2,k} = \frac{1}{\sqrt{C_k}}\left(B_{1,k} - \frac{2\sigma^2_{x_k|\theta,k}}{\sigma^2}\right),$$

where $w_{1,k}$, $w_{2,k}$, A_k, B_k, C_k is obtained above, and the condition $1 - 2B_{2,k}\kappa^2_{\theta,k} > 0$ is ensured by the existence of $F_{L_k|\boldsymbol{Y}_{1:k}}(l_k|\,\boldsymbol{Y}_{1:k}) = E_{\theta_k,x_k|\boldsymbol{Y}_{1:k}}[F_{L_k|\theta_k,x_k,\boldsymbol{Y}_{1:k}}(l_k|\,\theta_k, x_k, \boldsymbol{Y}_{1:k})]$ in this case, and can be formally proved by the mathematics induction principle.

This completes the proof part for $F_{L_k|\boldsymbol{Y}_{1:k}}(l_k|\,\boldsymbol{Y}_{1:k})$. Now we turn the attention to the mean, and variance of the RUL in this case, i.e. $E(L_k|\,\boldsymbol{Y}_{1:k})$, and $\mathrm{var}(L_k|\,\boldsymbol{Y}_{1:k})$, respectively.

Based on the property of the expectation operator, we can compute $E(L_k|\,\boldsymbol{Y}_{1:k})$, and $\mathrm{var}(L_k|\,\boldsymbol{Y}_{1:k})$ as

$$\begin{aligned} E(L_k|\,\boldsymbol{Y}_{1:k}) &= E_{\theta_k,x_k|\boldsymbol{Y}_{1:k}}[E(L_k|\,\theta_k, x_k, \boldsymbol{Y}_{1:k})] \\ &= E_{\theta_k|\boldsymbol{Y}_{1:k}}\left[E_{x_k|\theta_k,\boldsymbol{Y}_{1:k}}[E(L_k|\,\theta_k, x_k, \boldsymbol{Y}_{1:k})]\right] \\ &= E_{\theta_k|\boldsymbol{Y}_{1:k}}[E(L_k|\,\theta_k, \boldsymbol{Y}_{1:k})], \end{aligned} \tag{6.77}$$

$$\begin{aligned} \mathrm{var}(L_k|\,\boldsymbol{Y}_{1:k}) &= E_{\theta_k,x_k|\boldsymbol{Y}_{1:k}}[\mathrm{var}(L_k|\,\theta_k, x_k, \boldsymbol{Y}_{1:k})] + \mathrm{var}_{\theta_k,x_k|\boldsymbol{Y}_{1:k}}[E(L_k|\,\theta_k, x_k, \boldsymbol{Y}_{1:k})] \\ &= E_{\theta_k|\boldsymbol{Y}_{1:k}}\left[E_{x_k|\theta_k,\boldsymbol{Y}_{1:k}}[\mathrm{var}(L_k|\,\theta_k, x_k, \boldsymbol{Y}_{1:k})]\right] \\ &\quad + \mathrm{var}_{\theta_k,x_k|\boldsymbol{Y}_{1:k}}[E(L_k|\,\theta_k, x_k, \boldsymbol{Y}_{1:k})] \end{aligned} \tag{6.78}$$

At the same time, we have the following results for the mean and variance of the RUL conditional on θ_k, x_k, and $\boldsymbol{Y}_{1:k}$.

$$E(L_k|\,\theta_k, x_k, \boldsymbol{Y}_{1:k}) = \frac{\omega - x_k}{\theta_k}, \tag{6.79}$$

$$\mathrm{var}(L_k|\,\theta_k, x_k, \boldsymbol{Y}_{1:k}) = \frac{(\omega - x_k)\sigma^2}{\theta_k^3}. \tag{6.80}$$

It is therefore direct to get the result for the mean of the RUL estimation by the above derivation, and the fact in (73) as

$$\begin{aligned} E(L_k|\,\boldsymbol{Y}_{1:k}) &= E_{\theta_k|\boldsymbol{Y}_{1:k}}\left[E_{x_k|\theta_k,\boldsymbol{Y}_{1:k}}\left[\frac{\omega - x_k}{\theta_k}\right]\right] = E_{\theta_k|\boldsymbol{Y}_{1:k}}\left[\frac{\omega - \mu_{x_k|\theta,k}}{\theta_k}\right] \\ &= E_{\theta_k|\boldsymbol{Y}_{1:k}}\left[\frac{\omega - \hat{x}_{k|k} - \rho_k \frac{\kappa_{x,k}}{\kappa_{\theta,k}}(\theta_k - \hat{\theta}_{k|k})}{\theta_k}\right] \\ &= \left[\omega - \hat{x}_{k|k} + \rho_k \frac{\kappa_{x,k}}{\kappa_{\theta,k}}\hat{\theta}_{k|k})\right] E_{\theta_k|Y_{1:k}}\left[\frac{1}{\theta_k}\right] - \rho_k \frac{\kappa_{x,k}}{\kappa_{\theta,k}} \end{aligned}$$

$$= \left[\omega - \hat{x}_{k|k} + \rho_k \frac{\kappa_{x,k}}{\kappa_{\theta,k}} \hat{\theta}_{k|k})\right] \frac{\sqrt{2}}{\kappa_{\theta,k}} D\left(\frac{\hat{\theta}_{k|k}}{\sqrt{2}\kappa_{\theta,k}}\right) - \rho_k \frac{\kappa_{x,k}}{\kappa_{\theta,k}}$$

$$= w_{2,k} D_k - \rho_k \frac{\kappa_{x,k}}{\kappa_{\theta,k}} \tag{6.81}$$

where $D_k = \sqrt{2} D(\hat{\theta}_{k|k}/(\sqrt{2}\kappa_{\theta,k}))/\kappa_{\theta,k}$ and $D(\hat{\theta}_{k|k}/(\sqrt{2}\kappa_{\theta,k}))$ is the Dawson integral for $\hat{\theta}_{k|k}/(\sqrt{2}\kappa_{\theta,k})$.

As for the formulation of $\text{var}(L_k | \boldsymbol{Y}_{1:k})$, with the help of the facts in (73) and (74), we have the result for the first term of the last part in (95).

$$E_{\theta_k|\boldsymbol{Y}_{1:k}} \left[E_{x_k|\theta_k,\boldsymbol{Y}_{1:k}} \left[\text{var}(L_k | \theta_k, x_k, \boldsymbol{Y}_{1:k})\right]\right] = E_{\theta_k|\boldsymbol{Y}_{1:k}} \left[E_{x_k|\theta_k,\boldsymbol{Y}_{1:k}} \left[\frac{(\omega - x_k)\sigma^2}{\theta_k^3}\right]\right]$$

$$= E_{\theta_k|\boldsymbol{Y}_{1:k}} \left[\frac{(\omega - \mu_{x_k|\theta,k})\sigma^2}{\theta_k^3}\right]$$

$$= E_{\theta_k|\boldsymbol{Y}_{1:k}} \left[\frac{(\omega - \hat{x}_{k|k} - \rho_k \frac{\kappa_{x,k}}{\kappa_{\theta,k}} (\theta_k - \hat{\theta}_{k|k}))\sigma^2}{\theta_k^3}\right]$$

$$= E_{\theta_k|\boldsymbol{Y}_{1:k}} \left[\frac{(\omega - \hat{x}_{k|k} + \rho_k \frac{\kappa_{x,k}}{\kappa_{\theta,k}} \hat{\theta}_{k|k})\sigma^2}{\theta_k^3} - \frac{\rho_k \frac{\kappa_{x,k}}{\kappa_{\theta,k}} \sigma^2}{\theta_k^2}\right]$$

$$= w_{1,k} E_{\theta_k|\boldsymbol{Y}_{1:k}} \left[\frac{1}{\theta_k^3}\right] - \rho_k \sigma^2 \frac{\kappa_{x,k}}{\kappa_{\theta,k}} E_{\theta_k|\boldsymbol{Y}_{1:k}} \left[\frac{1}{\theta_k^2}\right]$$

$$= \frac{w_{1,k} \hat{\theta}_{k|k}^2}{\kappa_{\theta,k}^4} D_k - \frac{w_{1,k}}{2\kappa_{\theta,k}^2} D_k - \rho_k \sigma^2 \frac{\kappa_{x,k}}{\kappa_{\theta,k}^3} D_k + \rho_k \frac{\kappa_{x,k}}{\kappa_{\theta,k}^3} \sigma^2 - \frac{w_{1,k} \hat{\theta}_{k|k}}{2\kappa_{\theta,k}^4}. \tag{6.82}$$

At the same time, we have the result for the second term of the last part in (95) as

$$\text{var}_{\theta_k,x_k|\boldsymbol{Y}_{1:k}} \left[E(L_k | \theta_k, x_k, \boldsymbol{Y}_{1:k})\right] = \text{var}_{\theta_k,x_k|\boldsymbol{Y}_{1:k}} \left(\frac{\omega - x_k}{\theta_k}\right)$$

$$= E_{\theta_k,x_k|\boldsymbol{Y}_{1:k}} \left[\left(\frac{\omega - x_k}{\theta_k}\right)^2\right] - \left[E_{\theta_k,x_k|\boldsymbol{Y}_{1:k}} \left(\frac{\omega - x_k}{\theta_k}\right)\right]^2$$

$$= E_{\theta_k,x_k|\boldsymbol{Y}_{1:k}} \left[\frac{\omega^2 - 2\omega x_k + x_k^2}{\theta_k^2}\right] - [E(L_k | \boldsymbol{Y}_{1:k})]^2$$

$$= E_{\theta_k|\boldsymbol{Y}_{1:k}} \left[\frac{\omega^2 - 2\omega\mu_{x_k|\theta,k} + \mu_{x_k|\theta,k}^2 + \sigma_{x_k|\theta,k}^2}{\theta_k^2}\right] - [E(L_k | \boldsymbol{Y}_{1:k})]^2$$

$$= E_{\theta_k|\boldsymbol{Y}_{1:k}} \left[\frac{\omega^2 - 2\omega(\hat{x}_{k|k} + \rho_k \frac{\kappa_{x,k}}{\kappa_{\theta,k}} (\theta_k - \hat{\theta}_{k|k})) + (\hat{x}_{k|k} + \rho_k \frac{\kappa_{x,k}}{\kappa_{\theta,k}} (\theta_k - \hat{\theta}_{k|k}))^2 + \sigma_{x_k|\theta,k}^2}{\theta_k^2}\right]$$

$$- \left[E(L_k | \boldsymbol{Y}_{1:k})\right]^2$$

$$= \left[\omega^2 - 2\omega(\hat{x}_{k|k} - \rho_k \frac{\kappa_{x,k}}{\kappa_{\theta,k}} \hat{\theta}_{k|k}) + \sigma_{x_k|\theta,k}^2 + (\hat{x}_{k|k} - \rho_k \frac{\kappa_{x,k}}{\kappa_{\theta,k}} \hat{\theta}_{k|k})^2\right] E_{\theta_k|\boldsymbol{Y}_{1:k}} \left[\frac{1}{\theta_k^2}\right]$$

$$
\begin{aligned}
&+\rho_k^2\frac{\kappa_{x,k}^2}{\kappa_{\theta,k}^2}-[E(L_k|\,\boldsymbol{Y}_{1:k})]^2\\
&+2\left[\rho_k\frac{\kappa_{x,k}}{\kappa_{\theta,k}}(\hat{x}_{k|k}-\rho_k\frac{\kappa_{x,k}}{\kappa_{\theta,k}}\hat{\theta}_{k|k})-\omega\rho_k\frac{\kappa_{x,k}}{\kappa_{\theta,k}}\right]E_{\theta_k|\boldsymbol{Y}_{1:k}}\left[\frac{1}{\theta_k}\right]\\
=&\ \rho_k^2\frac{\kappa_{x,k}^2}{\kappa_{\theta,k}^2}+2H_kD_k-\frac{G_k}{\kappa_{\theta,k}^2}(1-\hat{\theta}_{k|k}D_k)-[E(L_k|\,\boldsymbol{Y}_{1:k})]^2,
\end{aligned}
\tag{6.83}
$$

with

$$
H_k=\rho_k\frac{\kappa_{x,k}}{\kappa_{\theta,k}}\left(\hat{x}_{k|k}-\rho_k\frac{\kappa_{x,k}}{\kappa_{\theta,k}}\hat{\theta}_{k|k}\right)-\omega\rho_k\frac{\kappa_{x,k}}{\kappa_{\theta,k}},
\tag{6.84}
$$

$$
G_k=\omega^2-2\omega\left(\hat{x}_{k|k}-\rho_k\frac{\kappa_{x,k}}{\kappa_{\theta,k}}\hat{\theta}_{k|k}\right)+\sigma_{x_k|\theta,k}^2+\left(\hat{x}_{k|k}-\rho_k\frac{\kappa_{x,k}}{\kappa_{\theta,k}}\hat{\theta}_{k|k}\right)^2.
\tag{6.85}
$$

In sum of above results, we finally obtain the result for the variance of the RUL estimation in this case as

$$
\begin{aligned}
\mathrm{var}(L_k|\,\boldsymbol{Y}_{1:k})=&\ E_{\theta_k,x_k|\boldsymbol{Y}_{1:k}}[\mathrm{var}(L_k|\,\theta_k,x_k,\boldsymbol{Y}_{1:k})]+\mathrm{var}_{\theta_k,x_k|\boldsymbol{Y}_{1:k}}[E(L_k|\,\theta_k,x_k,\boldsymbol{Y}_{1:k})]\\
=&\ \frac{w_{1,k}\hat{\theta}_{k|k}^2}{\kappa_{\theta,k}^4}D_k-\frac{w_{1,k}}{2\kappa_{\theta,k}^2}D_k-\rho_k\sigma^2\frac{\kappa_{x,k}}{\kappa_{\theta,k}^3}D_k+\rho_k\frac{\kappa_{x,k}}{\kappa_{\theta,k}^3}\sigma^2-\frac{w_{1,k}\hat{\theta}_{k|k}}{2\kappa_{\theta,k}^4}\\
&+\rho_k^2\frac{\kappa_{x,k}^2}{\kappa_{\theta,k}^2}+2H_kD_k-\frac{G_k}{\kappa_{\theta,k}^2}(1-\hat{\theta}_{k|k}D_k)-[E(L_k|\,\boldsymbol{Y}_{1:k})\\
=&\ Q_k+(J_k+2H_k+\frac{G_k}{\kappa_{\theta,k}^2}\hat{\theta}_{k|k})D_k-[E(L_k|\,\boldsymbol{Y}_{1:k})]^2.
\end{aligned}
\tag{6.86}
$$

where

$$
J_k=\frac{w_{1,k}\hat{\theta}_{k|k}^2}{\kappa_{\theta,k}^4}-\frac{w_{1,k}}{2\kappa_{\theta,k}^2}-\rho_k\frac{\kappa_{x,k}}{\kappa_{\theta,k}^3}\sigma^2,
\tag{6.87}
$$

$$
Q_k=\rho_k\frac{\kappa_{x,k}}{\kappa_{\theta,k}^3}\sigma^2-\frac{w_{1,k}\hat{\theta}_{k|k}}{2\kappa_{\theta,k}^4}+\rho_k^2\frac{\kappa_{x,k}^2}{\kappa_{\theta,k}^2}-\frac{G_k}{\kappa_{\theta,k}^2}.
\tag{6.88}
$$

This completes the proof of Theorem 6.5.

From the above results, once a new degradation measurement is available, we can obtain the estimation of z_k conditional on $\boldsymbol{Y}_{1:k}$ by the state-space model (6.54) with $z_k\sim N(\hat{\boldsymbol{z}}_{k|k},\boldsymbol{P}_{k|k})$. Therefore, the RUL of this monitored system can be adaptively estimated using Theorem 6.5. Comparing the results in Theorems 6.3 and 6.4 with the results in Theorem 6.5, we find that the uncertainties in the underlying stochastic degradation process, estimating the degradation state x_k, and random effect part, can propagate into the PDF of the RUL, $f_{L_k|\boldsymbol{Y}_{1:k}}(l_k|\,\boldsymbol{Y}_{1:k})$. Also, the parameters in $f_{L_k|\boldsymbol{Y}_{1:k}}(l_k|\,\boldsymbol{Y}_{1:k})$ can be updated using the Kalman filter such as $\hat{\theta}_{k|k}$, and $\kappa_{\theta,k}^2$. Thus, the estimated RUL by Theorem 6.5 accounts for three-source variability at the same time.

For the model (6.54) to be used for real-time estimation, several unknown initial parameters including σ^2, γ^2, μ_θ, and σ_θ^2 needed to be estimated from the historical data. Before determining these parameters in the next section, we have the following remark regarding the linkage of the RUL estimation results in Theorem 6.5 with the previous results.

Remark 6.3 Corresponding to Remark 6.1, it is not difficult to verify from the derivation process (see the online supplementary material) of Theorem 6.5 that, by selecting the parameters $\sigma^2, \gamma^2, \sigma_\theta^2$, the results in Theorem 6.5 can reduce to the RUL estimation results which only consider one- or two-source variability, as listed in Remark 6.1. ∎

6.4 Parameter Estimation

We now proceed to the issues regarding the estimation of $\sigma^2, \gamma^2, \mu_\theta, \sigma_\theta^2$ to initialize the degradation model and to implement the RUL estimation. Let $\boldsymbol{\Theta} = (\sigma^2, \gamma^2, \mu_\theta, \sigma_\theta^2)'$ be the unknown parameter vector, where $(\cdot)'$ denotes the vector transposition. To obtain the maximum likelihood estimation (MLE) of $\boldsymbol{\Theta}$ from the past degradation measurements, it is necessary to establish the identifiability of the degradation model under three-source variability. To do so, suppose there are N physically independently tested systems of the same kind, and the ith system is monitored at ordered times $t_1, \ldots, t_M$, with observed degradation measurements $\{Y_i(t_j) = y_{i,j}, i = 1, \ldots, N, j = 1, \ldots, M\}$ where M denotes the available number of the degradation measurements for each system. Due to the physical independence of the tested systems, we further suppose that the monitored degradation measurements from different systems are statistically independent. Therefore, from (6.2), the measurement at the jth time point t_j for the ith system is given by

$$Y_i(t_j) = \theta_i t_j + \sigma B(t_j) + \varepsilon_{ij}, \tag{6.89}$$

where θ_i are statistically independent and identically distributed following $N(\mu_\theta, \sigma_\theta^2)$, and ε_{ij} is the measurement error with $\varepsilon_{ij} \sim N(0, \gamma^2)$.

For convenience, we only use the degradation measurements of the ith system to investigate the identifiability of the degradation model. let $\boldsymbol{t} = (t_1, \ldots, t_M)'$, $\boldsymbol{y}_i = (y_{i,1}, \ldots, y_{i,M})'$. According to (6.89) and the statistically independent increments property of BM, we have that $\boldsymbol{y}_i$ is multivariate normally distributed with mean and variance as follows.

$$\boldsymbol{y}_i \sim N(\boldsymbol{\mu}, \boldsymbol{\Sigma}), \tag{6.90}$$

$$\boldsymbol{\mu}=\mu_\theta \boldsymbol{t}, \tag{6.91}$$

$$\boldsymbol{\Sigma} = \boldsymbol{\Omega} + \sigma_\theta^2 \boldsymbol{t}\boldsymbol{t}', \tag{6.92}$$

with

$$\boldsymbol{\Omega} = \sigma^2 \boldsymbol{Q} + \gamma^2 \boldsymbol{I}_M, \tag{6.93}$$

$$\boldsymbol{Q} = [\min\{t_i, t_j\}]_{1\le i,j\le M}, \tag{6.94}$$

where $\boldsymbol{I}_M$ is an identity matrix of order M.

To establish the identifiability of the degradation model, we first introduce the following lemma ([48], pp. 118).

Lemma 6.6 *Define a regression model* $\mathbf{y} \sim N(\boldsymbol{h}(\boldsymbol{\alpha}), \mathbf{V}(\boldsymbol{\alpha}, \boldsymbol{\beta}))$, *where* $\mathbf{y}$ *is the data vector,* $\boldsymbol{h}(\boldsymbol{\alpha})$ *is a linear or nonlinear vector function of the parameter vector* $\boldsymbol{\alpha}$, *and* $\boldsymbol{V}(\boldsymbol{\alpha}, \boldsymbol{\beta})$ *is the covariance matrix dependent on* $\boldsymbol{\alpha}$ *and another parameter vector* $\boldsymbol{\beta}$. *Then the regression model is identifiable iff* $\boldsymbol{h}(\boldsymbol{\alpha}_1) = \boldsymbol{h}(\boldsymbol{\alpha}_2)$, *and* $\boldsymbol{V}(\boldsymbol{\alpha}_1, \boldsymbol{\beta}_1) = \boldsymbol{V}(\boldsymbol{\alpha}_2, \boldsymbol{\beta}_2)$ *imply that* $\boldsymbol{\alpha}_1 = \boldsymbol{\alpha}_2$ *and* $\boldsymbol{\beta}_1 = \boldsymbol{\beta}_2$.

Based on Lemma 6.6, we can establish the identifiability theorem for the presented model under three-source variability in the following.

Theorem 6.6 *For the presented model with degradation measurements* $\boldsymbol{y}_i$, *the degradation model is identifiable iff the number of the degradation measurements is not less than 3, i.e.,* $M \ge 3$.

Proof It is known from (57) through (61) that $\boldsymbol{y}_i \sim N(\mu_\theta \boldsymbol{t}, \sigma_\theta^2 \boldsymbol{t}\boldsymbol{t}' + \sigma^2 \boldsymbol{Q} + \gamma^2 \boldsymbol{I}_M)$. Thus the sufficient and necessary condition for the identifiability of the degradation model under $\boldsymbol{y}_i$ is

$$\begin{cases} \mu_{\theta_1} \boldsymbol{t} = \mu_{\theta_2} \boldsymbol{t} \\ \sigma_{\theta_1}^2 \boldsymbol{t}\boldsymbol{t}' + \sigma_1^2 \boldsymbol{Q} + \gamma_1^2 \boldsymbol{I}_M = \sigma_{\theta_2}^2 \boldsymbol{t}\boldsymbol{t}' + \sigma_2^2 \boldsymbol{Q} + \gamma_2^2 \boldsymbol{I}_M \end{cases} \Leftrightarrow \begin{cases} \mu_{\theta_1} = \mu_{\theta_2}, \sigma_{\theta_1}^2 = \sigma_{\theta_2}^2 \\ \sigma_1^2 = \sigma_2^2, \gamma_1^2 = \gamma_2^2 \end{cases} \tag{6.95}$$

It is straightforward that $\mu_{\theta_1} \boldsymbol{t} = \mu_{\theta_2} \boldsymbol{t}$ means $\mu_{\theta_1} = \mu_{\theta_2}$. In addition, for the equality of the covariance function, we have

$$(\sigma_{\theta_1}^2 - \sigma_{\theta_2}^2)\boldsymbol{t}\boldsymbol{t}' + (\sigma_1^2 - \sigma_2^2)\boldsymbol{Q} + (\gamma_1^2 - \gamma_2^2)\boldsymbol{I}_M = 0. \tag{6.96}$$

It is noted that the off-diagonal elements of $\boldsymbol{I}_M$ are zeros, and thus we pay attention to the off-diagonal elements of (107). For the off-diagonal entry (i, j) of (107) with $i \ne j$, based on the formulations of $\boldsymbol{Q}$ and $\boldsymbol{t}\boldsymbol{t}'$, we have

$$(\sigma_{\theta_1}^2 - \sigma_{\theta_2}^2)t_j + (\sigma_1^2 - \sigma_2^2) = 0, 2 \le j \ne i \le M. \tag{6.97}$$

It is easy to verify that, if $M < 3$, the conditions in (106) do not hold. Instead, when $M = 3$, we have the unique solution $\sigma_{\theta_1}^2 - \sigma_{\theta_2}^2 = 0$ and $\sigma_1^2 - \sigma_2^2 = 0$, because the monitored times are ordered, i.e., $\begin{vmatrix} t_2 & 1 \\ t_3 & 1 \end{vmatrix} \neq 0$. This result implies that $\sigma_{\theta_1}^2 = \sigma_{\theta_2}^2$ and $\sigma_1^2 = \sigma_2^2$. Thus, from (107), we know that $\gamma_1^2 = \gamma_2^2$. From Lemma 6.6, the presented degradation model is identifiable under $\boldsymbol{y}_i$ when $M = 3$. It is not difficult to verify that the identifiability holds for the case $M > 3$. This completes the proof.

In the following, we assume $M \geq 3$, and utilize the degradation measurements of the N tested systems to estimate $\sigma^2, \gamma^2, \mu_\theta, \sigma_\theta^2$. For the ith system, we have known that

$$\boldsymbol{y}_i \sim N(\mu_\theta \boldsymbol{t}, \boldsymbol{\Omega} + \sigma_\theta^2 \boldsymbol{t}\boldsymbol{t}'). \tag{6.98}$$

Denote $\boldsymbol{Y} = (\boldsymbol{y}_1, \boldsymbol{y}_2, \ldots, \boldsymbol{y}_N)'$. Due to the statistical independence assumption of the degradation measurements of different systems, the log-likelihood function over parameter vector $\boldsymbol{\Theta}$ can be written as

$$\begin{aligned}\ell(\boldsymbol{\Theta} \,|\boldsymbol{Y}) = &-\frac{NM}{2}\ln(2\pi) - \frac{N}{2}\ln\left|\boldsymbol{\Omega} + \sigma_\theta^2 \boldsymbol{t}\boldsymbol{t}'\right| \\ &-\frac{1}{2}\sum_{i=1}^{N}\left(\boldsymbol{y}_i - \mu_\theta \boldsymbol{t}\right)'\left(\boldsymbol{\Omega} + \sigma_\theta^2 \boldsymbol{t}\boldsymbol{t}'\right)^{-1}\left(\boldsymbol{y}_i - \mu_\theta \boldsymbol{t}\right).\end{aligned} \tag{6.99}$$

To facilitate the MLE, we re-parameterize the parameters by $\tilde{\sigma}^2 = \sigma^2/\sigma_\theta^2, \tilde{\gamma}^2 = \gamma^2/\sigma_\theta^2, \tilde{\boldsymbol{\Omega}} = \boldsymbol{\Omega}/\sigma_\theta^2$. As such, (6.99) can be rewritten as

$$\begin{aligned}\ell(\boldsymbol{\Theta} \,|\boldsymbol{Y}) = &-\frac{NM}{2}\ln(2\pi) - \frac{NM}{2}\ln\sigma_\theta^2 - \frac{N}{2}\ln\left|\tilde{\boldsymbol{\Omega}} + \boldsymbol{t}\boldsymbol{t}'\right| \\ &-\frac{1}{2\sigma_\theta^2}\sum_{i=1}^{N}\left(\boldsymbol{y}_i - \mu_\theta \boldsymbol{t}\right)'\left(\tilde{\boldsymbol{\Omega}} + \boldsymbol{t}\boldsymbol{t}'\right)^{-1}\left(\boldsymbol{y}_i - \mu_\theta \boldsymbol{t}\right).\end{aligned} \tag{6.100}$$

Taking the first partial derivative of $\ell(\boldsymbol{\Theta} \,|\mathbf{Y})$ with respect to μ_θ and σ_θ^2 generates

$$\frac{\partial \ell(\boldsymbol{\Theta} \,|\boldsymbol{Y})}{\partial \mu_\theta} = \frac{1}{\sigma_\theta^2}\boldsymbol{t}'(\tilde{\boldsymbol{\Omega}} + \boldsymbol{t}\boldsymbol{t}')^{-1}\sum_{i=1}^{N}\left(\boldsymbol{y}_i - \mu_\theta \boldsymbol{t}\right), \tag{6.101}$$

and

$$\frac{\partial \ell(\boldsymbol{\Theta} \,|\boldsymbol{Y})}{\partial \sigma_\theta^2} = -\frac{NM}{2\sigma_\theta^2} + \frac{N}{2\sigma_\theta^4}\sum_{i=1}^{N}\left(\boldsymbol{y}_i - \mu_\theta \boldsymbol{t}\right)'\left(\tilde{\boldsymbol{\Omega}} + \boldsymbol{t}\boldsymbol{t}'\right)^{-1}\left(\boldsymbol{y}_i - \mu_\theta \boldsymbol{t}\right). \tag{6.102}$$

Then, for given values of $\tilde{\sigma}^2, \tilde{\gamma}^2$, and setting these two derivatives to zero, the results of the MLE for μ_θ, and σ_θ^2 can be expressed as

$$\hat{\mu}_\theta(\tilde{\boldsymbol{\Theta}}) = \frac{\boldsymbol{t}'(\tilde{\boldsymbol{\Omega}} + \boldsymbol{t}\boldsymbol{t}')^{-1}\sum_{i=1}^{N}\boldsymbol{y}_i}{N(\boldsymbol{t}'(\tilde{\boldsymbol{\Omega}} + \boldsymbol{t}\boldsymbol{t}')^{-1}\boldsymbol{t})}, \tag{6.103}$$

and

$$\hat{\sigma}_{\theta}^{2}(\tilde{\boldsymbol{\Theta}}) = \frac{1}{NM}\sum\nolimits_{i=1}^{N}\left(\boldsymbol{y}_i - \hat{\mu}_{\theta}(\tilde{\boldsymbol{\Theta}})\boldsymbol{t}\right)'(\tilde{\boldsymbol{\Omega}} + \boldsymbol{t}\boldsymbol{t}')^{-1}\left(\boldsymbol{y}_i - \hat{\mu}_{\theta}(\tilde{\boldsymbol{\Theta}})\boldsymbol{t}\right). \tag{6.104}$$

Note that $\hat{\mu}_{\theta}(\tilde{\boldsymbol{\Theta}})$, and $\hat{\sigma}_{\theta}^{2}(\tilde{\boldsymbol{\Theta}})$ are respectively the MLE of μ_{θ}, and σ_{θ}^{2} when other parameters are fixed at $\tilde{\boldsymbol{\Theta}}$. Then the profile likelihood function for $\tilde{\boldsymbol{\Theta}}$ can be obtained by substituting (6.103) and (6.104) into (6.100) as

$$\ell(\tilde{\boldsymbol{\Theta}}\,|\boldsymbol{Y}) = -\frac{NM}{2}\ln(2\pi) - \frac{NM}{2} - \frac{NM}{2}\ln\hat{\sigma}_{\theta}^{2}(\tilde{\boldsymbol{\Theta}}) - \frac{N}{2}\ln\left|\tilde{\boldsymbol{\Omega}} + \boldsymbol{t}\boldsymbol{t}'\right|. \tag{6.105}$$

The MLE of $\tilde{\sigma}^2, \tilde{\gamma}^2$ can be obtained by maximizing the profile log-likelihood function in (6.105) through a two-dimensional search. By substituting the MLE of $\tilde{\sigma}^2, \tilde{\gamma}^2$ into (6.103) and (6.104), we can obtain the MLE for μ_{θ} and σ_{θ}^2, and then the MLE for σ^2 and γ^2 can be achieved by inverting the relations $\tilde{\sigma}^2 = \sigma^2/\sigma_{\theta}^2$ and $\tilde{\gamma}^2 = \gamma^2/\sigma_{\theta}^2$, accordingly. Once the parameters $\sigma^2, \gamma^2, \mu_{\theta}, \sigma_{\theta}^2$ are estimated, the results in Sect. 6.2 can be used to estimate the RUL for a system in service on the basis of its real-time measurements.

Remark 6.4 It is noted that the above parameter estimation method is suitable for the balanced case, i.e., N tested systems are measured at common monitored times. However, the estimation method can be easily extended to the unbalanced case, i.e., different monitored times for different systems. In addition, the parameter estimation for the special cases listed in Remark 6.1 can be similarly implemented. This further validates the generality of the developed method, in terms of not only the RUL estimation but also the parameter estimation procedure. ∎

Remark 6.5 In general, to accurately estimate the parameters associated with the random effect, the number of systems N under test should be large enough. Otherwise, it may be wondered that the estimation error for μ_{θ} and σ_{θ} would be large. However, when the presented method in Sect. 6.2 is used to estimate the RUL of an operating system, the estimation error for μ_{θ} and σ_{θ} could be partially reduced even for small N. The presented method considers the updating of the random effect parameter θ based on real-time monitored degradation signals. This updating mechanism is achieved by state-space modeling and Kalman filtering techniques. On the other hand, the estimations of μ_{θ}, and σ_{θ} are only used as initial values of the Kalman filter; and instead, the posterior updates of μ_{θ}, and σ_{θ} are used in the RUL estimation, corresponding to $\hat{\theta}_{k|k}$, and $\kappa_{\theta,k}$, respectively. In the case study, we will illustrate the updates of $\hat{\theta}_{k|k}$, and $\kappa_{\theta,k}$, whose initial values are μ_{θ}, and σ_{θ}, respectively. ∎

6.5 Experimental Studies

In this section, a practical case study for gyros in an inertial platform is provided to illustrate the presented modeling framework. Here, three measures are employed to enable comparisons between the results with three-source variability, and the results only considering one- or two-source variability. The first measure, AIC [49], is used to compare the fitting of the models to the measured degradation data, calculated by

$$AIC = -2(\max \ell) + 2p, \tag{6.106}$$

where p is the number of estimated model parameters, and $\max \ell$ is the maximized likelihood. The AIC is frequently used in engineering and the statistical literature to guide model selection because the AIC can balance the tradeoff between model fitting and the over-parametrization problem. The smallest AIC value corresponds to the best fitting accuracy.

The second measure is the MSE about the actual RUL obtained at each measurement point, defined as [50]

$$MSE_k = E\left[(L_k - \tilde{L}_k)^2\right], \tag{6.107}$$

where $\tilde{L}_k$ denotes the actual RUL obtained at t_k, and the expectation is evaluated on the estimated RUL conditional on the available data.

Based on the MSE, we further define the total MSE (TMSE) as the sum of the MSE at each CM point over the whole life cycle, as

$$TMSE = \sum\nolimits_{k=1}^{M} MSE_k,$$

where M is the number of the total observations.

The third measure is the RE of the observed lifetime and the estimated lifetime. Let $\tilde{T}$ denote the lifetime corresponding to a path of the observed degradation signals, and let $\hat{T}_k$ be the estimated lifetime at t_k. If we use the mean lifetime as a point estimation of the lifetime, we have $\hat{T}_k = t_k + E[L_k]$, where $E[L_k]$ is the estimated mean RUL at t_k. Hence, the RE for the lifetime at t_k is

$$RE_k = \frac{\left|\tilde{T} - \hat{T}_k\right|}{\tilde{T}}. \tag{6.108}$$

In the following, we use these defined measures to evaluate the performance of the model fitting and RUL estimation for different cases based on the gyros' degradation signals.

6.5.1 Problem Description

A gyro fixed on an inertial platform is a key device of the inertial navigation system widely applied in weapon systems and space equipment. The gyro having two degrees of freedom from the driver axis, and sense axis is a mechanical structure to sense angular velocity, and linear acceleration of the inertial platform, respectively [32]. When the inertial platform is operating, the wheel of the gyro rotates at very high speeds, and can lead to rotation axis wear. As the wear is accumulated, the bearings on the gyro's electric motor will become deformed, and an increase of the gyro drift occurs. Past statistics data show that almost 70% of the failures of inertial platforms result from gyroscopic drift, and such drift is largely the result of bearing wear. It is therefore natural to consider that the gyro drift is related to the physical failure process of the inertial platform, and thus utilize the monitored gyro's drift data as the degradation signals to evaluate the health state of the inertial platform.

In this study, we use the drift-based degradation signals based on the fact that the values of the gyro's drift are correlated with the severity of the gyro's degradation. Generally, six drift coefficients can be obtained, including three constant drift coefficients, and three stochastic drift coefficients. However, the constant drift coefficients are largely due to the possible errors in fixing the gyro, and thus the associated signals are stationary; see Fig. 6.1 for an illustration of the measured data of one constant drift coefficient over time. Therefore, this kind of coefficient cannot reflect the degradation trend. Among stochastic drift coefficients, the drift coefficient along the sense axis generally plays a dominant role in the gyro's degradation failure. As such, we take the measured data along the sense axis as the degradation signals, and use them for degradation modeling and RUL estimation. We define the failure of the gyro as the first time of the underlying drift state hitting the threshold 0.36 °/h, which is determined at the design stage, and strictly enforced in practice.

From November 2011 to July 2012, we carried out the experiments on three gyros of the same type. In the experiments, the measured drift data were collected under the sampling interval of 2.5 hours, and a gravity acceleration of 9.794121 m/s^2, where a PC-based data acquisition system was used to acquire and store the drift data. After the experiments, we can collect all the drift data of three gyros. The data sets include the time-to-failure data, constant drift coefficients, and stochastic drift

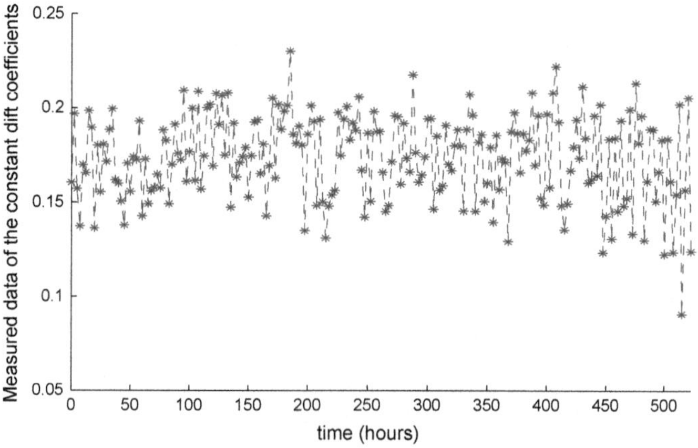

Fig. 6.1 Illustration of the measured constant drift coefficient

coefficients for three gyros. The first gyro was running to failure with the terminated life of 528 h, and 210 points of drift coefficients data were collected. The remaining two gyros were still functioning after testing 210 monitoring data, but we stopped the test to reduce experimental costs. The data of the drift coefficients along the sense axis for these three gyros are illustrated in Fig. 6.2, which contains the measured degradation signals of gyros contaminated by the measurement variability, i.e., $\boldsymbol{Y}_{1:k}$.

Figure 6.2 shows that the degradation paths of different gyros exhibit different degradation rates. In addition, it is observed that the degradation signals of gyros experience a lot of fluctuations. Such fluctuations are possibly the result of uncertainties in the underlying degradation process and measurement process. In physics, a Wiener process aims at modeling the movement of small particles in fluids and air with tiny fluctuations. The tiny increase or decrease in degradation over a small time interval behaves similarly to the random walk of small particles in fluids and air. This type of stochastic processes is appropriate to characterize the path of the gyro's degradation process [32, 33, 51]. Therefore, we use the proposed Wiener process based model to analyze the gyro's data in this chapter. Additionally, it is possible that the uncertain measurements before the failure time hit the threshold. However, the RUL of gyros defined in (4) is based on the underlying degradation states $\boldsymbol{X}_{1:k}$, i.e., the stochastic process $\{X(t), t \geq 0\}$. Thus, we cannot simply determine the failure time according to the degradation measurements. Rather, the failure time should be determined by the underlying degradation states $\boldsymbol{X}_{1:k}$, which are estimated from the degradation measurements. All the above observations may encourage that, for a

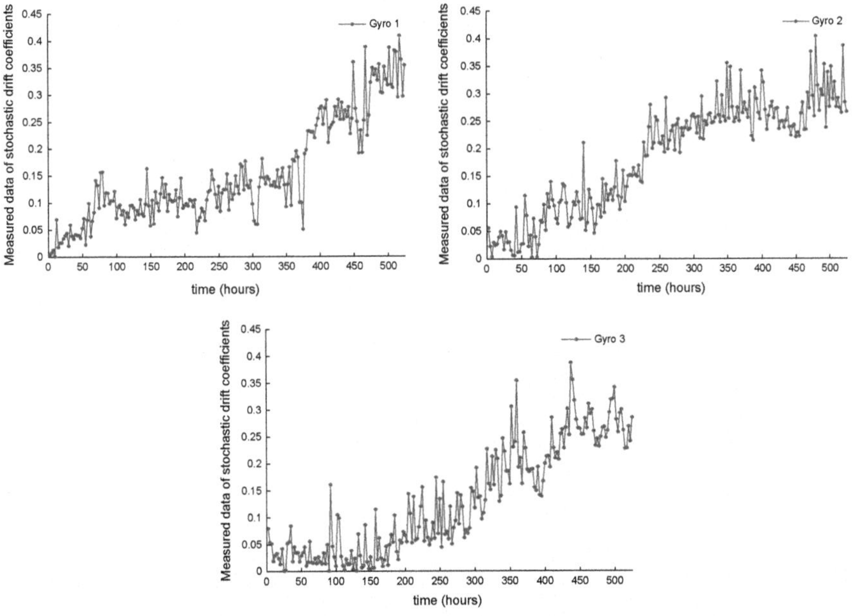

Fig. 6.2 Three measured gyro degradation signals for stochastic drift coefficients

degradation model to be realistic, it is more appropriate to incorporate three-source variability into the degradation modeling process. To show the superiority of incorporating three-source variability, we consider the following three competing cases for comparative studies.

1. Case 1: $\gamma^2 = 0$, the measurement uncertainty is ignored, and the results in Theorem 6.3 are used for RUL estimation.
2. Case 2: $\sigma_\theta^2 = 0$, the random effect is ignored, and the results in Theorem 6.4 are used for RUL estimation.
3. Case 3: $\gamma^2 = 0, \sigma_\theta^2 = 0$, only the temporal variability is considered, and the results in Theorem 6.2 are used for RUL estimation.

It is noted that Case 1 and Case 2 correspond to the cases only considering two-source variability, and Case 3 corresponds to the case only taking one-source variability. By contrast, we term our model with three-source case as Case 4. In the following, we use the degradation signals of the second and third gyros to estimate the model parameters and compare the model fitting for the above four cases, and the degradation signals of the first gyro are used for validation of the estimated RUL.

6.5.2 *Comparisons for Model Fitting*

Using the parameter estimation method presented in Sect. 6.4, we can obtain the estimated results of the parameters for four considered cases. The estimated parameters and the associated values of the log-likelihood function (log-LF) and AIC are shown in Table 6.2.

As shown in Table 6.2, the models with three-source variability have the greatest log-LF, and the least AIC. This result indicates that considering three-source variability in degradation modeling is necessary and can significantly improve the model fitting in terms of the AIC. The reason for this indication is clarified as follows. From the results in Case 1 and Case 4, we can learn that the unit-to-unit variability exists with the value greater than 0.0004. Such a value could affect the fitting of the degradation modeling because the estimate for μ_θ is about 0.0007, but the standard error σ_θ has the same order of magnitude as μ_θ. Additionally, we can find in Case 2 and Case 4 that the measurement uncertainty cannot be ignored because the estimated γ is

Table 6.2 Estimated parameters and associated values for log-LF and AIC

	μ_θ	σ_θ	σ	γ	log-LF	AIC
Case 1	0.000776	0.000479	0.011011	–	501.76	−997.52
Case 2	0.000769	–	0.010630	0.181364	409.76	−813.52
Case 3	0.000782	–	0.010945	–	391.17	−778.34
Case 4	0.000726	0.000487	0.010963	0.277260	576.65	−1145.3

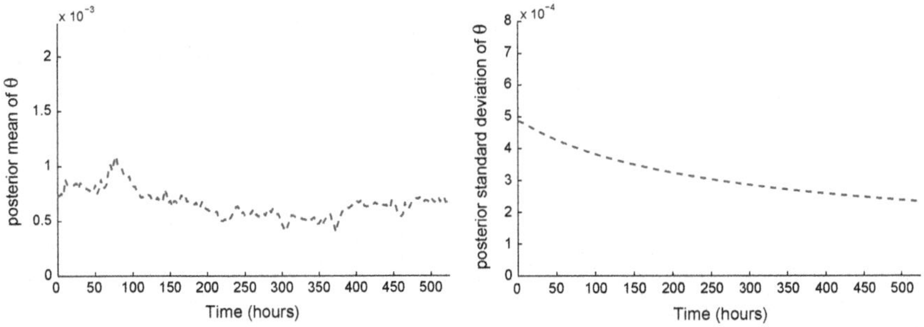

Fig. 6.3 Updates of $\hat{\theta}_{k|k}$ and $\kappa_{\theta,k}$ using degradation signals of the first gyro

relatively greater in the degradation signals. Therefore, it is not uncommon that Case 3 has the worst model fitting because both the measurement uncertainty and unit-to-unit variability are ignored in this case. As a result of the comparisons, it is suggested that the degradation model with three-source variability is the best fitting model.

On the other hand, it is observed that the estimated values of γ and σ are much larger than the values of μ_θ and σ_θ, which may indicate that the influence brought by the drift coefficient (represented by μ_θ and σ_θ) can be neglected. Our points regarding this phenomenon are as follows. According to the used underlying degradation model (1), and the measurement model (2), the mean of the degradation quantity at time t is $E(X(t)) = \mu_\theta t$, and the corresponding variance is $\mathrm{var}(X(t)) = \sigma_\theta^2 t^2 + \sigma^2 t$. Accordingly, the variance of the measured degradation signal at discrete time t_k is $\mathrm{var}(Y(t_k)) = \sigma_\theta^2 t_k^2 + \sigma^2 t_k + \gamma^2$. We can observe that the effect of μ_θ, σ_θ, σ in the degradation model is time-dependent, while γ not. Particularly, μ_θ, and σ_θ will have more influence over time. Therefore, the influence of the drift coefficient cannot be neglected. For example, using the parameters in Table 6.2 (Case 4), at $t_k = 500$, we have $E(X(t_k)) = \mu_\theta t_k = 0.363$, $\sigma_\theta^2 t_k^2 = 0.059$, $\sigma^2 t_k = 0.0601$, $\gamma^2 = 0.0769$. All the terms in the variance $\mathrm{var}(Y(t_k))$ (or $\mathrm{var}(X(t_k))$) have the same order of magnitude. Thus we cannot neglect the effect of μ_θ and σ_θ, particularly when the system's age t is large.

6.5.3 Comparisons for the Estimated RUL

Before going into the RUL estimation results, it is worthwhile to note that the number of systems under test used for estimating the parameters associated with the random effect is small because the gyro is very expensive, and funding is limited. According to Remark 6.5, we know that the estimation of the random effect parameter θ can be updated using real-time monitored degradation signals. Based on the estimated parameters in Sect. 6.5, and the degradation signals of the first gyro, the updates of $\hat{\theta}_{k|k}$ and $\kappa_{\theta,k}$, whose initial values are $\hat{\theta}_{0|0} = \mu_\theta = 0.000726$ and $\kappa_{\theta,0} = \sigma_\theta = 0.000487$, are illustrated in Fig. 6.3.

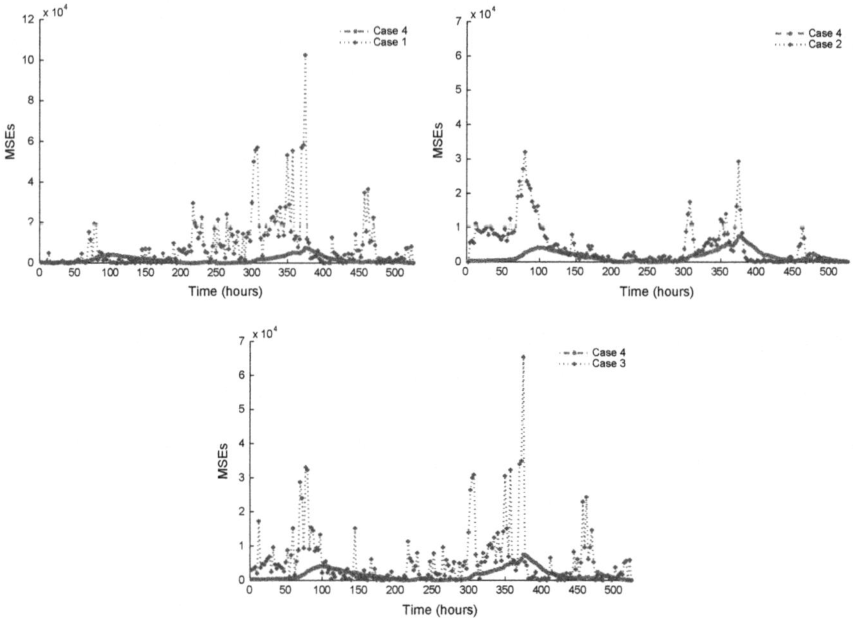

Fig. 6.4 Comparative results for RUL estimations in terms of MSE for the first gyro

As shown in Fig. 6.3, the posterior mean $\hat{\theta}_{k|k}$ and standard deviation $\kappa_{\theta,k}$ of θ are adaptively adjusted by the monitored degradation signals. Particularly, it is observed that $\kappa_{\theta,k}$ is decreasing as the degradation signals accumulated. This result implies that the variability caused by the random effect is reduced, and the estimation for θ tends to be tailored to this gyro under study. The above updating mechanism hopefully alleviates the estimation error for μ_θ and σ_θ resulted by the limit of the number of systems under test.

Now, we compare the performance of the estimated RUL of Case 4 against the results of Cases 1 through 3. First, the MSEs about the RUL associated with these cases can be obtained using the estimated PDF of the RUL and the definition of the MSE. We show the evolving paths of the MSEs of Cases 1–3 against Case 4 at each CM point in Fig. 6.4.

From Fig. 6.4, we find that the results with three-source variability maintain the MSEs of the estimated RUL at a relatively low level compared with other three cases. In Cases 1 through 3, the MSEs suffer from many large fluctuations which result from the uncertain degradation signals because these three cases do not fully consider the uncertainties in the degradation signals, i.e., taking either two-source or only one-source variability. By contrast, Case 4 avoids the large change of MSE in the estimated RUL because it takes all three-source variability simultaneously.

Another observation in Fig. 6.4 is that Case 2 behaves better than Cases 1 and 3 though the random effect is not considered. This behavior is the benefit of using the observation history of the degradation signals because the estimation in Case 2 is conditional on the real-time observed degradation signals up to the current CM point by Kalman filter, as implied by Theorem 6.4. Nevertheless, Case 4 is still better than Case 2 in that the random effect cannot be ignored in this case study, as discussed in Sect. 6.5.

Last, Case 1 and Case 3 have larger MSEs in most CM points because the measurement uncertainty is not taken into account. But the results in Table 6.2 shows that the measurement errors exist, and the variance is relatively large. Correspondingly, the TMSEs for Cases 1 through 4 are 1.5E+6, 1.8E+6, 9.8E+5, and 3.5E+5, respectively. These results are consistent with the above discussions, and further verify that the degradation modeling with three-source variability has a better RUL estimation, and can improve the estimation accuracy.

To check if the MSE of the RUL estimation is stable or not, we calculate the MSEs of the RUL estimation of the second gyro, as illustrated in Fig. 6.5.

In the computations corresponding to Fig. 6.5, we set the threshold to 0.3 °/h for numerical validation, which approximately corresponds to the lifetime of 525 h based on the underlying degradation state. From the results in Fig. 6.5, we can observe that the discussions for Fig. 6.4 also apply to Fig. 6.5. Thus, we verify that the MSE of the RUL estimation by the proposed method is stable for the gyros' data. Of course, we can also use the third gyro's data to do the validation. Those results are consistent with the results of the first and second gyros.

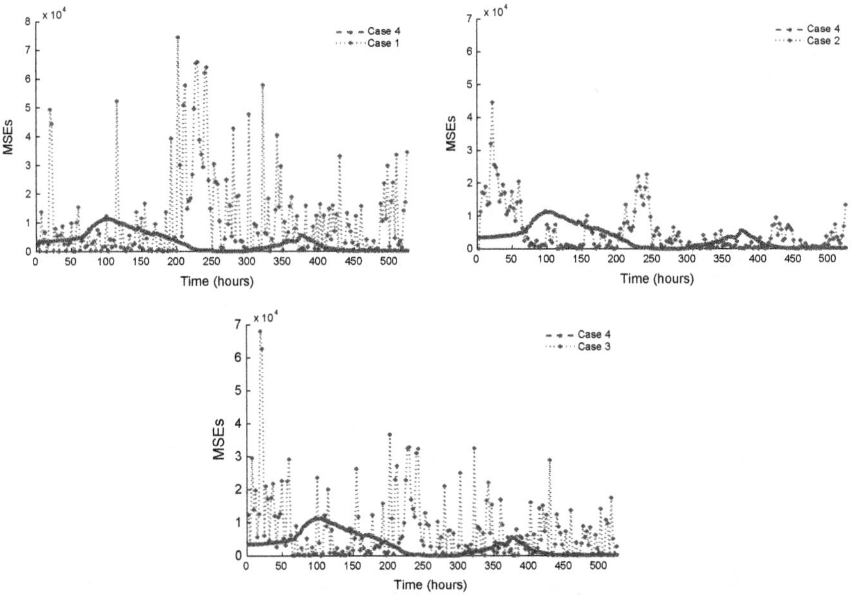

Fig. 6.5 Comparative results for RUL estimations in terms of MSE for the second gyro

Table 6.3 Estimated mean lifetimes and corresponding REs

	30th percentile	50th percentile	70th percentile	90th percentile
Case 1	636.45 (20.54%)	612.81 (16.06%)	600.52 (13.73%)	573.09 (8.54%)
Case 2	472.46 (10.52%)	567.17 (7.42%)	554.64 (5.05%)	546.24 (3.45%)
Case 3	473.74 (10.25%)	575.52 (9.0%)	599.69 (13.58%)	601.48 (13.92%)
Case 4	488.81 (7.42%)	505.63 (4.24%%)	548.93 (3.96%)	535.44 (1.41%)

The degradation signals can also be used for estimating the lifetime of the first gyro. For comparisons, we compute the estimated mean lifetimes and the corresponding REs (in the bracket) at the 30th, 50th, 70th, and 90th percentiles of the lifetime, as summarized in Table 6.3.

Observe that the results in Table 6.3 are consistent with the previous discussions. Therefore, Table 6.3 further provides the supporting evidence for the modeling motivation in this chapter; taking three-source variability simultaneously is necessary in the RUL estimation with the observation history dependency and real-time updating. It is indicated that taking three-source variability and incorporating the real-time degradation signals can improve the accuracy of the RUL estimation.

References

1. Haddad G, Sandborn PA, Pecht M (2012) An options approach for decision support of systems with prognostic capabilities. IEEE Trans Reliab 61(4):872–883
2. Pecht M (2008) Prognostics and health management of electronics. Wiley, New Jersey
3. Sheppard JW, Kaufman MA, Wilmering TJ (2009) IEEE standards for prognostics and health management. IEEE Aerosp Electron Syst Manag 22(9):34–41
4. Zio E, Peloni G (2011) Particle filtering prognostic estimation of the remaining useful life of nonlinear components. Reliab Eng Syst Saf 96(3):403–409
5. Elwany AH, Gebraeel NZ (2008) Sensor-driven prognostic models for equipment replacement and spare parts inventory. IIE Trans 40(7):629–639
6. Huynh KT, Barros A, Bérengure C (2012) Maintenance decision-making for systems operating under indirect condition monitoring: value of online information and impact of measurement uncertainty. IEEE Trans Reliab 61(2):410–425
7. Li R, Ryan JK (2011) A Bayesian inventory model using real-time condition monitoring information. Prod Oper Manag 20(5):754–771
8. Si XS, Wang W, Hu CH, Zhou DH (2011) Remaining useful life estimation-a review on the statistical data driven approaches. Eur J Oper Res 213(1):1–14
9. Zio E, Di Maio F (2010) A data-driven fuzzy approach for predicting the remaining useful life in dynamic failure scenarios of a nuclear system. Reliab Eng Syst Saf 95(1):49–57
10. Crowder M, Lawless J (2007) On a scheme for predictive maintenance. Eur J Oper Res 176(3):1713–1722
11. Meeker WQ, Escobar LA (1998) Statistical methods for reliability data. Wiley, New York
12. Pandey MD, Yuan XX, Van Noortwijk JM (2009) The influence of temporal uncertainty of deterioration on life-cycle management of structures. Struct Infrastruct Eng 5(2):145–156
13. Peng CY, Tseng ST (2009) Mis-specification analysis of linear degradation models. IEEE Trans Reliab 58(3):444–455

14. Whitmore GA (1995) Estimating degradation by a Wiener diffusion process subject to measurement error. Lifetime Data Anal 1(3):307–319
15. Ye ZS, Shen Y, Xie M (2012) Degradation-based burn-in with preventive maintenance. Eur J Oper Res 221(2):360–367
16. Ye ZS, Wang Y, Tsui KL, Pecht M (2013) Degradation data analysis using Wiener processes with measurement errors. IEEE Trans Reliab 62(4):772–780
17. Baraldi P, Mangili F, Zio E (2013) Investigation of uncertainty treatment capability of model-based and data-driven prognostic methods using simulated data. Reliab Eng Syst Saf 112(1):94–108
18. Liu R, Ma L, Kang R, Wang N (2011) The modeling method on failure prognostics uncertainties in maintenance policy decision process. In: Proceedings of the 9th international conference on reliability, maintainability and safety (ICRMS), Guiyang, China, pp 12–15
19. Saxena A, Celaya J, Saha B, Saha S, Goebel K (2010) Evaluating prognostics performance for algorithms incorporating uncertainty estimates. In: Proceedings of the IEEE aerospace conference 2010
20. Aalen OO, Borgan O, Gjessing HK (2008) Survival and event history analysis: a process point of view. Springer, New York
21. Giorgio M, Guida M, Pulcini G (2011) An age- and state-dependent Markov model for degradation processes. IIE Trans 43(9):621–632
22. Singpurwalla ND (1995) Survival in dynamic environments. Stat Sci 10(1):86–103
23. Gebraeel NZ, Lawley MA, Li R, Ryan JK (2005) Residual-life distributions from component degradation signals: a Bayesian approach. IIE Trans 37(6):543–557
24. Yuan XX, Pandey MD (2009) A nonlinear mixed-effects model for degradation data obtained from in-service inspections. Reliab Eng Syst Saf 94(2):509–519
25. Dieck RH (2004) Measurement uncertainty: methods and applications. The 4th edition, The Instrument Society of America, Research Triangular Park, North Carolina
26. Kolle C, O'Leary P (1998) Low-cost, high-precision measurement system, for capacitive sensors. Meas Sci Technol 9(3):510–517
27. Oxtoby NP, Sun HB, Wiseman HM (2003) Non-ideal monitoring of a qubit state using a quantum tunnelling device. J Phys Condens Matter 15(46):8055–8064
28. Chen N, Tsui KL (2013) Condition monitoring and residual life prediction using degradation signals: revisited. IIE Trans 45(9):939–952
29. Joseph VR, Yu IT (2006) Reliability improvement experiments with degradation data. IEEE Trans Reliab 55(1):149–157
30. Liao HT, Elsayed EA, Chan LY (2006) Maintenance of continuously monitored degrading systems. Eur J Oper Res 175(2):821–835
31. Lu CJ, Meeker WQ (1993) Using degradation measures to estimate a time-to-failure distribution. Technometrics 35(2):161–174
32. Si XS, Wang W, Chen MY, Hu CH, Zhou DH (2013) A Wiener-process-based degradation model with a recursive filter algorithm for remaining useful life estimation. Mech Syst Signal Process 35(1–2):219–237
33. Si XS, Wang W, Chen MY, Hu CH, Zhou DH (2013) A degradation path-dependent approach for remaining useful life estimation with an exact and closed-form solution. Eur J Oper Res 226(1):53–66
34. Tang J, Su TS (2008) Estimating failure time distribution and its parameters based on intermediate data from a Wiener degradation model. Nav Res Logist 55(3):265–276
35. Wang W, Carr M, Xu W, Kobbacy AKH (2011) A model for residual life prediction based on Brownian motion with an adaptive drift. Microelectron Reliab 51(2):285–293
36. Xu ZG, Ji YD, Zhou DH (2008) Real-time reliability prediction for a dynamic system based on the hidden degradation process identification. IEEE Trans Reliab 57(2):230–242
37. Zhou YF, Sun Y, Mathew J, Wolff R, Ma L (2011) Latent degradation indicators estimation and prediction: a monte carlo approach. Mech Syst Signal Process 25(1):222–236
38. Feng L, Wang HL, Si XS, Zou HX (2013) A state-space-based prognostic model for hidden and age-dependent nonlinear degradation process. IEEE Trans Autom Sci Eng. doi:10.1109/TASE.2012.2227960

39. Park JI, Bae SJ (2010) Direct prediction methods on lifetime distribution of organic light-emitting diodes from accelerated degradation tests. IEEE Trans Reliab 59(1):74–90
40. Wang X, Xu DH (2010) An inverse gaussian process model for degradation data. Technometrics 52(2):188–197
41. Ye ZS, Chen N (2013) The inverse Gaussian process as a degradation model. Technometrics. doi:10.1198/TECH.2009.08197
42. Ye ZS (2013) On the conditional increments of degradation processes. Stat Probab Lett 83(11):2531–2536
43. Tseng ST, Tang J, Ku LH (2003) Determination of optimal burn-in parameters and residual life for highly reliable products. Nav Res Logist 50(1):1–14
44. Sun JZ, Zuo HF, Wang W, Pecht MG (2012) Application of a state space modeling technique to system prognostics based on a health index for condition-based maintenance. Mech Syst Signal Process 28:585–596
45. Lee MLT, Whitmore GA (2006) Threshold regression for survival analysis: modeling event times by a stochastic process reaching a boundary. Stat Sci 21(4):501–513
46. Cox DR, Miller HD (1965) The theory of stochastic processes. Methuen and Company, London
47. Si XS, Zhou DH (2013) A generalized result for degradation model based reliability estimation. IEEE Trans Autom Sci Eng. doi:10.1109/TASE.2013.2260740
48. Demidenko E (2004) Mixed models: theory and applications. Wiley, New York
49. Akaike H (1974) A new look at the statistical model identification. IEEE Trans Autom Control 19(6):716–722
50. Carr MJ, Wang W (2011) An approximate algorithm for prognostic modelling using condition monitoring information. Eur J Oper Res 211(1):90–96
51. Si XS, Chen MY, Wang W, Hu CH, Zhou DH (2013) Specifying measurement errors for required lifetime estimation performance. Eur J Oper Res 231(3):631–644

Part III
Prognostic Techniques for Nonlinear Degrading Systems

Chapter 7
RUL Estimation Based on a Nonlinear Diffusion Degradation Process

7.1 Introduction

Because of limited natural resources, considerably increased safety and environmental concerns, and the drive to reduce operating costs, critical assets need to be managed over their entire life cycles—from design, manufacture, sale, and operation to their end of life in order to optimize life cycle management and reduce negative impact on the environment [1, 2]. For safety-critical equipment, such as aviation control systems and nuclear power generators, the accurate and early estimation of failure is critical in order to avoid catastrophic events that may cause severe damage to equipment, loss of human lives, and environmental disasters. A recent example was the explosion of the *Deepwater Horizon* oil well in the Gulf of Mexico, which resulted in the loss of lives, harm to the environment, and impact on average citizens. There were indications that this disaster indirectly resulted from the overuse of the system due to strong financial pressure. But the direct cause was the failure of the 'last line of defence', the so-called blowout preventer. The emerging discipline of prognostics and health management (PHM) could have addressed this problem [3].

PHM is a methodology that permits the assessment of the reliability of a system under its actual application conditions and exercises necessary management actions. Prognostics is a key step in PHM. Prognostics utilizes in situ monitoring and analysis to assess system degradation and determine the remaining useful life (RUL) of an asset. The RUL of an asset is defined as the length of time from the present time to the end of useful life. The need for RUL estimation is obvious, since it relates to a frequently asked question in industry, which is how long a monitored asset can survive based on the available information. Based on RUL estimation, appropriate actions can be planned. Especially for critical equipment, such as aircraft engines or inertial navigation platforms used in aerospace and weapon systems, determining if and when to take them out of service is important from both a cost-effective point of view and a safety point of view.

It is critically important to assess the RUL of an asset while it is in use, as this impacts the planning of maintenance activities, the supply chain, the replenishment of

X.-S. Si et al., *Data-Driven Remaining Useful Life Prognosis Techniques*,
Springer Series in Reliability Engineering, DOI 10.1007/978-3-662-54030-5_7

the inventory system, operational performance, and the profitability of the owner of an asset [4–12]. RUL estimation has also played an important role in the management of product reuse and recycling, which has an impact on human life, energy consumption, raw material use, pollution, and landfills [2, 13, 14]. The reused products must have sufficiently long lives to be able to be reused.

In the early 1980s, Derman et al. in [15] demonstrated the usefulness of lifetime distributions in the context of system life extension. Traditional failure-time analysis methods for estimating component lifetimes were heavily dependent on the time-to-failure data or lifetime data [16, 17]. However, some critical and valuable systems are not allowed to run to failure, or the tests to obtain failure information are very expensive. Therefore, lifetime data is often hard to obtain. In such cases, degradation data can be used as an alternate resource for lifetime analysis from an economical and practical viewpoint [18–23]. In many situations, such as the drift degradation of an inertial navigation system (INS) used in the aerospace industry, it is natural to view the failure event of interest as the result of a stochastic degradation process crossing a threshold level, i.e., to model the hitting time of the degradation as a time-dependent stochastic process. Singpurwalla [24], Cox [25], and Aalen and Gjessing [26] all advocated the development, adoption, and exploration of stochastic dynamic models in theory and practice for reliability estimation. On the other hand, dynamic environments induce changes in the physics of failure. Hence a stochastic process approach to model degradation provides flexibility with respect to describing the failure-generating mechanisms and the characteristics of the operating environment. A key idea behind this is that the lifetime can be defined as the first hitting time (FHT) of the degradation process reaching a failure threshold, and the probability density function (PDF) of the failure time is then modeled as the PDF of the FHT of the underlying stochastic process to mimic the true failure time. Lee and Whitmore in [27] have given a comprehensive review of a variety of FHT models and have discussed their potential. It is noted that even when the degradation process is observed to be at or above the threshold level it does not necessarily mean that the system has actually broken down. Such a threshold level is just a preset warning level that is set as a boundary for observed degradation for a variety of reasons.

Among stochastic process–based models, diffusion processes are types of random processes capable of describing random degradation [28, 29]. Brownian motion (BM) with a linear drift is a special diffusion process that has become very popular for degradation modeling in recent years. One of the most important advantages of modeling the degradation process using BM with a linear drift is that the distribution of the FHT of such a process crossing a constant threshold can be formulated analytically. This is known as the inverse Gaussian distribution; it has many merits and has been applied in reliability and lifetime analysis widely since the 1970s [30]. A fundamental problem related to BM with a linear drift is that it can only describe a linearly drifted diffusion process. However, nonlinearity exists extensively in practice and the linear model cannot trace the dynamic of such a degradation process. It is more likely that degradation may accelerate at a later stage of life. Therefore, for a model to be realistic it should incorporate some nonlinear structures. However, this issue has not been well documented in the literature.

Some nonlinear processes can be approximated to be near linear by some kind of transformation on the degradation data, such as log transformation [31, 32] or timescale transformation [33–35]. But these are limited to the cases that such transformations exist, and not many nonlinear processes can be transformed in these ways. There is also an implicit assumption used in the above transformation that the random part of the transformed process is still BM, which may not always be the case. This leads to the primary purpose of this chapter, which is to provide a useful model that can model general nonlinear degradation directly without the use of the transformations mentioned above. This would be particularly useful in cases where the nonlinear degradation process cannot be transformed to a linear degradation by manipulating the data. From a mathematical and application point of view, the difficulties encountered are mainly two kinds. First, as for a nonlinear-drifted diffusion process, the distribution of FHT is related to solving the Fokker–Planck–Kolmogorov (FPK) equation with boundary constraints [36]. For a nonlinear case, this task is rather difficult and even impossible. Second, the available closed forms of the PDFs of the FHT for nonlinear-drifted diffusion processes are very limited in special cases [37, 38]. Simulations can be used to deal with nonlinear cases, and a numerical solution approximation method was developed by Nardo et al. [39]. However, neither can provide a closed form of the PDF of the FHT, and both are subject to long computation times and require a lot of memory. However, in the context of PHM, it is necessary and valuable to derive and evaluate the PDF of FHT with a closed form from the degradation process and provide rapid online estimation for reliability analysis and maintenance scheduling.

From the above analysis, it is clear that nonlinear degradation processes have not been studied thoroughly. In order to formulate the distribution of RUL, we first transform the problem from calculating the FHT distribution of the nonlinear diffusion process crossing a constant threshold into a standard BM crossing a time-dependent boundary. This transformation is achieved by a well-known time-space transformation developed by Ricciardi [40]. It is noted that such a transformation is not on the degradation data; rather, it is on the model, which differs from the transformations in [31, 33–35]. Since there is no closed form for such a distribution, an analytical approximation of the distribution of the FHT of the transformed process is obtained in a closed form under a mild assumption. This is an important contribution to the literature, which has not been reported before. The unknown parameters in the degradation model are obtained using the maximum likelihood estimation (MLE) method, and we use two goodness-of-fit measures to compare the model fit. We demonstrate, using some real-world data sets, the usefulness of the proposed model and the necessity of incorporating a nonlinear structure into the degradation model, which has significantly improved the accuracy of RUL estimation.

The remaining parts of the chapter are organized as follows. Section 7.2 reviews the related literature with an emphasis on RUL modeling methods based on BM with drift. Section 7.3 presents two motivating examples and the modeling principle of our degradation model. Section 7.4 proposes the main theoretical results for calculating the PDF of the FHT. Section 7.5 discusses the procedure for parameter estimation. Three practical examples are presented to verify the proposed model in Sect. 7.6.

7.2 Literature Review

There are many methods for estimating RUL (see [3, 9, 41]). The methodologies are largely classified into physics-of-failure, machine learning, and statistics-based methods. Si et al. in [42] have presented a review on statistics-based methods for RUL estimation. In statistics-based methods, there are models relying on the availability of actual failure data as well as models needing a threshold for degradation to define the failure. BM with drift belongs to the latter.

BM with drift was originally used to depict the random walk of small particles in fluids and air in physics. The walk has a trend, so it is appropriate for modeling a dynamic process with an increasing or decreasing trend. The random term is normally distributed, so it is not a monotonic process. Due to its clarity in concept and its similarity to physical degradation processes, it has been widely applied in degradation modeling and further extended to RUL estimation. BM with drift is a type of stochastic process with Gaussian noise. The variance of the noise is a function of time, and, therefore, the cumulative degradation is infinitely divisible, as required by any physical degradation process [43]. This implies that the degradation described by BM with drift is free from the constraint of the sampling frequency and intervals. This is, to our understanding, a necessary property for a stochastic process to be able to model a physical degradation process. The only other stochastic process to possess such a property is the Gamma process [43].

Tseng et al. in [44] used BM with drift to determine the lifetime for the light intensity of LED lamps of contact image scanners. As an extension, Tseng and Peng in [45] proposed an integrated Wiener process to model the cumulative degradation path of a product's quality characteristics. Lee and Tang in [46] handled the failure-time prediction problem based on BM with drift under a time-censored degradation test, in which a modified expectation and maximization algorithm was used to estimate the mean failure time. A recent extension of the intermediate data for lifetime estimation can be found in [47]. Padgett and Tomlinson in [48] applied BM with drift to model accelerated degradation data and make an inference about the lifetime under a normal operating environment. Park and Padgett in [49, 50] applied BM with drift to model the initial damage under accelerated testing and infer the lifetime. Park and Padgett in [32] and Gebraeel et al. in [9] used log transformation to transform the exponential path into a linear path and then used BM with drift to conduct lifetime estimation. Joseph and Yu in [51] assumed that there existed some transformations that can transform a nonlinear degradation process into a linear process, and then they used BM with drift for degradation modeling and reliability improvement. Balka et al. in [52] reviewed some methods of cure models based on FHT for BM with drift. Though their review focused mainly on cure rate modeling in the biostatistics field, the principle is the same as our RUL modeling. Further, Peng and Tseng in [53] incorporated the random effect in the drift coefficient and measurement errors in BM with a drift-based degradation process for lifetime assessment. In a recent chapter, Wang et al. in [54] considered BM with drift using an adaptive random drift coefficient for RUL estimation. From above, we observe that though many extensions of RUL estimation methods using BM with drift have appeared, almost none of the previous models have explicitly considered the nonlinearity of the underlying degradation process without manipulating the degradation data. However, there is the exception of the work of Tseng and Peng [55]. In [55], Tseng and Peng addressed this problem using a stochastic differential equation for degradation modeling of LEDs with the restrictive assumption that the ratio between the expectation and the variance of the derivative of the degradation process was a constant. We note that such an assumption is difficult to justify in practice. Therefore, we can state that there are issues and challenges remaining to be solved.

In summary, the literature on RUL estimation based on BM with drift has treated only simple structures (linear or linearized by some kind of transformation on the degradation data, as discussed earlier). Nonlinearity has not been directly considered in the literature. In the next section, we will give two practical examples to show the necessity of incorporating a nonlinear structure into a degradation process model. Because of the nonlinearity introduced, the exact closed form of the PDF of the RUL does not exist, and we develop an approximate closed form under a mild assumption. To our knowledge, this chapter is the first to consider such nonlinearity with a closed form for RUL estimation based on BM with drift in the context of degradation modeling without using restrictive assumptions and data transformations. We start first with the motivating examples and RUL modeling principle.

7.3 Motivating Examples and RUL Modeling Principle

Example 1: *Drift coefficient in an inertial navigation platform.* The inertial platform is a key component in the inertial navigation systems in weapon systems and space equipment. Its operating state has a direct influence on navigation precision. The sensors fixed in an inertial platform mainly include three gyros and three accelerometers, which measure angular velocity and linear acceleration, respectively. Statistical analysis shows that almost 70% of failures of inertial platforms result from gyroscopic drift. In our case, the gyro fixed on an inertial platform is a mechanical structure having two degrees of freedom from the driver and sense axis (see [56, 57] for a general description of inertial navigation platforms and gyros). When the inertial platform is operating, the wheels of gyros rotate at very high speeds and can lead to rotation axis wear and finally result in the gyros' drift. With the accumulation of wear, the drift degrades and finally results in the failure of gyros. As such, the drift of gyros is used as a performance indicator to evaluate the health condition of an inertial platform. In our study, we only take the drift degradation measurement along the sense axis for illustrative purposes, since this variable plays a dominant role in the assessment of gyro degradation. The degradation data, including five tested items and nine measurements for each item, were obtained from inertial platforms' precision tests, the conditions which were similar to a field setting. The data are shown in Table 7.1 and are also shown graphically in Fig. 7.1. We can observe clearly nonlinear characteristics in drift degradation in Fig. 7.1.

Table 7.1 Drift degradation data from an inertial platform

Cm time (h)	Gyroscopic drift (°/h)				
	#1	#2	#3	#4	#5
2.5	0.050 7	0.130 0	0.013 5	0.105 2	0.092 8
5.0	0.178 9	0.255 4	0.030 9	0.199 6	0.263 3
7.5	0.205 9	0.315 3	0.007 7	0.292 7	0.301 0
10.0	0.254 8	0.396 6	0.151 0	0.306 0	0.351 5
12.5	0.311 7	0.242 4	0.116 2	0.286 0	0.281 8
15.0	0.523 6	0.347 3	0.023 7	0.291 8	0.437 5
17.5	0.611 4	0.499 4	0.268 1	0.354 0	0.335 7
20.0	0.661 6	0.710 5	0.493 4	0.432 6	0.441 3
22.5	1.816 6	2.639 9	0.651 3	0.504 2	0.815 7

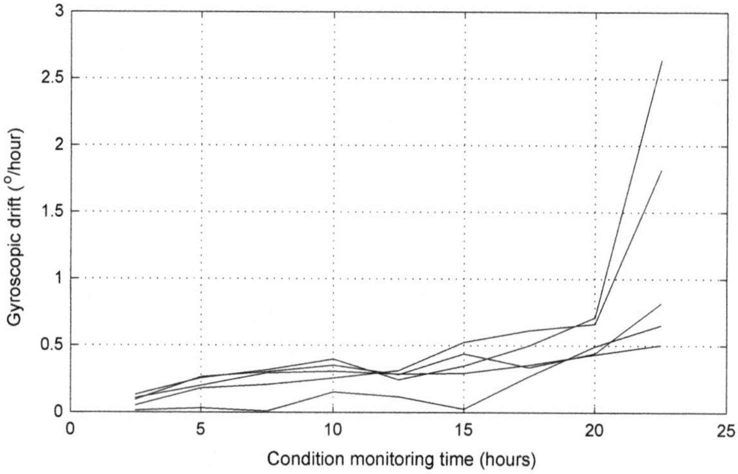

Fig. 7.1 Drift degradation path of gyros

Example 2: *Fatigue crack length in 2017-T4 aluminum alloy*. The 2017-T4 aluminum alloy is a common metallic material used in aircraft [58]. The quality of such material is assessed by the length of fatigue cracks. When the length of a fatigue crack is equal to or more than a predefined threshold level of 6 mm, a structure made of this material is considered to be in a very critical state and is defined as failed. The obtained data represent crack length propagation in four test specimens of 2017-T4 aluminum alloy under a stress level of 200 MPa, which simulates actual operating conditions. For each item, the fatigue crack length is recorded per 0.1 million cycles until the end of the experiment. During testing, ten crack levels were recorded for each item. The data are given in Table 7.2 and shown graphically in Fig. 7.2. Similarly, we can see that nonlinearity exists in the degradation process of 2017-T4 aluminum alloy.

Table 7.2 A2017-T4 aluminum alloy fatigue crack data

Rotating cycle ($\times 10^5$)	Fatigue crack length (mm)			
	#1	#2	#3	#4
1.5	0.04	0.34	0.32	0.38
1.6	0.32	0.40	0.40	0.50
1.7	0.40	0.58	0.50	0.60
1.8	0.48	0.80	0.70	0.75
1.9	0.50	1.20	1.25	1.10
2.0	1.80	1.80	1.40	1.25
2.1	2.00	2.80	2.00	1.50
2.2	2.60	2.40	3.30	2.00
2.3	5.00	3.40	4.70	3.00
2.4	6.00	7.00	5.60	6.80

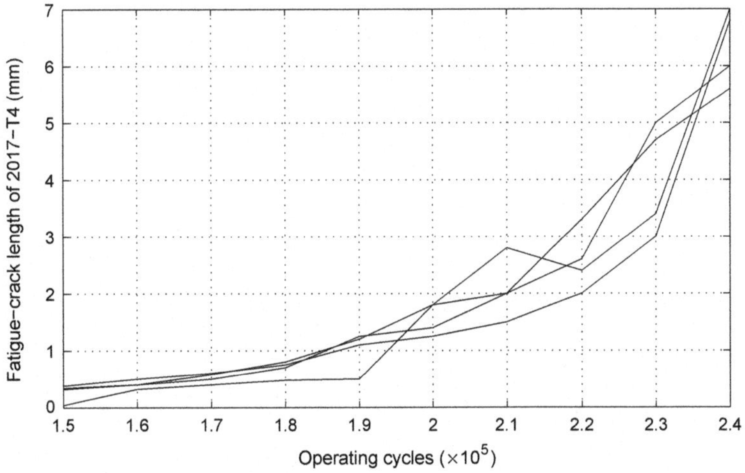

Fig. 7.2 Degradation measures of A2017-T4 fatigue crack growth data

From the reasons given in Sects. 7.1 and 7.3, and further motivated by the above practical examples, it is natural to model both fatigue crack length and drift degradation as a stochastic process with a nonlinear path, since this can capture the dynamics of the degradation process and lead to a better understanding of the nature of the failure event.

Let $X(t)$ denote the degradation at time t. Then,

$$X(t) = X(0) + \int_0^t \mu(t; \boldsymbol{\theta})dt + \sigma_B B(t), \tag{7.1}$$

where the degradation process $X(t)$ is driven by a standard Brownian motion $B(t)$ with a nonlinear drift of $\mu(t;\boldsymbol{\theta})$. In Eq. (7.1), $\mu(t;\boldsymbol{\theta})$ and σ_B are the drift and diffusion coefficients, respectively; $\mu(t;\boldsymbol{\theta})$is a nonlinear function with t and parameter vector $\boldsymbol{\theta}$. Equation (7.1) is presumed to satisfy the regularity conditions (in Ito's sense) that guarantee a weakly unique global solution of Eq. (7.1) [59]. Clearly, if $\mu(t;\boldsymbol{\theta}) = \mu$, Eq. (7.1) becomes the conventional linear drifted model in [9, 47, 53].

Considering that each item possibly experiences different sources of variations during its operation, for a degradation model to be realistic, it is more appropriate to incorporate item-to-item variability in the degradation process. As such, we treat the parameter vector $\boldsymbol{\theta}$ to be a function of two parameters: $\boldsymbol{\theta} = (\boldsymbol{a}, \boldsymbol{b})$, where $\boldsymbol{a}$ is a random effect representing between-item variation, and $\boldsymbol{b}$ is a fixed effect that is common to all items. For simplicity, we assume that $\boldsymbol{\theta}$ and $B(t)$ are s-independent and that $\boldsymbol{a}$ follows $N\left(\boldsymbol{\mu}_a, \boldsymbol{\Sigma}_a\right)$. The ideas of random effects and Gaussian assumptions are widely used in degradation modeling (see [9, 20, 21, 53, 60]). It is worth noting that if we set $\sigma_B = 0$ in Eq. (7.1), then the proposed degradation model reduces to the true degradation path model adopted in conventional random-effect regression models [21, 60, 61].

We have established the degradation model using a general diffusion process. We now illustrate how to estimate the RUL based on the established model. Along the line of the work by Aalen and Gueless in [26] and Lee and Whitmore in [27], we use the concept of the FHT to define the lifetime and then infer the RUL.

Figure 7.3 illustrates the RUL modeling principle. When degradation $X(t)$ reaches a preset critical level w, the plant is declared to be nonusable, and, therefore, it is natural to view the event of lifetime termination as the point when the degradation process $X(t)$ crosses the threshold level w for the first time. This FHT requirement may be considered to be restrictive to some cases, since the degradation may go back after the first hit. However, for critical equipment it is usually mandatory to put this

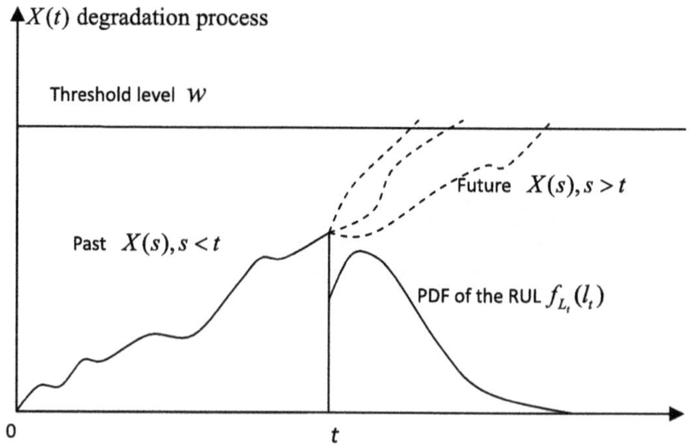

Fig. 7.3 Illustration of the degradation modeling and RUL estimation principle

into practice; once the observed degradation is equal to or above the set threshold level, the equipment must be stopped operating. From the FHT concept, the lifetime Tcan be defined as

$$T = \inf\{t : X(t) \geq w|\, X(0) < w\}. \tag{7.2}$$

Without loss of generality, we first consider the case $X(0) = 0$ below, and then we proceed to the case $X(t), t > 0$.

As mentioned previously, the RUL can be written as $L_t = \{l_t : T - t|\, T > t\}$ at current time t. Therefore, if $f_T(t)$ is known, then the PDF of the RUL can be formulated as

$$f_{L_t}(l_t) = f_T(t + l_t)/R(t), \tag{7.3}$$

where $f_T(t+l_t)$ is the PDF of the lifetime at $t+l_t$ and $R(t)$ is the reliability function at t [62]. Equation (7.3) is the model for a population of identical items without taking into account the individually observed $X(s)$, $s < t$, for each item. Such $f_{L_t}(l_t)$, particularly when $t = 0$, is needed at the design or testing stage before the actual item is put into use in order to have an estimated designed life PDF or to recommend a maintenance schedule for a future planning purpose. As for users, it should be noted that once the observed degradations become available, then an individually predicted RUL is required to make a dynamic decision to determine if and when the item monitored should be repaired or replaced. In the next section, we consider situations where $t = 0$ and $t > 0$ for $X(t)$. Clearly, the key for estimating the RUL is to derive the PDF of lifetime $f_T(t)$; and thus we first focus on deriving $f_T(t)$ in the following section.

7.4 Lifetime Distribution and Parameter Estimation of the Proposed Degradation Model

7.4.1 Derivation of the Lifetime Distribution

It is difficult to derive a general analytical form of the PDF of the FHT when $\mu(t; \boldsymbol{\theta})$ is not a constant. In the following, we develop an approximate lifetime distribution in a closed form.

For simplicity, we begin without considering the random effects of the parameters in the following derivations. In order to derive the lifetime distribution function, we use Lemma 7.1 to transform the degradation process into a standard BM, which has an explicit form of the PDF of the FHT.

Lemma 7.1 *Let a diffusion process $X(t)$ with the drift and the diffusion coefficients $\mu(x, t)$ and $\sigma(x, t)$, respectively, and $c_1(t)$ and $c_2(t)$ be arbitrary functions of the time. Then, if and only if*

$$\mu(x,t) = 4\cdot\frac{\partial\sigma(x,t)}{\partial x} + \frac{[\sigma(x,t)]^{1/2}}{2}\left\{c_1(t) + \int_z^x \frac{c_2(t)\sigma(t,u) + \partial\sigma(u,t)/\partial t}{[\sigma(u,t)]^{3/2}}du\right\}, \tag{7.4}$$

there exists a transformation $\tilde{x} = \psi(t,x), \tilde{t} = \phi(t)$ *that can change the original Kolmogorov equation of the diffusion process into the Kolmogorov equation of the standard BM. This transformation can be specified as*

$$\begin{aligned}\psi(t,x) &= (k_1)^{1/2}\exp\left\{-\frac{1}{2}\int_{t_0}^t c_2(\tau)d\tau\right\}\cdot\int_z^x \frac{1}{(\sigma(y,t))^{1/2}}dy - \\ &\quad \frac{(k_1)^{1/2}}{2}\int_{t_2}^t c_1(\tau)\exp\left\{-\frac{1}{2}\int_{t_0}^\tau c_2(u)du\right\}d\tau + k_2 \\ \phi(t) &= k_1\int_{t_2}^t \exp\left\{-\frac{1}{2}\int_{t_0}^\tau c_2(u)du\right\}d\tau + k_3,\end{aligned} \tag{7.5}$$

where z *is an arbitrary value in the diffusion path of* $X(t)$*. Here,* $t_i \in [0,\infty), i = 0,1,2$ *and* k_1, k_2, k_3 *are arbitrary constants with the only restriction being that* $k_1 > 0$*.*

The definition of $\mu(x,t)$ and $\sigma(x,t)$ can be found in [36]. The proof of Lemma 7.1 has been well documented by Ricciardi in [40] and is thus omitted here. In this way, formulating the PDF of the FHT of $X(t)$ crossing a constant threshold w is equivalent to calculating the PDF of the FHT of a standard BM crossing a transformed boundary using $\tilde{x}$ and $\tilde{t}$. Since t_1, t_2, k_1, k_2, k_3 are arbitrary constants with the only restriction that $k_1 > 0$, we set $k_1 = 1, k_2 = k_3 = 0$ in the following. In our modeling process, we set $X(0) = 0$, and thus we have $t_0 = 0$ and $z = 0$ as well for notation convenience. Now, in order to facilitate deriving the PDF of the FHT of the proposed degradation process crossing a constant threshold w, we first transform this problem into the PDF of the FHT of a standard BM crossing a time-dependent threshold. The result is summarized as follows:

Theorem 7.1 *For the degradation process* $X(t)$*, if* $\mu(t;\boldsymbol{\theta})$ *is a continuous function of time* t *in* $[0,\infty)$*, then the PDF of the FHT of* $X(t)$ *crossing a critical level* w *can be formulated as*

$$p_{X(t)}(w,t) = p_{B(t)}\left(S_B(t),t\right)\frac{d\phi(t)}{dt}, \tag{7.6}$$

where $p_{B(t)}\left(S_B(t),t\right)$ *is the PDF of the FHT of the standard BM* $B(t)$ *crossing the corresponding threshold* $S_B(t)$*. In such a case, the transformation can be written as*

$$\begin{aligned}&\psi(t,x) = \tfrac{1}{\sigma_B}\left(x - \int_0^t \mu(\tau;\boldsymbol{\theta})d\tau\right), \phi(t) = t \\ &S_B(t) = \tfrac{1}{\sigma_B}\left(w - \int_0^t \mu(\tau;\boldsymbol{\theta})d\tau\right)\end{aligned}. \tag{7.7}$$

Proof From Lemma 7.1, we can see that the sufficient and necessary condition for transforming the degradation process $X(t)$ to a standard BM $B(t)$ is that there exist

$c_1(t)$ and $c_2(t)$, which make (7.4) hold. Since $\mu(x,t) = \mu(t;\boldsymbol{\theta}), \sigma(x,t) = \sigma_B^2$, we directly have from (7.4)

$$\mu(t;\boldsymbol{\theta}) = \frac{\sigma_B}{2}\left\{c_1(t) + \int_0^x \frac{c_2(t)}{\sigma_B}dy\right\}. \tag{7.8}$$

Clearly, for Eq. (7.7) to hold, we can set $c_1(t) = 2f(t;\boldsymbol{\theta})/\sigma_B, c_2(t) = 0$, where $f(t,\boldsymbol{\theta})$ is a function of time and $\boldsymbol{\theta}$. Thus, we can see that the conditions of Lemma 7.1 can be satisfied.

Therefore, using the transformation $\tilde{x}$ and $\tilde{t}$ in Lemma 7.1, we can change the original Kolmogorov equation of diffusion $X(t)$ into the Kolmogorov equation of a standard BM. From (7.5), such transformation can be specified as $\psi(t,x) = \frac{1}{\sigma_B}\left(x - \int_0^t \mu(\tau;\boldsymbol{\theta})d\tau\right), \phi(t) = t$. Using the transformation $\tilde{x} = \psi(t,x), \tilde{t} = \phi(t)$, the boundary $S_B(\tilde{t})$ for the standard BM can be calculated by

$$\begin{aligned} S_B(\tilde{t}) &= \psi\left(\varphi^{-1}(\tilde{t}), w\left[\varphi^{-1}(\tilde{t})\right]\right) - \psi(t_0, x_0) = \psi\left(t, w\right) \\ &= \frac{1}{\sigma_B}\left(w - \int_0^t \mu(\tau;\boldsymbol{\theta})d\tau\right) = S_B\left(t\right). \end{aligned} \tag{7.9}$$

It is noted that we use the result $\psi(t_0, x_0) = 0$, since $t_0 = 0$ and $x_0 = 0$, as given previously. Then we can obtain the PDF of the FHT of the diffusion $X(t)$ directly:

$$\begin{aligned} p_{X(t)}(w,t) &= p_{B(t)}\left(S_B(\tilde{t}), \tilde{t}\right)\frac{d\varphi(t)}{dt} = p_{B(t)}\left(S_B\left[\varphi(t)\right], \varphi(t)\right)\frac{d\varphi(t)}{dt} \\ &= p_{B(t)}\left(S_B(t), t\right)\frac{d\varphi(t)}{dt}. \end{aligned} \tag{7.10}$$

This completes the proof of Theorem 7.1.

Remark 7.1 We note from Eq. (7.8) that there is a unique pair of functions, $c_1(t)$ and $c_2(t)$. When the degradation process is time homogeneous, i.e., $\mu(t;\boldsymbol{\theta})$ is independent of the time variable, then $c_1(t)$ is a constant value.

It is clear that the problem under consideration can be transformed to formulate the FHT distribution of a standard BM crossing time-varying boundary $S_B(t)$. As mentioned in the introduction, an analytical and explicit form is more desirable than numerical approximation in reliability and RUL estimation. To achieve this aim, we present the following analytical approximation. First, we give one of the lemmas presented by Durbin [63].

Lemma 7.2 *For a Gaussian process $W(t)$ with $E\left[W(t)\right] = 0$ and covariance function $\rho(s,t)$ for $0 \le s \le t$, where $E\left[\cdot\right]$ is the expectation operator, if the following assumptions—*

1. *The boundary function, $S(s)$, is continuous in $0 \le s < t$, and is left differentiable at t.*

2. *The covariance, $\rho(s,t)$, is positive definite and has continuous first-order partial derivatives on the set $\{(s,t): 0 \le s < t\}$, where appropriate left or right derivatives are taken at $s = 0$, $s = t$.*
3. $\lim_{s \to t} \left[\frac{\partial \rho(s,t)}{\partial s} - \frac{\partial \rho(s,t)}{\partial t} \right] = \lambda_t$, *with* $0 < \lambda_t < \infty$

—can be satisfied, then the PDF of the FHT of the process $W(t)$ crossing the boundary $S(t)$ can be written as

$$p_{W(t)}(S(t), t) = b(t) h_{W(t)}(t), \tag{7.11}$$

where $h_{W(t)}(t)$ denotes the PDF of $W(t)$ on boundary $S(t)$. That is,

$$h_{W(t)}(t) = \frac{1}{\sqrt{2\pi\rho(t,t)}} \exp\left[-\frac{S^2(t)}{2\rho(t,t)} \right], \tag{7.12}$$

and $b(t)$ can be written as

$$b(t) = \lim_{s \to t} (t-s)^{-1} E_{W(s)|W(t)} \left[I(s,W)\left(S(s) - W(s)\right) \middle| W(t) = S(t) \right], \tag{7.13}$$

where $I(s,W)$ is an indicator defined to equal 1 if the sample path of the Gaussian process of interest does not cross the boundary prior to time s and to equal 0 otherwise.

Now we introduce one important assumption that will be used to prove Theorem 7.2 as follows:

Assumption 7.1 If the degradation process is hitting the threshold at a certain time t exactly, then the probability that such a process crossed the threshold level before time t is assumed to be negligible.

Assumption 7.1 implies that $I(s,W) \cong 1$ at the time of hitting the threshold and is used for obtaining an approximated lifetime distribution. It requires some clarification. Indeed, there will be a nonzero probability that the process will cross the threshold before t in a strict sense since t is not the FHT. However, such a probability should be small because of the drift. Also, the probability of the process hitting the threshold several times should be small depending on the values of the drift and diffusion parameters. We know that degradation data are always collected at discrete time points, so we can only stop the item when we first observe that $X(t)$ has crossed w. So whether the degradation process had crossed w before is irrelevant, since the exact FHT was not observed. There is also an additional advantage of using this assumption, which is that the PDF of the FHT defined here is not a strict FHT by definition, but rather a hitting time not far from it. There are concerns that using the FHT as the time of failure is conservative. For example, Barker and Newby in [64] argued that a system could hit the critical threshold and then return back to or below the critical level. Hence, they defined the lifetime as the last exit time from the threshold. Wang and Xu in [23] recently expressed a similar idea. The same idea can also be found in the statistical control chart literature [65]. This shows that our formulation and consideration have a practical meaning. It is difficult to theoretically assess the accuracy of the approximation, but we present several comparisons using the simulation after obtaining the main results to show the closeness of our approximation and empirically validate this assumption.

Based on Lemmas 7.1 and 7.2, Theorem 7.1 and Assumption 7.1, we present the following theorem for constructing the PDF of the FHT of our established degradation process crossing a constant threshold w.

Theorem 7.2 *For the degradation process $X(t)$ given by Eq. (7.1), if $\mu(t;\boldsymbol{\theta})$ is a continuous function of time t in $[0,\infty)$, then the PDF of the FHT of $X(t)$ crossing a constant boundary w can be approximated with an explicit form under Assumption 7.1 as follows:*

$$p_{X(t)}(w,t) \cong \frac{1}{\sqrt{2\pi t}}\left(\frac{S_B(t)}{t}+\frac{1}{\sigma_B}\mu(t;\boldsymbol{\theta})\right)\exp\left[-\frac{S_B^2(t)}{2t}\right]. \tag{7.14}$$

Proof From Theorem 7.1 and Lemma 7.2, the PDF of the FHT of $X(t)$ crossing a constant threshold w can be written as

$$p_{X(t)}(w,t) = p_{B(t)}\left(S_B(t),t\right)\frac{d\phi(t)}{dt} = b(t)h_{B(t)}(t)\frac{d\phi(t)}{dt}, \tag{7.15}$$

where $h_{B(t)}(t)$ denotes the PDF of the standard BM at boundary $S_B(t)$. Then from Eqs. (7.11) and (7.12), and noting that $\rho(t,t)=t$ for a standard BM, $b(t)$ and $h_{B(t)}(t)$ can be formulated as follows:

$$h_{B(t)}(t) = \frac{1}{\sqrt{2\pi t}}\exp\left[-\frac{S_B^2(t)}{2t}\right], \tag{7.16}$$

$$b(t) = \lim_{s\to t}(t-s)^{-1}E\left[I(s,B)\left(S_B(s)-B(s)\right)\middle|\, B(t)=S_B(t)\right]. \tag{7.17}$$

Under the transformation $\tilde{t}=\phi(t)$ in Theorem 7.1, we directly have $\tilde{t}=\phi(t)=t$. This means that there is no timescale transformation in the process of transforming the diffusion process $X(t)$ into a standard BM. As a result, using Assumption 7.1, we assumed that the probability of the transformed BM, $B(t)$, reaching boundary $S_B(t)$ before t, given that $B(t)$ crossed $S_B(t)$ at time t, can be neglected; as such, we can approximately have $I(s,B)\cong 1$. Then, using the property of a standard BM and the L'Hospital rule [66], $b(t)$ can be formulated as

$$\begin{aligned} b(t) &\cong \lim_{s\to t}(t-s)^{-1}\mathbb{E}_{B(s)|B(t)}\left[\left(S_B(s)-B(s)\right)\middle|\, B(t)=S_B(t)\right] \\ &= \lim_{s\to t}\frac{S_B(s)-\mathbb{E}_{B(s)|B(t)}\left[B(s)\middle|\, B(t)=S_B(t)\right]}{t-s} \\ &= \lim_{s\to t}\frac{S_B(s)-sS_B(t)/t}{t-s} = \frac{S_B(t)}{t}-\frac{\mathrm{d}S_B(t)}{\mathrm{d}t} = \frac{S_B(t)}{t}+\frac{1}{\sigma_B}\mu(t;\boldsymbol{\theta}). \end{aligned} \tag{7.18}$$

This completes the proof of Theorem 7.2.

If the degradation is monotonic, such as the gamma processes, then a simple FHT model using $\Pr(T \le t) = \Pr(X(t) \ge w)$ can be directly obtained. However, our process is not monotonic, and Assumption A does not imply a monotonic property. The process can hit the threshold before t or go back below the threshold after t, but this is just assumed to be a small probability. This is consistent with the characteristics of diffusion processes. In this sense, our assumption is weaker than the monotonic assumption.

Below, we consider two special cases of $\mu(t; \boldsymbol{\theta})$ and give the corresponding results using Theorem 7.2. In addition, these results will be used for the examples in Sect. 7.5. Of course, other forms of $\mu(t; \boldsymbol{\theta})$ can be selected, but we only consider these two special cases for illustrative purposes.

Corollary 7.1 *Consider that the unknown parameters are fixed and that there are no random effects among them. With the drift coefficients* $\mu(t; \boldsymbol{\theta}) = abt^{b-1}$ *and* $\mu(t; \boldsymbol{\theta}) = ab\exp(bt)$ *corresponding to Model 1 (M_1) and Model 2 (M_2), respectively, we can obtain two different nonlinear diffusion degradation models. The PDFs of FHT under M_1 and M_2 can be formulated, respectively, from Eqs. (7.9) and (7.14), as*

$$f_{T|M_1,\boldsymbol{\theta}}\left(t|\, M_1, \boldsymbol{\theta}\right) \cong \frac{w - at^b(1-b)}{\sigma_B\sqrt{2\pi t^3}} \exp\left\{-\frac{\left(w - at^b\right)^2}{2\sigma_B^2 t}\right\}, \tag{7.19}$$

$$f_{T|M_2,\boldsymbol{\theta}}\left(t|\, M_2, \boldsymbol{\theta}\right) \cong \frac{w - a\left(\exp\{bt\} - bt\exp\{bt\} - 1\right)}{\sigma_B\sqrt{2\pi t^3}} \exp\left\{-\frac{\left(w - a\exp\{bt\} + a\right)^2}{2\sigma_B^2 t}\right\}, \tag{7.20}$$

where $\boldsymbol{\theta} = (a, b)$.

In the following, the obtained results are compared with the results under the monotone assumption.

Under the monotone assumption, the lifetime of the system can be obtained by $\Pr(T \le t) = \Pr(X(t) \ge w)$. This implies that the process can only hit the threshold once and cannot go back. However, for BM-based models, the degradation path is not monotonic so the simple approximation as above equation cannot be used, or if used can only be considered as a crude approximation or will yield some unexpected results.

First, based on the monotone assumption, the lifetime distribution can be calculated as

$$\Pr(T \le t) = \Pr(X(t) \ge w) = 1 - \Phi\left(\frac{w - h(t; \boldsymbol{\theta})}{\sigma_B^2 t}\right), \tag{7.21}$$

where $\Phi(\cdot)$ is the CDF of the standard normal variable, and $h(t; \boldsymbol{\theta}) = \int_0^t \mu(\tau; \boldsymbol{\theta})\mathrm{d}\tau$. In this case, the obtained PDF $f_T(t)$ from the nonlinear model in this chapter can be formulated as

$$f_T(t) = -\frac{\mathrm{d}}{\mathrm{d}t}\Phi\left(\frac{w - h(t;\boldsymbol{\theta})}{\sigma_B^2 t}\right) = -\frac{1}{\sqrt{2\pi}}\exp\left\{-\frac{(w - h(t;\boldsymbol{\theta}))^2}{2\sigma_B^2 t}\right\}\cdot\frac{\mathrm{d}}{\mathrm{d}t}\frac{w - h(t;\boldsymbol{\theta})}{\sqrt{\sigma_B^2 t}}. \tag{7.22}$$

Since

$$\frac{\mathrm{d}}{\mathrm{d}t}\frac{w - h(t;\boldsymbol{\theta})}{\sqrt{\sigma_B^2 t}} = \frac{1}{\sigma_B}\left[\frac{0.5h(t;\boldsymbol{\theta}) - 0.5w}{t^{3/2}} - \frac{\mu(t;\boldsymbol{\theta})}{\sqrt{t}}\right], \tag{7.23}$$

we further have

$$f_T(t) = \frac{1}{\sigma_B\sqrt{2\pi}}\exp\left\{-\frac{(w - h(t;\boldsymbol{\theta}))^2}{2\sigma_B^2 t}\right\}\cdot\left[\frac{0.5h(t;\boldsymbol{\theta}) - 0.5w}{t^{3/2}} - \frac{\mu(t;\boldsymbol{\theta})}{\sqrt{t}}\right]. \tag{7.24}$$

Take the model M_1 for an example, under the monotone assumption, following (7.24), the estimated PDF of the lifetime is

$$f_T(t) = \frac{0.5w - at^b(0.5 - b)}{\sigma_B\sqrt{2\pi t^3}}\exp\left\{-\frac{(w - at^b)^2}{2\sigma_B^2 t}\right\}. \tag{7.25}$$

When $b = 1$, we have

$$f_T(t) = \frac{0.5w + 0.5at}{\sigma_B\sqrt{2\pi t^3}}\exp\left\{-\frac{(w - at)^2}{2\sigma_B^2 t}\right\}. \tag{7.26}$$

When $b = 0$, we have

$$f_T(t) = \frac{0.5w - 0.5a}{\sigma_B\sqrt{2\pi t^3}}\exp\left\{-\frac{(w - a)^2}{2\sigma_B^2 t}\right\}. \tag{7.27}$$

Clearly, these results cannot be consistent with the exact results for these two special cases. In contrast, the method in this chapter can obtain the exact result as follows:

If $b = 1$, there is

$$f_{T|M_1,\boldsymbol{\theta}}(t|\,M_1,\boldsymbol{\theta}) = \frac{w}{\sigma_B\sqrt{2\pi t^3}}\exp\left\{-\frac{(w - at)^2}{2\sigma_B^2 t}\right\}. \tag{7.28}$$

If $b = 0$, there is

$$f_{T|M_1,\boldsymbol{\theta}}(t|\,M_1,\boldsymbol{\theta}) = \frac{w - a}{\sigma_B\sqrt{2\pi t^3}}\exp\left\{-\frac{(w - a)^2}{2\sigma_B^2 t}\right\}. \tag{7.29}$$

Regarding the above comparative results, we have the following remark:

Remark 7.2 In this chapter, we do not assume that degradation process is monotonic and we just say the probability of hitting the threshold before t is small enough. This allows the degradation process to be non-monotonic and is consistent with the characteristics of diffusion process. There are two advantages to utilize our assumption compared with the monotonic assumption. First, in deriving the lifetime distribution, we actually used the process information before t. Look at the following equation:

$$b(t) = \lim_{s \to t}(t-s)^{-1} \cdot \mathrm{E}_{W(s)|W(t)}\left[I(s, W)(S(s) - W(s))|W(t) = S(t)\right].$$

Because there is no transformation in timescale and so the timescale in the $X(t)$ and the transformed process is the same. Therefore, in Eq. (7.18), using the information before t is equivalent to use the process information of $X(t)$ before t. However, applying $F(t) = \Pr(X(t) \geq \omega)$ to approximate the lifetime, only the degradation $X(t)$ at time t is used. In this sense, the proposed model can utilize more information. Second, from (7.19) and (7.20), $f_{T|M_1,\boldsymbol{\theta}}(t|\, M_1, \boldsymbol{\theta})$ can be reduced to the inverse Gaussian distribution exactly when $b = 1$ and both $f_{T|M_1,\boldsymbol{\theta}}(t|\, M_1, \boldsymbol{\theta})$ and $f_{T|M_2,\boldsymbol{\theta}}(t|\, M_2, \boldsymbol{\theta})$ can be reduced to the FHT distribution of the diffusion process with zero drift when $b = 0$, also exactly. This is as expected since any properly developed nonlinear model should cover a linear model as a special case. However, if we use $\Pr(T \leq t) = \Pr(X(t) \geq w)$ to obtain the FHT distribution for M_1 and M_2 as an approximation under the monotonic assumption, it has been shown without difficulty that the obtained results cannot go back to the linear and zero drift cases exactly. These discussions reveal the rationality of the proposed model and method.

7.4.2 Lifetime Distribution Under Random Effects

In the above derivations, we assume that there is no random effect in the model parameter space. However, different items have variability in their degradation paths. This can be interpreted as the item-to-item variability. In the current literature, considering the random effect in a parameter is a common way to characterize this variability. For simplicity, we consider that a is the random effect representing between-item variation and that it follows a normal distribution with mean μ_a and variance σ_a^2, while b is the fixed effect and common to all items. In order to facilitate the derivation in the case of considering the random effect in unknown parameters, we give the following lemma:

Lemma 7.3 *If $Z \sim N(\mu, \sigma^2)$, and $w, A, B, C \in \boldsymbol{R}$, then the following holds:*

$$E_Z\left[(w - AZ) \cdot \exp\left(-\frac{(w-BZ)^2}{2C}\right)\right] = \sqrt{\frac{C}{B^2\sigma^2 + C}}\left(w - A\frac{B\sigma^2 w + \mu C}{B^2\sigma^2 + C}\right) \times \exp\left(-\frac{(w-B\mu)^2}{2\left(B^2\sigma^2 + C\right)}\right). \quad (7.30)$$

Based on Theorem 7.3, we can obtain the PDF of the FHT by the law of total probability. The main results are summarized in the following corollary:

Corollary 7.2 *If b is fixed and $a \sim N(\mu_a, \sigma_a^2)$, the PDFs of the FHT of M_1 and M_2 can be formulated as*

$$f_{T|M_1}(t|\,M_1) \cong \frac{1}{\sqrt{2\pi t^3\left(\sigma_a^2 t^{2b-1}+\sigma_B^2\right)}}\left(w-(t^b-bt^b)\frac{w\sigma_a^2 t^{b-1}+\mu_a\sigma_B^2}{\sigma_a^2 t^{2b-1}+\sigma_B^2}\right)\times$$
$$\exp\left\{-\frac{\left(w-\mu_a t^b\right)^2}{2t\left(\sigma_a^2 t^{2b-1}+\sigma_B^2\right)}\right\}, \tag{7.31}$$

$$f_{T|M_2}(t|\,M_2) \cong \frac{1}{\sqrt{2\pi t^2\left(\sigma_a^2\gamma(t)^2+\sigma_B^2 t\right)}}\left(w-\beta(t)\frac{w\sigma_a^2\gamma(t)+\mu_a\sigma_B^2 t}{\sigma_a^2\gamma(t)^2+\sigma_B^2 t}\right)\times$$
$$\exp\left\{-\frac{\left(w-\mu_a\gamma(t)\right)^2}{2\left(\sigma_a^2\gamma(t)^2+\sigma_B^2 t\right)}\right\}, \tag{7.32}$$

with $\gamma(t)=\exp(bt)-1$, $\beta(t)=\exp(bt)-bt\exp(bt)-1$.

Proof For M_1, from Corollary 7.1 and using the law of total probability, we obtain

$$f_{T|M_1}(t|\,M_1) = \frac{1}{\sigma_B\sqrt{2\pi t^3}}E_a\left\{\left(w-at^b(1-b)\right)\exp\left[-\left(w-at^b\right)^2\Big/2\sigma_B^2 t\right]\right\}. \tag{7.33}$$

Let $A=t^b(1-b)$, $B=t^b$, $C=\sigma_B^2 t$; the result can be obtained straightforwardly using Theorem 7.3.

The proof for M_2 is similar and is thus omitted. It is noted that the model in [53] can be incorporated into our modeling framework when $b=1$ for M_1.

In order to check the appropriateness of Assumption 7.1 and the approximation accuracy of our proposed models, we develop an FHT simulation algorithm to generate the FHT from $\{X(t), t\geq 0\}$ defined as Eq. 7.1) to compare with the PDF of the FHT from our model. The fundamental principle of this simulation algorithm is that we can approximate $\{X(t), t\geq 0\}$ with the following so-called *Euler* approximation [59, 70]

$$X_{(k+1)\Delta t} = X_{k\Delta t}+\mu(k\Delta t;\boldsymbol{\theta})\Delta t+\sigma_B Y\sqrt{\Delta t},$$

where $Y\sim N(0,1)$ and Δt is the discretization step.

If parameter $\boldsymbol{\theta}$ in the above equation is unknown, we should first use a parameter estimation algorithm to estimate $\boldsymbol{\theta}$. For simplicity, but without loss of generality, we assume that the parameter set $\boldsymbol{\theta}$ or its distribution is known in our simulation. Therefore, according to the definition of the FHT, we present an algorithm for simulating the FHT as follows:

Algorithm 7.1 (FHT simulation algorithm)

Step 1: Initialize the number of sampling paths M, discretization step Δt, threshold w, and initial state X_0.

Step 2: Launch the mth sampling path from the initial setting in Step 1 and let $k = 0$.

Step 3: At time instant $k\Delta t$ for the mth sampling path, sample random numbers $\boldsymbol{\theta}$ and Y from its distribution, respectively.

Step 4: Calculate $X^{(m)}_{(k+1)\Delta t}$ using a Euler approximation from $X^{(m)}_{k\Delta t}$. If $X^{(m)}_{(k+1)\Delta t} \geq w$, the FHT of the mth sampling path can be calculated by $T^{(m)} = (k+1)\Delta t$, the mth sampling path is terminated. As such, set $m = m + 1$ and return to Step 2. Otherwise, set $k = k + 1$ and return to Step 3 to continue the mth sampling path until condition $X^{(m)}_{(k+1)\Delta t} \geq w$ is satisfied and $T^{(m)}$ is obtained.

Step 5: Repeat Steps 2–4 until M FHTs are simulated, i.e., $\boldsymbol{T} = \{T^{(1)}, T^{(2)}, \ldots T^{(M)}\}$.

As an illustration, we consider a diffusion process with $dX(t) = 1.5at^{1/2}dt + \sigma_B dB(t)$, where $a \sim N(1, 0.001)$. In order to start our simulation algorithm, we set the threshold $w = 2.5$, initial state $X_0 = 0$, simulation sample path number $M = 10000$, and discretization step $\Delta t = 0.01$. Since σ_B has a dominant role for governing the process uncertainty, we give three different cases for σ_B to compare our results with the simulated FHTs, i.e., 1) *Case* 1$---\sigma_B$ *has a small value such as* $\sigma_B = 0.05$; 2) *Case* 2$---\sigma_B$ *has a moderate value such as* $\sigma_B = 0.2$; 3) *Case* 3$---\sigma_B$ *has a relatively large value such as* $\sigma_B = 2$. The main results are summarized in Fig. 7.4.

As shown in Fig. 7.4, it is obvious that our method can generate an accurate approximation of the PDF of the FHT of the given process. It should be noted that simulated FHTs are the realizations of the random FHT, so when the sample size is large, the histograms approach the true PDF of the FHT. Therefore, from the above comparisons, they provide evidence that the approximate method presented in this chapter is satisfactory. This empirically validates our Assumption 7.1.

It can be seen that the correspondence between the histograms generated via simulations and the PDF curves produced by our models are very close in Appendix-B. Of course, such a simulation can be used to generate an empirical PDF of the FHT, but our method can obtain an explicit parametric form of the estimated FHT, which is desired for online realization and real-time engineering applications.

7.4.3 The Distribution of the RUL Estimation

Now we proceed to our main aim, which is the estimation of the RUL at a particular point of time t_i, which may be the ith monitoring point from the starting time. It is not a straightforward transformation of w by $w - X(t_i)$, so we present another theorem as follow:

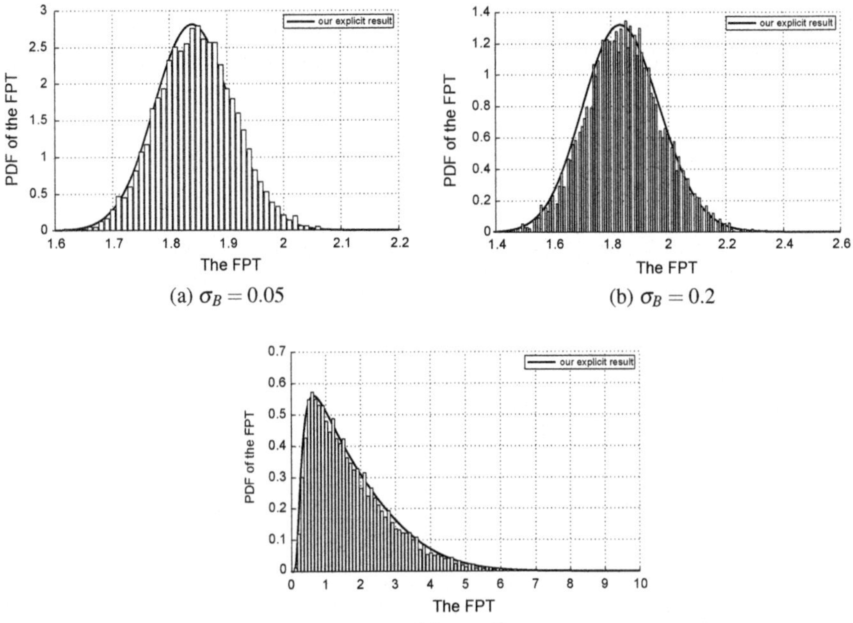

(a) $\sigma_B = 0.05$
(b) $\sigma_B = 0.2$
(c) $\sigma_B = 2$

Fig. 7.4 Comparison of the simulated FHT and our obtained explicit result

Theorem 7.3 *Under the same conditions as in Corollary 7.2, the PDFs of the RUL of M_1 and M_2 can be formulated at time t_i with the available degradation measurement $X(t_i)$ as*

$$f_{L_i|x_i,M_1}(l_i|x_i, M_1) \cong \frac{1}{\sqrt{2\pi l_i^2\left(\sigma_a^2\eta(l_i)^2+\sigma_B^2 l_i\right)}} \exp\left\{-\frac{(w_i-\mu_a\eta(l_i))^2}{2\left(\sigma_a^2\eta(l_i)^2+\sigma_B^2 l_i\right)}\right\} \times \left(w_i-\left(\eta(l_i)-bl_i(l_i+t_i)^{b-1}\right)\frac{\sigma_a^2\eta(l_i)w_i+\mu_a\sigma_B^2 l_i}{\sigma_a^2\eta(l_i)^2+\sigma_B^2 l_i}\right), \tag{7.34}$$

with $\eta(l_{t_i}) = (l_{t_i}+t_i)^b - t_i^b$, $w_{t_i} = w - X(t_i)$, and

$$f_{L_i|x_i,M_2}(l_i|x_i, M_2) \cong \frac{1}{\sqrt{2\pi l_i^2\left(\sigma_a^2\gamma(l_i)^2+\sigma_B^2 l_i\right)}} \exp\left\{-\frac{(w_i-\mu_a\gamma(l_i))^2}{2\left(\sigma_a^2\gamma(l_i)^2+\sigma_B^2 l_i\right)}\right\} \times \left(w_i-\beta(l_i)\frac{\sigma_a^2 w_i\gamma(l_{t_i})+\mu_a\sigma_B^2 l_i}{\sigma_a^2\gamma(l_i)^2+\sigma_B^2 l_i}\right), \tag{7.35}$$

with $\gamma(l_{t_i}) = \exp\left(b(l_{t_i}+t_i)\right) - \exp(bt_i)$, and $\beta(l_{t_i}) = (1-bl_{t_i})\exp\left(b(l_{t_i}+t_i)\right) - \exp(bt_i)$.

Proof We only prove the case for M_1; the proof for M_2 is similar and thus is omitted. Since we observe $X(t_i)$ at t_i; then for $t \geq t_i$, the degradation process can be written as $X(t) = X(t_i) + a(t^b - t_i^b) + \sigma_B B(t - t_i)$. In such a case, the residual $t - t_i$ corresponds to the realization of the RUL at time t_i if t is the FHT of $\{X(t), t \geq t_i\}$. Having this in mind, we take the transformation $l_{t_i} = t - t_i$ with $l_{t_i} \geq 0$, and then the process $\{X(t), t \geq t_i\}$ can be transformed into

$$X(l_{t_i} + t_i) - X(t_i) = a\left((l_{t_i} + t_i)^b - t_i^b\right) + \sigma_B B(l_{t_i}), with\ l_{t_i} \geq 0. \tag{7.36}$$

As a result, the RUL at time t_i is equal to the FHT of the process $\left\{Y(l_{t_i}), l_{t_i} \geq 0\right\}$ crossing the threshold $w_{t_i} = w - X(t_i)$, where $Y(l_{t_i}) = X(l_{t_i} + t_i) - X(t_i)$ and $Y(0) = 0$. That is to say,

$$Y(l_{t_i}) = a\left((l_{t_i} + t_i)^b - t_i^b\right) + \sigma_B B(l_{t_i}).$$

It is easy to verify that $\left\{Y(l_{t_i}), l_{t_i} \geq 0\right\}$ satisfies all the conditions of Theorems 7.1, 7.2 and Lemma 7.2; so we directly have $\mu(l_{t_i}; \boldsymbol{\theta}) = ab\left(l_{t_i} + t_i\right)^{b-1}$ and $S_B(l_{t_i}) = \frac{1}{\sigma_B}\left[w_{t_i} - a\left((l_{t_i} + t_i)^b - t_i^b\right)\right]$. Similar to the deriving process of Corollary 7.2, we can obtain the PDF of the RUL of M_1, as summarized in (7.34) using the results of Theorems 7.2 and 7.3 after some complicated manipulations.

For parameter estimation, the usual MLE approach can be used to estimate the unknown parameters. In the following, we only consider M_1 and M_2 to show how to estimate the unknown parameters from the available degradation data.

7.5 Parameters Estimation

We now present a parameter estimation algorithm for the degradation model based on M_1 and M_2 in order to implement the derived models. We assume two cases here: one is that all items are measured at the same time, and the other is that the items are measured at different time points. The detailed presentations are summarized below.

In order to achieve parameter estimation, we assume that there are N tested items, and the degradation measurements of the $n$thitem are available at time $t_{n,1}, \ldots, t_{n,m_n}$, where m_n denotes the available number of degradation measurements of the nth item, $n = 1, \ldots, N$. Therefore, the sample path of the nth item at the jth time point $t_{n,j}$ is, from Eq. (7.1), given by

$$X_n(t_{n,j}) = \varphi(t_{n,j})a_n + \sigma_B B(t_{n,j}), \tag{7.37}$$

where $j = 1, \ldots, m_n$ and a_n are s-independent and identically distributed following $N(\mu_a, \sigma_a^2)$, as in Corollary 7.2.

For simplicity, we define $\varphi(t)$ as $\varphi(t) = t^b$ and $\varphi(t) = \exp(bt) - 1$ for M_1 and M_2, respectively, and let $\boldsymbol{T}_n = \left(t_{n,1}, \ldots, t_{n,m_n}\right)'$, $T_{n,j} = \varphi(t_{n,j})$, $\boldsymbol{X}_n =$

$\left(x_n(t_{n,1}), \ldots, x_n(t_{n,m_n})\right)'$, where $(\cdot)'$ denotes the vector transposition, and $\boldsymbol{X}$ denotes all the degradation data, consisting of $\boldsymbol{X}_n, n = 1, \ldots, N$. According to Eq. (7.37) and the independent assumption of BM, $\boldsymbol{X}_n$ follows a multivariate normal with mean and variance as follows:

$$\tilde{\boldsymbol{\mu}}_n = \mu_a \boldsymbol{T}_n, \quad \boldsymbol{\Sigma}_n = \boldsymbol{\Omega}_n + \sigma_a^2 \boldsymbol{T}_n \boldsymbol{T}_n', \tag{7.38}$$

where

$$\boldsymbol{Q}_n = \begin{bmatrix} t_{n,1} & t_{n,1} & \cdots & t_{n,1} \\ t_{n,1} & t_{n,2} & \cdots & t_{n,2} \\ \vdots & \vdots & \vdots & \vdots \\ t_{n,1} & t_{n,2} & \cdots & t_{n,m_n} \end{bmatrix}, \quad \boldsymbol{\Omega}_n = \sigma_B^2 \boldsymbol{Q}_n. \tag{7.39}$$

Due to the independent assumption of the degradation measurements of different items, the log-likelihood function over parameter set $\boldsymbol{\Theta} = \left(\mu_a, \sigma_a^2, \sigma_B^2, b\right)^T$ can be written as

$$\begin{aligned} \ell(\boldsymbol{\Theta} \mid \boldsymbol{X}) = & -\ln(2\pi) \sum\nolimits_{n=1}^{N} m_n - \frac{1}{2} \sum\nolimits_{n=1}^{N} \ln |\boldsymbol{\Sigma}_n| \\ & - \frac{1}{2} \sum\nolimits_{n=1}^{N} (\boldsymbol{X}_n - \mu_a \boldsymbol{T}_n)' \boldsymbol{\Sigma}_n^{-1} (\boldsymbol{X}_n - \mu_a \boldsymbol{T}_n), \end{aligned} \tag{7.40}$$

where

$$|\boldsymbol{\Sigma}_n| = |\boldsymbol{\Omega}_n| \left(1 + \sigma_a^2 \boldsymbol{T}_n' \boldsymbol{\Omega}_n^{-1} \boldsymbol{T}_n\right), \tag{7.41}$$

$$\boldsymbol{\Sigma}_n^{-1} = \boldsymbol{\Omega}_n^{-1} - \frac{\sigma_a^2}{1 + \sigma_a^2 \boldsymbol{T}_n' \boldsymbol{\Omega}_n^{-1} \boldsymbol{T}_n} \boldsymbol{\Omega}_n^{-1} \boldsymbol{T}_n \boldsymbol{T}_n' \boldsymbol{\Omega}_n^{-1}. \tag{7.42}$$

Taking the first partial derivative of the log-likelihood function of Eq. (7.40) with respect to μ_a, σ_a gives

$$\frac{\partial \ell(\boldsymbol{\Theta} \mid \boldsymbol{X})}{\partial \mu_a} = \sum\nolimits_{n=1}^{N} \boldsymbol{T}_n' \boldsymbol{\Sigma}_n^{-1} \boldsymbol{X}_n - \mu_a \sum\nolimits_{n=1}^{N} \boldsymbol{T}_n' \boldsymbol{\Sigma}_n^{-1} \boldsymbol{T}_n, \tag{7.43}$$

$$\begin{aligned} \frac{\partial \ell(\boldsymbol{\Theta} \mid \boldsymbol{X})}{\partial \sigma_a} = & -\sum\nolimits_{n=1}^{N} \frac{\sigma_a \boldsymbol{T}_n' \boldsymbol{\Omega}_n^{-1} \boldsymbol{T}_n}{1 + \sigma_a^2 \boldsymbol{T}_n' \boldsymbol{\Omega}_n^{-1} \boldsymbol{T}_n} + \\ & \frac{\sigma_a \sum_{n=1}^{N} (\boldsymbol{X}_n - \mu_a \boldsymbol{T}_n)' \boldsymbol{\Omega}_n^{-1} \boldsymbol{T}_n \boldsymbol{T}_n' \boldsymbol{\Omega}_n^{-1} (\boldsymbol{X}_n - \mu_a \boldsymbol{T}_n)}{\left(1 + \sigma_a^2 \boldsymbol{T}_n' \boldsymbol{\Omega}^{-1} \boldsymbol{T}_n\right)^2}. \end{aligned} \tag{7.44}$$

Case 1: Degradation measurements are available for all paths at the same time, and the number of measurements of each item is the same, i.e., m_n is a constant for all items and $t_{n,j} = t_{l,j}$ for $n, l = 1, \ldots, N$.

Along the line of the work presented in [53], the subscript of $\boldsymbol{T}_n$, $\boldsymbol{\Omega}_n$, $\boldsymbol{\Sigma}_n$ in Eqs. (7.38)–(7.44) can be removed. Thus, using Eqs. (7.41)–(7.44) can be reduced to

$$\frac{\partial \ell(\boldsymbol{\Theta}|\boldsymbol{X})}{\partial \mu_a} = \frac{\sum_{n=1}^N \boldsymbol{T}'\boldsymbol{\Omega}^{-1}\boldsymbol{X}_n - N\mu_a\boldsymbol{T}'\boldsymbol{\Omega}^{-1}\boldsymbol{T}}{1+\sigma_a^2\boldsymbol{T}'\boldsymbol{\Omega}^{-1}\boldsymbol{T}}, \tag{7.45}$$

$$\frac{\partial \ell(\boldsymbol{\Theta}|\boldsymbol{X})}{\partial \sigma_a} = -\frac{N\sigma_a\boldsymbol{T}'\boldsymbol{\Omega}^{-1}\boldsymbol{T}}{1+\sigma_a^2\boldsymbol{T}'\boldsymbol{\Omega}^{-1}\boldsymbol{T}} + \frac{\sigma_a\sum_{n=1}^N (\boldsymbol{X}_n - \mu_a\boldsymbol{T})'\boldsymbol{\Omega}^{-1}\boldsymbol{T}\boldsymbol{T}'\boldsymbol{\Omega}^{-1}(\boldsymbol{X}_n - \mu_a\boldsymbol{T})}{\left(1+\sigma_a^2\boldsymbol{T}'\boldsymbol{\Omega}^{-1}\boldsymbol{T}\right)^2}. \tag{7.46}$$

Then, for specific values of σ_B^2, b, and setting these two derivatives to zero, the results of MLE for μ_a, σ_a^2 can be expressed as

$$\hat{\mu}_a = \frac{\sum_{n=1}^N \boldsymbol{T}'\boldsymbol{\Omega}^{-1}\boldsymbol{X}_n}{N\boldsymbol{T}'\boldsymbol{\Omega}^{-1}\boldsymbol{T}}, \tag{7.47}$$

$$\hat{\sigma}_a = \left\{\frac{1}{N\left(\boldsymbol{T}'\boldsymbol{\Omega}^{-1}\boldsymbol{T}\right)^2}\sum_{n=1}^N (\boldsymbol{X}_n - \hat{\mu}_a\boldsymbol{T})'\boldsymbol{\Omega}^{-1}\boldsymbol{T}\boldsymbol{T}'\boldsymbol{\Omega}^{-1}(\boldsymbol{X}_n - \hat{\mu}_a\boldsymbol{T}) - \frac{1}{\boldsymbol{T}'\boldsymbol{\Omega}^{-1}\boldsymbol{T}}\right\}^{1/2}. \tag{7.48}$$

Further, the profile likelihood function for σ_B, b in terms of estimated μ_a and σ_a^2 can be written as

$$\begin{aligned}\ell(\sigma_B, b\,|\boldsymbol{X}, \hat{\mu}_a, \hat{\sigma}_a) = &-\frac{Nm}{2}\ln(2\pi) - \frac{N}{2} - \frac{N}{2}\ln|\boldsymbol{\Omega}| - \\ &\frac{1}{2}\left\{\sum_{n=1}^N \boldsymbol{X}_n'\boldsymbol{\Omega}^{-1}\boldsymbol{X}_n - \frac{\sum_{n=1}^N\left(\boldsymbol{T}'\boldsymbol{\Omega}^{-1}\boldsymbol{X}_n\right)^2}{\boldsymbol{T}'\boldsymbol{\Omega}^{-1}\boldsymbol{T}}\right\} - \\ &\frac{N}{2}\ln\left\{\frac{\sum_{n=1}^N\left(\boldsymbol{T}'\boldsymbol{\Omega}^{-1}\boldsymbol{X}_n\right)^2}{N\boldsymbol{T}'\boldsymbol{\Omega}^{-1}\boldsymbol{T}} - \frac{\left(\sum_{n=1}^N \boldsymbol{T}'\boldsymbol{\Omega}^{-1}\boldsymbol{X}_n\right)^2}{N^2\boldsymbol{T}'\boldsymbol{\Omega}^{-1}\boldsymbol{T}}\right\}.\end{aligned} \tag{7.49}$$

The MLE of σ_B, b can be obtained by maximizing the profile log-likelihood function in Eq. (7.49) through a two-dimensional search. Then, substituting σ_B, b into Eqs. (7.47) and (7.48), we can obtain the MLE for μ_a, σ_a^2 accordingly.

Case 2: Degradation measurements are available for all paths at different times with different numbers of measurements for each item.

It is clear that Eq. (7.44) may not have an explicit solution through setting the right-hand side of Eq. (7.44) to 0. Thus, for the specific value of σ_a, σ_B, b, and setting the first derivatives of Eq. (7.40) with respect to μ_a to zero, the result of MLE for μ_a can be expressed as

$$\hat{\mu}_a = \frac{\sum_{n=1}^N \boldsymbol{T}_n'\boldsymbol{\Sigma}_n^{-1}\boldsymbol{X}_n}{\sum_{n=1}^N \boldsymbol{T}_n'\boldsymbol{\Sigma}_n^{-1}\boldsymbol{T}_n}. \tag{7.50}$$

Then the profile log-likelihood function of σ_a, σ_B, and b in terms of estimated μ_a can be written as

$$\ell(\sigma_a, \sigma_B, b\,|\boldsymbol{X}, \hat{\mu}_a) = -\frac{\ln(2\pi)}{2}\sum_{n=1}^{N} m_n - \frac{1}{2}\sum_{n=1}^{N}\ln|\boldsymbol{\Sigma}_n| - \frac{1}{2}\left\{\sum_{n=1}^{N}\boldsymbol{X}'_n\boldsymbol{\Sigma}_n^{-1}\boldsymbol{X}_n - 2\frac{\sum_{n=1}^{N}\boldsymbol{T}'_n\boldsymbol{\Sigma}_n^{-1}\boldsymbol{X}_n}{\sum_{n=1}^{N}\boldsymbol{T}'_n\boldsymbol{\Sigma}_n^{-1}\boldsymbol{T}_n}\sum_{n=1}^{N}\boldsymbol{T}'_n\boldsymbol{\Sigma}_n^{-1}\boldsymbol{X}_n + \left(\frac{\sum_{n=1}^{N}\boldsymbol{T}'_n\boldsymbol{\Sigma}_n^{-1}\boldsymbol{X}_n}{\sum_{n=1}^{N}\boldsymbol{T}'_n\boldsymbol{\Sigma}_n^{-1}\boldsymbol{T}_n}\right)^2\sum_{n=1}^{N}\boldsymbol{T}'_n\boldsymbol{\Sigma}_n^{-1}\boldsymbol{T}_n\right\}. \tag{7.51}$$

The MLE of σ_a, σ_B, b, can be obtained by maximizing the profile log-likelihood function in Eq. (7.51) through a three-dimensional search. In this chapter, we use the MATLAB function "fminsearch" for this aim. The function "fminsearch" is a MATLAB function for multidimensional searching using the simplex search method; details can be found in [67]. Now, substituting σ_a, σ_B, b into Eq. (7.50), we obtain the MLE for μ_a.

7.6 Examples of the Applications of the Models

In this section, we illustrate three real-data examples: laser data used in [53, 61], drift degradation data from the INS, and fatigue crack data for 2017-T4 aluminum alloy, as shown in Sect. 7.3. The data analysis was performed using MATLAB.

To compare the fitting of the proposed models, the Akaike information criterion (AIC) [68], and overall mean squared errors (MSEs) of the fitted model compared with the empirical distribution obtained directly from the data for each point in $t_{n,1}, \ldots, t_{n,m_n}$ were both used. AIC balances the log-likelihood with the number of parameters estimated to overcome the problem of overparameterization. The AIC is calculated

$$AIC = -2(\max \ell) + 2p, \tag{7.52}$$

where p is the number of estimated model parameters and $\max \ell$ is the maximized likelihood.

It is noted, however, that MSE directly assesses the fit to the data, so it is another useful measure of goodness of fit [49], though it was acknowledged that AIC is frequently used in engineering and statistical literature to give a guideline for model selection [32, 50]. Let $\hat{F}(t_{n,j}, \hat{\boldsymbol{\Theta}})$ denote the estimated value of the CDF of the lifetime at time $t_{n,j}$ for the nth item with the MLE of the parameters, $\hat{\boldsymbol{\Theta}}$, $\tilde{F}(t_{n,j})$ denote the empirical CDF value at time $t_{n,j}$ for the nth item. This value can be estimated by the median rank method, due to Wilk and Gnanadesikan [69] then we have

$$MSE = \frac{1}{N}\sum\nolimits_{n=1}^{N}\frac{1}{m_n}\sum\nolimits_{j=1}^{m_n}\left(\hat{F}(t_{n,j},\hat{\Theta}) - \tilde{F}(t_{n,j})\right)^2. \tag{7.53}$$

In both criteria, the smallest AIC and MSE values correspond to the best fitting accuracy, and thus they can serve as criteria for model selection.

7.6.1 Laser Data

First, we use the laser data in [61] to compare our methods with the work of Peng and Tseng in [53], in which BM with drift was used for modeling the degradation. The random effect in the drift coefficient was considered as well. We chose their model as a reference for comparison primarily because it is a relatively general one in the literature of BM with drift for degradation modeling. We refer to the model presented by Peng and Tseng in [53] as M_0 below. A detailed data description can be found in [61]. Since the data show a clearly linear path, our objective is to show that our models are more general and can fit this linear data set well. Similar to [53], we set the threshold as $w = 10$. Then, under M_0, M_1, and M_2 separately, we obtain the MLE of the unknown parameters in these models based on the method presented in Sect. 7.4.2. For comparison, we summarize the corresponding estimation results of parameters, the log-likelihood function value (log-LF), and the mean time to failure (MTTF), and we calculate the AIC and MSE from the fitted models (see Table 7.3).

From Table 7.3, we can see that our resulting log-LF and MTTF estimations are slightly different from the results of M_0. The model M_0 displays a marginally better fitting than models M_1 and M_2 in terms of the log-LF. This better performance comes from the linear nature of the laser degradation data. In addition, for M_1, we have $b = 1.0178$. Obviously, b approaches 1 in model M_1. This shows that our models can reduce to model M_0 if the data is linear, which validates our statement that our models are more general, and model M_0 is a special case of our models in the linear case. There are marginal differences among AICs, but in terms of MSE,

Table 7.3 Comparisons of three degradation models with laser data

	M_0	M_1	M_2
μ_a	0.002 036 6	0.001 757 2	132.36
σ_a	0.000 421 48	0.000 360 96	27.144
b	–	1.017 8	1.493 5e-5
σ_B	0.010 115 5	0.010 899	2.763 0
log-LF	65.858 5	65.780 2	65.615 4
MTTF	4912.4	4934.2	4921.4
AIC	−125.717 0	−123.560 4	−123.230 8
MSE$\times 10^{-3}$	2.4	0.032 34	0.283 12

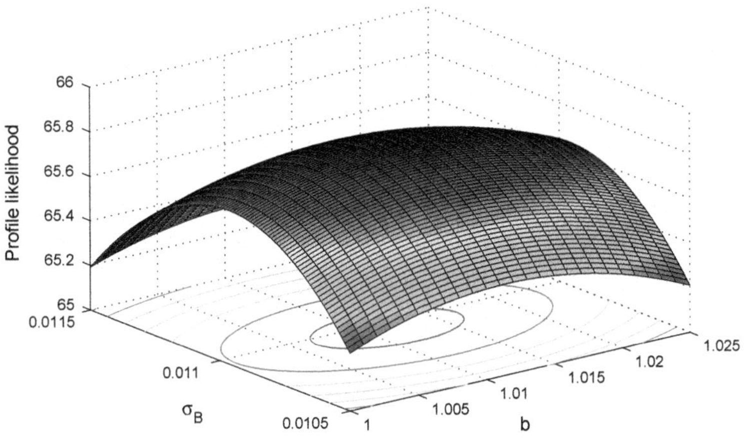

Fig. 7.5 Profile likelihood function of (σ_B, b) for laser data under M_1

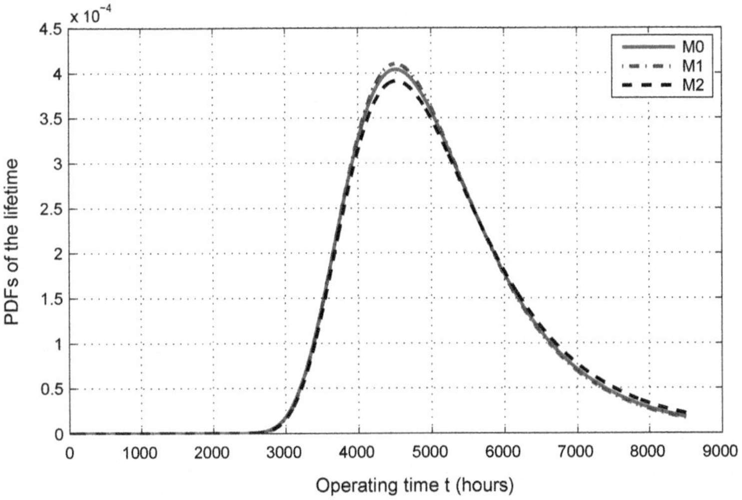

Fig. 7.6 Comparison of the PDFs of the FHT under M_0, M_1, and M_2 with laser data at time zero

M_1 and M_2 show a significant improvement compared with model M_0. It is difficult to prove the uniqueness of the MLE in this case, but we draw a contour plot with a 3-D perspective surface for M_1, as shown in Fig. 7.5, which empirically demonstrates that the profile likelihood function is convex and the MLE is unique in this case. For the following examples, similar results can be obtained but are not shown here due to the limited space. It is noted that "fminsearch" is stable in all cases tested.

Correspondingly, we obtain the PDFs of lifetime at time 0 for M_0, M_1, and M_2, respectively, as shown in Fig. 7.6.

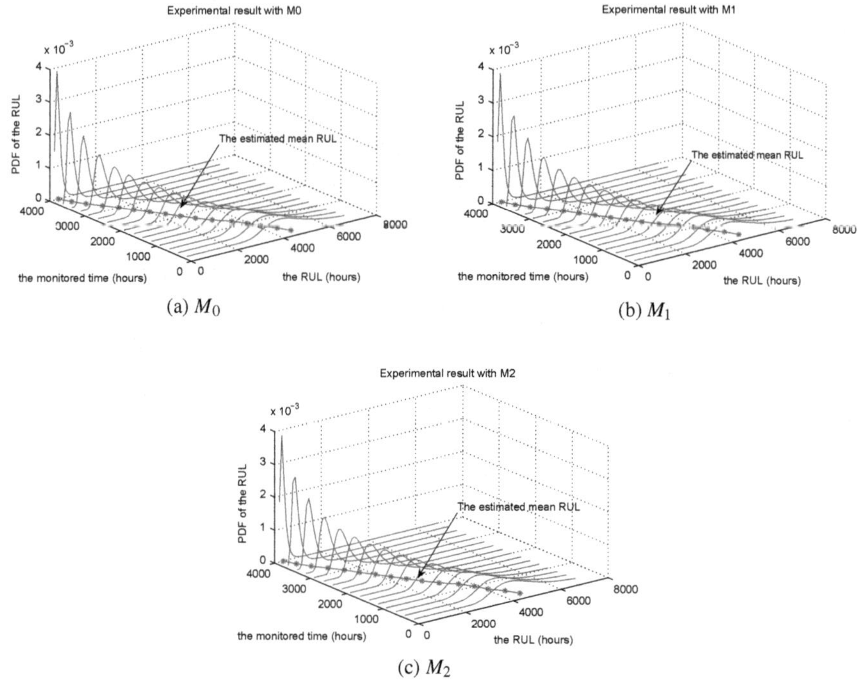

Fig. 7.7 Comparison of the PDFs and means of the RULs under M_0, M_1, and M_2 with the laser data (the observed RULs are marked with $*$)

From Fig. 7.6 we can see that the PDFs of the estimated lifetime for M_0, M_1, and M_2 are almost the same and their differences are trivial. In order to illustrate the usefulness of our method in RUL estimation, we select the 2nd sample in the laser data set to show the estimated RUL curves from M_0, M_1, and M_2 at each measuring point, see Fig. 7.7.

Figure 7.7 shows that the differences of the PDFs of the RULs among these three models are small visually. This, in turn, implies that our methods provide at least as good a fit as the linear method for this case study.

7.6.2 Drift Degradation Data of INS

The gyroscopes used in our example are very expensive, and therefore only limited tests can be performed to obtain the degradation data. Further, because of the influence of the gyroscope's drift on the precision of a navigation system, once the observed drift value is beyond a preset threshold, it must be replaced by a new one to maintain the precision of the INS. In our experiment, the threshold is set as $0.6(^\circ/\text{h})$. In our experiment, the drift data are collected automatically, and the failure times are

Table 7.4 Comparisons of three degradation models with drift degradation data of INS

	M_0	M_1	M_2
μ_a	0.055 705	2.938 6e-25	9.335 8e-9
σ_a	0.024 954	2.732 9e-25	8.667 1e-9
b	–	18.088	0.814 82
σ_B	0.202 89	0.065 709 3	0.065 746
log-LF	−13.898	28.376	28.540
MTTF	13.401 4	23.744	22.557
AIC	33.796 0	−48.752 0	−49.080 0
MSE$\times 10^{-3}$	0.325 2	0.006 2	0.005 5

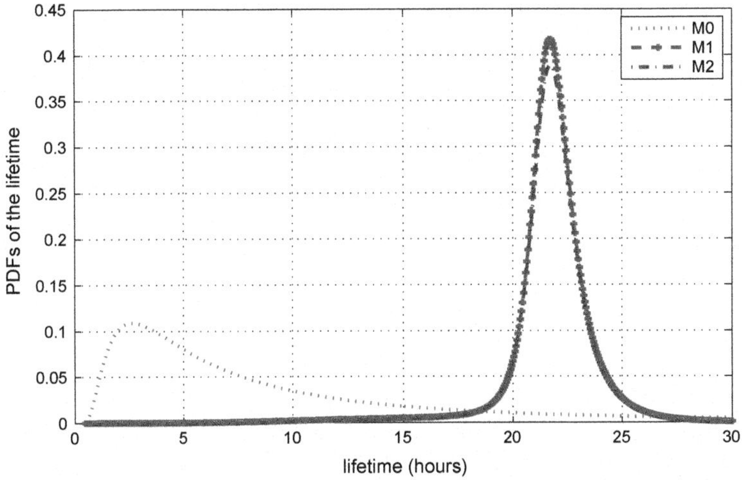

Fig. 7.8 Comparison of the PDFs of the lifetime under M_0, M_1, and M_2 with gyroscope drift data

recorded as the times that the observed values cross over the threshold. After the experiment, the MTTF is about 21.5 h. The drift data of the four gyroscopes are shown in Table 7.2. Using the parameter estimation method, we obtain the corresponding estimation results of the parameters, as shown in Table 7.4. For comparison, we also summarize the estimated log-LF, the MTTF, the AIC, and MSE in Table 7.4.

As shown in Table 7.4, the estimated values of b in M_1 and M_2 clearly confirm the nonlinear characteristics. As in the fatigue crack data case, Table 7.4 shows that our models clearly outperform model M_0 in terms of log-LF, MTTF, and the AIC and MSE. Similar to the previous two studies, the PDFs of the lifetimes of M_0, M_1, and M_2 at time zero are shown in Fig. 7.8 for comparison.

Since the degradation paths of the gyroscopes' drifts are nonlinear, as illustrated in Fig. 7.1, it is natural that models M_1 and M_2 should achieve a better fit of the PDFs of the lifetimes and the associated PDFs of RULs. Compared to M_0, the uncertainty in

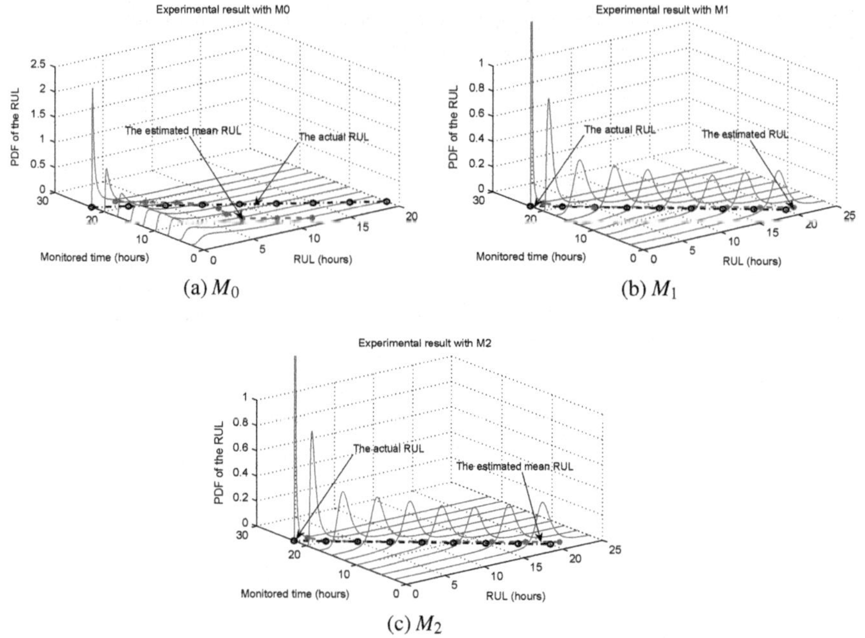

Fig. 7.9 Comparison of the PDFs of the RULs under M_0, M_1, and M_2 with drift degradation data from INS

the estimated PDFs of the lifetimes under M_1 and M_2 is smaller, as seen in Fig. 7.8. Also, the location of the lifetime distribution is not far from the observed MTTF. Therefore, the validity of the proposed methods is further demonstrated.

We now use an individual item to test our obtained models and demonstrate their results in RUL estimation at run time. We select the degradation data of the 4th item in this data set, as shown in Table 7.1, to show the RUL estimation results under M_0, M_1, and M_2, respectively, at different measuring points based on actual observed degradation data. According to Theorem 7.3, the corresponding PDFs of the RULs, the estimated mean RULs, and the actual RULs for the 4th item are shown graphically in Fig. 7.9 for comparison.

As shown in Fig. 7.9, as expected, model M_0 cannot compare with the performance of models M_1 and M_2. The reason is clearly due to the nonlinearity in this case, which further demonstrates the importance of considering the nonlinearity if the degradation process is nonlinear. The prediction errors produced by the linear model can lead to misleading recommendations, which could result in premature replacements in the last two nonlinear examples.

7.6.3 Fatigue Crack Data of 2017-T4

The obtained data are from measured crack length propagation from four test specimens of 2017-T4 aluminum alloy (see Table 7.2 and Fig. 7.2). For all specimens, the crack length is measured at ten measuring points. In general, the time to cross the threshold cannot be recorded precisely. The obtained lifetime data are often interval-censored between two consecutive sampling points. Therefore, we use the data of the first time when the value is observed to be over the threshold as the approximate FHT (failure). In our experiment, the mean time to failure was about 2.4×10^5 cycles. It should be noted that although the sample size is relatively small, the estimated results using our methods are still satisfactory in terms of the AIC and MSE, as shown below. We summarize the main estimation results of interest in Table 7.5.

We now add one model which transforms the original data by $\log X(t)$, which is a common way to transform nonlinear data to linear. Through this data transformation, we treat the data as linear. M_0 in [53] is used. We refer to this model as M_3.

As expected, parameter b in M_1 is far from 1, and this confirms a nonlinear degradation path. In addition, Table 7.4 shows that our models outperform models M_0 and M_3 in terms of log-LF, MTTF, the AIC, and MSE. Specifically, our estimations show significantly better fit in terms of both the AIC and MSE compared with M_0 and M_3. This case study demonstrates the better performance of the presented models over M_0 and M_3 using data transformation in the case of nonlinear degradation. We show the PDFs of lifetimes for M_0, M_1, M_2, and M_3 at time zero based on this data set in Fig. 7.10.

It is obvious that there are significant differences among the PDFs of the lifetimes under M_0, M_1, M_2, and M_3 with this fatigue crack data. The reason is that the fatigue crack data used in this chapter display a clearly nonlinear pattern (see Fig. 7.2). Therefore, model M_0 has limited modeling capability in this case. This weakness can be seen from Fig. 7.10, since the estimated PDF under M_0 of the lifetime covers a wide range so that its uncertainty is very large compared with the results of models M_1 and M_2. The estimated result of M_3 is similar to M_0, although the data is transformed

Table 7.5 Comparison of three degradation models with Fatigue crack data

	M_0	M_1	M_2	M_3
μ_a	2.640 3	7.464 5e-5	0.000 149 55	0.767 16
σ_a	2.105 5	1.440 3e-5	3.531e-5	1.113 5
b	–	12.803	4.440 2	–
σ_B	3.281 7	1.762	0.007 788	1.725 7
log-LF	–63.252	–38.942	–14.366	–42.157
MTTF$\times 10^5$ (cycles)	1.168 6	2.291 2	2.394 1	0.819 2
AIC	132.504 0	85.884 0	36.732 0	90.314 0
MSE$\times 10^{-3}$	276.0	6.6	0.157 95	390.6

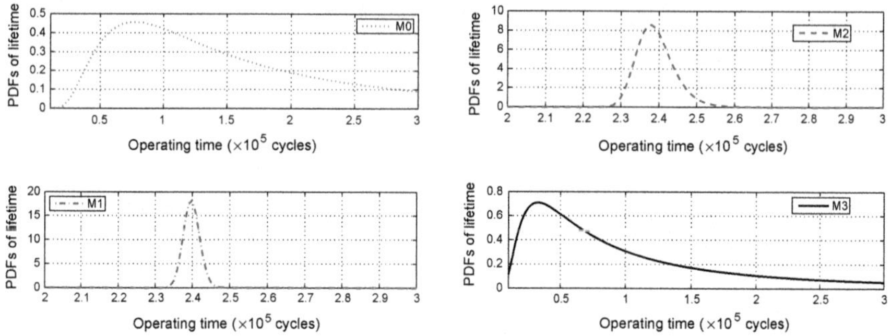

Fig. 7.10 Comparison of the PDFs of the lifetime under M_0, M_1, M_2, and M_3 at time zero with the fatigue crack data

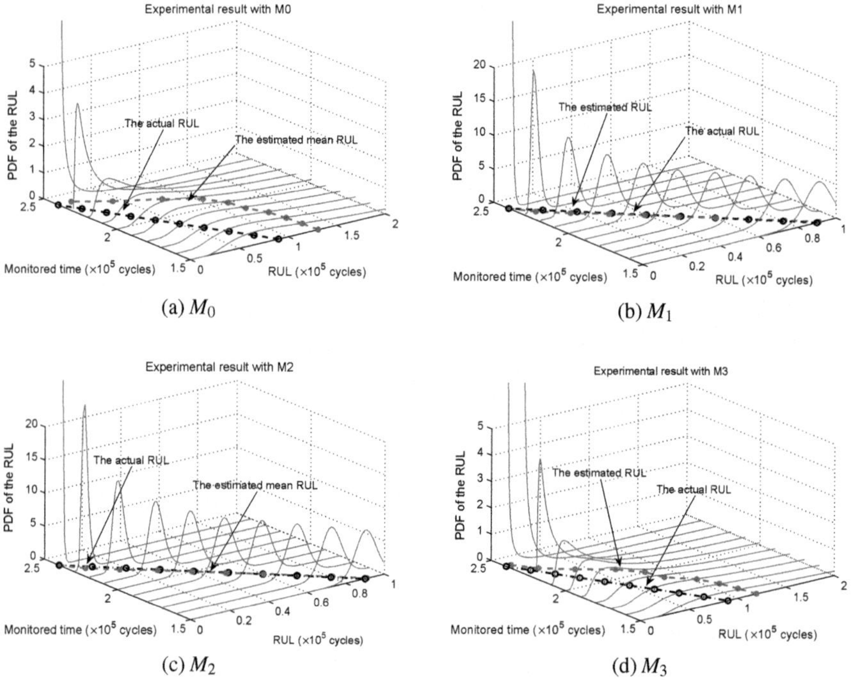

(a) M_0 (b) M_1 (c) M_2 (d) M_3

Fig. 7.11 Comparison of the PDFs and mean of the RULs under M_0, M_1, M_2, and M_3 with the fatigue crack data to take the nonlinearity into account. The results show that such transformation is limited in solving the nonlinear problem in this case.

Similarly, we use the degradation data of the 3rd item in this data set to show the estimated PDFs of the RULs under M_0, M_1, and M_2 at run time. The corresponding PDFs of the RULs, the estimated mean RULs, and the actual RULs are shown graphically in Fig. 7.11 for comparison.

We can see clearly from Fig. 7.11 that the predictions produced by models M_1 and M_2 significantly outperform those of M_0 and M_3 both in terms of the PDFs of the RULs and the mean RULs. In Fig. 7.11, the observed RULs are plotted by a straight line with circle marks while the predicted mean RULs are marked with asterisk signs. From Fig. 7.11, we observe that the estimated results of M_3 are better than M_0, but they are still not satisfactory. Instead, the actual RULs and predicted mean RULs from M_1 and M_2 almost overlap. This shows the necessity of considering the nonlinearity in degradation processes if the process is or appears to be nonlinear.

References

1. Bayus BL (1998) An analysis of product lifetimes in a technologically dynamic industry. Manag Sci 44:763–775
2. Mazhar MI, Kara S, Kaebernick H (2007) Remaining life estimation of used components in consumer products life cycle data analysis by Weibull and artificial neural networks. J Oper Manag 25:1184–1193
3. Pecht M (2008) Prognostics and health management of electronics. John Wiley, New Jersey
4. Altay N, Green WGIII (2006) OR/MS research in disaster operations management. Eur J Oper Res 175:475–493
5. Block HW, Savits TH, Singh H (2002) A criterion for burn-in that balances mean residual life and residual variance. Oper Res 50:290–296
6. Elwany AH, Gebraeel NZ (2008) Sensor-driven prognostic models for equipment replacement and spare parts inventory. IIE Trans 40:629–639
7. Fan CY, Chang PC, Fan PS (2010) A system dynamics modeling approach for a military weapon maintenance supply system. Int J Prod Econ (2010). doi:10.1016/j.ijpe.2010.07.015
8. Huh WT, Janakiraman G, Muckstadt JA, Rusmevichientong P (2009) Asymptotic optimality of order-up-to policies in lost sales inventory systems. Manag Sci 55(3):404–420
9. Jardine AKS, Lin D, Banjevic D (2006) A review on machinery diagnostics and prognostics implementing condition-based maintenance. Mech Syst Signal Process 20(7):1483–1510
10. Papakostas N, Papachatzakis P, Xanthakis V, Mourtzis D, Chryssolouris G (2010) An approach to operational aircraft maintenance planning. Decis Support Syst 48:604–612
11. Tomlin B (2006) On the value of mitigation and contingency strategies for managing supply chain disruption risks. Manag Sci 52(5):639–657
12. Wang W (2007) A two-stage prognosis model in condition based maintenance. Eur J Oper Res 182:1177–1187
13. Maillart LM, Ivy JS, Ransom S, Diehl K (2008) Assessing dynamic breast cancer screening policies. Oper Res 56:1411–1427
14. Ryu YU, Chandrasekaran R, Jacob V (2004) Prognosis using an isotonic prediction technique. Manag Sci 50:777–785
15. Derman C, Lieberman GJ, Ross SM (1984) On the use of replacements to extend system life. Oper Res 32:616–627
16. Leemis LM (1987) Variate generation for accelerated life and proportional hazards models. Oper Res 35:892–894

17. Shen Y, Tang LC, Xie M (2009) A model for upside-down bathtub-shaped mean residual life and its properties. IEEE Trans Reliab 58(3):425–431
18. Chen Z, Zheng S (2005) Lifetime distribution based degradation analysis. IEEE Trans Reliab 54:3–10
19. Escobar LA, Meeker WQ (2006) A review of accelerated test models. Stat Sci 21(4):552–577
20. Gebraeel N, Pan J (2008) Prognostic degradation models for computing and updating residual life distributions in a time-varying environment. IEEE Trans Reliab 57(4):539–550
21. Lu CJ, Meeker WQ (1993) Using degradation measures to estimate a time-to-failure distribution. Technometrics 35(2):161–174
22. Nelson W (1990) Accelerated testing: statistical models, test plans, and data analysis. Wiley, New York
23. Wang X, Xu D (2010) An inverse Gaussian process model for degradation data. Technometrics 52(2):188–197
24. Singpurwalla ND (1995) Survival in dynamic environments. Stat Sci 10(1):86–103
25. Cox DR (1999) Some remarks on failure-times, surrogate markers, degradation, wear, and the quality of life. Lifetime Data Anal 5:307–314
26. Aalen OO, Gjessing HK (2001) Understanding the shape of the hazard rate: a process point of view (with discussion). Stat Sci 16(1):1–22
27. Lee M-LT, Whitmore GA (2006) Threshold regression for survival analysis: modeling event times by a stochastic process reaching a boundary. Stat Sci 21(4):501–513
28. Karlin S, Taylor HM (1981) A second course in stochastic processes. Academic press, California
29. Lefebvre M, Aoudia DA (2011) Two-dimensional diffusion processes as models in lifetime studies. Int J Syst Sci. doi:10.1080/00207721.2011.563870
30. Chhikara RS, Folks JL (1977) The inverse Gaussian distribution as a lifetime model. Technometrics 19(4):461–468
31. Gebraeel N, Lawley MA, Li R, Ryan JK (2005) Residual-life distributions from component degradation signals: a Bayesian approach. IIE Trans 37:543–557
32. Park C, Padgett WJ (2005) Accelerated degradation models for failure based on geometric Brownian motion and gamma processes. Lifetime Data Anal 11(4):511–527
33. Doksum KA, Hoyland A (1992) Models for variable-stress accelerated life testing experiments based on Wiener processes and the inverse Gaussian distribution. Technometrics 34(1):74–82
34. Wang X (2010) Wiener processes with random effects for degradation data. J Multivar Anal 101(2):340–351
35. Whitmore GA, Schenkelberg F (1997) Modeling accelerated degradation data using Wiener diffusion with a time scale transformation. Lifetime Data Anal 3:27–45
36. Cox DR, Miller HD (1965) The theory of stochastic processes. Methuen and Company, London
37. Buonocore A, Caputo L, Pirpzzi E, Ricciardi LM (2011) The first passage time problem for Gauss-diffusion processes: algorithmic approaches and applications to LIF neuronal model. Methodol Comput Appl Prob 13(1):29–57
38. Mehr CB, McFadden JA (1965) Certain property of Gaussian processes and their first-passage times. J R Stat Soc Ser B 27:505–522
39. Nardo ED, Nobile AG, Pirozzi E, Ricciardi LM (2001) A computational approach to first-passage-time problems for Gauss-Markov processes. Adv Appl Prob 33:453–482
40. Ricciardi LM (1976) On the transformation of diffusion processes into the Wiener process. J Math Anal Appl 54:185–199
41. Heng A, Zhang S, Tan CC, Mathew J (2009) Rotating machinery prognostics: state of the art, challenges and opportunities. Mech Syst Signal Process 23:724–739
42. Si XS, Wang W, Hu CH, Zhou DH (2011) Remaining useful life estimation-A review on the statistical data driven approaches. Eur J Oper Res 213(1):1–14
43. Bondesson L (1979) A general result on infinite divisibility. Ann Prob 7(6):965–979
44. Tseng ST, Tang J, Ku LH (2003) Determination of optimal burn-in parameters and residual life for highly reliable products. Naval Res Logist 50:1–14

45. Tseng ST, Peng CY (2004) Optimal burn-in policy by using an integrated Wiener process. IIE Trans 36:1161–1170
46. Lee MY, Tang J (2007) A modified EM-algorithm for estimating the parameters of inverse Gaussian distribution based on time-censored Wiener degradation data. Statistica Sinica 17:873–893
47. Tang J, Su TS (2008) Estimating failure time distribution and its parameters based on intermediate data from a Wiener degradation model. Naval Logist Res 55:265–276
48. Padgett WJ, Tomlinson MA (2004) Inference from accelerated degradation and failure data based on Gaussian process models. Lifetime Data Anal 10:191–206
49. Park C, Padgett WJ (2005) New cumulative damage models for failure using stochastic processes as initial damage. IEEE Trans Reliab 54(3):530–540
50. Park C, Padgett WJ (2006) Stochastic degradation models with several accelerating variables. IEEE Trans Reliab 55(2):379–390
51. Joseph VR, Yu IT (2006) Reliability improvement experiments with degradation data. IEEE Trans Reliab 55(1):149–157
52. Balka J, Desmond AF, McNicholas PD (2009) Review and implementation of cure models based on first hitting times for Wiener processes. Lifetime Data Anal 15:147–176
53. Peng CY, Tseng ST (2009) Mis-specification analysis of linear degradation models. IEEE Trans Reliab 58(3):444–455
54. Wang W, Carr M, Xu W, Kobbacy AKH (2011) A model for residual life prediction based on Brownian motion with an adaptive drift. Microelectron Reliab 51(2):285–293
55. Tseng ST, Peng CY (2007) Stochastic diffusion modeling of degradation data. J Data Sci 5:315–333
56. Si XS, Hu CH, Yang JB, Zhou ZJ (2011) A new prediction model based on belief rule base for system's behavior prediction. IEEE Trans Fuzzy Syst. doi:10.1109/TFUZZ.2011.2130527
57. Woodman OJ (2007) An introduction to inertial navigation, Technical report, Published by the University of Cambridge Computer Laboratory, http://www.cl.cam.ac.uk/techreport/
58. Kharoufeh JP, Cox SM (2005) Stochastic models for degradation-based reliability. IIE Trans 37:533–542
59. Kloeden P, Platen E (1995) Numerical solution of stochastic differential equations. Springer, New York
60. Park JI, Bae SJ (2010) Direct prediction methods on lifetime distribution of organic light-emitting diodes from accelerated degradation tests. IEEE Trans Reliab 59(1):74–90
61. Meeker WQ, Escobar LA (1998) Statistical methods for reliability data. Wiley, New York
62. Kalbfleisch JD, Prentice RL (2002) The statistical analysis of failure time data. Wiley, New York
63. Durbin J (1985) The first-passage density of a continuous Gaussian process to a general boundary. J Appl Prob 22:99–122
64. Barker CT, Newby MJ (2009) Optimal non-periodic inspection for a multivariate degradation model. Reliab Eng Syst Safety 94:33–43
65. Oakland JS (2008) Stat Process Control, 6th edn. Butterworth-Heinemann, Woburn
66. Taylor AE (1952) L'Hospital's rule. Am Math Mon 59(1):20–24
67. Lagarias JC, Reeds JA, Wright MH, Wright PE (1998) Convergence properties of the nelder-mead simplex method in low dimensions. SIAM J Optim 9(1):112–147
68. Akaike H (1974) A new look at the statistical model identification. IEEE Trans Autom Control 19(6):716–722
69. Wilk MB, Gnanadesikan R (1968) Probability plotting methods for the analysis data. Biometrika 55(1):1–17
70. Beskos A, Papaspiliopoulos O, Roberts GO, Fearnhead P (2006) Exact and computationally efficient likelihood-based estimation for discretely observed diffusion processes. J R Stat Assoc Ser B 68(3):333–382

Chapter 8
Prognostics for Age- and State-Dependent Nonlinear Degrading Systems

8.1 Introduction

Prognostics and health management (PHM) has been proved to be an effective methodology for improving reliability and reducing the operation risk of technological systems via management actions [1]. As a key part of PHM, prognosis for degrading systems has drawn much attention from both researchers and engineers in recent years [2–6]. In particular, the remaining useful life (RUL) estimation of systems serves as one of the most important ingredients of prognosis, and lays the groundwork for sequential decision-making activities, such as condition based maintenance, optimal inspection, and spare part ordering.

Therefore, how to estimate the RUL of systems has been attached great importance in practice. The purpose of this chapter is to estimate the RUL for degrading systems, with emphasis on the systems experiencing stochastic but age- and state-dependent nonlinear deteriorating processes. We will elaborate on this issue later.

Due to limited funds and time-consuming testing processes, traditional prognostic methods always suffer a huge problem that the failure time data of the concerned degrading systems are absent [7–9].

On the other hand, the system deteriorates over time, inevitably resulting from its operating load and environment impacts. In this circumstance, most failures of systems arise from a degradation mechanism, and there are characteristics that deteriorate over the system operation. Examples include bearings, batteries, motor drives, and capacitors in analog electronic circuits [10–14]. The past decade has witnessed a growing research interest in various aspects of stochastic degradation modeling-based prognostic methods. Si et al. in [15] provided a comprehensive review on various degradation models for estimating the RUL of the system, including Gamma processes [16–18], Wiener processes [19–22], inverse Gaussian processes [23, 24], and Markovian-based models [25, 26]. All the reviewed models in [16–27] are age-based. Specifically, the degradation level in these models is assume to be physically dependent only on the system age, but not on the current system degradation state. However, in many situations, it would be more appropriate to consider an age- and

X.-S. Si et al., *Data-Driven Remaining Useful Life Prognosis Techniques*,
Springer Series in Reliability Engineering, DOI 10.1007/978-3-662-54030-5_8

state-dependent degradation model so as to account for the s-dependence of the degradation process on both the system age and state.

These age- and state-dependent degradation phenomena are frequently encountered in practice. For example, as the degradation level increases, the system degradation rate might increase, and the system resistance to failure might reduce.

To our knowledge, the studies on age- and state-dependent degradation modeling are very limited in the literature, as opposed to a great deal of effort made to age-dependent degradation models. The only exceptions are the works appearing in [28–30]. Most recently, Giorgio et al. in [31, 32] made the first attempt along this direction, and presented some Markov chain based degradation models whose transition probabilities between the process states depend on the current state and the current age of the system under study. It is actually worth pointing out that their developed models are only suitable to represent strictly monotonic degradation processes. However, in many industrial systems, a non-monotonic degradation process, e.g. resulting from minor repair or reduced intensity of use, can provide a good description of the system's degradation signals, including rotating bearings [33], bridge beam degradation [20], batteries [9, 11, 34], light emitting diode (LED) lamps [35], carbon film resistors [36, 37], etc. On the other side, in [29–32], the continuous degradation process is approximated by a Markov chain with discrete degradation states. This approximating process introduces many context-dependent parameters, which might pose difficulty in applications. Together with these discussions, we can conclude that it is still desirable to develop age- and state-dependent degradation models for continuously degrading systems, whose degradation progression might be non-monotonic.

Due to its favorable mathematical properties and physical interpretations, Wiener processes have been widely adopted to characterize continuously degrading systems with non-monotonic degradation progressions, such as the examples in [19–22]. It is known, for complicated systems, that nonlinearity and stochasticity are two important factors contributing to their degradation processes, and thus have to be taken into account in stochastic degradation-modeling-based RUL estimation. However, most works on Wiener-process-based RUL estimation methods focused on linear models or models that can be made linear by logarithmic or timescale transformations [21]. The research on RUL estimation using nonlinear models is still very limited. Recently, Si et al. in [38] studied how to model the nonlinear degradation process, and presented a more general nonlinear Wiener-process-based model for RUL estimation. Based on this model, unlike the usually used monotonic assumption, they formulated an analytical approximation of the RUL distribution under a mild assumption. Feng et al. in [39] extended the above work to the case of partially observed degradation signals. More comparisons between linear models and nonlinear models can be found in [40–44]. It is worth noting that, even in rarely reported nonlinear cases, the current research is still limited to the case of modeling the degradation as an age-dependent stochastic process.

The above survey of the related works and discussions pose an interesting challenge for prognostic studies through age- and state-dependent nonlinear degradation modeling. Still unsolved is an important and practical problem of how to achieve

RUL estimation for age- and state-dependent nonlinear degradation processes. In this chapter, we propose a general age- and state-dependent nonlinear degradation model for prognostics of continuously degrading systems. In this model, we present a diffusion process with age- and state-dependent nonlinear drift coefficients, as well as age- and state-dependent volatility coefficients, to represent the dynamics and nonlinearity of the degradation progression.

To estimate the RUL distribution, the proposed diffusion process is first converted into a diffusion process with age- and state-dependent nonlinear drift but constant volatility through a Lamperti transformation.

Further, based on a well-known time-space transformation, we obtain an analytical approximated RUL distribution in the concept of the first passage time. To implement the presented model, a maximum likelihood estimation (MLE) method for unknown model parameters is presented based on closed-form approximated transition density functions of the degradation states through the Hermite expansion method.

Overall, the key contribution of this work is investigating the RUL estimation for age- and state-dependent nonlinear degrading systems, which to the best of our knowledge is the first effort on this topic. As a by-product, the results for pure age-dependent models are nested within the results obtained in this chapter. An illustrative example is provided to show how the main results of this chapter can be applied to a specific age- and state-dependent nonlinear degradation model. Finally, a case study is provided by fitting the bearing degradation data from the PHM data challenge of 2012 to the presented model [45]. Comparative results suggest the necessity of investigating age- and state-dependent nonlinear degradation modeling in prognostics.

The remaining parts of this chapter are organized as follows. Section 8.2 describes the problem. Section 8.3 derives the main results of the RUL estimation under age- and state-dependent nonlinear degradation models. Section 8.4 presents the framework of the MLE for unknown model parameters. A particular degradation model is proposed in Sect. 8.5 for illustrative purposes. Section 8.6 provides a case study for the bearings data from the PHM data challenge of 2012 [45].

8.2 Problem Formulation

Let $\{X(t), t \geq 0\}$ denote the stochastic process describing the degradation of the system over its operating time. In this chapter, $\{X(t), t \geq 0\}$ is modeled as a nonlinear diffusion process.

In general, a nonlinear diffusion process can be represented in the form of a stochastic differential equation (SDE), which is expressed as

$$dX(t) = \mu(X(t), t; \boldsymbol{\theta})dt + \sigma(X(t), t; \boldsymbol{\theta})dB(t), \tag{8.1}$$

where $\{B(t), t \geq 0\}$ is a standard Brownian motion process, $\mu(X(t), t; \boldsymbol{\theta})$ is the drift coefficient function, and $\sigma(X(t), t; \boldsymbol{\theta})$ is the volatility coefficient function. $\mu(x, t; \boldsymbol{\theta})$, and $\sigma(x, t; \boldsymbol{\theta})$ can be arbitrary nonlinear functions with the parameter vector $\boldsymbol{\theta}$. Without loss of generality, we suppose $t_0 = 0, X(0) = 0$ for simplicity throughout the chapter.

Our model differs from existing methods for degradation modeling in that it uses nonlinear drift coefficient functions and volatility coefficient functions depending not only on time (or age) t, but also on the system state $X(t)$. In addition, the presented model extends the dediffusion-process-based degradation model, one of whose special cases is the Wiener-process-based degradation model, to a more general situation. Particularly, (8.1) can be simplified to the age-dependent nonlinear drifted model in [38], and a state-dependent degradation model as mentioned in [35–37], by setting $\mu(x, t; \boldsymbol{\theta}) = \mu(t; \boldsymbol{\theta})$, $\sigma(x, t; \boldsymbol{\theta}) = \sigma_B$, and $\mu(x, t; \boldsymbol{\theta}) = \mu(X(t); \boldsymbol{\theta})$, $\sigma(x, t; \boldsymbol{\theta}) = \sigma(X(t); \boldsymbol{\theta})$, respectively. A further simplification by letting $\mu(x, t; \boldsymbol{\theta})$ and $\sigma(x, t; \boldsymbol{\theta})$ be constants reduces both age-dependent and state-dependent models to conventional linear degradation models in [29, 31, 32]. These observations reflect the generality of the presented model.

To estimate the lifetime of the degrading system, the lifetime is usually defined as the first time that the degradation process crosses the failure threshold.

Thus, by the concept of the first passage time (FPT), the lifetime T of the degrading system can be formally defined as

$$T := \inf\{t : X(t) \geq w | X(0) < w\}, \tag{8.2}$$

where the probability density function (PDF) of T can be described as $f_T(t)$, and w is a preset constant failure threshold level determined by the required performance of a specific system (see, e.g., ISO 2372 and ISO 10816 for vibration level).

To utilize the degradation monitoring information, suppose the degradation is inspected at some discrete time points $0 = t_0 < t_1 < t_1 < \cdots < t_k$, and let $x_k = X(t_k)$ denote the degradation observation of the system at time t_k. The set of observations of the degradation states up to time t_k is presented as $\boldsymbol{X}_0^k = \{x_0, x_1, \ldots, x_k\}$. Thus, using the concept of the FPT, we define the RUL L_k of the system at time t_k as

$$L_k = \inf\{l_k > 0 : X(l_k + t_k) \geq w\}, \tag{8.3}$$

with the PDF $f_{L_k}(l_k)$.

Based on the definition of L_k, $f_{L_k}(l_k)$ is consequently decided by the drift and volatility coefficient functions of the degradation process, together with the failure threshold w, and the state of the system. Therefore, the primary goal of this chapter is to derive the PDF $f_{L_k}(l_k)$ of the RUL under the above model formulation. The novelty of the presented model is to allow us to take age- and state-dependent impacts into account simultaneously to estimate the RUL distribution. This approach distinguishes our model from the existing models that only consider the age-dependent impact on the estimated RUL. In the following section, we provide the solution to $f_{L_k}(l_k)$ in the presence of age- and state-dependent degradation processes.

8.3 RUL Estimation by Degradation Modeling

To derive the estimated RUL distribution, we employ the Lamperti transformation proposed in [46, 47] to change the stochastic process $\{X(t), t \geq 0\}$ described in (8.1) into a stochastic process $\{Y(t), t \geq 0\}$ with unit volatility. A Larmperti transformation intends to standardize the variance of a diffusion process so that it has unit variance. The rationale of this transform is that it is achieved by Ito's formula, which is widely accepted and used (see, e.g., [46–48]). Besides, we suppose that $\mu(x, t)$ and $\sigma(x, t)$ are differentiable, which is the premise required to use such a transformation. There are two advantages performing Lamperti transformation for $\{X(t), t \geq 0\}$. On the one hand, after such a transformation, some conclusions in [38] can be utilized to derive an analytical approximate PDF of the RUL. On the other hand, after performing a Lamperti transformation, the existence of the degradation state's closed-form approximated log-transition density expansions will facilitate the parameters estimation of (8.1).

These issues will be shown in the following.

For the studied model (8.1), the Lamperti transformation is represented as

$$y = \gamma(x, t) = \int^{x} \frac{1}{\sigma(w, t; \boldsymbol{\theta})} dw, \tag{8.4}$$

and the SDE corresponding to the transformed process $\{Y(t), t \geq 0\}$ is

$$dY(t) = \mu_Y(Y(t), t; \boldsymbol{\theta})dt + dB(t), \tag{8.5}$$

where the drift coefficient function $\mu_Y(Y(t), t; \boldsymbol{\theta})$ is obtained by Ito's formula as

$$\mu_Y(Y(t), t; \boldsymbol{\theta}) = \frac{\mu[\gamma^{-1}(Y(t), t), t; \boldsymbol{\theta}]}{\sigma[\gamma^{-1}(Y(t), t), t; \boldsymbol{\theta}]} - \frac{1}{2}\frac{\partial\sigma[\gamma^{-1}(Y(t), t), t; \boldsymbol{\theta}]}{\partial x} + \frac{\partial\gamma[\gamma^{-1}(Y(t), t), t; \boldsymbol{\theta}]}{\partial t}, \tag{8.6}$$

in which $\gamma^{-1}(y, t)$ denotes the inverse function of $\gamma(y, t)$.

Obviously, $\mu_Y(Y(t), t; \boldsymbol{\theta})$ is still an age- and state-dependent nonlinear function. To guarantee that there is a unique stochastic process satisfying (8.6), it is assumed that $\mu_Y(Y(t), t; \boldsymbol{\theta})$ satisfies the conditions given in Lemma 8.1. Actually, Lemma 8.1 is a particular case of the proved theory ensuring a unique solution to a general SDE [49].

Lemma 8.1 *There exists a unique stochastic process* $\{X(t), t \geq 0\}$ *satisfying the SDE of*

$$dX(t) = g(X(t), t; \boldsymbol{\theta})dt + \sigma_B dB(t), \tag{8.7}$$

if the following conditions hold.

(i) $g(x, t; \boldsymbol{\theta})$ *is measurable about* x, t, *and* $|g(x, t; \boldsymbol{\theta})|^{\frac{1}{2}} \in L^2_{T\times R}$.

(ii) (Lipschitz conditions) There is a positive constant $K \in R$ letting

$$|g(x, t; \boldsymbol{\theta}) - g(y, t; \boldsymbol{\theta})| < K|x - y|, \quad \forall t \in [0, T], \forall x, y \in R.$$

(iii) (Linear growth conditions) There is a constant $C > 0$ letting

$$|g(x, t; \boldsymbol{\theta})| \leq C(1 + |x|), \quad \forall t \in [0, T], \forall x, y \in R.$$

Based on the above discussions, we conclude that, under certain conditions, a general stochastic process $\{X(t), t \geq 0\}$ can be equivalently transformed into a kind of stochastic process with fixed volatility such as (8.5) and (8.7). Therefore, we mainly consider the type of stochastic degradation processes $\{X(t), t \geq 0\}$ in (8.7) for simplicity, but without loss of generality.

Inspired by [38], to achieve the lifetime distribution under the degradation process $\{X(t), t \geq 0\}$ in (8.7), we first change the problem calculating the PDF of the FPT crossing a constant threshold into the problem of a standard BM crossing a time-varying threshold. This change can be done by the well-known time-space transformation. The corresponding conditions and transformations for the presented model (8.7) are summarized as Lemma 8.2.

Lemma 8.2 *The degradation process $\{X(t), t \geq 0\}$ described by (8.7) can be changed into a standard BM $\{B(\tilde{t}), \tilde{t} \geq 0\}$ by a time-space transformation $\tilde{x} = \Psi(t, x), \tilde{t} = \varphi(t)$, iff there exist the functions $c_1(t)$ and $c_2(t)$ of time t satisfying*

$$c_1(t) = \frac{2g(x, t; \boldsymbol{\theta}) - xc_2(t)}{\sigma_B}, \tag{8.8}$$

$$c_2(t) = 2\frac{\partial g(x, t; \boldsymbol{\theta})}{\partial x}. \tag{8.9}$$

Then $\Psi(t, x)$, and $\varphi(t)$ can be written as

$$\Psi(x, t) = \exp\left[-\frac{1}{2}\int_0^t c_2(u)du\right]\frac{x}{\sigma_B} - \frac{1}{2}\int_0^t c_1(\tau)\exp\left[-\frac{1}{2}\int_0^\tau c_2(u)du\right]d\tau,$$

$$\varphi(t) = \int_0^t \exp\left[-\frac{1}{2}\int_0^\tau c_2(u)du\right]d\tau. \tag{8.10}$$

Lemma 8.2 can be proved by substituting $\mu(x, t, \boldsymbol{\theta}) = f(x, t, \boldsymbol{\theta})$ into Lemma 1 in [38]. After this time-space transformation, the degradation process defined in (x, t) can be described as a standard BM in $(\tilde{x}, \tilde{t})$.

Now, based on the transformed standard BM, the purpose to obtain the PDF of the FPT of the degradation process in $\{X(t), t \geq 0\}$ crossing a constant critical threshold w can be achieved through formulating the PDF of the FPT of a standard BM crossing a time-dependent threshold. Thus, for a degradation process $\{X(t), t \geq 0\}$ in the form of (8.7), the PDF of lifetime T can be formulated under the time-space transformation $\tilde{x} = \Psi(t, x)$ and $\tilde{t} = \varphi(t)$ as

$$f_T(t) = p_{B(\tilde{t})}(S(\tilde{t}), \tilde{t})\frac{d\varphi(t)}{dt}, \tag{8.11}$$

where $p_{B(\tilde{t})}(S(\tilde{t}), \tilde{t})$ is the PDF of the standard BM crossing a $\tilde{t}$-dependent critical level $S(\tilde{t})$. The corresponding transformations are $S(\tilde{t}) = \Psi(w[\varphi^{-1}(\tilde{t})], \varphi^{-1}(\tilde{t})) = \Psi(w, \varphi^{-1}(\tilde{t}))$, and $\tilde{t} = \varphi(t)$.

At present, through the Lamperti transformation and time-space transformation, the lifetime distribution of the system described by an age- and state- dependent degradation process (8.7) has been transformed into the PDF of the FPT of a standard BM crossing a time-dependent boundary $S(\tilde{t})$. It is noted that an analytical form of $f_T(t)$ is desirable for decision-making under prognostic information, especially in the circumstance of online estimation analysis and maintenance scheduling. However, under the sense of the FPT, it is often really difficult to derive a closed-form $f_T(t)$, no matter what method is utilized.

Fortunately, for such a kind of diffusion with age- and state-dependent drift coefficient but constant volatility coefficient, we can use the method proposed in [38] to achieve an analytical approximation for the PDF of lifetime. To get an analytical form of the lifetime distribution, we make the same assumption as [38] that if the degradation process is hitting the threshold at a certain time t exactly, then the probability that such a process crossed the threshold before time t is negligible. This assumption has been well demonstrated by recent prognostics studies in [50–52].

Together with Lemma 8.2, (8.11), and the above assumption, we present the following theorem for constructing the PDF of the lifetime associated with the degradation process (8.7).

Theorem 8.1 *For the degradation process $\{X(t), t \geq 0\}$ given by (8.7), if $g(x, t, \boldsymbol{\theta})$ is a continuous function of the time t and the state x, then the PDF of the FPT of $\{X(t), t \geq 0\}$ crossing a constant threshold w can be approximated with an explicit form as*

$$f_T(t) \cong \frac{1}{\sqrt{2\pi\tilde{t}}}\left[\frac{S(\tilde{t})}{\tilde{t}} - \frac{dS(\tilde{t})}{d\tilde{t}}\right]\exp\left[-\frac{S^2(\tilde{t})}{2\tilde{t}}\right]\frac{d\varphi(t)}{dt} \tag{8.12}$$

where $S(\tilde{t}) = \Psi(w, \varphi^{-1}(\tilde{t}))$ and $\tilde{t} = \varphi(t)$ are determined by (8.10).

The proof of Theorem 8.1 is given in the Appendix.

As mentioned in the section above, (8.1) is a generalization of many degradation models in literature. Due to the Lamperti transformation, such a generality statement can also be applied to (8.7). As a result, the method calculating the lifetime distribution in this section works in a series of cases by specifying different forms of $g(x, t, \boldsymbol{\theta})$, and the according results for several special cases are summarized in Table 8.1.

In Table 8.1, the model M_1 is an age- and state-dependent nonlinear degradation model; M_2, and M_4 are age-dependent nonlinear models which become linear models when $b_T = 0$, and $b_T = 1$ respectively; and M_3 is a state-dependent degradation model. Using the obtained result in Theorem 8.1, we can formulate the lifetime

Table 8.1 Lifetime distribution estimation results of some degradation models

Model	M_1	M_2	M_3	M_4
$\mu(x,t;\boldsymbol{\theta})$	$b_X x + a_T b_T e^{b_T t}$	$a_T b_T e^{b_T t}$	$a_X + b_X x$	$a_T b_T t^{b_T - 1}$
$\sigma(x,t;\boldsymbol{\theta})$	σ_B	σ_B	σ_B	σ_B
$c_2(t)$	$2b_X$	0	$2b_X$	0
$c_1(t)$	$\frac{2a_T b_T e^{b_T t}}{\sigma_B}$	$\frac{2a_T b_T e^{b_T t}}{\sigma_B}$	$\frac{2a_X}{\sigma_B}$	$\frac{2a_T b_T t^{b_T - 1}}{\sigma_B}$
$\varphi(t)$	$\frac{1-e^{-b_X t}}{b_X}$	t	$\frac{1-e^{-b_X t}}{b_X}$	t
$\frac{d\varphi(t)}{dt}$	$e^{-b_X t}$	1	$e^{-b_X t}$	1
$\psi(x,t;\boldsymbol{\theta})$	$\frac{xe^{-b_X t}}{\sigma_B} + \frac{a_T b_T[1-e^{(b_T - b_X)t}]}{\sigma_B(b_T - b_X)}$	$\frac{x+a_T-a_T e^{b_T t}}{\sigma_B}$	$\frac{(xb_X+a_X)e^{-b_X t}-a_X}{\sigma_B}$	$\frac{x-a_T t^{b_T}}{\sigma_B}$
$S(\tilde{t})$	$\frac{w(1-b_X\tilde{t})}{\sigma_B} + \frac{a_T b_T[1-(1-b_X\tilde{t})^{(b_X - b_T)/b_X}]}{\sigma_B(b_T - b_X)}$	$\frac{w+a_T-a_T e^{b_T t}}{\sigma_B}$	$\frac{w(1-b_X\tilde{t})-a_X\tilde{t}}{\sigma_B}$	$\frac{w-a_T t^{b_T}}{\sigma_B}$
$\frac{dS(\tilde{t})}{d\tilde{t}}$	$-\frac{wb_X}{\sigma_B} - \frac{a_T b_T(1-b_X\tilde{t})^{-b_T/b_X}}{\sigma_B(b_T - b_X)}$	$-\frac{a_T b_T e^{b_T t}}{\sigma_B}$	$-\frac{a_T b_T e^{b_T t}}{\sigma_B}$	$-\frac{a_T b_T \tilde{t}^{b_T}}{\sigma_B}$

distributions for all models in Table 8.1. All the required quantities in (8.12) for lifetime estimation are summarized in Table 8.1. It is worth mentioning that the models M_1 and M_2 will be used in the subsequent illustrative example and case study.

Remark 8.1 In (8.7), if the drift coefficient function $g(x,t,\boldsymbol{\theta})$, obtained through Lamperti transformation from (8.1), is s-independent of the degradation state, then the time-space transformation in Lemma 8.2 becomes a space-only transformation, which is demonstrated by the results of the models M_2 and M_4 with $c_2(t) = 0$ and $\varphi(t) = t$ in Table 8.1. It is straightforward to verify that the results in Table 8.1 can include current results of linear models, and age-dependent nonlinear models as special cases. This result further shows the generality of the obtained results by Theorem 8.1. ■

Remark 8.2 The results in Theorem 8.1 and Table 8.1 do not take into account the individual-to-individual variability of degrading systems.

When the individual-to-individual variability is involved, we can use a popular method of introducing some random effects in the model parameters [53]. Then, the PDF of the lifetime can be estimated by the law of total probability. ■

Until now, we have achieved the goal of calculating the lifetime distribution $f_T(t)$ for systems with age- and state-dependent nonlinear degradation models. However, the results in Theorem 8.1 and (8.12) are obtained under the condition of $t_0 = 0$, $X(0) = 0$, and thus fail to incorporate the degradation observation of the system into the estimated lifetime. For a practical system, a more common situation is that the system is monitored at time t_k, and the state is then evaluated as $x_k = X(t_k)$. Based on the degradation modeling, and the observed degradation state x_k, the RUL of the system at the monitoring time t_k can be estimated by the following results.

Theorem 8.2 *For the degradation process described by (8.7), if the degradation observation at t_k is $x_k = X(t_k)$, then the PDF of the estimated RUL at time t_k equals the FPT of the stochastic process $\{Y(l_k), l_k \geq 0\}$ crossing a constant threshold $w_k = w - x_k$, where $Y(l_k) = X(l_k + t_k) - x_k$, and $Y(0) = 0$. The process $\{Y(l_k), l_k \geq 0\}$ can be described by*

$$dY(l_k) = g(x_k + Y(l_k), (l_k + t_k); \boldsymbol{\theta})dl_k + \sigma_B dB(l_k), \tag{8.13}$$

and the according PDF of the RUL at t_k can be approximately formulated as

$$f_{L_k}(l_k) \cong \frac{l_k}{\sqrt{2\pi \tilde{l}_k}} \left[\frac{S(\tilde{l}_k)}{\tilde{l}_k} - \frac{dS(\tilde{l}_k)}{d\tilde{l}_k} \right] \exp\left[-\frac{S^2(\tilde{l}_k)}{2\tilde{l}_k} \right] \frac{d\varphi(l_k)}{dl_k}, \tag{8.14}$$

where $S(\tilde{l}_k) = \Psi^*(w_k, \varphi^{*-1}(\tilde{l}_k))$, and $\tilde{l}_k = \varphi^*(l_k)$ are determined as

$$\Psi^*(w_k, l_k) = \exp\left[-\frac{1}{2} \int_0^{l_k} c_2^*(u)du \right] \frac{w_k}{\sigma_B} - \frac{1}{2} \int_0^{l_k} c_1^*(\tau) \exp\left[-\frac{1}{2} \int_0^{\tau} c_2^*(u)du \right] d\tau,$$

$$\varphi^*(l_k) = \int_0^{l_k} \exp\left[-\frac{1}{2} \int_0^{\tau} c_2^*(u)du \right] d\tau. \tag{8.15}$$

The proof of Theorem 8.2 is given in the Appendix, together with the forms of $c_1^*(u)$ and $c_2^*(u)$.

As observed from Theorem 8.2, there are shifts in the forms of the drift and volatility coefficient functions, which make the PDF of the RUL $f_{L_k}(l_k)$ no longer just be a simple substitution of w by w_k into $f_T(t)$. At the monitoring time t_k with the degradation state x_k, applying Theorem 8.2 to the model $M_j, j = 1, 2, 3, 4$ in Table 8.1 can provide an efficient approach to calculate the PDF of the estimated RUL L_k at t_k. To distinguish the models considering the current degradation observation from the models in Table 8.1 for lifetime estimation, let M_j^* denote the corresponding model for the process $\{Y(l_k), l_k \geq 0\}$ obtained from the model M_j through Theorem 8.2. The main results associated with the estimated PDF of the RUL for model M_j^* are summarized in Table 8.2.

By comparing the results in Table 8.1 with those of Table 8.2, we can find the differences in models used to derive the lifetime distribution $f_T(t)$ and the PDF of the RUL $f_{L_k}(l_k)$ at t_k clearly, because the current degradation observation has been incorporated into the estimated RUL. It is not surprising that the result for model M_1^* can include the results of the other three models as special cases. Actually, this observation holds also true for several current results under linear models and age-

Table 8.2 PDFs of the estimated RUL results for some degradation models

Model	M_1^*	M_2^*	M_3^*	M_4^*
$\mu(y, l_k; \boldsymbol{\theta})$	$b_X x_k + b_X y + a_T e^{b_T t_k} b_T e^{b_T l_k}$	$a_T e^{b_T t_k} b_T e^{b_T l_k}$	$a_X + b_X x_k + b_X y$	$a_T b_T (t_k + l_k)^{b_T - 1}$
$\sigma(y, l_k; \boldsymbol{\theta})$	σ_B	σ_B	σ_B	σ_B
$c_2(l_k)$	$2b_X$	0	$2b_X$	0
$c_1(l_k)$	$\frac{2(a_T b_T e^{b_T t_k} e^{b_T l_k} + b_X x_k)}{\sigma_B}$	$\frac{2a_T e^{b_T t_k} b_T e^{b_T l_k}}{\sigma_B}$	$\frac{2(a_X + b_X x_k)}{\sigma_B}$	$\frac{2a_T b_T (t_k + l_k)^{b_T - 1}}{\sigma_B}$
$\varphi(l_k)$	$\frac{1 - e^{-b_X l_k}}{b_X}$	l_k	$\frac{1 - e^{-b_X l_k}}{b_X}$	l_k
$\frac{d\varphi(l_k)}{d\tilde{l}_k}$	$e^{-b_X l_k}$	1	$e^{-b_X l_k}$	1
$\psi(y, l_k; \boldsymbol{\theta})$	$\frac{x e^{-b_X l_k}}{\sigma_B} + + \frac{x_k (1 - e^{-b_X l_k})}{\sigma_B} + \frac{a_T e^{b_T t_k} b_T [1 - e^{(b_T - b_X) l_k}]}{\sigma_B (b_T - b_X)}$	$\frac{x + a_T e^{b_T t_k} - a_T e^{b_T t_k} e^{b_T l_k}}{\sigma_B}$	$-\frac{a_X + b_X x_k}{\sigma_B} + \frac{(y b_X + a_X + b_X x_k) e^{-b_X l_k}}{\sigma_B}$	$\frac{x_k - a_T (t_k + l_k)^{b_T}}{\sigma_B}$
$S(\tilde{l}_k)$	$\frac{w_k (1 - b_X \tilde{l}_k)}{\sigma_B} + \frac{b_X x_k \tilde{l}_k}{\sigma_B} + \frac{a_T e^{b_T t_k} b_T [1 - (1 - b_X \tilde{l}_k)^{\frac{b_X - b_T}{b_X}}]}{\sigma_B (b_T - b_X)}$	$\frac{w + a_T e^{b_T t_k} - a_T e^{b_T t_k} e^{b_T l_k}}{\sigma_B}$	$\frac{w_k (1 - b_X \tilde{l}_k) - (a_X + b_X x_k) \tilde{l}_k}{\sigma_B}$	$\frac{w_k - a_T (t_k + l_k)^{b_T}}{\sigma_B}$
$\frac{dS(\tilde{l}_k)}{d\tilde{l}_k}$	$-\frac{w_k b_X}{\sigma_B} + \frac{b_X x_k}{\sigma_B} - \frac{a_T e^{b_T t_k} b_T (1 - b_X \tilde{l}_k)^{-\frac{b_T}{b_X}}}{\sigma_B (b_T - b_X)}$	$-\frac{a_T e^{b_T t_k} b_T e^{b_T l_k}}{\sigma_B}$	$-\frac{w b_X + a_X + b_X x_k}{\sigma_B}$	$-\frac{a_T b_T (t_k + l_k)^{b_T - 1}}{\sigma_B}$

dependent nonlinear models because the structure of model (8.7) has a very general form.

To end this section, we summarize the main steps deriving the RUL distribution of the system with an age- and state-dependent degradation process.

Step 1: A general age- and state-dependent nonlinear degradation process $\{X(t), t \geq 0\}$ is transformed into a process $\{Y(t), t \geq 0\}$ with a drift function $\mu_Y(Y(t), t; \boldsymbol{\theta})$ and a unit volatility coefficient through Lamperti transformation.

Step 2: By the time-space transformation, $\{Y(t), t \geq 0\}$ is further transformed into a standard BM $\{B(\tilde{t}), \tilde{t} \geq 0\}$.

Step 3: Under a weak assumption, we obtain an analytical approximate PDF of the FPT of $B(\tilde{t})$ crossing a $\tilde{t}$-dependent boundary as the lifetime distribution corresponding to $\{X(t), t \geq 0\}$.

Step 4: The PDF of the RUL at time t_k is derived by utilizing the current degradation observation with x_k according to Theorem 8.2.

By now, for a specific degradation process in the form (8.1), the lifetime distribution $f_T(t)$ and the PDF of RUL $f_{L_k}(l_k)$ at the monitoring time t_k are accessible by applying Theorems 8.1 and 8.2. As for the model parameters, when the degradation observations are available, a parameter estimation procedure is needed to identify unknown parameters in (8.1). This procedure is the focus of the next section.

8.4 Model Parameter Estimation Framework

Regarding the problem of estimating unknown parameters $\boldsymbol{\theta}$ in (8.1), it is worth noting that the observation of the degradation process is usually discrete, while the model specification of the stochastic degradation process is continuous. Consequently, a closed-form of the degradation state transition function is unavailable for most cases. This unavailability leads to the difficulty of directly using maximum likelihood estimation (MLE) for parameter estimation. In this chapter, to get the MLE of unknown parameters $\boldsymbol{\theta}$, then we use the Hermite expansion method, which is seminally proposed in [46, 54], to find a closed-form approximate transition density function of the degradation states, and then to construct the log-likelihood function. Due to the nature of the studied model (8.1), only the case of univariate degradation processes is considered here for simplicity. For multivariate cases, see [48, 55] for references.

Now, we provide a brief summary of the parameter estimation framework using the method of Hermite expansion proposed in [47]. First, an age- and state-dependent degradation model in (8.1) is changed into a diffusion process (8.5) by a Lamperti transformation. Let $\Delta = t - t_0$, and define another transformation for $\{Y(t), t \geq 0\}$ in (8.5) as

$$z = \frac{y - y_0}{\sqrt{\Delta}}. \tag{8.16}$$

Then, the transition density of $\{Y(t), t \geq 0\}$, denoted by $P_Z(t, z|t_0, y_0)$, can be approximated as a J-th order Hermite expansion

$$p_Z^{(J)}(t, z|t_0, y_0) = \phi(z) \sum_{j=0}^{J} \eta_Z^{(j)}(t, t_0, y_0) H_j(z), \tag{8.17}$$

where $\phi(z) = \frac{1}{\sqrt{2\pi}} e^{-(z^2)/(2)}$, and the Hermite polynomials are a series of orthogonal base functions defined as $H_j(z) = \phi(z)^{(-1)}(d^j)(dz^j)\phi(z)$. In (8.17), the coefficients $\eta_Z^{(j)}(t, t_0, y_0)$ can be obtained using the orthogonality of Hermite polynomials and Taylor expansion, and the transition function $P_Z(t, z|t_0, y_0)$ is thus approximated as

$$p_Y^{(J,K)}(t, y|t_0, y_0) = \frac{1}{\sqrt{\Delta}} \phi\left(\frac{y - y_0}{\sqrt{\Delta}}\right) \left\{ \sum_{j=0}^{J} \frac{1}{j!} \left[\sum_{i=0}^{K} \frac{\Delta^i}{i!} A_Y^i \circ H_j\left(\frac{y - y_0}{\sqrt{\Delta}}\right) |_{y=y_0} \right] H_j\left(\frac{y - y_0}{\sqrt{\Delta}}\right) \right\}, \tag{8.18}$$

where the infinitesimal operator A_Y is defined by

$$A_Y \circ f(t, y, t_0, y_0) = \frac{\partial f(t, y, t_0, y_0)}{\partial t} + \mu_Y(t, y) \frac{\partial f(t, y, t_0, y_0)}{\partial y} + \frac{1}{2} \frac{\partial^2 f(t, y, t_0, y_0)}{\partial y^2}. \tag{8.19}$$

A general definition of the infinitesimal operator for a univariate time-inhomogeneous diffusion process can be found in [47, 55]. In degradation modeling practice, only the observations of $\{X(t), t \geq 0\}$ are available. Based on (8.18), we can obtain the approximate transition density of $\{X(t), t \geq 0\}$ as

$$p_X^{(J,K)}(t, x|t_0, x_0) = \frac{1}{\sigma(x,t)} p_Y^{(J,K)}(t, \gamma(x,t)|t_0, \gamma(x_0, t_0)), \tag{8.20}$$

and formulate the log-likelihood $\ell(\boldsymbol{\theta}|\mathbf{X})$ as

$$\ell(\boldsymbol{\theta}|\mathbf{X}) = \sum_{n=1}^{k} \ln P_X^{J,K}(t_n, x_n|t_{n-1}, x_{n-1}). \tag{8.21}$$

After choosing suitable orders J and K according to the requirements for the modeling accuracy and the complexity of calculation, we obtain the MLE $\hat{\boldsymbol{\theta}}$ of $\boldsymbol{\theta}$ by maximizing $\ell(\boldsymbol{\theta}|\mathbf{X})$ with respect to $\boldsymbol{\theta}$. As described in [55], this maximization is relatively faster, because the simulation of many paths from the degradation process $\{X(t), t \geq 0\}$, which dominates many other estimation procedures for diffusion processes, is not required here to approximate the transition density.

8.5 An Illustrative Example

To illustrate the proposed approach, we consider a new state- and age-dependent nonlinear diffusion-process-based degradation model. This model will be applied subsequently to the bearing degradation of the PHM data challenge of 2012 [45]. In this model, only the drift coefficient is considered to be state- and age- dependent for the purposes of simplicity and comparability with several existing degradation models. The diffusion coefficient is set constant, as done in [38].

8.5.1 Degradation Model and Lifetime Estimation

Specifically, the concerned model is expressed as

$$dX(t) = [b_X X(t) + a_T b_T \exp(b_T t)]dt + \sigma_B dB(t). \tag{8.22}$$

It can be observed that the drift coefficient in (8.22) is both age- and state-dependent, which makes this model different from all diffusion-process-based degradation models in literature. For descriptive convenience, we represent this degradation model as M_1. Accordingly, we consider its reduced version as the age-dependent nonlinear diffusion model M_2, expressed by

$$dX(t) = a_T b_T e^{b_T t} dt + \sigma_B dB(t). \tag{8.23}$$

It is straightforward to find that the model M_2 (8.23) is a special case of the degradation model M_1 when $b_X = 0$. Here, $b_X = 0$ means that the drift coefficient is s-independent of the degradation state. Note that M_2 was first proposed in [38], and has indicated very good fitting performance in a series of degradation data, such as fatigue growth data, laser data, and the degradation data of the inertial navigation system. In the following, the lifetime distribution, and the RUL distribution can be respectively derived by applying Theorems 8.1 and 8.2.

Firstly, from Theorem 8.1, we can formulate the PDF of the lifetime corresponding to the degradation process described in (8.22) as

$$f_{T|M_1}(t; w, \boldsymbol{\theta}_{M_1}) = \frac{b_X^{3/2} e^{-b_X t}}{\sigma_B \sqrt{2\pi \left(1 - e^{-b_X t}\right)^3}} \left(w + \frac{a_T b_T}{b_T - b_X} - \frac{a_T b_T^2}{(b_T - b_X) b_X} e^{(b_T - b_X)t} + \frac{a_T b_T}{b_X} e^{b_T t} \right) \times \exp\left[- \frac{\left(w b_X e^{-b_X t} + \frac{a_T b_T b_X}{b_T - b_X} (1 - e^{(b_T - b_X)t})] \right)^2}{2(1 - e^{-b_X t}) b_X \sigma_B^2} \right]. \tag{8.24}$$

Further, at a specific monitoring time t_k with the degradation observation x_k, the PDF of the estimated RUL can be calculated according to Theorem 8.2 as

$$f_{L_k|M_1}(l_k; w, \boldsymbol{\theta}_{M_1}) = \frac{b_X^{3/2} e^{-b_X l_k}}{\sigma_B \sqrt{2\pi \left(1 - e^{-b_X l_k}\right)^3}} \left(w_k + \frac{a_T b_T e^{b_T t_k}}{b_T - b_X} - \frac{a_T b_T^2 e^{b_T t_k}}{(b_T - b_X) b_X} e^{(b_T - b_X) l_k} + \frac{a_T b_T e^{b_T (t_k + l_k)}}{b_X} \right) \times \exp\left\{ - \frac{\left[w_k b_X e^{-b_X l_k} - x_k b_X (1 - e^{-b_X l_k}) + \frac{a_T b_T b_X}{b_T - b_X} (1 - e^{(b_T - b_X) l_k})] \right]^2}{2(1 - e^{-b_X l_k}) b_X \sigma_B^2} \right\} \tag{8.25}$$

where $w_k = w - x_k$. What should be pointed out is that, in (8.24) and (8.25), for notation convenience, we use the formulation

$$\frac{b_X^{3/2}}{\sqrt{(1 - e^{-b_X l_k})^3}} = \sqrt{\frac{b_X}{(1 - e^{-b_X l_k})} \times \frac{b_X}{(1 - e^{-b_X l_k})}}. \tag{8.26}$$

Similarly, from Theorems 8.1 and 8.2, the PDFs of the lifetime T and the RUL L_k under model M_2 can be respectively formulated as

$$f_{T|M_2}(t, w; \boldsymbol{\theta}_{M_2}) = \frac{w + a_T + a_T e^{b_T t}(b_T t - 1)}{\sigma_B \sqrt{2\pi t^3}} \exp\left[- \frac{(w + a_T - a_T e^{b_T t})^2}{2\sigma_B^2 t} \right], \tag{8.27}$$

and

$$f_{L_k|M_2}(l_k, w; \boldsymbol{\theta}_{M_2}) = \frac{w_k + a_T e^{b_T t_k} + a_T e^{b_T(t_k+l_k)}(b_T l_k - 1)}{\sigma_B \sqrt{2\pi l_k^3}} \exp\left[-\frac{(w_k + a_T e^{b_T t_k} - a_T e^{b_T(t_k+l_k)})^2}{2\sigma_B^2 l_k}\right]. \tag{8.28}$$

By comparing the distributions of the lifetime T and the RUL L_k between M_1 and M_2, we can find that the introduction of state dependency in degradation modeling has a great influence on the lifetime estimation. During the process of conducting time-space transformation, $c_2(t)$ equals 0 in M_2 due to Lemma 8.2, which indicates that no time transformation will be made for only age-dependent nonlinear degradation models. However, such a time transformation must be performed for M_1 because the drift coefficient function is s-dependent on the degradation state and $c_2(t) = 2b_X$. Furthermore, it can also be concluded that there is a need for a time-space transformation in the situation of state-independent nonlinear degradation models.

8.5.2 Parameters Estimation

Suppose that up to time t_k the degradation process has been measured discretely at an ordered time series $\boldsymbol{T}_0^k = [t_0, \ldots, t_n, \ldots, t_k]$ with $n < k$, and the corresponding observations are collected as $\boldsymbol{X}_0^k = [x_0, \ldots, x_n, \ldots, x_k]'$. The interval between observation time t_n and t_{n-1} is denoted as Δ_n, where $\Delta_n = t_n - t_{n-1}$. For different n, Δ_n can be either equal or unequal, depending on the specific observation schedule.

Now we illustrate how to apply the MLE method proposed in Sect. 8.4 to estimate unknown parameters in the above degradation model. Here we only illustrate the case in M_1, as the estimating procedure in M_2 is quite similar and simpler. When the degradation observations of degrading systems are obtained, the Hermite expansion method described in Sect. 8.4 is used to find the MLE of unknown parameters in the age- and state-dependent nonlinear degradation model (8.22). As for the model described by (8.22), we have the Lamperti transformation of (8.22) as

$$dY(t) = \mu_Y(Y(t), t; \boldsymbol{\theta})dt + dB(t) = \left[b_X Y(t) + \frac{a_T b_T}{\sigma_B} \exp(b_T t)\right] dt + dB(t), \tag{8.29}$$

where $Y(t) = \gamma(X(t), t) = X(t)/\sigma_B$.

By setting $J = 4$, and $K = 2$ in (8.18) and (8.20), we can obtain an approximation of the transition density accurate enough for parameter estimation with a relatively low cost of computation. Consecutively, the approximated transition density from t_{n-1} with state x_{n-1} to t_n with x_n can be written as

$$p_X^{(4,2)}(t_n, x_n | t_{n-1}, x_{n-1}) = \frac{1}{\sigma_B} p_Y^{(4,2)}(t_l, \gamma(x_n, t_n) | t_{n-1}, \gamma(x_{n-1}, t_{n-1}) \qquad (8.30)$$

where $p_Y^{(4,2)}(t_n, y_n | t_{n-1}, y_{n-1})$ is

$$p_Y^{(4,2)}(t_n, y_n | t_{n-1}, y_{n-1}) = \frac{1}{\sqrt{\Delta_n}} \phi\left(\frac{y_n - y_{n-1}}{\sqrt{\Delta_n}}\right) \sum_{j=0}^{4} \beta_j^2(y_n, y_{n-1}, t_n, t_{n-1}) H_j\left(\frac{y_n - y_{n-1}}{\sqrt{\Delta_n}}\right). \qquad (8.31)$$

The particular forms of $\beta_j^2(y_n, y_{n-1}, t_n, t_{n-1})$ for both M_1 and M_2 are summarized in the Appendix. Then Substituting (8.31) into (8.21) yields the log-likelihood function as

$$\begin{aligned} \ell(\boldsymbol{\theta} | \boldsymbol{X}_0^k) &= \sum_{n=1}^{k} \ln P_X^{4,2}(t_n, x_n | t_{n-1}, x_{n-l}) \\ &= \sum_{n=1}^{k} \ln p_Y^{(4,2)}(t_n, y_n | t_{n-1}, y_{n-1}) - k \ln \sigma_B. \end{aligned} \qquad (8.32)$$

Therefore, maximizing (8.32) will generate the MLE of unknown parameters b_X, a_T, b_T, and σ_B in the model M_1. In the case study of this chapter, through taking $-\ell(\boldsymbol{\theta} | \boldsymbol{X}_0^k)$ as the objective function, we use the well-known fminsearch function in MATLAB to achieve the goal of maximizing $\ell(\boldsymbol{\theta} | \boldsymbol{X}_0^k)$.

8.5.3 *Verifying the Accuracy of the Proposed Method*

Observe that there are two approximations used in the proposed method: the approximation in Theorem 1, and the approximation related with the transition density in the parameters estimation method. In the presented RUL estimation method, these two approximations are implemented in series, and thus it is inevitable to wonder whether the approximation errors will be accumulated. Hence, the demonstration of the accuracy of the proposed approximation method will be desired, as suggested by the reviewer. In this chapter, to verify the accuracy of the proposed approximation method against the true value, we carried out a numerical experiment based on M_1 and the widely used Euler–Maruyama discretization policy [56].

First, we generate M degradation paths using the true parameters $\boldsymbol{\theta}$ by the *Euler* discretization

$$X_{(k+1)\Delta t} = X_{k\Delta t} + \mu(X_{k\Delta t}, k\Delta t; \boldsymbol{\theta})\Delta t + \sigma_B Y \sqrt{\Delta t}, \qquad (8.33)$$

where $Y \sim N(0, 1)$, and Δt is the discretizing size.

Then, the degradation data of some randomly selected paths are used to estimate the parameters through the MLE method proposed in Sect. 8.4. The estimated parameters $\check{\boldsymbol{\theta}}$ are not only compared with the true parameters $\boldsymbol{\theta}$, but also used to calculate the analytical lifetime PDF $f_{T|M}(t; w, \check{\boldsymbol{\theta}})$. Using the algorithm in the appendix of [38], we also simulate the FPT of the degradation paths $\boldsymbol{T} = \{T^{(1)}, T^{(2)}, \ldots, T^{(M)}\}$, where $T^{(m)}$ is the FPT of the $m-$th degradation path.

We compare the simulation results, namely the histogram of $\boldsymbol{T}$, the analytical lifetime PDF using the true parameters $f_{T|M}(t; w, \check{\boldsymbol{\theta}})$, and the analytical lifetime PDF using the estimated parameters $f_{T|M}(t; w, \boldsymbol{\theta})$ with each other, to verify the accuracy of the proposed approximation. As such, the accuracy of the presented approximated lifetime PDF, the accuracy of the proposed MLE method, and the influence of these series approximation will be jointly considered.

Specifically, the drift coefficient function is specified as $\mu(X_{k\Delta t}, k\Delta t; \boldsymbol{\theta}) = b_X X_{k\Delta t} + a_T b_t \exp(b_T k\Delta t)$ with parameters $\boldsymbol{\theta} = [b_X, a_T, b_T, \sigma_B]'$ in the illustrative example. To initialize the FPT simulation algorithm, we set the failure threshold $w = 1$, the sample size $M = 10\,000$, the initial degradation value $X_{0\Delta T} = 0$, and the discretization step $\Delta t = 0.01$. The true values of parameters are given as $b_x = 0.01$, $a_T = 2$, and $b_T = 0.2$. Additionally, because σ dominates the uncertainty of the degradation process, three different values of σ_B are used for comparisons: a small value $\sigma_B = 0.02$, a medium value $\sigma_B = 0.1$, and a large value $\sigma_B = 0.4$. The main results are shown in Fig. 8.1, in which the true, and estimated analytical lifetime distributions are obtained by substituting $\boldsymbol{\theta}$, and $\check{\boldsymbol{\theta}}$ into (8.25), respectively.

The corresponding estimated parameters for three different cases are $\check{\boldsymbol{\theta}} = [0.0201, 1.9953, 0.1984, 0.0261]'$, $\check{\boldsymbol{\theta}} = [0.0202, 2.012, 0.2232, 0.0935]'$, and $\check{\boldsymbol{\theta}} = [0.0096, 1.7257, 0.2329, 0.4046]'$, respectively. These results indicate that an accurate estimate for the parameters can be obtained by maximizing the analytical approximation of the likelihood function. The solid lines in Fig. 8.1, presenting the analytical approximation of the lifetime PDF using the true parameters, verify that the approximation in Theorem 8.1 matches the simulated lifetime well. In addition, the dashed lines, denoting the analytical approximated PDF of the lifetime using the estimated parameters, show that the proposed method can also provide a very close approximation to the simulated lifetime. Thus, the effectiveness of two approximations in the presented method is verified numerically.

8.6 Case Study

In this section, we fit the bearing data from the PHM data challenge of 2012 [45] to the proposed age- and state-dependent nonlinear degradation model. Note that the failure times of bearings are known in the used PHM challenge data set, and thus the actual RUL at each monitoring time is available. This information will facilitate the model verification.

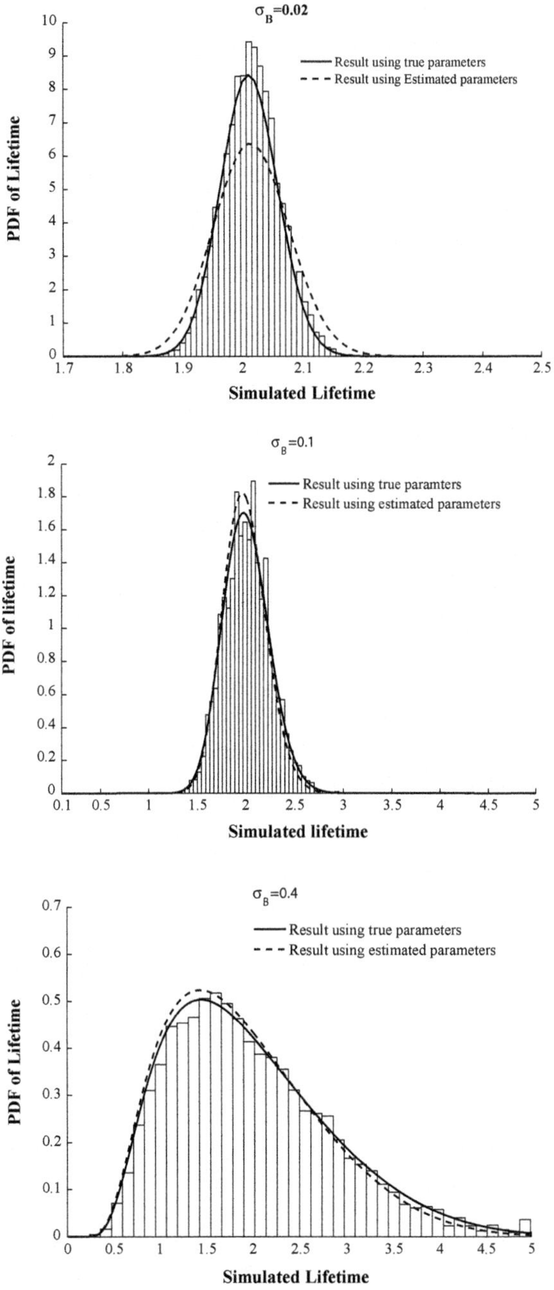

Fig. 8.1 Comparison of the histogram for the simulated FPT and the analytical approximation of the PDF

Specifically, the models M_1 and M_2 in Sect. 8.5 are used in this case study for illustration and demonstration purposes. Besides the maximized log-likelihood ℓ_{max}, we use a performance measures such as Akaike Information Criterion (AIC) [57], and the score of accuracy (SOA) of the estimated RUL [45, 51], for model comparisons. The smallest AIC corresponds to the best fitting accuracy, while the highest SOA value indicates the best estimation of RUL. We introduce the definitions of AIC, and SOA as follows.

Compared with existing models, an age- and state-dependent model generally needs more parameters. Thus, AIC is used here to overcome the problem of overparameterization, which is defined as

$$AIC = -2\ell_{max} + 2p \tag{8.34}$$

where p is the number of the estimated model parameters, and ℓ_{max} is the maximized likelihood.

The reasons why SOA is used arise from the following two considerations. First, in practice, the underestimate of the RUL only leads to conservative decision-making in maintenance and spare ordering for the degrading system, while the overestimated RUL might result in disasters and immeasurable losses due to unexpected failures of the system. In this circumstance, the SOA of the estimated RUL can differentiate between underestimates and overestimates due to the fact that the good performance of the estimation model relates to early predictions of the RUL. Second, SOA is first defined and used in PHM data challenges, which have been held successfully many times, and the SOA measure has indicated its superiority (see, e.g. [45, 51]). Specifically, the SOA of the estimated RUL for experiment i is defined as

$$SOA_{i,t_k} = \begin{cases} e^{-\ln(0.5)(Er_{k,t_i}/5)} & , Er_{i,t_k} \leq 0 \\ e^{+\ln(0.5)(Er_{k,t_i}/20)} & , Er_{i,t_k} > 0 \end{cases}, \tag{8.35}$$

where Er_{i,t_k} is the prediction error at time t_k on experiment i defined by

$$Er_{i,t_k} = 100 \times \frac{ActL_k - \widehat{L}_k}{ActL_k}. \tag{8.36}$$

In (8.36), $ActL_k$ is the actual RUL at time t_k, and $\widehat{L}_k$ denotes the estimated RUL which can be the s-measures mode, mean, or median of the distribution of the RUL. Note that SOA is applied to the case that a point estimation of the RUL is available. In this chapter, we use the mode of the estimated PDF of the RUL as the point estimate because the distribution of the RUL is often highly skewed, and thus it is reasonable to use the mode as a measure of central tendency rather than the mean.

After extracting degradation features using the approach described in [58], we choose the band-pass filtered kurtosis minus 3 as the degradation path. The extracted degradation data are illustrated in Fig. 8.1. The purpose of drawing Fig. 8.2 is to illustrate the considered degradation processes, and the practical degradation data. As

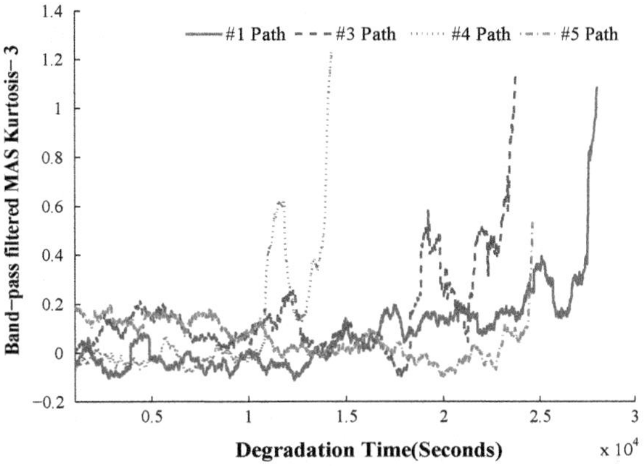

Fig. 8.2 Some degradation paths of ball bearings under operating condition 1

Table 8.3 Estimated parameters and associated values for log-LF and AIC

Model	Path	b_X	a_T	b_T	σ_B	ℓ_{max}	AIC	SOA
M_1	3	−2.357e−4	1.496e−26	2.488e−3	6.449e−3	5847.481	−11686.962	23.794
	4	−6.543e−4	3.172e−4	0.645e−4	7.523e−3	3257.247	−6506.494	23.635
	5	−3.825e−4	5.312e−90	8.325e−3	3.623e−3	7611.754	−15215.508	23.915
M_2	3	–	5.080e−33	3.0854e−3	6.358e−3	5845.908	−11685.816	23.169
	4	–	−1.212e−1	−4.873e−4	6.355e−3	3249.143	−6492.286	23.531
	5	–	2.129e−93	8.642e−3	3.6228e−3	7608.516	−15211.032	23.890

shown in Fig. 8.1, most of the degradation paths begin near zero, which corresponds to a healthy state at the very beginning of the system's degradation. As the degradation is accumulated, the degradation path approaches a known preset failure threshold $w = 1$, according to the training sets. In addition, the degradation paths exhibit nonlinear trends. This result necessitates nonlinear degradation modeling for bearing deterioration.

In this case study, the MLEs of unknown model parameters based on the training data can serve as an initial solution for the MATLAB optimization function fminsearch.

For each degradation path $\#3 through \#5$ under the first operation condition with 1800 rpm and 4000 N, the last 25 observations were regarded as the test set, while the remaining observations of each path were treated as the training set. Then, we model the degradation data with the models M_1, and M_2, respectively. The main parameter estimation results for different models were summarized in Table 8.3, together with comparative results regarding ℓ_{max}, AIC, and $SOA_k = \sum_{k=1}^{24} SOA_{i,t_k}$.

As shown in Table 8.3, the estimated parameters of b_X in M_1 confirm the existence of the state dependency in the degradation process, and the age- and state-dependent model M_1 outperforms the age-dependent model M_2 in terms of the ℓ_{max}, AIC, and SOA for all degradation paths #3, #4, and #5. Further, the SOA at each observation

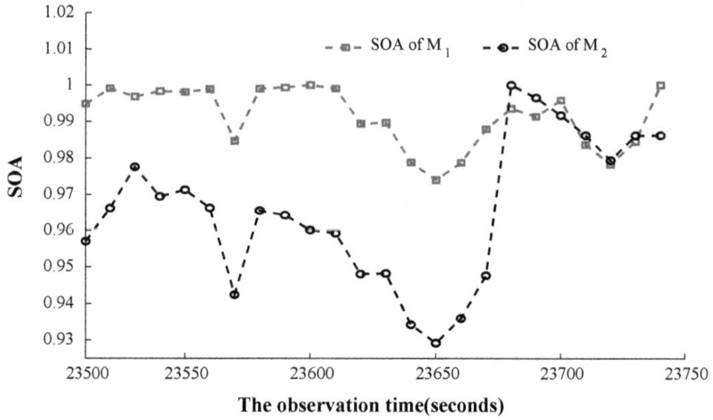

Fig. 8.3 Comparison of SOA of the RUL Estimation under M_1 and M_2 with degradation path #3

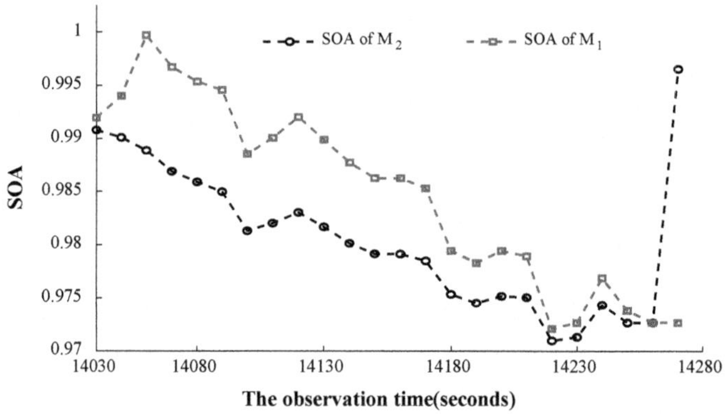

Fig. 8.4 Comparison of SOA of the RUL Estimation under M_1 and M_2 with degradation path #4

times of the test set are shown graphically in Figs. 8.3, 8.4, and 8.5 for the degradation paths #3, #4, and #5, respectively.

The results illustrated in Figs. 8.3, 8.4 and 8.5 demonstrate that model M_1 has more accurate RUL estimations at almost all the observation times than does model M_2.

In addition, the performance of the model M_1 exhibits a stable character. These observations also suggest the advantage of considering the age- and state dependency in degradation modeling and prognostics.

To have further comparisons, Figs. 8.6, 8.7 and 8.8 show the estimated PDFs of the RULs, the estimated modes of the RULs, and the actual RULs under the models M_1 and M_2 at the last 25 observations with degradation paths #3, #4, and #5, respectively.

It is clear in Figs. 8.6 through 8.8 that the estimated RULs of model M_1 are closer to the actual RULs than those of model M_2, especially for degradation paths #3 and #4. Actually, degradation paths #3 and #4 show more complex dynamics than does

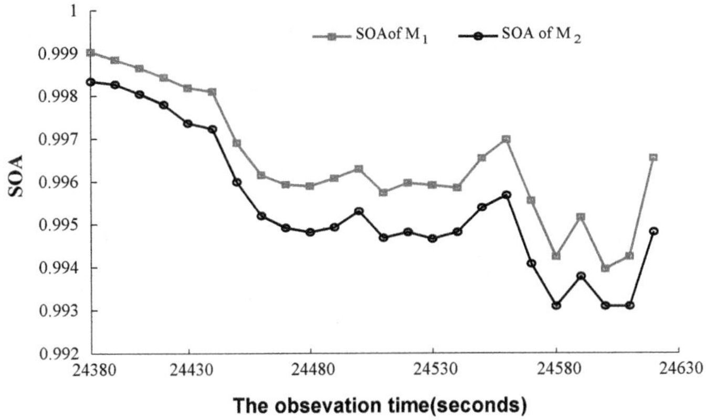

Fig. 8.5 Comparison of SOA of the RUL Estimation under M_1 and M_2 with degradation path #5

path #5. The degradation trends of paths #3 and #4 are not obvious because of large fluctuations. In such a case, using age- and state-dependent nonlinear degradation models can improve the accuracy of the RUL estimation significantly.

The measures used in this chapter for model comparisons deserve some comments. First, AIC and the maximized log-likelihood function ℓ_{max} are both criteria used for measuring the modeling fitness rather than for evaluating the accuracy of the prognosis, because AIC and the likelihood function are calculated from the degradation data that have been observed. However, in the prognosis, the failure event does not occur at the current time, and the predicted RUL is at a future time. In this case, though the improvement assessed by the AIC and the likelihood function in Table III is marginal, the results in Figs. 8.6, 8.7 and 8.8 for prognosis indicate the superiority of the proposed method. Second, the observation, that the improvement by comparing SOAs is marginal, arises from the fact that only the point estimate of the RUL is used in calculating SOA. However, the RUL is a random variable describing the future time, and thus the estimated RUL is inevitable to include the uncertainty. As a result, the PDF of the RUL includes all information of the random variable L_k, i.e., the RUL. Therefore, the estimated PDFs of the RUL used in Figs. 8.6, 8.7 and 8.8 are more appropriate to reflect the prognostic ability of the proposed method because more information is involved. As expected, Figs. 8.6, 8.7 and 8.8, and the corresponding results, show some improvement of the presented method. Together with these comparative studies, we conclude that it is necessary to consider the state dependency in modeling nonlinear degradation processes of degrading systems, and the performance of the estimated RUL is expected to be improved as a result.

Proof of Theorem 8.1

Proof As formulated in (8.11), the PDF of the FPT of $X(t)$ crossing a constant threshold W equals the PDF of the transformed standard BM crossing a $\tilde{t}$-dependent critical level $S(\tilde{t})$. From Lemma 8.2 in [38], $p_{B(\tilde{t})}(S(\tilde{t}), \tilde{t})$ can be formulated as

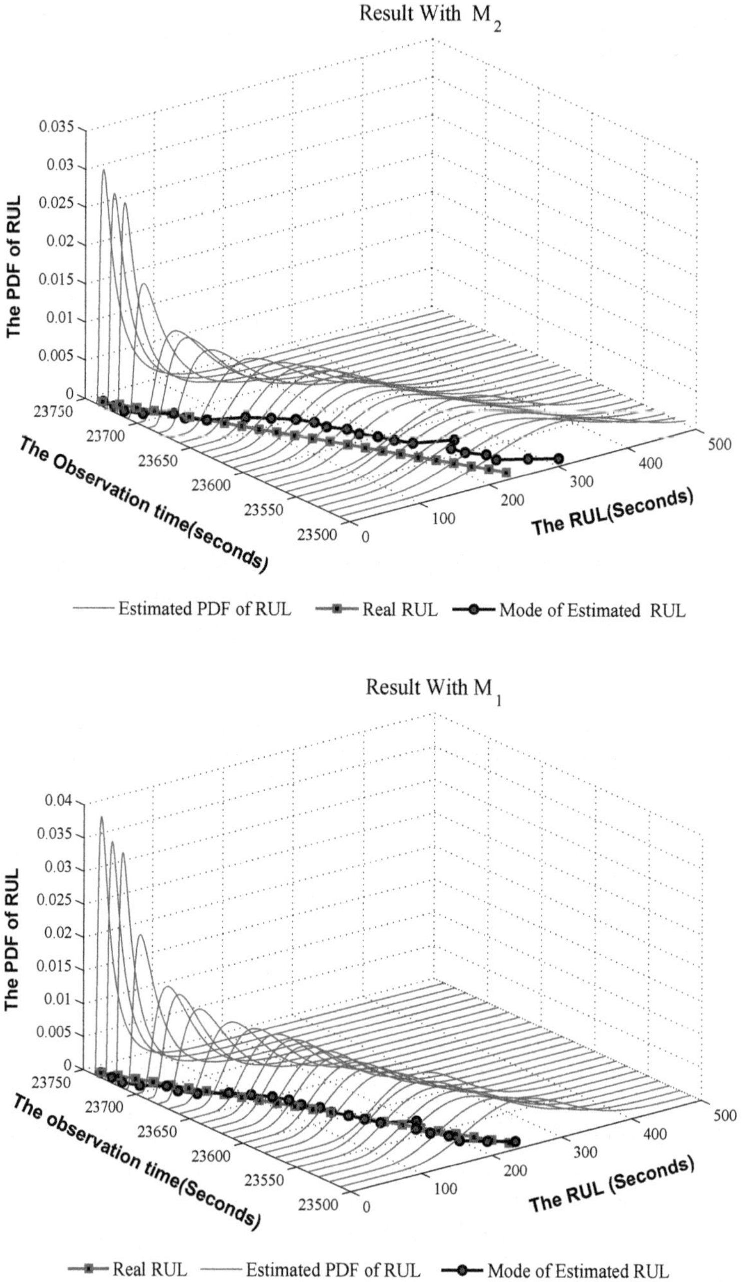

Fig. 8.6 Comparison of PDF and Mode of the RUL under M_1 and M_2 with degradation path #3

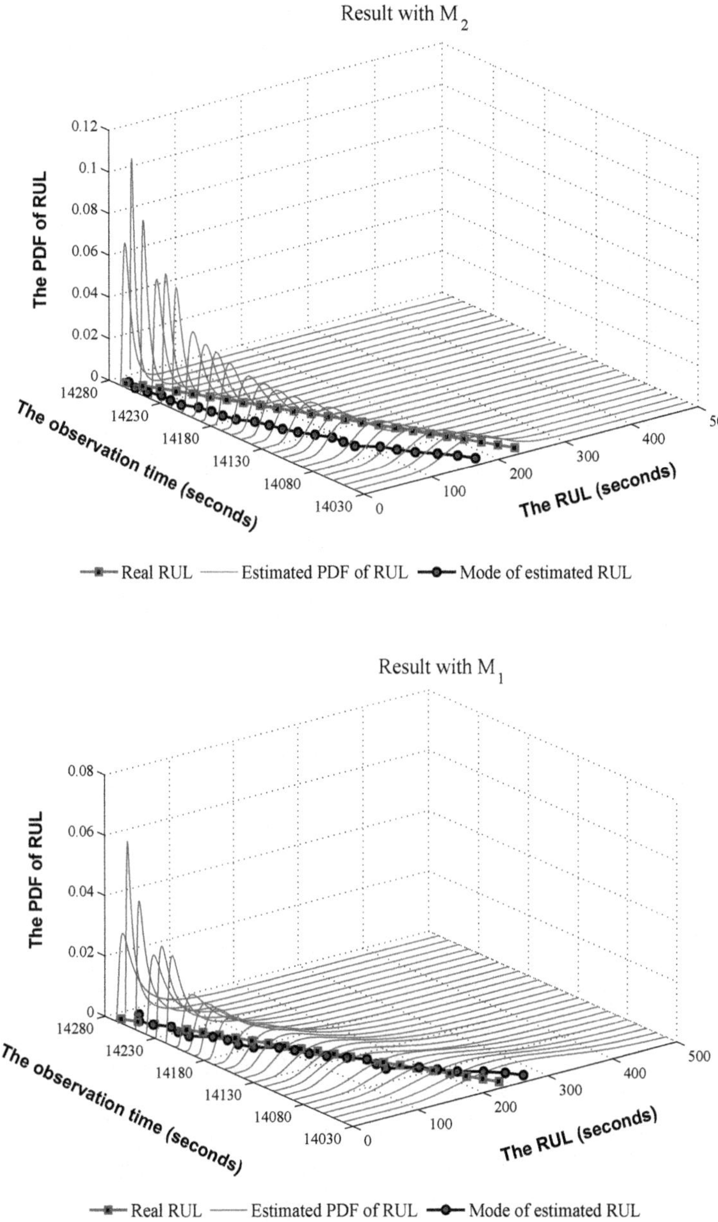

Fig. 8.7 Comparison of PDF and Mode of the RUL under M_1 and M_2 with degradation path #4

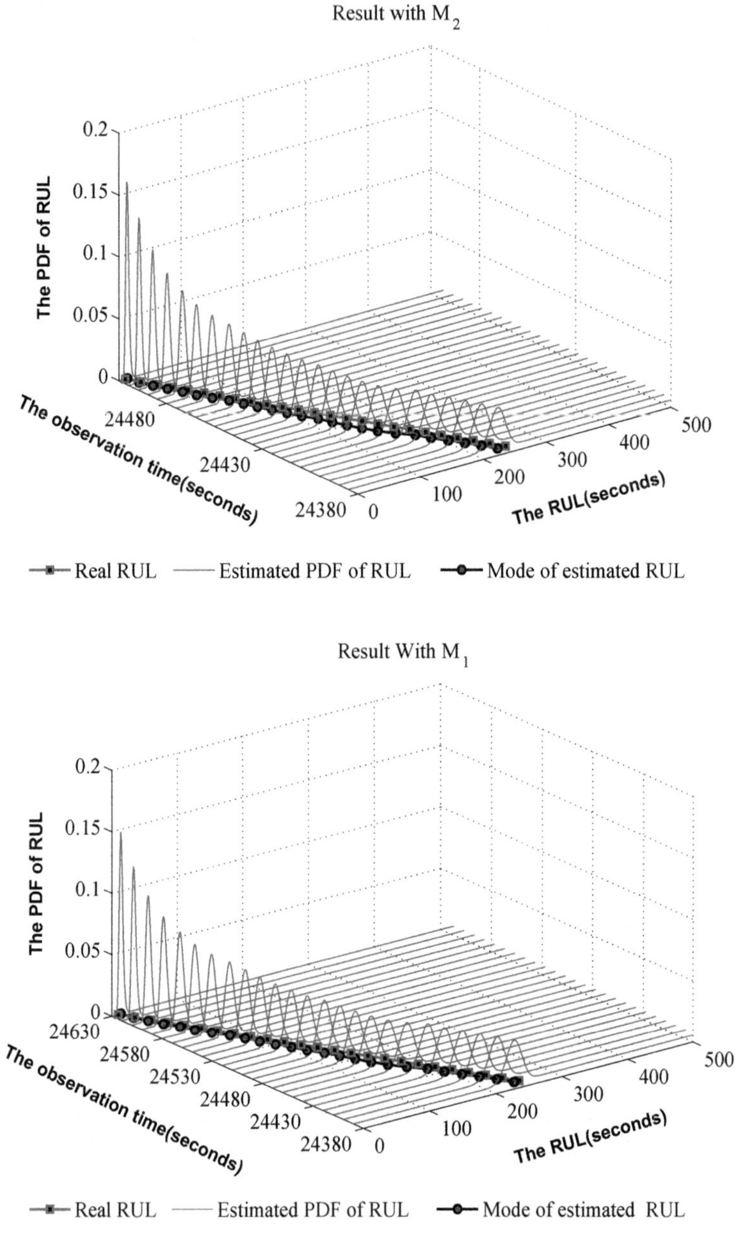

Fig. 8.8 Comparison of PDF and Mode of the RUL under M_1 and M_2 with degradation path #5

$$p_{B(\tilde{t})}(S(\tilde{t}), \tilde{t}) = b(\tilde{t}) h_{B(\tilde{t})}(\tilde{t}), \tag{8.37}$$

where $h_{B(\tilde{t})}(\tilde{t})$ is the PDF of $B(\tilde{t})$ on boundary $S(\tilde{t})$, expressed as

$$h_{B(\tilde{t})}(\tilde{t}) = \frac{1}{\sqrt{2\pi \tilde{t}}} \exp\left[-\frac{S^2(\tilde{t})}{2\tilde{t}}\right]; \tag{8.38}$$

and $b(\tilde{t})$ can be written as

$$b(\tilde{t}) = \lim_{s \to \tilde{t}} (\tilde{t} - s)^{-1} E_{B(s)|B(\tilde{t})} \left[I(s, B(\tilde{t}))(S_B(s) - B(s)) | B(\tilde{t}) = S_B(\tilde{t}) \right]. \tag{8.39}$$

In the above formula, the indicator function $I(s, B(\tilde{t})) \cong 1$ under the assumption that the probability of the transformed BM $B(\tilde{t})$ reaching boundary $S_B(\tilde{t})$ before $\tilde{t}$ can be neglected. Then, $b(\tilde{t})$ will be formulated as

$$\begin{aligned}
b(\tilde{t}) &\cong \lim_{s \to \tilde{t}} \frac{E_{B(s)|B(\tilde{t})} \left[(S_B(s) - B(s)) | B(\tilde{t}) = S_B(\tilde{t}) \right]}{\tilde{t} - s} \\
&= \lim_{s \to \tilde{t}} \frac{S_B(s) - E_{B(s)|B(\tilde{t})} \left[B(s) | B(\tilde{t}) = S_B(\tilde{t}) \right]}{\tilde{t} - s} = \lim_{s \to \tilde{t}} \frac{S_B(s) - s S_B(\tilde{t})/\tilde{t}}{\tilde{t} - s} \\
&= \lim_{s \to \tilde{t}} \frac{S_B(s) - S_B(\tilde{t}) + S_B(\tilde{t}) - s S_B(\tilde{t})/\tilde{t}}{\tilde{t} - s} = \lim_{s \to \tilde{t}} \frac{S_B(\tilde{t})(\tilde{t} - s)/\tilde{t} - (S_B(\tilde{t}) - S_B(s))}{\tilde{t} - s} \\
&= \frac{S_B(\tilde{t})}{\tilde{t}} - \frac{d S_B(\tilde{t})}{d\tilde{t}}.
\end{aligned} \tag{8.40}$$

This completes the proof of Theorem 8.1.

***Proof of Theorem* 8.1**

Proof After observing $X(t_k)$ at t_k, we can write the degradation process $\{X(t), t \geq t_k\}$ for $t > t_k$ as $dX(t) = g(X(t), t; \boldsymbol{\theta})dt + \sigma dB(t - t_k)$, with initial value x_k. Let

$$t = t_k + l_k, \tag{8.41}$$

and

$$Y(l_k) = X(l_k + t_k) - x_k. \tag{8.42}$$

Then, we have $Y(0) = 0$, and the following SDE with regard to l_k.

$$dX(t_k + l_k) = g(X(t_k + l_k), t_k + l_k)d(t_k + l_k) + \sigma_B dB(l_k + t_k) \tag{8.43}$$

From the linkages between t with t_k and $X(t)$ with $Y(l_k)$, the stochastic process $\{X(t), t \geq t_k\}$ can be transformed into the process $\{Y(l_k), l_k \geq 0\}$ in l_k according to the properties of BM as

$$dY(l_k) = g(x_k + Y(l_k), l_k + t_k; \boldsymbol{\theta})dl_k + \sigma_B dB(l_k). \tag{8.44}$$

As a result, the RUL at time t_k is equal to the FPT of the process $\{Y(l_k), l_k \geq 0\}$ crossing the threshold $w_k = w - x_k$. It is not difficult to verify that $\{Y(l_k), l_k \geq 0\}$ satisfies all the conditions of Lemmas 8.1 and 8.2. For simplicity, set $g^*(y, l_k; \boldsymbol{\theta}) = g(y+x_k, l_k+t_k; \boldsymbol{\theta})$. According to Lemma 8.2, both $c_1^*(l)$, and $c_2^*(l)$ can be determined as

$$c_1^*(l_k) = \frac{2g^*(y, l_k; \boldsymbol{\theta}) - yc_2(l_k)}{\sigma_B}, \tag{8.45}$$

$$c_2^*(l_k) = \frac{\partial g^*(y, l_k; \boldsymbol{\theta})}{\partial y}. \tag{8.46}$$

And at the same time the time-space transformation $\Psi^*(l_k, y)$, and $\phi^*(l_k)$ can be obtained by

$$\Psi^*(y, l_k) = \exp\left[-\frac{1}{2}\int_0^{l_k} c_2^*(u)du\right]\frac{y}{\sigma_B} - \frac{1}{2}\int_0^{l_k} c_1^*(\tau)\exp\left[-\frac{1}{2}\int_0^{\tau} c_2^*(u)du\right]d\tau,$$

$$\varphi^*(l_k) = \int_0^{l_k} \exp\left[-\frac{1}{2}\int_0^{\tau} c_2^*(u)du\right]d\tau. \tag{8.47}$$

Therefore, the conclusion in Theorem 8.1 can be used conveniently to calculate the PDF of the RUL l_k when x_k is available at time t_k. As such, we directly have $S(\tilde{l}_k) = \Psi^*(w_k, \varphi^{*-1}(\tilde{l}_k))$, and the PDF of the RUL at t_k as (8.16).

This completes the proof of Theorem 8.2.

The particular forms of $\beta_j^2(y_n, y_{n-1}, t_n, t_{n-1})$

The particular forms of $\beta_j^2(y_n, y_{n-1}, t_n, t_{n-1})$ in (8.31) are summarized herein as follows.

$$\beta_0^2(y_n, y_{n-1}, t_n, t_{n-1}) = 1$$

$$\beta_1^2(y_n, y_{n-1}, t_n, t_{n-1}) = -\sqrt{\Delta_n}\psi - \frac{\sqrt{\Delta_n^3}}{4}(2\psi_{10} + 2\psi\psi_{01} + \psi_{20})$$

$$\beta_2^2(y_n, y_{n-1}, t_n, t_{n-1}) = \frac{\Delta_n}{2}(\psi^2 + \psi_{10}) + \frac{\Delta_n^2}{12}(6\psi\psi_{01} + 6\psi^2\psi_{10} + 4\psi_{10}^2 + 4\psi_{11} + 7\psi\psi_{20} + 2\psi_{30})$$

$$\beta_3^2(y_n, y_{n-1}, t_n, t_{n-1}) = -\frac{\sqrt{\Delta_n^3}}{6}(\psi^3 + 3\psi\psi_{10} + \psi_{20})$$

$$\beta_4^2(y_n, y_{n-1}, t_n, t_{n-}) = -\frac{\Delta_n^2}{24}(\psi^4 + 6\psi^2\psi_{10} + 3\psi_{10}^2 + 4\psi\psi_{20} + \psi_{30})$$

where ψ, ψ_{10}, ψ_{01}, ψ_{11}, ψ_{20}, ψ_{20} are expressed as

$$
\begin{aligned}
\psi &= \mu_Y(y_{n-1}, t_{n-1}) = b_X y_{n-1} + \frac{a_T b_T}{\sigma_B} \exp(b_T t_{n-1}) \\
\psi_{01} &= \frac{\partial \mu_Y(y,t)}{\partial t}|_{y=y_{n-1}} = \frac{a_T b_T^2}{\sigma_B} \exp(b_T t_{n-1}) \\
\psi_{10} &= \frac{\partial \mu_Y(y,t)}{\partial y}|_{y=y_{n-1}} = b_X \\
\psi_{11} &= \frac{\partial \mu_Y(y,t)}{\partial y \partial t}|_{y=y_{n-1}} = 0 \\
\psi_{20} &= \frac{\partial^2 \mu_Y(y,t)}{\partial y^2}|_{y=y_{n-1}} = 0 \\
\psi_{30} &= \frac{\partial^3 \mu_Y(y,t)}{\partial y^3}|_{y=y_{n-1}} = 0
\end{aligned}
$$

for M_1, and

$$
\begin{aligned}
\psi &= \mu_Y(y_{n-1}, t_{n-1}) = \frac{a_T b_T}{\sigma_B} \exp(b_T t_{n-1}) \\
\psi_{01} &= \frac{\partial \mu_Y(y,t)}{\partial t}|_{y=y_{n-1}} = \frac{a_T b_T^2}{\sigma_B} \exp(b_T t_{n-1}) \\
\psi_{10} &= \frac{\partial \mu_Y(y,t)}{\partial y}|_{y=y_{n-1}} = 0 \\
\psi_{11} &= \frac{\partial \mu_Y(y,t)}{\partial y \partial t}|_{y=y_{n-1}} = 0 \\
\psi_{20} &= \frac{\partial^2 \mu_Y(y,t)}{\partial y^2}|_{y=y_{n-1}} = 0 \\
\psi_{30} &= \frac{\partial^3 \mu_Y(y,t)}{\partial y^3}|_{y=y_{n-1}} = 0
\end{aligned}
$$

for M_2.

References

1. Pecht M (2008) Prognostics and health management of electronics. Wiley, New Jersey
2. Wang W (2007) A prognosis model for wear prediction based on oil-based monitoring. J Oper Res Soc 58:887–893
3. Sikorska JZ, Hodkiewicz M, Ma L (2011) Prognostic modelling options for remaining useful life estimation by industry. Mech Syst Signal Process 25:1803–1836
4. Hu C, You BD, Wang P, Yoon JT (2012) Ensemble of data-driven prognostic algorithms for robust prediction of remaining useful life. Reliab Eng Syst Saf 103:120–135
5. Hu J, Zhang L, Ma L, Liang W (2011) An integrated safety prognosis model for complex system based on dynamic Bayesian network and ant colony algorithm. Expert Syst Appl 38:1431–1446

6. Lorton A, Fouladirad M, Grall A (2013) Methodology for probabilistic model-based prognosis. Eur J Oper Res 225:443–454
7. Lu CJ, Meeker WQ (1993) Using degradation measures to estimate a time-to-failure distribution. Technometrics 35(2):161–174
8. Chen Z, Zheng S (2005) Lifetime distribution based degradation analysis. IEEE Trans Reliab 54:3–10
9. Sun JW, Li L, Xi LF (2012) Modified two-stage degradation model for dynamic maintenance threshold calculation considering uncertainty. IEEE Trans Autom Sci Eng 09(1):209–212
10. Chen N, Tsui KL (2013) Condition monitoring and residual life prediction using degradation signals: revisited. IIE Trans 45(9):939–952
11. Saha B, Goebel K, Poll S, Christophersen J (2009) Prognostics methods for battery health monitoring using a Bayesian framework. IEEE Trans Instrum Meas 58(2):291–296
12. Nuhic A, Terzimehic T, Soczka-Guth T, Buchholz M, Dietmayer K (2013) Health diagnosis and remaining useful life prognostics of lithium-ion batteries using data-driven methods. J Power Sources 239:680–688
13. Zhao JY, Liu F (2007) Reliability assessment of the metalized film capacitors from degradation data. Microelectron Reliab 47:434–436
14. Kimura K, Iwabuch T, Morooka K, Ishikawa Y (1990) A useful index for estimating residual life of motor insulation. IEEE Electr Insul Mag 6(2):29–34
15. Si XS, Wang W, Hu CH, Zhou DH (2011) Remaining useful life estimation-a review on the statistical data driven approaches. Eur J Oper Res 213(1):1–14
16. van Noortwijk JM (2009) A survey of the application of gamma processes in maintenance. Reliab Eng Syst Saf 94:2–21
17. Lawless J, Crowder M (2004) Covariates and random effects in a gamma process model with application to degradation and failure. Lifetime Data Anal 10:213–227
18. Ye ZS, Xie M, Tang LC, Chen N (2014) Efficient semi-parametric estimation of Gamma processes for deteriorating products. Technometrics. doi:10.1080/00401706.2013.869261
19. Si XS, Wang W, Chen MY, Hu CH, Zhou DH (2013) A Wiener-process-based degradation model with a recursive filter algorithm for remaining useful life estimation. Mech Syst Signal Proces 35:219–237
20. Wang X (2010) Wiener processes with random effects for degradation data. J Multivar Anal 101:340–351
21. Whitmore GA, Schenkelberg F (1997) Modelling accelerated degradation data using wiener diffusion with a time scale transformation. Lifetime Data Anal 3:27–45
22. Ye ZS, Wang Y, Tsui KL, Pecht M (2013) Degradation data analysis using Wiener process with measurement errors. IEEE Trans Reliab 57(4):539–550
23. Wang X, Xu D (2010) An inverse Gaussian process model for degradation data. Technometrics 52(2):188–197
24. Ye ZS, Chen N (2014) The Inverse Gaussian process as a degradation model. Technometrics. doi:10.1080/00401706.2013.830074
25. Kharoufeh JP, Solo CJ, Ulukus MY (2010) Semi-Markov models for degradation-based reliability. IIE Trans 42:599–612
26. Dong M, He D (2007) Hidden semi-Markov model-based methodology for multi-sensor equipment health diagnosis and prognosis. Eur J Oper Res 178:858–878
27. Ye ZS, Xie M (2014) Stochastic modelling and analysis of degradation for highly reliable products. Appl Stoch Models Bus Ind. doi:10.1002/asmb.2063
28. Bogdanoff JL, Kozin F (1985) Probabilistic models of cumulative damage. Wiley, New York
29. Giorgio M, Guida M, Pulcini G (2010) A state-dependent wear model with an application to marine engine cylinder liner. Technometrics 52(2):172–189
30. Giorgio M, Guida M, Pulcini G (2010) A parametric markov chain to model age- and state-dependent wear processes. In: Pietro M, Piercesare S (eds) Complex data modelling and computationally intensive statistical methods. Springer, Milan
31. Giorgio M, Guida M, Pulcini G (2011) An age- and state-dependent Markov model for degradation processes. IIE Trans 43(9):621–632

32. Guida M, Pulcini G (2011) A continuous-state Markov model for age- and state-dependent degradation processes. Struct Saf 33(6):354–366
33. Gebraeel N, Lawley MA, Li R, Ryan JK (2005) Residual-life distributions from component degradation signals: a Bayesian approach. IIE Trans 37:543–557
34. Si XS (2015) An adaptive prognostic approach via nonlinear degradation modelling: Application to battery data. IEEE Trans Ind Electron. doi:10.1109/TIE.2015.2393840
35. Chiao CH, Hamada M (1996) Using degradation data from an experiment to achieve robust reliability for light emitting diodes. Qual Reliab Eng Int 12:89–94
36. Park C, Padgett WJ (2005) Accelerated degradation models for failure based on geometric Brownian motion and Gamma processes. Lifetime Data Anal 11:511–527
37. Meeker WQ, Escobar LA, Lu CJ (1998) Accelerated degradation tests: modeling and analysis. Technometrics 40(2):89–99
38. Si XS, Wang W, Hu CH, Zhou DH, Pecht MG (2012) Remaining useful life estimation based on a nonlinear diffusion degradation process. IEEE Trans Reliab 61(1):50–67
39. Feng L, Wang HL, Si XS, Zhou HX (2013) A state space based prognostic model for hidden and age-dependent nonlinear degradation process. IEEE Trans Autom Sci Eng 10(4):1072–1086
40. Peng CY, Tseng ST (2009) Mis-specification analysis of linear degradation models. IEEE Trans Reliab 58(3):444–455
41. Whitmore GA (1995) Estimating degradation by a Wiener diffusion process subject to measurement error. Lifetime Data Anal 1:307–319
42. Yuan XX, Pandey MD (2009) A nonlinear mixed-effects model for degradation data obtained from in-service inspections. Reliab Eng Syst Saf 94:509–519
43. Wu SJ, Shao J (1999) Reliability analysis using the least squares method in nonlinear mixed-effect degradation models. Statistica Sinica 9:855–877
44. Si XS, Wang W, Chen MY, Hu CH, Zhou DH (2013) A degradation path-dependent approach for remaining useful life estimation with an exact and closed-form solution. Eur J Oper Res 226:53–66
45. Nectoux P, Gouriveau R, Medjaher K, Ramasso E, Morello B, Zerhouni N, Varnier C (2012) PRONOSTIA: An experimental platform for bearings accelerated life test. IEEE International conference on prognostics and health management, Denver, Colorado, USA
46. AÏt-Sahalia Y (2002) Maximum-likelihood estimation of discretely sampled diffusions: a closed-form approach. Econometrica 70(1): 223–262
47. Egorov AV, Li H, Xu Y (2003) Maximum likelihood estimation of time-inhomogeneous diffusions. J Econom 114:107–139
48. AÏt-Sahalia Y (2008) Closed-form likelihood expansions for multivariate diffusions. Ann Stat 36(2): 906–937
49. Mao X (2007) Stochastic differential equations and application. International Publishers in Science and Technology, Chichester
50. Wang ZQ, Wang WB, Hu CH, Si XS (2014) An additive Wiener process-based prognostic model for hybrid deteriorating systems. IEEE Trans Reliab 63(1):208–222
51. Son KL, Fouladirad Mi, Barros A, Levrat E, Iung B (2013) Remaining useful life estimation based on stochastic deterioration models: a comparative study. Reliab Eng Syst Saf 112:165–175
52. Wang XL, Balakrishnan N, Guo B (2014) Residual life estimation based on a generalized Wiener degradation. Reliab Eng Syst Saf 124:13–23
53. Si XS, Zhou DH (2013) A generalized result for degradation model-Based reliability estimation. IEEE Trans Autom Sci Eng. doi:10.1109/TASE.2013.2260740
54. Kalbfleisch JD, Prentice RL (2002) The statistical analysis of failure time data. Wiley, New York
55. Choi S (2013) Closed-form likelihood expansions for multivariate time-inhomogeneous diffusions. J Econom 174:45–65
56. Kloeden P, Platen E (1995) Numerical solution of stochastic differential equations. Springer, New York

57. Akaike H (1974) A new look at the statistical model identification. IEEE Trans Autom Control 19(6):716–722
58. Sutrisno E, Oh H, Vasan ASS, Pecht M (2012) Estimation of remaining useful life of ball bearings using data driven methodologies. IEEE International Conference on Prognostics and Health Management, Denver, Colorado, USA

Chapter 9
Adaptive Prognostic Approach via Nonlinear Degradation Modeling

9.1 Introduction

With the ever-increased high requirement of reliability and safety for critical systems, accurately assessing the pending failure of a system has become an active research area over the past decades. This also leads to an emerging concept called prognostics and health management (PHM) [1–3]. PHM is an enabling discipline consisting of technologies and methods to assess the reliability of a system in its actual life-cycle conditions, to determine the advent of failure, and to mitigate system risk. The past decade has witnessed an increasing research interest on various aspects of PHM due primarily to the fact that PHM have been extensively applied in a variety of fields including electronics, smart grid, batteries, bearings, motor drives, electromechanical structures, analog electronic circuits, power industry, aerospace, and military application, public health management [4–11].

The prognostic part of PHM is typically characterized by estimating the remaining useful life (RUL) of a system, conditional on the available information at hand. In fact, the estimation of the RUL has been considered as one of the most central components in PHM [12–19]. Effective and accurate RUL estimation can avoid catastrophic events, extend life cycle, and schedule timely healthcare actions [20–22]. Many technological systems are subject, during their operating life, to a gradual deterioration process that progressively degrades their performance until a failure occurs [13, 23, 24]. For example, lithium-ion batteries have been widely used in many fields, including consumer electronics, electric vehicles, marine systems, aircrafts, satellites, etc., due to their high power density, low weight, long lifetime, no memory effect, and other advantages. Nevertheless, an inevitable problem is that lithium-ion battery performance degrades with cycling and aging. Such functionality degradation of lithium-ion batteries may cause reduced performance and even catastrophic failure. For example, in 2013, all Boeing 787 Dreamliners were indefinitely grounded due to battery failures that occurred on two planes. However, the progression of degradation of systems such as lithium-ion batteries and electromechanical systems is typically stochastic in nature, resulting in the difficulty to estimate the RUL with certainty. To

X.-S. Si et al., *Data-Driven Remaining Useful Life Prognosis Techniques*,
Springer Series in Reliability Engineering, DOI 10.1007/978-3-662-54030-5_9

characterize the uncertainty of the estimated RUL, the form of the probability density function (PDF) of the RUL is often required. Thus, this chapter pays a particular attention to the estimation of the PDF of the required RUL.

The degradation processes of technological systems are usually affected by two kind of variability, namely individual variability and temporal variability [25, 26]. The individual variability determines heterogeneity among the degradation paths of different systems. The temporal uncertainty is referred to as the inherent uncertainty associated with the progression of the degradation over time. For the individual variability, it can be usually modeled by introducing system-specific random effects by some model parameters [27, 28]. Some efforts in this respect have been made by proposing Bayesian procedures for real-time estimation of the RUL of degrading units on the basis of sensor-based degradation models [29]. However, they assumed that the variance parameter of the model is fixed for different units and thus is estimated by offline data. In practice, different units have different degradation magnitudes and hence different variance parameters. In addition, the fixed variance parameter dominates the variance of the estimated RUL for a particular system since random effects on model parameters of this kind of models are very small for a particular system [25].

For each single system, the temporal variability determines the random fluctuation over time of the degradation process. As advocated by [30, 31], this can be described by an appropriate stochastic process which provides flexibility on describing the characteristics of the temporal effects. Therefore, this chapter mainly considers the stochastic process-based modeling approaches for prognostics. The central idea in this kind of approaches is that the lifetime can be defined as the first hitting time (FHT) of the degradation process crossing a failure threshold [32], i.e., soft failure.

If the degradation process is monotonic, the relationship between the degradation and the FHT can be easily established [33]. For non-monotonic degradation process, a Wiener process is a commonly used stochastic process-based model with a constant degradation drift and with the dynamic part driven by the standard Brownian motion. However, most of the works on the RUL estimation based on the Wiener process assume that the degradation process has a linear path or can be linearized by logarithmic or time-scale transformations [16, 24, 28, 34–36]. Such assumption restricts the applications of these prognostic methods in several complex circumstances. For example, during the aircraft flight process, the engine load varies as the aircraft is exposed to several distinct flight conditions including takeoff, maximum climb, maximum cruise, ground idle, etc. Moreover, if inclement weather is encountered during flight, the aircraft may change altitude, accelerate, or decelerate to avoid turbulence. Another example is the lithium-ion batteries which contains three different operational profiles (charge, discharge, and impedance). These environmental/operational conditions affect the degradation processes of systems and lead to the varying degradation rate of systems, i.e., nonlinear stochastic degradation.

As for nonlinear stochastic deteriorating systems, a nonlinear stochastic degradation model for RUL estimation was presented in [26] by introducing a time-varying degradation rate function. In [26], the RUL distribution was derived under the assumption that if the degradation process hits the threshold at a certain time

t, then the probability that such a process crossed the threshold level before time t can be negligible. However, only the current degradation observation is considered in the estimated RUL and the parameters in the model cannot be adaptively updated via newly monitored degradation data. There are reported models utilizing the degradation data to date for evaluating and updating the RUL [29, 34, 37]. Using the history of observed degradation, degradation models based on linear Wiener processes were proposed by [34, 37] with an adapted drift to estimate the RUL. The work [29] presented an exponential-like degradation model with a Brownian error for RUL estimation, where the historical data to date of individual equipment are incorporated by Bayesian mechanism. Following [29], many variants have been reported (see a review in [38]). However, such models are limited to those which can transform the exponential path into a linear path. The work in [26] was extended by [39], but the result for RUL estimation was limited to the nonlinear model with the power function. Recently, Zhou et al. in [40] proposed a promising prognostic framework for individual units subject to hard failure, based on joint modeling of degradation signals and time-to-event data. In this work, the degradation signals are modeled using a mixed-effects regression model and time-to-event data are modeled using the Cox Proportional Hazard (PH) model where the degradation signals are used as covariate information. Then, by updating the parameters in the degradation model, the RUL prediction is achieved by formulating the relationship between the survival function and the hazard rate function. Therefore, the proposed prediction method in [40] focuses mainly on the case of hard failure. In other words, both the degradation signals and time-to-event data are required to be available. However, time-to-event data might be scarce or even non-existent for systems which are costly or time-consuming to collect time-to-event data. In this case, the hazard rate and its associated parameters are difficult to determine. In addition, the mixed-effects regression models might not be effectively model the temporal variability in stochastic degradation processes [30, 31].

There are also prognostic studies with nonlinear models, but under simulation-based framework [41–43]. For example, Zhou et al. in [42] proposed a nonlinear state-space model to estimate the RUL, where the RUL was estimated by Monte Carlo method and the unknown parameters were estimated through expectation-maximization (EM) algorithm. Orchard et al. in [43] proposed the concept of artificial evolution to estimate unknown model parameters within a particle-filtering-based framework and an adaptation mechanism to modify the variance of the uncertainty source that defines the performance of the artificial evolution algorithm. To do so, the health state and model parameters can be jointly estimated by particle filter. Similar ideas can also be found in [44–46]. However, the particle filter is a Monte Carlo-based approach, and thus only the numerical result of the estimated RUL distribution can be obtained and an explicit PDF estimate of the RUL is difficult to achieve. In the context of PHM, it is necessary and valuable to derive the PDF of the RUL with an explicit form so as to provide real-time RUL estimation for the subsequent maintenance scheduling.

The purpose of this chapter is to develop a general nonlinear degradation process model that can generate an explicit RUL distribution and adapt to the history of observed degradation data.In particular, the goal is to shed light on three fundamental issues frequently encountered in soft failure prognosis: (i) nonlinear degradation modeling without requiring data transformation and time-to-event data, (ii) obtaining and adaptively updating the explicit expression of the RUL distribution derived from the degradation process with newly observed data, and (iii) updating estimated parameters of the degradation model.

To address the above issues, an adaptive and nonlinear prognostic model is presented for RUL estimation, in which the dynamics and nonlinearity of the degradation process are modeled by a time-dependent drift coefficient. This degradation model can cover conventionally linear models. In order to make the RUL estimation depending on the history of the observations, a state-space model is constructed and Bayesian filtering is applied to update one parameter in the drifting function through treating this parameter as a hidden state variable. The PDF of the RUL is derived with an explicit form and some commonly used linear model-based methods can be shown to be special cases of the proposed model. In addition, the EM algorithm in conjunction with the Kalman filter is utilized to update the drifting parameter and other parameters in the state-space model simultaneously and recursively. Finally, several numerical examples and an application to lithium-ion batteries data are provided to validate the proposed approach.

The remainder parts are organized as follows. Section 9.2 develops an adaptive and nonlinear prognostic model for RUL estimation. In Sect. 9.3, the parameter estimation/updating approach is presented. An illustrative example to the power model is given in Sect. 9.4. In Sect. 9.5, numerical examples and an application are provided to demonstrate the developed method.

9.2 Nonlinear Model Description and RUL Estimation

9.2.1 Modeling Description

Let $\{X(t), t \geq 0\}$ denote the stochastic process describing the progression of degradation over operating time t when starting from a known initial state $X(0) = x_0$. In order to model the nonlinear stochastic process of systems such as lithium-ion batteries, the degradation at time t is denoted as follows

$$X(t) = x_0 + \lambda \cdot \int_0^t \mu(\tau; \vartheta) \mathrm{d}\tau + \sigma_B B(t), \tag{9.1}$$

where the degradation process $X(t)$ is driven by a standard Brownian motion (BM) $B(t)$with a nonlinear drift $\lambda \cdot \mu(t; \vartheta)$. In (9.1), $\lambda \cdot \mu(t; \vartheta)$ and σ_B are the drift and diffusion coefficients, respectively; $\mu(t; \vartheta)$ is a nonlinear function over t with unknown parameter vector ϑ, which can be used to characterize time-variable and nonlinear performance of systems such as lithium-ion batteries and electromechanical systems;

λ is a proportional parameter controlling the speed of the nonlinear degradation while ϑ is used to determine the shape of the degradation progression. Without loss of generality, it is assumed $X(0) = x_0 = 0$. Further, let $\boldsymbol{\phi} = [\lambda, \vartheta, \sigma_B]$ denote the model parameters.

The motivation of using degradation model (9.1) is twofold. First, the engineering systems possibly experiences different operating conditions such as charge, discharge, and impedance conditions for lithium-ion batteries, and thus the degradation paths of systems exhibit varying degradation rates. Second, the model (1) uses a relatively general Wiener process with a nonlinear drift part. It is general since it can cover linear models used in the literature and include a variety of nonlinear paths through selecting the forms of $\mu(t; \vartheta)$. For example, if $\mu(t; \vartheta)$ is a constant, (9.1) is reduced to the Wiener process with a constant drift, which has been widely used to model degradation processes with linear paths [16, 24, 34, 37].

Now, we illustrate the main modeling principle of estimating the RUL based on (9.1). From [31, 32], the concept of the FHT is used to define the lifetime and then infer the RUL. In other words, when the degradation $X(t)$ modeled as (9.1) reaches a preset critical level w, the system can be declared to be failed and thus there has no useful lifetime left. Therefore, it is natural to view the event of lifetime termination as the point that $X(t)$ exceeds the threshold level w for the first time. For critical equipments, it is usually mandatory for putting this into practice and once the observed degradation is equal or above the set threshold level that the system must be stopped for inspecting.

Under model expressed by (9.1), and based on the concept of the FHT, the FHT T of $\{X(t), t \geq 0\}$ crossing threshold w can be defined as

$$T = \inf\{t : X(t) \geq w | X(0) < w\}. \tag{9.2}$$

As a result, the key for estimating the RUL is to derive the PDF of lifetime T, denoted as $f_{T|\boldsymbol{\phi}}(t|\boldsymbol{\phi})$.

9.2.2 Derivation of the RUL Distribution

From [26], the approximated distribution of T can be obtained by the following lemma.

Lemma 9.1 *For the degradation process $\{X(t), t \geq 0\}$ given by (9.1), if $\mu(t; \vartheta)$ is a continuous function of time t in $[0, \infty)$, then the PDF of the FHT T of $X(t)$ defined by (9.2) can be approximated with an explicit form as follows:*

$$f_{T|\boldsymbol{\phi}}(t|\boldsymbol{\phi}) \cong \frac{1}{\sqrt{2\pi t}} \left(\frac{S_B(t)}{t} + \frac{\lambda}{\sigma_B} \mu(t; \vartheta) \right) \exp\left[-\frac{S_B^2(t)}{2t} \right], \tag{9.3}$$

where $S_B(t) = (w - \lambda \cdot \int_0^t \mu(\tau; \vartheta) d\tau)/\sigma_B$.

Lemma 9.1 is obtained in the context of using a Wiener process for degradation modeling under an assumption that, if the degradation process is hitting the threshold w at a certain time t exactly, then the probability that such a process crossed the threshold level before time t is negligible. This assumption has been well demonstrated by recent prognostics studies in [26, 47–49]. From Lemma 9.1, it can be easily found that most of the current models based on the Wiener process with a linear drift can be covered by the proposed model. That is to say, $f_{T|\boldsymbol{\phi}}(t|\boldsymbol{\phi})$ can be reduced to the inverse Gaussian distribution provided that $\lambda \cdot \mu(t;\vartheta)$ is a constant.

However, in (9.3), the estimated PDF of the FHT does not consider the real-time observations of the degradation process. Considering the potential for updating the knowledge of the process when new degradation observation $X(t_i) = x_i$ become available, for $t \geq t_i$, the degradation process over time since t_i can be revised as

$$\begin{aligned} X(t) = x_i + \lambda & \left(\int_0^t \mu(\tau;\vartheta)\mathrm{d}\tau - \lambda \cdot \int_0^{t_i} \mu(\tau;\vartheta)\mathrm{d}\tau \right) + \\ & \sigma_B \left(B(t) - B(t_i) \right) \\ = & x_i + \lambda \cdot \int_{t_i}^t \mu(\tau;\vartheta)\mathrm{d}\tau + \sigma_B B(t - t_i) \end{aligned} \tag{9.4}$$

for $t \geq t_i$.

In such case, the residual $t - t_i$ corresponds to the realization of the RUL at time t_i if t is the FHT of $\{X(t), t \geq t_i\}$. Given $X(t_i) = x_i$, the RUL at t_i can be defined as

$$L_i = \inf \{ l_i : X(t_i + l_i) \geq w | X(t_i) = x_i, \boldsymbol{\phi} \}, \tag{9.5}$$

with the PDF $f_{L_i|x_i,\boldsymbol{\phi}}(l_i|x_i,\boldsymbol{\phi})$.

Taking the transformation $l_i = t - t_i$ with $l_i \geq 0$ for (9.4), the process $\{X(t), t \geq t_i\}$ can be transformed with timescale over the residual time l_i, i.e., RUL, as

$$X(l_i + t_i) = x_i + \lambda \cdot \int_{t_i}^{l_i + t_i} \mu(\tau;\vartheta)\mathrm{d}\tau + \sigma_B \left(B(l_i + t_i) - B(t_i) \right). \tag{9.6}$$

As a result, the RUL at time t_i is equal to the FHT of the process $\{Y(l_i), l_i \geq 0\}$ crossing threshold $w_i = w - x_i$, where $Y(l_i) = X(l_i + t_i) - x_i$ and $Y(0) = 0$. That is to say, given t_i,

$$Y(l_i) = \lambda \cdot \int_{t_i}^{l_i + t_i} \mu(\tau;\vartheta)\mathrm{d}\tau + \sigma_B W(l_i), \tag{9.7}$$

where $W(l_i) = B(l_i + t_i) - B(t_i)$.

In order to derive $f_{L_i|x_i,\boldsymbol{\phi}}(l_i|x_i,\boldsymbol{\phi})$, it is necessary to prove that the stochastic process, $\{W(l_i), l_i \geq 0\}$, with $W(l_i) = B(l_i + t_i) - B(t_i)$ is still a BM. This is guaranteed by the following lemma.

Lemma 9.2 *Given t_i, the stochastic process, $\{W(t), t \geq 0\}$, with $W(t) = B(t + t_i) - B(t_i)$ for any $t \geq 0$ is still a standard BM, where $\{B(t), t \geq 0\}$ is a standard BM.*

The proof of Lemma 9.2 can be easily achieved by checking the properties of the standard BM. As such, the estimated PDF of the RUL, $f_{L_i|x_i,\boldsymbol{\phi}}(l_i|x_i, \boldsymbol{\phi})$, conditional on the current observation and $\boldsymbol{\phi}$ can be obtained by the following theorem.

Theorem 9.1 *Under the same conditions as Lemma 9.1, the PDF of the RUL can be formulated at time t_i with the available current degradation measurement x_i as*

$$f_{L_i|x_i,\boldsymbol{\phi}}(l_i|x_i, \boldsymbol{\phi}) \cong \frac{w_i - \lambda\left(\upsilon(l_i) - l_i\mu(l_i + t_i; \vartheta)\right)}{\sigma_B\sqrt{2\pi l_i^3}} \times \exp\left[-\frac{(w_i - \lambda\upsilon(l_i))^2}{2\sigma_B^2 l_i}\right], \tag{9.8}$$

with $\upsilon(l_i) = \int_{t_i}^{l_i+t_i} \mu(\tau; \vartheta)d\tau$ and $w_i = w - x_i$.

Proof Based on Lemma 9.1, (9.6) and (9.7), it is known that $\{Y(l_i), l_i \geq 0\}$ is still a nonlinear degradation process satisfying (9.1) with a drift part $\lambda \cdot \mu(l_i + t_i; \vartheta)$ and initial value $Y(0) = 0$. Therefore, the RUL estimation at t_i can be calculated as the FHT of $\{Y(l_i), l_i \geq 0\}$ crossing w_i. In addition, it is easy to verify that $\{Y(l_i), l_i \geq 0\}$ satisfies all the conditions of Lemma 9.1; so it is obtained $S_B(l_i) = \left[w_i - \int_{t_i}^{l_i+t_i} \mu(\tau; \vartheta)\mathrm{d}\tau\right]/\sigma_B$. From (9.3), the proof of Theorem 9.1 is completed.

9.2.3 Adaptive RUL Estimation

It is noted that (9.11) only uses the current monitoring value x_i when making extrapolations about the future trend of the degradation process. Now, consider making the estimated RUL depend on $\boldsymbol{X}_{0:i} = \{x_0, x_1, \ldots, x_i\}$, which is the history of degradation observations for the system up to t_i. Toward this end, from the FHT concept, the definition of the RUL L_i at t_i in (9.7) can be revised as

$$L_i = \inf\left\{l_i : X(t_i + l_i) \geq w | \boldsymbol{X}_{0:i}\right\}, \tag{9.9}$$

with the PDF $f_{L_i|\boldsymbol{X}_{0:i}}(l_i|\boldsymbol{X}_{0:i})$.

To incorporate the history of the observations, the concept of combined parameter and state estimation, suggested in [50], has been frequently adopted in prognostic studies such as [43–46]. This chapter also utilizes this concept to achieve joint state and parameter estimation and to make the history of the observations be incorporated into the estimation. Specifically, an updating procedure for the parameter λ in the drifting part $\lambda \cdot \mu(t;\vartheta)$ is introduced by a random walk model $\lambda_i = \lambda_{i-1} + \eta$ over time, where $\eta \sim N(0, Q)$. In this case, the parameter λ evolves as a time-dependent variable, conditional on the observed data up to t_i. The degradation equation can be reconstructed with a state-space model as

$$\begin{cases} \lambda_i = \lambda_{i-1} + \eta \\ x_i = x_{i-1} + \lambda_{i-1}\Omega_i(\vartheta) + \sigma_B\varepsilon_i \end{cases}, \tag{9.10}$$

where $\Omega_i(\vartheta) = h(t_i;\vartheta) - h(t_{i-1};\vartheta)$, $h(t_i;\vartheta) = \int_0^{t_i} \mu(\tau;\vartheta)d\tau$, the error term in the state equation is distributed as $\eta \sim N(0, Q)$, and $\varepsilon_i = \left[B(t_k) - B(t_{k-1})\right] \sim N(0, t_i - t_{i-1})$. The use of $t_i - t_{i-1}$ as the variance of ε_i is required by the property of BM. Here it is assumed that the initial drift λ_0 follows a normal distribution with mean μ_0 and variance P_0. As such, the drift parameter is considered as a hidden 'state' and can only be estimated from the historical information to date, $\boldsymbol{X}_{0:i}$. In the first equation of (9.10), λ_i follows a Gaussian distribution which can be estimated by a recursive filter based on $\boldsymbol{X}_{0:i}$. In the following, denote its mean as $\hat{\lambda}_i = E(\lambda_i|\boldsymbol{X}_{0:i})$ and its variance as $P_{i|i} = \mathrm{var}(\lambda_i|\boldsymbol{X}_{0:i})$. In the framework of Bayesian filtering, $\hat{\lambda}_i$ and $P_{i|i}$ can be easily obtained by Kalman filtering in a recursive manner, as summarized below.

Algorithm 9.1 (*Kalman filtering algorithm for estimating* λ_i)
Step 1: Initialize μ_0, P_0.
Step 2: State estimation at time t_i

$\hat{\lambda}_{i|i-1} = \hat{\lambda}_{i-1}$
$P_{i|i-1} = P_{i-1|i-1} + Q$
$K_i = P_{i|i-1}\Omega_i(\vartheta)(\Omega_i(\vartheta)^2 P_{i|i-1} + \sigma_B^2(t_i - t_{i-1}))^{-1}\cdot$
$\hat{\lambda}_i = \hat{\lambda}_{i-1} + K_i\left(x_i - x_{i-1} - \hat{\lambda}_{i-1}\Omega_i(\vartheta)\right)$

Step 3: Updating variance $P_{i|i} = P_{i|i-1} - \Omega_i(\vartheta)K_i P_{i|i-1}$.

Based on (9.10) and the Gaussian nature of the Kalman filter, the PDF of λ_i conditional on $\boldsymbol{X}_{0:i}$ is still Gaussian with

$$p(\lambda_i|\boldsymbol{X}_{0:i}) = \frac{1}{\sqrt{2\pi P_{i|i}}}\exp\left[-\frac{(\lambda_i - \hat{\lambda}_i)^2}{2P_{i|i}}\right], \tag{9.11}$$

where the dependence between λ_i and $\boldsymbol{X}_{0:i}$ is contained in $\hat{\lambda}_i$ and $P_{i|i}$.

Remark 9.1 In above updating mechanism, we only make λ adaptive with the observed data to date. For a more realistic implementation, making both λ and ϑ adapted with the history may be more appropriate, but this consideration will require more complicated estimating procedure such as extended Kalman filter or particle filter. This consideration deserves future research.

From the state equation in (9.10), it is known that λ_i is random and characterized by the PDF $p(\lambda_i|\boldsymbol{X}_{0:i})$, as denoted by (9.11). Consequently, to derive $f_{L_i|\boldsymbol{X}_{0:i}}(l_i|\boldsymbol{X}_{0:i})$, it is obtained

$$f_{L_i|X_{0:i}}(l_i|X_{0:i}) = \int f_{L_i|X_{0:i},\lambda_i}(l_i|X_{0:i},\lambda_i)p(\lambda_i|X_{0:i})\mathrm{d}\lambda_i, \tag{9.12}$$

where $f_{L_i|X_{0:i},\lambda_i}(l_i|X_{0:i},\lambda_i)$ denotes the estimated RUL conditional on λ_i and $X_{0:i}$. It is worth noting that other parameters are omitted in the conditional part of $f_{L_i|X_{0:i},\lambda_i}(l_i|X_{0:i},\lambda_i)$ for notation simplicity since they are not random. In order to calculate $f_{L_i|X_{0:i},\lambda_i}(l_i|X_{0:i},\lambda_i)$, the following result is given.

Theorem 9.2 *Once $X_{0:i}$ is available at t_i, the following holds,*

$$f_{L_i|X_{0:i},\lambda_i}(l_i|X_{0:i},\lambda_i) = f_{L_i|X(t_i)=x_i,\lambda_i}(l_i|\lambda_i,x_i). \tag{9.13}$$

Proof Due to the Markov property of the standard BM, it is obtained

$$\begin{aligned}
f_{L_i|X_{0:i},\lambda_i}(l_i|X_{0:i},\lambda_i) &= \frac{\mathrm{d}}{\mathrm{d}l_i}\Pr(L_i \le l_i|X_{0:i},\lambda_i)\\
&= \frac{\mathrm{d}}{\mathrm{d}l_i}\Pr(\inf\{l_i : X(l_i+t_i) \ge w|X_{0:i}\} \le l_i|X_{0:i},\lambda_i)\\
&= \frac{\mathrm{d}}{\mathrm{d}l_i}\Pr(\inf\{l_i : X(l_i+t_i) \ge w|X_{0:i}\} \le l_i|X(t_i)=x_i,\lambda_i)\\
&= f_{L_i|X(t_i)=x_i,\lambda_i}(l_i|\lambda_i,x_i).
\end{aligned} \tag{9.14}$$

This completes the proof.

In order to calculate $f_{L_i|X_{0:i}}(l_i|X_{0:i})$, it is required to compute the integration of (9.12). To avoid the lengthy derivation, the following lemma is directly given [16].

Lemma 9.3 *If $Z \sim N(\mu,\sigma^2)$ and $w, A, B \in \mathbb{R}$, $C \in \mathbb{R}^+$, then the following holds:*

$$\begin{aligned}
&E_Z\left[(A-Z)\cdot\exp\left(-\frac{(B-Z)^2}{2C}\right)\right]\\
&= \sqrt{\frac{C}{\sigma^2+C}}\left(A-\frac{\sigma^2 B+\mu C}{\sigma^2+C}\right)\times\exp\left(-\frac{(B-\mu)^2}{2\left(\sigma^2+C\right)}\right).
\end{aligned} \tag{9.15}$$

From (9.12) and Theorem 9.2, the following conclusion is obtained to derive the PDF of the RUL conditional on $X_{0:i}$, i.e., $f_{L_i|X_{0:i}}(l_i|X_{0:i})$.

Theorem 9.3 *As for the degradation process defined in (9.1), given $X_{0:i}$, the PDF of the RUL conditional on $X_{0:i}$ can be formulated as*

$$\begin{aligned}
f_{L_i|X_{0:i}}(l_i|X_{0:i}) \cong & \frac{w_i\Lambda(l_i)-\alpha_i(l_i;\vartheta)\Delta(l_i)}{\sqrt{2\pi l_i^2\Lambda^3(l_i)}}\times\\
&\exp\left[-\frac{\left(w_i-\hat{\lambda}_i\upsilon(l_i)\right)^2}{2\Lambda(l_i)}\right]
\end{aligned} \tag{9.16}$$

where $\alpha_i(l_i;\vartheta)=\upsilon(l_i)-l_i\mu(l_i+t_i;\vartheta)$, $\upsilon(l_i)=\int_{t_i}^{l_i+t_i}\mu(\tau;\vartheta)d\tau$, $\Lambda(l_i)=P_{i|i}\upsilon(l_i)^2+\sigma_B^2 l_i$, *and* $\Delta(l_i)=P_{i|i}\upsilon(l_i)w_i+\hat{\lambda}_i\sigma_B^2 l_i$.

Proof In order to complete the proof, $f_{L_i|X_{0:i},\lambda_i}(l_i|X_{0:i},\lambda_i)$ should be first derived from (9.13). As a result of Theorem 9.1, $f_{L_i|X_{0:i},\lambda_i}(l_i|X_{0:i},\lambda_i)$ can be derived from (9.8) through replacing λ with λ_i. From (9.11)–(9.13), it is obtained

$$\begin{aligned}
&f_{L_i|X_{0:i}}(l_i|X_{0:i})=\int f_{L_i|X(t_i)=x_i,\lambda_i}(l_i|\lambda_i,x_i)p(\lambda_i|X_{0:i})d\lambda_i\\
&=E_{\lambda_i|X_{0:i}}\left[f_{L_i|X(t_i)=x_i,\lambda_i}(l_i|\lambda_i,x_i)\right]\\
&=\frac{1}{\sigma_B\sqrt{2\pi l_i^3}}E_{\lambda_i|\mathbf{X}_{0:i}}\left[(w_i-\lambda_i\alpha_i(l_i;\vartheta))\exp\left[-\frac{(w_i-\lambda_i\upsilon(l_i))^2}{2\sigma_B^2 l_i}\right]\right]\\
&=\frac{\alpha_i(l_i;\vartheta)}{\sigma_B\sqrt{2\pi l_i^3}}E_{\lambda_i|\mathbf{X}_{0:i}}\left[\left(\frac{w_i}{\alpha_i(l_i;\vartheta)}-\lambda_i\right)\exp\left[-\frac{\left(\frac{w_i}{\upsilon(l_i)}-\lambda_i\right)^2}{2\sigma_B^2 l_i/(\upsilon^2(l_i))}\right]\right]
\end{aligned}\tag{9.17}$$

where $\alpha_i(l_i;\vartheta)=\upsilon(l_i)-l_i\mu(l_i+t_i;\vartheta)$.

Therefore, setting the followings and applying Lemma 9.3 will complete the proof after necessary simplification.

$$A=\frac{w_i}{\alpha_i(l_i;\vartheta)},\ B=\frac{w_i}{\upsilon(l_i)},\ C=\frac{\sigma_B^2 l_i}{\upsilon^2(l_i)}.\tag{9.18}$$

From (9.16), once a new degradation observation is available and parameter λ is updated, the RUL of this monitored system can be adaptively updated. Comparing (9.16) with (9.8), it is found that the uncertainty in estimating λ_i, i.e., $P_{i|i}$, propagates into the PDF of the RUL $f_{L_i|X_{0:i}}(l_i|X_{0:i})$ and $P_{i|i}$ can be updated by Kalman filter as well.

For the model (9.16) to be used in real-time estimation, several unknown parameters, including σ_B^2, ϑ, μ_0, P_0 and Q, need to be estimated. The selection of these parameters will surely affect the performance of the RUL estimation. For instance, the selected variance Q of the random walk noise determines both the rate of the convergence of λ and the estimation performance once convergence is achieved. A large Q will yield quick convergence but tracking with too wide a variance, whereas too small a Q may yield a very slow convergence. Therefore, it is desirable to estimate these parameters to obtain the best possible performance. The mechanism to estimate these parameters will be discussed in the next section.

9.3 Adaptive Parameter Estimation

In this section, $\boldsymbol{\theta}$ is used to denote the unknown parameter vector, $\boldsymbol{\theta}=\left[\sigma_B^2,\vartheta,\mu_0,P_0,Q\right]$, and $\Upsilon_i=\{\lambda_0,\lambda_1,\ldots,\lambda_i\}$ is the history of the proportional parameter up to t_i. In order to achieve simultaneous estimation of the proportional parameter λ and $\boldsymbol{\theta}$, this chapter considers to calculate the maximum likelihood estimation (MLE) of $\boldsymbol{\theta}$ once the new observation x_i is available. In this case, the log-likelihood function for the degradation data $X_{0:i}$ can be written as

$$\iota(\boldsymbol{\theta}) = \log[p(\boldsymbol{X}_{0:i}|\boldsymbol{\theta})] = \sum_{j=1}^{i} \log[p(x_j|\boldsymbol{X}_{0:j-1}, \boldsymbol{\theta})]. \tag{9.19}$$

In the following, $\hat{\boldsymbol{\theta}}_i = \left[\sigma_{B,i}^2, \vartheta_i, \mu_{0,i}, P_{0,i}, Q_i\right]$ is used to represent the MLE of $\boldsymbol{\theta}$ conditional on $\boldsymbol{X}_{0:i}$ at t_i. Under model (9.10), $p(\boldsymbol{X}_{0:i}|\,\boldsymbol{\theta})$ is induced by the missing data, since the data Υ_i about parameter λ is used as the state in state-space model (9.10). In general, it is difficult to write the function $p(\boldsymbol{X}_{1:i}|\,\boldsymbol{\theta})$ with an explicit form when there exists a hidden variable. As such, it is natural to consider the effect of missing data Υ_i to search for the optimum of the log-likelihood function of the parameters. The EM algorithm [51] provides a natural framework for approaching a parameter estimation problem involved hidden variables.

In the EM algorithm, by exploiting the relationship between $p(\boldsymbol{X}_{0:i}\,|\boldsymbol{\theta})$and $p(\boldsymbol{X}_{0:i}, \Upsilon_i|\,\boldsymbol{\theta})$, it is possible to generate a sequence of parameter estimates that converge to the MLE of the parameters. The complete-data log-likelihood is defined as $\ell(\boldsymbol{\theta}) = \log p(\boldsymbol{X}_{0:i}, \Upsilon_i|\,\boldsymbol{\theta})$. Then the estimation procedure consists of two steps E-step and M-step,

- ***E-step***: Calculate

$$\ell(\boldsymbol{\theta}|\,\hat{\boldsymbol{\theta}}_i^{(k)}) = \mathrm{E}_{\Upsilon_i|\boldsymbol{X}_{0:i},\hat{\boldsymbol{\theta}}_i^{(k)}}\left\{\log p(\boldsymbol{X}_{0:i}, \Upsilon_i|\,\boldsymbol{\theta})\right\}, \tag{9.20}$$

where $\hat{\boldsymbol{\theta}}_i^{(k)} = \left[\sigma_{B,i}^{2(k)}, a_i^{(k)}, \mu_{0,i}^{(k)}, P_{0,i}^{(k)}, Q_i^{(k)}\right]$ denotes the parameter estimation in the kth iteration conditional on $\boldsymbol{X}_{0:i}$.
- ***M-step***: Calculate

$$\hat{\boldsymbol{\theta}}_i^{(k+1)} = \arg\max_{\boldsymbol{\theta}} \ell(\boldsymbol{\theta}|\,\hat{\boldsymbol{\theta}}_i^{(k)}). \tag{9.21}$$

To be more precise, in the E-step, to obtain the expected value of $\ell(\boldsymbol{\theta})$ with respect to the kth iteration of $\hat{\boldsymbol{\theta}}_i^{(k)}$ and the observed data $\boldsymbol{X}_{0:i}$, there is

$$\begin{aligned}\ell(\boldsymbol{\theta}|\,\hat{\boldsymbol{\theta}}_i^{(k)}) &= \mathrm{E}_{\Upsilon_i|\boldsymbol{X}_{0:i},\hat{\boldsymbol{\theta}}_i^{(k)}}\left\{\log p(\boldsymbol{X}_{0:i}, \Upsilon_i|\,\boldsymbol{\theta})\right\}\\ &= \mathrm{E}_{\Upsilon_i|\boldsymbol{X}_{0:i},\hat{\boldsymbol{\theta}}_i^{(k)}}\left\{\log p(\boldsymbol{X}_{0:i}|\,\Upsilon_i, \boldsymbol{\theta})\right\} + \mathrm{E}_{\Upsilon_i|\boldsymbol{X}_{0:i},\hat{\boldsymbol{\theta}}_i^{(k)}}\left\{\log p(\Upsilon_i|\,\boldsymbol{\theta})\right\}.\end{aligned} \tag{9.22}$$

Following (9.13), the full likelihood function can be formulated as

$$\begin{aligned}&\log \ell(\boldsymbol{\theta}) = \log p(\boldsymbol{X}_{0:i}|\,\Upsilon_i, \boldsymbol{\theta}) + \log p(\Upsilon_i|\,\boldsymbol{\theta}) = \log p(\lambda_0|\,\boldsymbol{\theta}) +\\ &\log \prod_{j=1}^{i} p(\lambda_j|\,\lambda_{j-1}, \boldsymbol{\theta}) + \log \prod_{j=1}^{i} p(x_j|\,\lambda_{j-1}, \boldsymbol{\theta})\end{aligned} \tag{9.23}$$

From state-space model (9.10), it is obtained $\lambda_j | \lambda_{j-1} \sim N(\lambda_{j-1}, Q)$, $x_j | \lambda_{j-1} \sim N\left(x_{j-1} + \lambda_{j-1}\int_{t_{j-1}}^{t_j} \mu(\tau;\vartheta)\mathrm{d}\tau, \sigma_B^2(t_j - t_{j-1})\right)$ and $\lambda_0 \sim N(\mu_0, P_0)$. Using (9.23) and ignoring the constant terms, the complete likelihood function can be written as

$$\begin{aligned}2\log\ell(\boldsymbol{\theta}) = &-\log P_0 - \tfrac{(\lambda_0-\mu_0)^2}{P_0} - \sum_{j=1}^{i}\left(\log Q + \tfrac{(\lambda_j-\lambda_{j-1})^2}{Q}\right)\\ &-\sum_{j=1}^{i}\left(2\log\sigma_B + \frac{\left(x_j - x_{j-1} - \lambda_{j-1}\int_{t_{j-1}}^{t_j}\mu(\tau;\vartheta)\mathrm{d}\tau\right)^2}{\sigma_B^2(t_j-t_{j-1})}\right)\end{aligned} \tag{9.24}$$

The conditional expectation, $\ell(\boldsymbol{\theta}|\hat{\boldsymbol{\theta}}_i^{(k)})$, given in (9.20) can be written as

$$\begin{aligned}\ell(\boldsymbol{\theta}|\hat{\boldsymbol{\theta}}_i^{(k)}) &= E_{\Upsilon_i|\boldsymbol{X}_{0:i},\hat{\boldsymbol{\theta}}_i^{(k)}}\left[2\log\ell(\boldsymbol{\theta})\right]\\ &= E_{\Upsilon_i|\boldsymbol{X}_{0:i},\hat{\boldsymbol{\theta}}_i^{(k)}}\left[-\log P_0 - \frac{(\lambda_0-\mu_0)^2}{P_0} - \sum_{j=1}^{i}\left(\log Q + \frac{(\lambda_j-\lambda_{j-1})^2}{Q}\right)\right.\\ &\left.-\sum_{j=1}^{i}\left(2\log\sigma_B + \frac{\left(x_j - x_{j-1} - \lambda_{j-1}\int_{t_{j-1}}^{t_j}\mu(\tau;\vartheta)\mathrm{d}\tau\right)^2}{\sigma_B^2(t_j-t_{j-1})}\right)\right].\end{aligned} \tag{9.25}$$

Clearly, to derive the expectation implies to work out $E_{\Upsilon_i|\boldsymbol{X}_{0:i},\hat{\boldsymbol{\theta}}_i^{(k)}}(\lambda_j)$, $E_{\Upsilon_i|\boldsymbol{X}_{0:i},\hat{\boldsymbol{\theta}}_i^{(k)}}(\lambda_j^2)$ and $E_{\Upsilon_i|\boldsymbol{X}_{0:i},\hat{\boldsymbol{\theta}}_i^{(k)}}(\lambda_j\lambda_{j-1})$, which are the conditional expectation given $\boldsymbol{X}_{0:i}$. In this chapter, Rauch-Tung-Striebel (RTS) smoother is used to provide an optimal estimate of the above conditional expectations [52]. The smoothing algorithm is summarized as follows, where $M_{j|i} = \mathrm{cov}(\lambda_j, \lambda_{j-1}|\boldsymbol{X}_{0:i})$.

Algorithm 9.2 (*RTS smoothing algorithm*)
Step 1: Forwards iteration by Kalman filter algorithm
Step 2: Backwards iteration

$S_j = P_{j|j}P_{j+1|j}^{-1}$
$\hat{\lambda}_{j|i} = \hat{\lambda}_j + S_j(\hat{\lambda}_{j+1|i} - \hat{\lambda}_{j+1|j}) = \hat{\lambda}_j + S_j(\hat{\lambda}_{j+1|i} - \hat{\lambda}_j)$
$P_{j|i} = P_{j|j} + S_j^2(P_{j+1|i} - P_{j+1|j})$

Step 3: Initialize

$M_{i|i} = \left(1 - K_i\int_{t_{j-1}}^{t_j}\mu(\tau;\vartheta)\mathrm{d}\tau\right)P_{i-1|i-1}$

Step 4: Backwards iteration for smoothing covariance

$M_{j|i} = P_{j|j}S_{j-1} + S_j(M_{j+1|i} - P_{j|j})S_{j-1}$

From the RTS smoothing algorithm, the conditional expectations, $E_{\Upsilon_i|\boldsymbol{X}_{0:i},\hat{\boldsymbol{\theta}}_i^{(k)}}(\lambda_j)$, $E_{\Upsilon_i|\boldsymbol{X}_{0:i},\hat{\boldsymbol{\theta}}_i^{(k)}}(\lambda_j^2)$, and $E_{\Upsilon_i|\boldsymbol{X}_{0:i},\hat{\boldsymbol{\theta}}_i^{(k)}}(\lambda_j\lambda_{j-1})$ can be calculated as follows:

$$\begin{aligned}&E_{\Upsilon_i|\boldsymbol{X}_{0:i},\hat{\boldsymbol{\theta}}_i^{(k)}}(\lambda_j) = \hat{\lambda}_{j|i},\\ &E_{\Upsilon_i|\boldsymbol{X}_{0:i},\hat{\boldsymbol{\theta}}_i^{(k)}}(\lambda_j^2) = \hat{\lambda}_{j|i}^2 + P_{j|i},\\ &E_{\Upsilon_i|\boldsymbol{X}_{0:i},\hat{\boldsymbol{\theta}}_i^{(k)}}(\lambda_j\lambda_{j-1}) = P_{j|j}S_{j-1} +\\ &\qquad S_j(M_{j+1|i} - P_{j|j})S_{j-1} + \hat{\lambda}_{j|i}\hat{\lambda}_{j-1|i}\end{aligned} \tag{9.26}$$

Together with (9.25) and (9.26), $\ell(\boldsymbol{\theta} \mid \hat{\boldsymbol{\theta}}_i^{(k)})$ can be written as

$$\begin{aligned}
&2\ell(\boldsymbol{\theta} \mid \hat{\boldsymbol{\theta}}_i^{(k)}) = E_{\Upsilon_i|\boldsymbol{X}_{0:i},\hat{\boldsymbol{\theta}}_i^{(k)}}\left[2\ell_i(\boldsymbol{\theta})\right] \\
&= E_{\Upsilon_i|\boldsymbol{X}_{0:i},\hat{\boldsymbol{\theta}}_i^{(k)}}\Bigg[-\log P_0 - \frac{(\lambda_0-\mu_0)^2}{P_0} - \sum_{j=1}^i\left(\frac{(\lambda_j-\lambda_{j-1})^2}{Q}\right) \\
&-\sum_{j=1}^i\left(\log\sigma_B^2 + \frac{\left(x_j-x_{j-1}-\lambda_{j-1}\int_{t_{j-1}}^{t_j}\mu(\tau;\vartheta)\mathrm{d}\tau\right)^2}{\sigma_B^2(t_j-t_{j-1})}\right)\Bigg] - i\log Q \\
&= -\log P_0 - \frac{C_{0|i}-2\hat{\lambda}_{0|i}\mu_0+\mu_0^2}{P_0} - \\
&\sum_{j=1}^i\left(\log Q + \frac{C_{j|i}-2C_{j,j-1|i}+C_{j-1|i}}{Q}\right) - \\
&\sum_{j=1}^i\left(\log\sigma_B^2 + \frac{(x_j-x_{j-1})^2+(\int_{t_{j-1}}^{t_j}\mu(\tau;\vartheta)\mathrm{d}\tau)^2C_{j-1|i}}{\sigma_B^2(t_j-t_{j-1})}\right) + \\
&\sum_{j=1}^i\left(\frac{2\hat{\lambda}_{j-1|i}(x_j-x_{j-1})\cdot\int_{t_{j-1}}^{t_j}\mu(\tau;\vartheta)\mathrm{d}\tau}{\sigma_B^2(t_j-t_{j-1})}\right)
\end{aligned} \tag{9.27}$$

After obtaining $\ell(\boldsymbol{\theta} \mid \hat{\boldsymbol{\theta}}_i^{(k)})$, the unknown parameters $\hat{\boldsymbol{\theta}}_i^{(k+1)}$ can be obtained by maximizing $\ell(\boldsymbol{\theta} \mid \hat{\boldsymbol{\theta}}_i^{(k)})$ with respect to $\boldsymbol{\theta}$. However, it is difficult to obtain the explicit form of $\hat{\boldsymbol{\theta}}_i^{(k+1)}$ through maximizing (9.27) due to the effect of the nonlinear part involved by the parameter ϑ. In order to tackle this problem, $\hat{\boldsymbol{\theta}}_i^{(k+1)}$ is calculated through the technique of a profile log-likelihood function as follows.

Firstly, given ϑ, the results of the estimated parameter $\hat{\boldsymbol{\theta}}_i^{(k+1)}$ in the $(k+1)$th step can be summarized in the following.

$$\begin{aligned}
\mu_{0,i}^{(k+1)} &= \hat{\lambda}_{0|i}, \\
P_{0,i}^{(k+1)} &= C_{0|i} - \hat{\lambda}_{0|i}^2 = P_{0|i}, \\
Q_i^{(k+1)} &= \frac{1}{i}\sum_{j=1}^i\left(C_{j|i} - 2C_{j,j-1|i} + C_{j-1|i}\right), \\
\sigma_{B,i}^{2(k+1)}(\vartheta) &= \frac{1}{i}\sum_{j=1}^i\left(\frac{(x_j-x_{j-1})^2+(\int_{t_{j-1}}^{t_j}\mu(\tau;\vartheta)\mathrm{d}\tau)^2C_{j-1|i}}{t_j-t_{j-1}}\right) - \\
&\quad \frac{1}{i}\sum_{j=1}^i\left(\frac{2\hat{\lambda}_{j-1|i}(x_j-x_{j-1})\int_{t_{j-1}}^{t_j}\mu(\tau;\vartheta)\mathrm{d}\tau}{t_j-t_{j-1}}\right).
\end{aligned} \tag{9.28}$$

with $C_{j|i} = E_{\Upsilon_i|\boldsymbol{X}_{0:i},\hat{\boldsymbol{\theta}}_i^{(k)}}(\lambda_j^2)$, $C_{j,j-1|i} = E_{\Upsilon_i|\boldsymbol{X}_{0:i},\hat{\boldsymbol{\theta}}_i^{(k)}}(\lambda_j\lambda_{j-1})$. It is noted here that $\sigma_{B,i}^{2(k+1)}(\vartheta)$ is a function of ϑ, while $\mu_{0i}^{(k+1)}$, $P_{0i}^{(k+1)}$, $Q_i^{(k+1)}$ are free of the effect of ϑ provided that $\hat{\boldsymbol{\theta}}_i^{(k)}$ is given.

Secondly, substituting (9.27) into (9.28) gives the profile log-likelihood function as

$$\begin{aligned}
&2\ell(\vartheta \mid \hat{\boldsymbol{\theta}}_i^{(k)}) = -\log P_{0|i} - 1 - \\
&\sum_{j=1}^i\left(\log Q_i^{(k+1)} + \frac{C_{j|i}-2C_{j,j-1|i}+C_{j-1|i}}{Q_i^{(k+1)}}\right) \\
&-\sum_{j=1}^i\left(\log\sigma_{B,i}^{2(k+1)}(\vartheta) + \frac{(x_j-x_{j-1})^2+(\int_{t_{j-1}}^{t_j}\mu(\tau;\vartheta)\mathrm{d}\tau)^2C_{j-1|i}}{\sigma_{B,i}^{2(k+1)}(\vartheta)(t_j-t_{j-1})}\right). \\
&+\sum_{j=1}^i\left(\frac{2\hat{\lambda}_{j-1|i}(x_j-x_{j-1})\int_{t_{j-1}}^{t_j}\mu(\tau;\vartheta)\mathrm{d}\tau}{\sigma_{B,i}^{2(k+1)}(\vartheta)(t_j-t_{j-1})}\right)
\end{aligned} \tag{9.29}$$

As such, the estimate of ϑ, i.e., $\vartheta_i^{(k)}$, can be obtained by maximizing the profile log-likelihood function in (9.29) through a search algorithm. In this chapter, 'fminsearch' function in MATLAB is utilized to optimize (9.29), which is relatively easy to perform. Now, substituting $\vartheta_i^{(k)}$ into (9.28), the estimates of $\sigma_{B,i}^{2(k+1)}(\vartheta_i^{(k)})$, $\mu_{0,i}^{(k+1)}$, $P_{0,i}^{(k+1)}$, $Q_i^{(k+1)}$ are obtained, respectively. In contrast with directly maximizing (9.27) to obtain $\hat{\boldsymbol{\theta}}_i^{(k+1)}$, the number of the unknown parameters in the search algorithm is reduced by adopting the above profile log-likelihood function technique. Then, the algorithm iterates the E-step and M-step until a criterion of convergence is satisfied.

9.4 An Illustrative Example

For simplicity and an illustration purpose, this section considers the following power model to illustrate the implementation of the presented approach,

$$X(t) = \lambda t^a + \sigma_B B(t). \tag{9.30}$$

It is noted that (9.30) is a nonlinear model fallen into the type defined in (9.1) with $\mu(t;\vartheta) = at^{a-1}$ and $\boldsymbol{\phi} = [\vartheta, \sigma_B] = [\lambda, a, \sigma_B^2]$. This kind of power model has been found in many applications such as modeling crack length and battery data [53–55]. Of course, different forms of $\mu(t;\vartheta)$ could be used, if it is suggested by the specific application of interest.

By Lemma 9.1, the PDF of the FHT corresponding to (9.30) can be formulated directly as

$$f_{T|\boldsymbol{\phi}}(t|\,\boldsymbol{\phi}) \cong \frac{w - \lambda t^a(1-a)}{\sigma_B\sqrt{2\pi t^3}} \exp\left[-(w-\lambda t^a)^2/2\sigma_B^2 t\right]. \tag{9.31}$$

Once the history of degradation observations $\boldsymbol{X}_{0:i}$ up to t_i is available, the PDF of the estimated RUL can be formulated by Theorem 9.3 as

$$f_{L_i|\boldsymbol{X}_{0:i}}(l_i|\,\boldsymbol{X}_{0:i}) \cong \frac{w_i \Lambda(l_i) - \Delta(l_i)}{\sqrt{2\pi l_i^2 \Lambda^3(l_i)}} \exp\left[-\frac{\left(w_i - \hat{\lambda}_i \upsilon(l_i)\right)^2}{2\Lambda(l_i)}\right], \tag{9.32}$$

with $\alpha_i(l_i;\vartheta) = \upsilon(l_{t_i}) - al_i(l_i+t_i)^{a-1}$, $\upsilon(l_i) = (l_i+t_i)^a - t_i^a$, $\Lambda(l_i) = P_{i|i}\upsilon(l_i)^2 + \sigma_B^2 l_i$, $\Delta(l_i) = P_{i|i}\upsilon(l_i)w_i + \hat{\lambda}_i\sigma_B^2 l_i$ and $w_i = w - x_i$.

It is noted that $P_{i|i}$ and $\hat{\lambda}_i$ can be obtained via Algorithm 9.1 by setting $h(t_i;\vartheta) = t_i^a$. Accordingly, the parameter estimation algorithm in Sect. 9.3 can be applied to the model (9.30) so as to implement adaptive RUL estimation.

9.5 Numerical Example and Case Study

In this section, a numerical simulation and a case study for battery data are provided to compare the performance of the presented method with some existing models. Here a loss function is employed to enable a direct comparison with any alternative modeling approach that produces a conditional PDF for the underlying RUL estimation. The loss function is the mean-squared error (MSE) about the actual RUL obtained at each observation point [16], defined as

$$\mathrm{MSE}_i = \int_0^{\infty} (l_i - \tilde{l}_i)^2 f_{L_i|\boldsymbol{X}_{0:i}}(l_i|\boldsymbol{X}_{0:i})\,\mathrm{d}l_i, \tag{9.33}$$

where $\tilde{l}_i$ is the actual RUL obtained at t_i and $f_{L_i|\boldsymbol{X}_{0:i}}(l_i|\boldsymbol{X}_{0:i})$ is the according conditional PDF of the RUL estimated by (9.16).

The total mean-squared error (TMSE) is the sum of the MSE at each CM point over the whole life cycle. If there are N observations, the TMSE of the RUL estimation based on these observations can be calculated as

$$\mathrm{TMSE} = \sum_{i=1}^{N} \mathrm{MSE}_i. \tag{9.34}$$

Due to the above definition and the effect of the integral operator, MSE_i is actually a metric to characterize the accumulated estimation error between the actual RUL and the estimated RUL described by a probabilistic distribution. This metric will be used in the following numerical example for model comparisons.

9.5.1 Numerical Example

The purpose of this numerical example is twofold. The first is to illustrate the implementation of the proposed method by comparing the estimation results with existing models. The second is to verify the usefulness of the proposed parameters estimation algorithm. Specifically, a nonlinear Wiener process as $X(t) = \lambda t^a + \sigma_B B(t)$ is directly considered to generate the simulation data by the so-called *Euler* approximation. Hence the simulated data are generated by $X_{(i+1)\Delta t} = X_{i\Delta t} + \lambda a\,(i\Delta t)^{a-1}\,\Delta t + \sigma_B Y\sqrt{\Delta t}$ where $Y \sim N(0, 1)$ and Δt is the discretization step. Here, set $\lambda = 0.05$, $a = 2$, $\sigma_B^2 = 0.16$ and $\Delta t = 0.1$. When the threshold is $w = 10.36$, one simulated path is obtained as shown in Fig. 9.1, where the FHT can be approximated as 14.8 after 148 sampling.

It can be found that there is no simple method which can transform the above nonlinear simulation model into a linear process under the same timescale. In order to show the superiority of the proposed approach, the following four competing

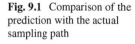

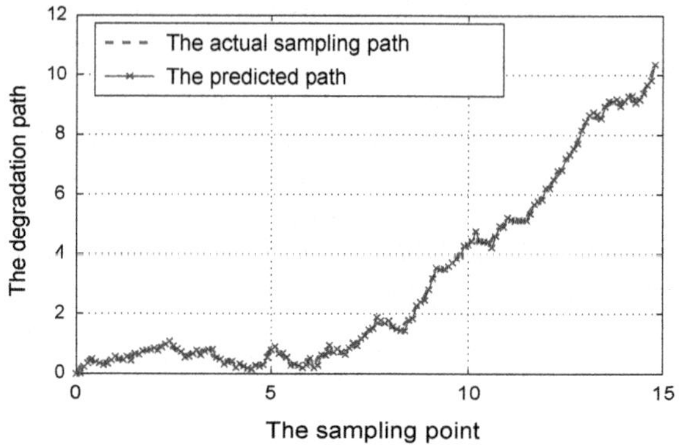

Fig. 9.1 Comparison of the prediction with the actual sampling path

models, which have been widely used in degradation modeling fields, are considered for a comparative purpose.

(i) Model 1. It is a Wiener process with a linear drift such as $X(t) = \lambda t + \sigma_B B(t)$. To achieve a fair comparison, all the sampling data are used to obtain the MLE of λ and σ_B, denoted by $\hat{\lambda}$ and $\hat{\sigma}_B$. Once these parameters are obtained, they are fixed in the process of estimating the RUL distribution.

(ii) Model 2. This model also uses the same process as Model 1. However, at each sampling point all the sampling data to date are used to compute the MLE $\hat{\lambda}$ and $\hat{\sigma}_B$, and then the estimated RUL distribution at this sampling point is obtained. Namely, $\hat{\lambda}$ and $\hat{\sigma}_B$ are updated in a sequential fashion.

(iii) Model 3. Based on the same process as Model 1 and Model 2, the method developed in [37] is adopted to estimate λ and σ_B, in which a Kalman filter and the EM algorithm are used together.

(iv) Model 4. Here, the presented model and parameters estimation method developed in this chapter are used, but the RUL distribution is calculated by (9.8), where the unknown parameters, λ and σ_B, are replaced by $\hat{\lambda}_i$ and the estimate of σ_B in Sect. 9.3. In this model, the PDF of the RUL is not fully dependent on the observation history to date since this model only replaces λ with its mean $\hat{\lambda}_i$ and does not consider its distribution. Thus, this model is different from the proposed model which calculates the RUL distribution from (9.16).

In addition, the prediction equation for degradation in Models 1–3 is $\hat{x}_i = x_{i-1} + \hat{\lambda} \cdot \Delta t$, while the prediction equation for the presented approach and Model 4 is $\hat{x}_i = x_{i-1} + \hat{\lambda}_i a \left(i\Delta t\right)^{a-1} \Delta t$ where Δt is the sampling interval and a is estimated by the method developed in Sect. 9.3.

The one step prediction of the degradation using the proposed method is illustrated by Fig. 9.1. It can be seen that the proposed method can recover the degradation path very well and the MSE between the predictions and the actual data is 0.0113. In contrast, the MSEs using Models 1–3 are 0.0179, 0.0175, 0.0158, respectively. This shows that the presented method produces better prediction accuracy for

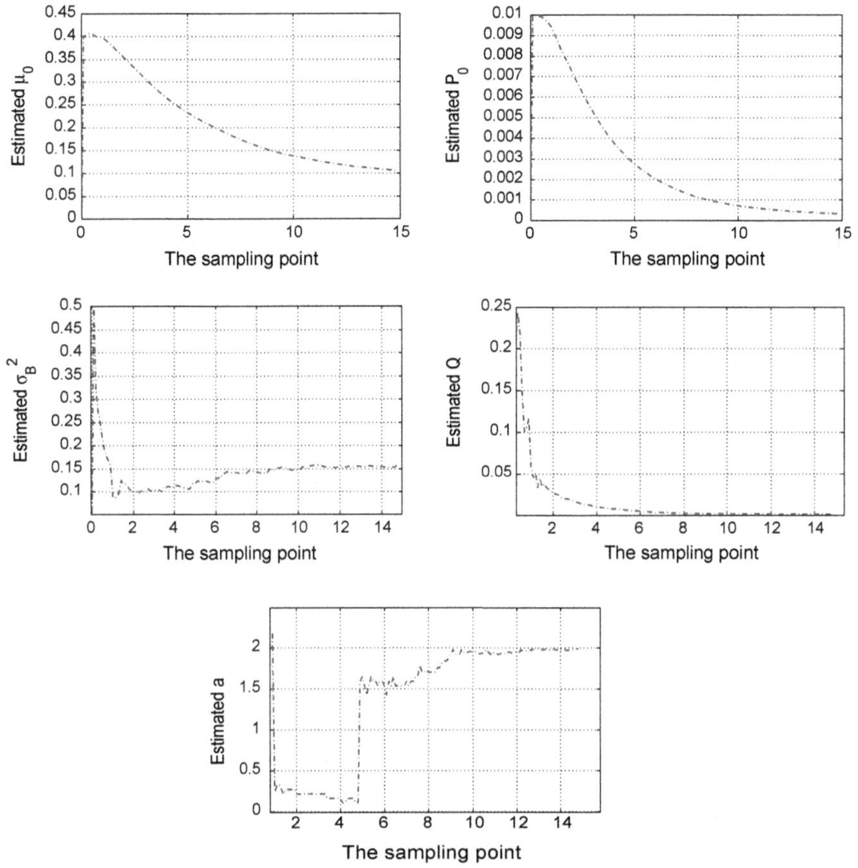

Fig. 9.2 Estimated parameters μ_0, P_0, σ_B^2, Q, a with the sampling data accumulated

degradation data. The evolving path of the estimated parameter vector $\boldsymbol{\theta}$, consisting of σ_B^2, a, μ_0, P_0, Q, is illustrated in Fig. 9.2.

From Fig. 9.2, it is observed that the parameter estimation method developed in Sect. 9.3 works well. In particular, the estimates of σ_B^2 and a can respectively approach the true values 0.16 and 2 closely as the sampling data are accumulated. In addition, λ is made be adapted to the observed data and be estimated from the Kalman filter, but the sample mean of the estimated $\hat{\lambda}_i$ based on all sampling points is 0.049 in the simulation, which is also close to the true value 0.05. These results verify the usefulness of the developed parameters estimation method. For the results for RUL estimation, the comparisons with Models 1–4 are shown in Fig. 9.3 from the 144th sampling point to the 147th point.

It can be seen from Fig. 9.3 that the proposed method has obvious advantage over linear Wiener process-based Models 1–3 and Model 4 that does not consider the observation history. In particular, the estimated PDF of the RUL from the presented

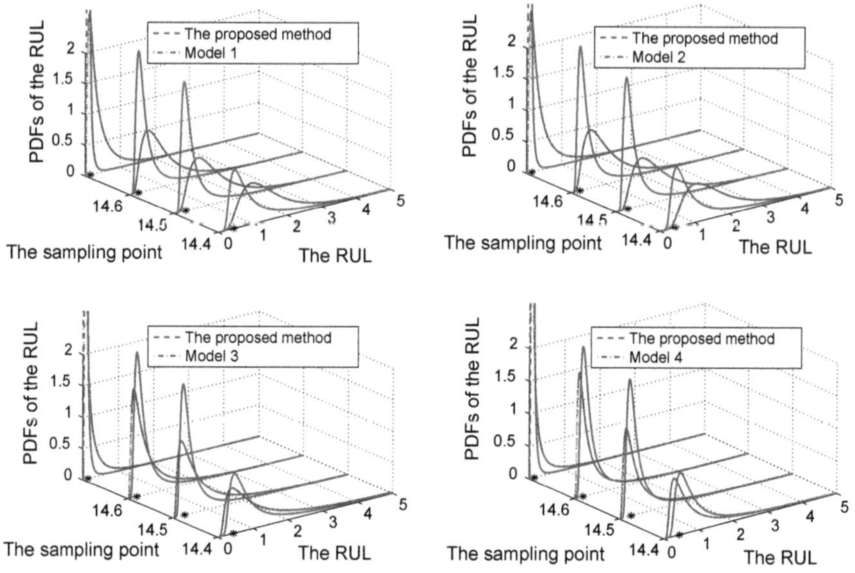

Fig. 9.3 Comparative results of the proposed method with Models 1–4 in RUL estimation (the asterisk denotes the end of life)

method is much tighter than other competing models. This is the desired result since all the data have been used for estimating the RUL in the proposed approach and all model parameters can be updated once new observation is available. Thus, the uncertainty in the estimated RUL can be reduced naturally.

To further compare the performance of the proposed method against Models 1–4, the MSE and TMSE about the RUL associated with these models are calculated. The evolving paths of the MSE at each observation point for all methods are shown in Fig. 9.4. Note that only the results of the last 88 observation points (i.e., from the 61th sampling point to the 148th point) are plotted since Fig. 9.2 indicates that the changes of the estimation will be smaller after the 60th observation point.

From Fig. 9.4, it is found that the proposed method maintains the MSE of the RUL estimation in a relatively low level compared with linear Wiener process based Models 1–3 and approaches a small MSE more quickly than Models 1–3. In addition, the proposed method avoids the abrupt change of MSE occurring in Model 4 since the presented method fully uses the observation history. Furthermore, the TMSEs of the presented method and Models 1–4 are 249.2, 1789.4, 2348.3, 910.8, 513.4, respectively. Obviously, the presented method has the least TMSE and thus has a better RUL estimation using the measure MSE than the competing models.

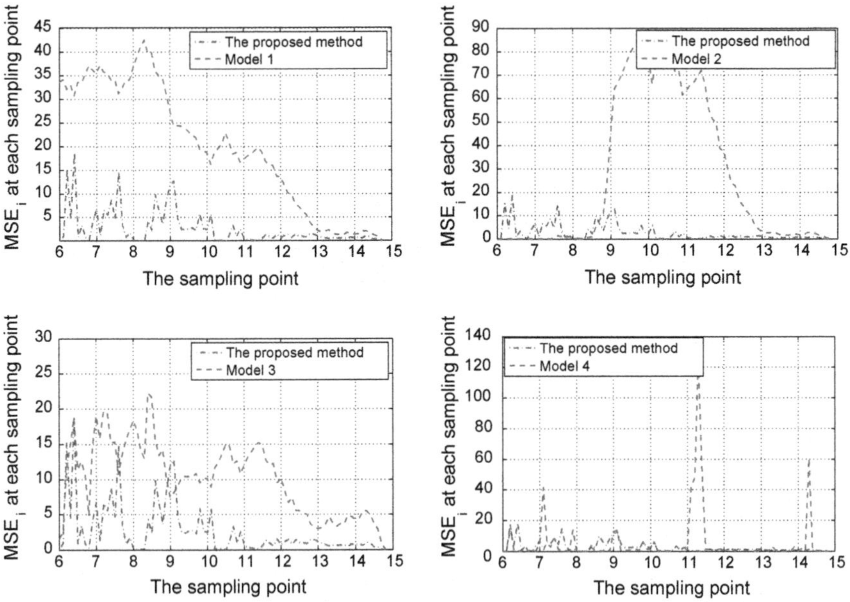

Fig. 9.4 Comparative results of the proposed method with Models 1–4 in MSE of the RUL

9.5.2 Lithium-Ion Battery Life Prognosis

In this experiment, the presented method is applied for RUL estimation of lithium-ion batteries to demonstrate the effectiveness. The used degradation datasets are provided by the NASA Ames Prognostics Center of Excellence [56]. In the used dataset, a set of four lithium-ion batteries (#5, #6, #7 and #18) were run through three different operational profiles (charge, discharge, and impedance) at room temperature. Charging was carried out in a constant current mode at 1.5 A until the battery voltage reached 4.2 V and then continued in a constant voltage mode until the charge current dropped to 20 mA. Discharge was carried out at a constant current level of 2 A until the battery voltage fell to 2.7, 2.5, 2.2, and 2.5 V for batteries #5, #6, #7, and #18 respectively. Impedance measurement was carried out through an electrochemical impedance spectroscopy frequency sweep from 0.1 Hz to 5 kHz. Repeated charge and discharge cycles result in accelerated aging of the batteries while impedance measurements provide insight into the internal battery parameters that change as aging progresses. Based on the analysis of the available performance measures of the Lithium-ion batteries, the capacity can be used to characterize the long-term degradation process induced by the charge-discharge operational cycle [56–58]. The experiments were stopped when the batteries reached end-of-life criteria, which was a 30% fade in rated capacity (from 2 to 1.4 Ah). The capacity data of four batteries are illustrated in Fig. 9.5. In this case, the lifetime of battery #5 is about 125 cycles when the failure threshold is 1.4 Ah.

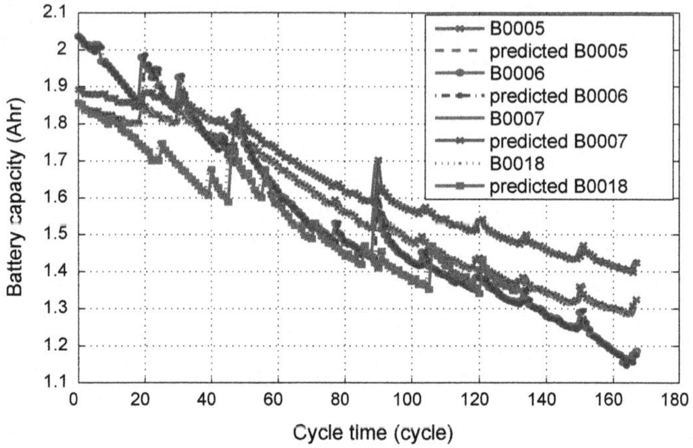

Fig. 9.5 Degradation paths and predicted paths of four batteries

Figure 9.5 shows that the degradation signals of batteries experience a lot of fluctuations. In physics, Brownian motion process aims at modeling the movement of small particles in fluids and air with tiny fluctuations. The tiny increase or decrease in degradation over a small time interval behaves similarly to the random walk of small particles in fluids and air. In addition, it is observed that the degradation path of battery exhibits a trend with variable degradation rate. Thus, a degradation model with a time-varying degradation rate could be suitable. All these observations encourage that the type of stochastic processes in (9.1), driven by BM with a nonlinear drift, is appropriate to characterize the path of the battery degradation process. Therefore, the presented model (9.1) is used to analyze the batteries data in this chapter.

For comparative purpose, this chapter also considers the case that the power law model is directly fitted to the battery data, termed as Model 5. In Model 5, the parameters are estimated by a Bayesian method as [40] and the lifetime is estimated by formulating the distribution of the degradation quantity as [29]. In order to verify the presented method, this chapter first fits the four batteries data to the presented model and compare the fitting accuracy with Models 1–5 in Sect. 9.5.1. Note that the proposed method is derived based on the case that the degradation process has an increasing trend. However, the capacity of the battery has a decreasing trend over time. Thus, to apply the proposed method, the original data are transformed by make the initial capacity minus all the capacity data of battery. Accordingly, the critical failure threshold should be changed as the initial capacity minus 1.4 Ah. Based on the transformed batteries data, the predicted paths of the batteries capacity by the proposed method can be obtained after the inverse transformation, which are illustrated in Fig. 9.5. For battery #5, the MSEs between the predicted degradation and the actual degradation for the presented method and Models 1–5 are 1.03E-4, 1.92E-4, 1.65E-4, 1.62E-4, 1.53E-4, 1.58E-4, respectively. This shows the better accuracy in the degradation prediction of the developed method. Accordingly, the

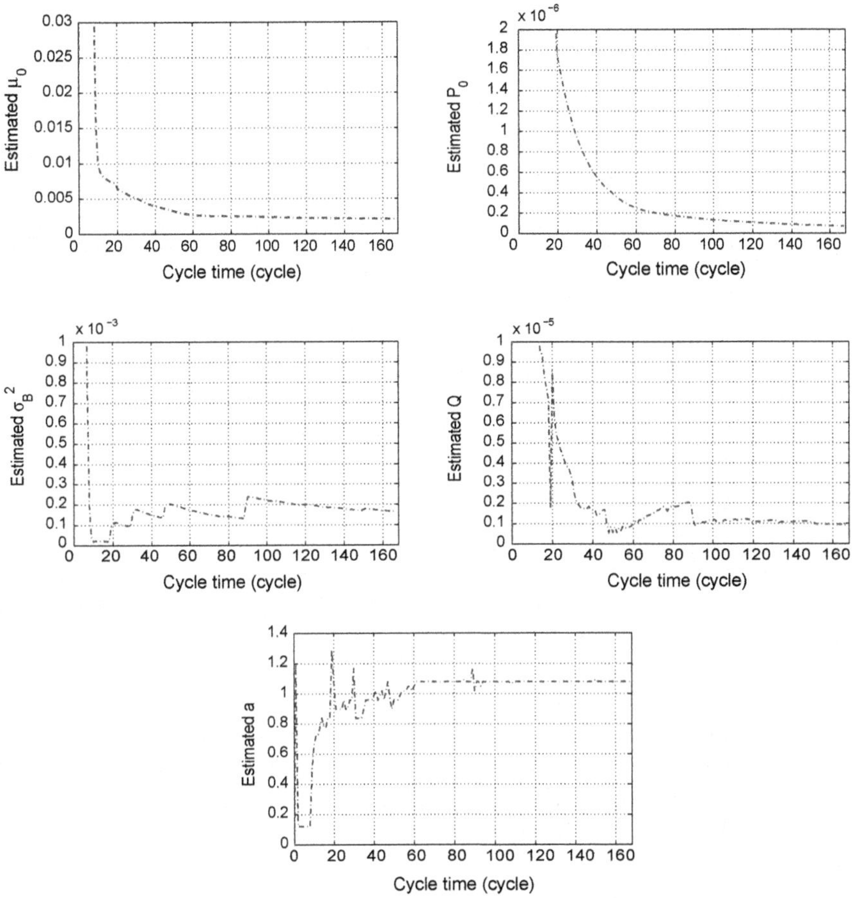

Fig. 9.6 Estimated parameters $\mu_0, P_0, \sigma_B^2, Q, a$ with the data of Battery #5

estimated parameters evolving with the cycle for battery #5 are shown in Fig. 9.6. It is observed from Fig. 9.6 that, as the battery degradation data are accumulated, the model parameters will converge quickly. In addition, if there are some great changes in the degradation data, the estimated parameters can reflect such data changes as expected (see σ_B^2, Q, and a for examples). These observations reflect the adaptive ability of the developed parameters estimation algorithm.

Based on the updated model parameters, the degradation signals of the battery capacity are further used for estimating the RUL of batteries. In this chapter, the data of the battery #5 are taken for example. For comparisons, the mediums of the RUL, the associated 95% confidence intervals (CI) and relative errors (RE) of the estimated RUL for #5 are computed at the 30th, 60th, and 90th percentiles of the lifetime, as summarized in Table 9.1. Here, the median is chosen as an approximate value for the point estimate of the RUL since the distribution of the RUL is often

Table 9.1 MRUL, 95%CI, RE of the estimated parameters

	30th (%) percentile			60th (%) percentile			90th (%) percentile		
	MRUL	95%CI	RE	MRUL	95%CI	RE	MRUL	95%CI	RE
Model 1	68.8	(63.5,96.3)	21.4	39.5	(37.2,56.1)	21.0	14.2	(11.2,17.9)	13.6
Model 2	67.1	(62.4,95.8)	23.3	39.1	(35.3,57.4)	21.8	14.5	(12.1,19.4)	16.0
Model 3	69.4	(64.2,95.4)	20.7	40.8	(37.8,55.3)	18.4	13.9	(10.1,16.8)	11.2
Model 4	72.5	(69.3,92.8)	17.1	43.8	(40.2,53.5)	12.4	11.4	(9.4,14.1)	8.8
Model 5	70.2	(65.2,93.5)	19.9	42.6	(38.1,54.2)	14.8	11.2	(8.1,14.5)	10.4
Proposed model	74.8	(71.1,91.5)	14.5	45.8	(41.3,52.1)	8.4	11.9	(10.2,13.6)	4.8

highly skewed and thus it is reasonable to use the median as a measure of central tendency rather than the mean. In addition, let $\tilde{l}_i$ denote the actual RUL at t_i, and let $\hat{l}_i$ be the estimated RUL by its median at t_i. Then, the RE for the RUL at t_i is defined as $RE_i = \left(\left|\tilde{l}_i - \hat{l}_i\right| / \tilde{l}_i\right) \times 100$. This measure is used for model comparisons.

The results in Table 9.1 indicate that all estimation results for nonlinear models are superior to the results of linear models. In nonlinear models, Model 5 is a mixed-effect regression model while Model 4 and the presented model are stochastic process-based models. As expected, the results imply that stochastic process-based models are more superior in the estimation accuracy. In stochastic process-based models, Model 4 calculates the PDF of the RUL through (9.8) and is not fully dependent on the observation history to date since this model only replaces λ with its mean $\hat{\lambda}_i$ and does not consider its distribution. Therefore, the accuracy of the RUL estimation by the presented method is much improved as opposed to Models 1–5 and the superiority of the presented method is validated.

Overall, the numerical example and case study effectively provides the supporting evidence of the presented modeling development—nonlinear degradation modeling and RUL estimation with the observation history dependency. On one hand, the results demonstrate that the conventional linear Wiener process-based model cannot give satisfactory results in the nonlinear case even though the updating mechanism is incorporated in Model 3. But the proposed method works well. On the other hand, the comparative results verify that incorporating the observation history to date can improve the accuracy of the RUL estimation indeed. Both are evaluated by the measures of MSE and TMSE.

References

1. Pecht M (2008) Prognostics and health management of electronics. Wiley, New Jersey
2. Sheppard JW, Kaufman MA, Wilmering TJ (2009) IEEE standards for prognostics and health management. IEEE Aerosp Electron Syst Mag 22(9):34–41

3. Lall P, Lowe R, Goebel K (2012) Prognostics health management of electronic systems under mechanical shock and vibration using Kalman filter models and metrics. IEEE Trans Ind Electron 59(11):4301–4314
4. Singleton RK II, Strangas EG, Aviyente S (2014) Extended Kalman filtering for remaining useful life estimation of bearings. IEEE Trans Ind Electron 62(3):1781–1790. doi:10.1109/TIE.2014.2336616
5. Brown ER, McCollom NN, Moore EE, Hess A (2007) Prognostics and health management-a data-driven approach to supporting the F-35 lightning II. IEEE Aerosp Conference, 3–10(March):1–12
6. Saha B, Goebel K, Poll S, Christophersen J (2009) Prognostics methods for battery health monitoring using a Bayesian framework. IEEE Trans Instrum Meas 58(2):291–296
7. Javed K, Gouriveau R, Zerhouni N, Nectoux P (2015) Enabling health monitoring approach based on vibration data for accurate prognostics. IEEE Trans Ind Electron 62(1):647–656
8. Tsui KL, Wong SY, Jiang W, Lin CJ (2011) Recent research and developments in temporal and spatiotemporal surveillance for public health. IEEE Trans Reliab 60(1):49–58
9. Zio E, Compare M (2013) Evaluating maintenance policies by quantitative modeling and analysis. Rel Eng Syst Saf 109:53–65
10. Climente-Alarcon V, Antonino-Daviu JA, Strangas E, Riera-Guasp M (2014) Rotor bar breakage mechanism and prognosis in an induction motor. IEEE Trans Ind Electron 62(3):1814–1825. doi:10.1109/TIE.2014.2336604
11. Vasan ASS, Long B, Pecht M (2013) Diagnostics and prognostics method for analog electronic circuits. IEEE Trans Ind Electron 60(11):5277–5291
12. Baraldi P, Mangili F, Zio E (2013) Investigation of uncertainty treatment capability of model-based and data-driven prognostic methods using simulated data. Rel Eng Syst Saf 112(1):94–108
13. Chen N, Tsui KL (2013) Condition monitoring and residual life prediction using degradation signals: revisited. IIE Trans 45(9):939–952
14. DiMaio F, Tsui KL, Zio E (2012) Combining relevance vector machines and exponential regression for bearing residual life estimation. Mech Syst Signal Process 31:405–427
15. Mazhar MI, Kara S, Kaebernick H (2007) Remaining life estimation of used components in consumer products life cycle data analysis by Weibull and artificial neural networks. J Oper Manag 25:1184–1193
16. Si XS, Wang W, Chen MY, Hu CH, Zhou DH (2013) A degradation path-dependent approach for remaining useful life estimation with an exact and closed-form solution. Eur J Oper Res 226(1):53–66
17. Orchard ME, Hevia-Koch P, Zhang B, Tang L (2013) Risk measures for particle-filtering-based state-of-charge prognosis in lithium-ion batteries. IEEE Trans Ind Electron 60(11):5260–5269
18. Soualhi A, Razik H, Clerc G, Doan D (2014) Prognosis of bearing failures using hidden markov models and the adaptive neuro-fuzzy inference system. IEEE Trans Ind Electron 61(6):2864–2874
19. Si XS, Hu CH, Kong XY, Zhou DH (2014) A residual storage life prediction approach for systems with operation state switches. IEEE Trans Ind Electron 61(11):6304–6315
20. Zio E, Peloni G (2011) Particle filtering prognostic estimation of the remaining useful life of nonlinear components. Rel Eng Syst Saf 96(3):403–409
21. Wang W, Pecht M (2011) Economic analysis of canary-based prognostics and health management. IEEE Trans Ind Electron 58(7):3077–3089
22. Strangas EG, Aviyente S, Neely JD, Zaidi SSH (2013) The effect of failure prognosis and mitigation on the reliability of permanent-magnet AC motor drives. IEEE Trans Ind Electron 60(8):3519–3528
23. Ye ZS, Tang LC, Xu HY (2011) A distribution-based systems reliability model under extreme shocks and natural degradation. IEEE Trans Rel 60(1):246–256
24. Ye ZS, Wang Y, Tsui KL, Pecht M (2013) Degradation data analysis using Wiener processes with measurement errors. IEEE Trans Rel 62(4):772–780

25. Pandey MD, Yuan X-X, van Noortwijk JM (2009) The influence of temporal uncertainty of deterioration on life-cycle management of structures. Struct Infrastruct Eng 5(2):145–156
26. Si XS, Wang W, Hu CH, Zhou DH, Pecht MG (2012) Remaining useful life estimation based on a nonlinear diffusion degradation process. IEEE Trans Rel 61(1):50–67
27. Yuan X-X, Pandey MD (2009) A nonlinear mixed-effects model for degradation data obtained from in-service inspections. Rel Eng Syst Saf 94:509–519
28. Eker FO, Camci F, Guclu A, Yilboga H, Sevkli M, Sain S (2011) A simple state-based prognostic model for railway turnout systems. IEEE Trans Ind Electron 58(5):1718–1726
29. Gebraeel N, Lawley MA, Li R, Ryan JK (2005) Residual-life distributions from component degradation signals: a Bayesian approach. IIE Trans 37(5):543–557
30. Singpurwalla ND (1995) Survival in dynamic environments. Stat Sci 10(1):86–103
31. Aalen OO, Borgan O, Gjessing HK (2008) Survival and event history analysis: a process point of view. Springer, New York
32. Lee MLT, Whitmore GA (2006) Threshold regression for survival analysis: modeling event times by a stochastic process reaching a boundary. Stat Sci 21(4):501–513
33. Park JI, Bae SJ (2010) Direct prediction methods on lifetime distribution of organic light-emitting diodes from accelerated degradation tests. IEEE Trans Rel 59(1):74–90
34. Si XS, Wang W, Hu CH, Chen MY, Zhou DH (2013) A Wiener-process-based degradation model with a recursive filter algorithm for remaining useful life estimation. Mech Syst Signal Process 35(1–2):219–237
35. Wang X (2010) Wiener processes with random effects for degradation data. J Multivar Anal 101(2):340–351
36. Whitmore GA, Schenkelberg F (1997) Modelling accelerated degradation data using Wiener diffusion with a time scale transformation. Lifetime Data Anal 3(1):27–45
37. Wang W, Carr M, Xu W, Kobbacy AKH (2011) A model for residual life prediction based on Brownian motion with an adaptive drift. Microelectron Rel 51(2):285–293
38. Si XS, Wang W, Hu CH, Zhou DH (2011) Remaining useful life estimation-a review on the statistical data driven approaches. Eur J Oper Res 213(1):1–14
39. Si XS, Hu CH, Wang W, Chen MY (2011) An adaptive and nonlinear drift-based Wiener process for remaining useful life estimation, In Proceedings Prognostics and Health Management Conference, Shenzhen, China, 2011, IEEE Xplore
40. Zhou Q, Zhou S, Mao X, Salman M (2014) Remaining useful life prediction of individual units subject to hard failure. IIE Trans 46(10):1017–1030
41. Olivares B, Cerda M, Silva J (2013) Particle-filtering-based prognosis framework for energy storage devices with a statistical characterization of state-of-health regeneration phenomena. IEEE Trans Instrum Meas 32(2):364–376
42. Zhou YF, Sun Y, Mathew J, Wolff R, Ma L (2011) Latent degradation indicators estimation and prediction: a Monte Carlo approach. Mech Syst Signal Process 25(1):222–236
43. Orchard M, Tobar F, Vachtsevanos G (2009) Outer feedback correction loops in particle filtering-based prognostic algorithms: statistical performance comparison. Stud Inform Control 18(4):295–304
44. Orchard M (2007) A particle filtering based framework for fault diagnosis and failure prognosis, Ph.D. thesis, Department of Electrical and Computer Engineering, Georgia Institute of Technology
45. Daigle MJ, Goebel K (2013) Model-based prognostics with concurrent damage progression processes. IEEE Trans Syst Man Cybern Syst 43(3):535–546
46. Miao Q, Xie L, Cui HJ, Liang W, Pecht M (2013) Remaining useful life prediction of lithium-ion battery with unscented particle filter technique. Microelectron Rel 53(6):805–810
47. Wang ZQ, Wang WB, Hu CH, Si XS (2014) An additive Wiener process-based prognostic model for hybrid deteriorating systems. IEEE Trans Rel 63(1):208–222
48. Son KL, Fouladirad Mi, Barros A, Levrat E, Iung B (2013) Remaining useful life estimation based on stochastic deterioration models: a comparative study. Rel Eng Syst Saf 112(2):165–175

49. Wang XL, Balakrishnan N, Guo B (2014) Residual life estimation based on a generalized Wiener degradation. Rel Eng Syst Saf 124(1):13–23
50. Liu J, West M (2001) Combined parameter and state estimation in simulation-based filtering. In: Doucet A, de Freitas N, Gordon N (eds) Sequential Monte Carlo methods in practice. Springer-Verlag, New York
51. Dempster AP, Laird NM, Rubin DB (1977) Maximum likelihood from incomplete data via the EM algorithm. J R Stat Soc Ser B 39(1):1–38
52. Rauch HE, Tung F, Striebel CT (1965) Maximum likelihood estimates of linear dynamic systems. AIAA J 3(8):1145–1450
53. Svensson J (2007) Survival estimation for opportunistic maintenance. Ph.D. thesis, Mathematical Statistics, Chalmers University of Technology, Göteborg
54. Xing Y, Ma EWM, Tsui K-L, Pecht M (2013) An ensemble model for predicting the remaining useful performance of lithium-ion batteries. Microelectron Rel 53:811–820
55. Lorén S, de Maré J (2012) Maintenance for reliability–a case study. Ann Oper Res 224(1):111–119. doi:10.1007/s10479-011-0873-8
56. Saha B, Goebel K (2007) "Battery data set", NASA Ames Prognostics Data Repository, Moffett Field, CA: NASA Ames [Online]. http://ti.arc.nasa.gov/project/prognostic-data-repository/
57. Jin G, Matthews DE, Zhou Z (2013) A Bayesian framework for on-line degradation assessment and residual life prediction of secondary batteries in spacecraft. Rel Eng Syst Saf 113(1):7–20
58. Shahriari M, Farrokhi M (2013) Online state-of-health estimation of VRLA batteries using state of charge. IEEE Trans Ind Electron 60(1):191–202

Chapter 10
Prognostics for Hidden and Age-Dependent Nonlinear Degrading Systems

10.1 Introduction

10.1.1 Motivation

With the rapid development of modern condition monitoring (CM) techniques, condition-based maintenance (CBM) which implements maintenance actions based on the CM information has become an active research area for reducing operation and maintenance costs. As a very important step of CBM, prognosis plays a critical role in CBM practice [1–3]. According to the available monitoring information, the main purpose of the prognosis is to assess how long the equipment can operate with a satisfactory reliability level from the present time, i.e., remaining useful life (RUL). As a result, the accurate RUL estimation can provide a sufficient and efficient decision support for the subsequent maintenance scheduling [4]. Besides, such prognostic information has impacts on the other aspects of management activities, such as spare parts provision, operational performance, the profitability of the owner of an equipment, and the management of product reuse and recycle [5].

The traditional methods for estimating equipment lifetime depend on the time-to-failure data or lifetime data. However, most of critical equipments and new products are forbidden to run to failure, or the cost of obtaining the failure data through the accelerated life test is very high. On the other hand, the equipment deteriorates over time inevitably since it operates with certain load under various environments. In contrast to the failure data, the stochastic degradation signals related to the health state of the equipment can be obtained relatively easily through the CM techniques, such as the vibration monitoring and oil analysis [6, 7]. Therefore, stochastic models for degradation data-based RUL estimation have been studied by more and more researchers (see a review [5]).

In general, nonlinearity and stochasticity are important factors which have to be considered when investigating RUL estimation in the framework of stochastic modeling. In literature, most of works on RUL estimation using stochastic models focused on linear models or models that can be linearized by logarithmic [8–10] or time-scale

X.-S. Si et al., *Data-Driven Remaining Useful Life Prognosis Techniques*,
Springer Series in Reliability Engineering, DOI 10.1007/978-3-662-54030-5_10

transformations [11–13]. The research on RUL estimation using nonlinear models is still very limited. Even in nonlinear case, the current research is limited to the case of modeling the degradation as a directly observable process for a population of "identical" components [14, 15]. However, it is well-identified that hidden or partially observable degradation process is frequently encountered in practice, because of the complexity of the equipment, or the high cost to monitor the degradation state directly [5]. In addition, adaptive updating RUL estimation for a particular component in service using the real-time data is increasingly important in recent decade. Because utilizing the CM information to date can make the RUL estimation sharper and more tailored to an individual component than only using the current data or not using at all. Overall, it is fair to say that, though many approaches for RUL estimation are available, an important and practical problem remains open, namely, how to achieve adaptive RUL estimation for hidden and nonlinear stochastic degradation process since nonlinearity and partial observability are commonly encountered in practice. The consideration of this problem will make the estimation tailored to an individual component through its real-time monitoring data which relate to its operational and environmental characteristics. It is worth noting that the above discussions regarding linearity and nonlinearity for degradation processes are with respect to the operation time of the equipment, i.e., age, rather than the degradation state itself. Thus, the term 'nonlinear degradation process' means 'age-dependent nonlinear degradation process' in this chapter. To our knowledge, there is no report on adaptive RUL estimation approach based on hidden and age-dependent nonlinear degradation process in literature, which can provide an explicit RUL distribution. This is the primary motivation of this chapter.

10.1.2 Related Works

There has been a significant volume of research in stochastic models to model the degradation process for RUL estimation in the past decades, such as random coefficient regression model [14], Wiener process [15–17] and Gamma process [18–21].Si et al. [5] classified stochastic models-based RUL estimation approaches into two main categories according to the characteristics of the degradation data: direct data-driven methods and indirect data-driven methods. The direct data-driven methods utilize the data which can describe the degradation state of the equipment directly. Lu and Meeker in [14] considered the item-to-item variability and proposed a general random coefficient regression model to characterize the degradation process of a population of equipments. Along the line of [14], many extensions have appeared, such as [22, 23]. However, the random coefficient regression model can only estimate the lifetime distribution for a class of equipment. Based on the model of [14], Gebraeel et al. [8–10] conducted many excellent works and presented some random coefficient models focused on estimating the RUL of single unit under Bayesian framework [24]. However, for each in-service equipment, its degradation process is generally stochastic and uncertain in nature because of the effect of the so-called temporal variability [25]. The above random coefficient models do not take the temporal variability into account. Due to the temporal variability of a degradation process, it is appropriate to adopt a stochastic process to model the degradation progression [25]. In the most recent, Si et al. [15] studied how to model the nonlinear degradation process and presented a more general nonlinear Wiener process-based model for RUL estimation. Based on this model, they formulated an analytical approximation of the RUL distribution under a mild assumption instead of the monotonic assumption. But they only considered that the degradation state can be observed directly as the random coefficient regression model, and obtained the results for two specific nonlinear degradation forms. Moreover, the method of [15] cannot update the RUL estimation in a real-time manner for an individual operating equipment, i.e., the parameters are estimated off-line by the historical data of a population of identical equipments rather than an operating equipment degradation observations up to date, and once determined, they are fixed.

The indirect data-driven methods consider that the degradation process is hidden or partially observable but there is a stochastic relationship between the observable CM variable and the actual degradation state. Therefore, the central idea of the indirect data-driven methods is to appropriately model the hidden degradation process and establish the relationship between the observable CM data and the hidden degradation state. Among various models for establishing such relationship, the state-space model is an efficient approach since it can describe both the hidden state of the equipment and the uncertain relationship between the hidden state and the direct observation. Furthermore, using state-space model can realize real-time estimation naturally according to the updating and prediction equations. There are many researches on using the state-space model to estimate the RUL. However, most of the literature only considered that the hidden degradation process evolved linearly over time or assumed that both the state equation and the observation equation followed a linear formulation [26–31]. For example, Xu et al. [26] proposed a real-time reliability prediction method for a dynamic system, and they adopted the Brownian Motion (BM) with linear drift to model the hidden state. Zhou et al. [31] utilized the BM with drift as the hidden degradation process, and established the state-space model to estimate the RUL of the equipment. But, they only considered the BM with a linear drift. We know that nonlinearity exists extensively in practice and the linear model cannot characterize the dynamic of such a degradation process. However, there is scarce literature appeared for nonlinear cases. The main cause for such scarcity is that it is rather difficult to obtain the analytical form of the probability density function (PDF) of the RUL for the nonlinear case. But, in the context of CBM, it is necessary and valuable to derive the PDF of the RUL with an analytical form so as to provide rapid real-time RUL estimation for the subsequent maintenance scheduling. Cadini et al. [32] considered that the hidden fatigue crack growth process was nonlinear and non-Gaussian. They employed the state-space model and particle filter to estimate the RUL of the equipment. Similar idea can also be found in [33, 34], where particle filter was used to predict the hidden degradation state. However, particle filter is a Monte Carlo-based approach, and thus only numerical result of the estimated RUL distribution can be obtained and the heavy computation may be incurred in most cases. In addition, these researches assumed that the model parameters were known or fixed once determined by the historical data. Zhou et al. [18] proposed a Gamma process-based state-space model to estimate the RUL for the nonlinear and hidden degradation process, where the RUL was estimated by Monte Carlo method and the unknown parameters were estimated through Expectation Maximization (EM) algorithm. But they did not concern the updating issue for parameters and the explicit RUL distribution cannot be obtained due to the Monte Carlo nature of their method. Similar state-space model also appeared in [19]. Actually, Gamma process based models are only appropriate to represent strictly monotonic degradation processes. However, In many engineering practices, a non-monotonic degradation process, e.g., as a result of minor repair or a reduced intensity of use [35], can provide a good description of the system's behavior, such as rotating bearings [8], bridge beam degradation [12], drift coefficient in an inertial navigation platform [15], the LED lamps [36], the Carbon-film resistor [37], etc. More importantly, since the Gamma distribution is very complicate, most literature utilizes the numerical simulation methods to estimate the RUL.

From the above survey over the related works, we can observe that the issues about how to model the hidden and nonlinear degradation process for RUL estimation and how to formulate the analytical form of the RUL distribution have not been well solved. This leads to the first goal of this chapter, i.e., to develop a state-space-based prognostic model for RUL estimation, in which the hidden degradation process is modeled by a nonlinear state equation, and the relationship between the hidden state and the measurement is characterized by an observation equation, to derive the RUL distribution. As far as we know, real-time RUL estimation is desirable in practice so that the RUL of monitored equipment is repeatedly updated during its operation to ensure the most recently calculated RUL value accurately reflects the current reality of the equipment. This consists of our second objective.

10.1.3 Main Works of This Chapter

In this chapter, motivated by the above observations, we use a nonlinear drift-based BM to characterize the nonlinearity of the hidden degradation state. Since the state is hidden, we construct a state-space model by linking the hidden state with the measurement and utilize the Extended Kalman Filter (EKF) and the EM algorithm jointly to evaluate the degradation state and the unknown parameters. Then, we incorporate the distribution of the estimated state from the measurements into the RUL estimation and derive the parameterized analytical form of the RUL distribution approximately. As a result, we can re-estimate the parameter and the state once the new observation data are available. Accordingly, we can update the RUL distribution in line with newly available data and realize real-time RUL estimation.

Our work is different from the previous works, such as [8, 11, 12, 14, 15, 18], and [26] in several major aspects: (1) we explicitly consider the effect of hidden degradation and incorporate the uncertainty of the hidden state estimation from the measurements into the RUL estimation, thus we account for both the uncertainty in degradation process and the uncertainty of the state estimation; (2) under the condition that the degradation is hidden, we consider a general age-dependent nonlinear degradation model and obtain the analytical approximation of the RUL distribution, which is free of the assumptions that the degradation process is linear or can be linearized by logarithmic or time-scale transformations; (3) the real-time RUL estimation can be realized by re-estimating the model parameters and the hidden state jointly with the incoming new data, in which the uncertainty of the estimation for hidden state is incorporated into the updated RUL distribution.

Finally, we provide a numerical example and a practical case study for NASA battery to illustrate the application of the developed approach. With simulated/realistic data, we analyze and compare the RUL estimation results of the developed approach with the results of linear model. Particularly, we consider three different kinds of nonlinear degradation functions, which are often used in literature. In the case study, we utilize the Akaike information criterion (AIC) and mean squared error (MSE) as measures to determine the most appropriate nonlinear model by comparing the fitness to the monitoring data. In contrast with the linear model, it is found that our proposed method is superior to the linear model in terms of both the AIC and the MSE.

The chapter is organized as follows: In Sect. 10.2, we formulate the problem and derive the RUL distribution. Section 10.3 develops the parameter estimation approach. Section 10.4 provides several illustrative examples of theoretical results. In Sect. 10.5, a numerical example is provided to demonstrate the approach. Section 10.6 gives a case study on a NASA battery set to show the effectiveness of our proposed approach.

10.2 Problem Formulation and RUL Estimation

10.2.1 Problem Formulation

Now let us consider a dynamic system in which the hidden degradation state at time t is denoted by $X(t)$. In this chapter, we describe the hidden degradation process as an unobservable BM with nonlinear drift, i.e.,

$$X(t) = X(0) + \int_0^t \mu(\tau; \boldsymbol{\vartheta})\,\mathrm{d}\tau + \sigma_B B(t) \tag{10.1}$$

where $X(0)$ denotes the initial state and is assumed to be zero, which is frequently used in literature. $\int_0^t \mu(\tau; \boldsymbol{\vartheta})\,\mathrm{d}\tau$ and σ_B are the drift part and diffusion coefficient, respectively; $B(t)$ is the standard BM. Here we assume that (10.1) satisfies the regularity conditions that guarantee a weakly unique global solution [38]. From the above degradation model, we can find that if $\mu(\tau; \boldsymbol{\vartheta})$ is a nonlinear function with t and unknown parameter vector $\boldsymbol{\vartheta}$, the model can capture the nonlinear characteristics of the degradation process. Moreover, it can contain various nonlinear processes by selecting the different forms of $\mu(\tau; \boldsymbol{\vartheta})$. If $\int_0^t \mu(\tau; \boldsymbol{\vartheta})\,\mathrm{d}\tau = \eta t$, (10.1) will reduce to the traditional BM with constant drift which has been widely used to model the degradation process with the linear path.

Before proceeding to the construction of the measurement equation in the dynamic system, a few remarks about the model for the nonlinear degradation process are summarized as follows.

Remark 10.1 The term "nonlinear" here means the hidden degradation state evolves nonlinearly over time and the mean degradation path is a nonlinear function of time t, i.e., age, rather than the state $X(t)$. Thus, the term 'nonlinear degradation process' means 'age-dependent nonlinear degradation process' in this chapter. For the age-dependent or age-and state-dependent degradation process modeling, The interested reader may refer to [39–42].

Remark 10.2 In literature on degradation modeling, many researches assumed that the nonlinear process can be linearized approximately by some kinds of transformations on the degradation data, such as log-transformation [8–10] or time-scale transformation [11–13]. However, not many nonlinear degradation processes can be transformed to be nearly linear in these ways. Rather, our model can characterize the general nonlinear degradation process directly without the use of the transformations mentioned above. If the nonlinear degradation process cannot be transformed to a linear degradation by manipulating the data, our model would be particularly useful.

The measurement equation in the dynamic system which describes the relationship between the observable measurement obtained from the available sensors and the hidden degradation state is formulated as

$$Y(t) = g(X(t); \boldsymbol{\xi}) + \varepsilon(t) \tag{10.2}$$

where $g(X(t); \boldsymbol{\xi})$ is a nonlinear function of $X(t)$ with the unknown parameter vector $\boldsymbol{\xi}$, and $\varepsilon(t)$ is assumed to be independent and identically distributed with $\varepsilon(t) \sim N(0, \sigma^2)$ at any time point t. It is further assumed that $\varepsilon(t)$ and $B(t)$ are mutually independent.

For deriving the RUL estimation according to the constructed state-space model (1) and (2), we adopt the concept of the first hitting time (FHT) to define the lifetime and then proceed to deducing the RUL. The lifetime can be interpreted as the FHT of the degradation state to a threshold level [43]. This definition of lifetime according to the FHT may be considered to be restrictive to some cases, since the degradation may go back after the first crossing the threshold. However, for the critical equipment, whose failure may lead to severe consequence to the equipment, or to the human, or even to the environment, once the degradation process $X(t)$ is equal to or beyond the predefined threshold level, the equipment will be considered to be failed and has to be repaired. According to the concept of FHT, the lifetime T can be defined as

$$T = \inf\{t : X(t) \geq \omega \mid X(0) < \omega\} \tag{10.3}$$

where ω is the predefined threshold level.

10.2.2 RUL Estimation

As defined in (10.3), we can observe that the key issue to estimate the RUL in the sense of the FHT is to derive the PDF of T, i.e., $f_T(t)$. Under the monotone assumption of degradation process, the lifetime distribution $F_T(t)$ can be formulated

$$F_T(t) = \Pr(T \leq t) = \Pr(X(t) \geq \omega). \tag{10.4}$$

then a simple FHT model can be directly obtained. However, for the BM-based models, the degradation path is not monotonic so the simple model cannot be used, or if used can only be considered as a crude approximation.

As stated in [15], the distribution of the FHT for a nonlinear-drifted diffusion process is related to solving the Fokker–Planck–Kolmogorov (FPK) equation with boundary constraints. This task is rather difficult. Thus, they first utilized a well-known time-space transformation on the model to calculate the FHT distribution of a standard BM crossing a time-dependent boundary instead of solving this problem from the nonlinear-drifted BM crossing a constant threshold. Then, in such case, they obtained an analytical approximation of the distribution of the FHT in a closed form under a mild assumption compared with the monotone assumption. Following [15], we make the following assumption to derive the approximated distribution of T.

Assumption: If the degradation process is hitting the threshold at a certain time t exactly, then the probability that such a process crossed the threshold level before time t is assumed to be negligible.

The above assumption implies that the process can hit the threshold before t or go back below the threshold after t, but this is just assumed to be a small probability. Thus, the above assumption is different from the monotone assumption. As for the validation of the above assumption, Si et al. [15] performed extensive simulations and thus we will not discuss this issue here. Based on the above assumption, the following lemma can be established.

Lemma 10.1 *For the degradation process* $\{X(t),\ t \geq 0\}$, *if* $\mu(t; \boldsymbol{\vartheta})$ *is a continuous function of time* t *in* $[0, \infty)$, *then the PDF of the FHT defined in (10.3) can be approximated with an explicit form as*

$$f_T(t) \approx \frac{1}{\sqrt{2\pi t}} \left(\frac{S_B(t)}{t} + \frac{1}{\sigma_B} \mu(t; \boldsymbol{\vartheta}) \right) \exp\left(-\frac{S_B^2(t)}{2t} \right) \tag{10.5}$$

where $S_B(t)$ denotes the time-dependent boundary corresponding to the standard BM, and

$$S_B(t) = \left(\omega - \int_0^t \mu(\tau; \boldsymbol{\vartheta})\,\mathrm{d}\tau\right)/\sigma_B. \tag{10.6}$$

The proof of Lemma 7.1 has been presented in Chap. 7. It is not difficult to verify that the obtained PDF of the lifetime T in Lemma 10.1 can cover the FHT distribution of the linear model and Brownian motion with zero drift. This is important since the exact results for the linear case and zero drift case exist and the nonlinear model should go back to these cases. However, the results derived from (10.4) cannot be consistent with the exact results for these two special cases. This further brings out the usefulness of our assumption and the derived result. We summarized the detailed comparisons of our results with the results obtained by (10.4) in Sect. 10.2.3.

We have already obtained the PDF of the lifetime through Lemma 10.1, and now for formulating the PDF of the RUL at the ith monitoring point $t_i > 0$ by incorporating the real-time degradation state $X(t_i)$, we first give the following remark.

Remark 10.3 Suppose that the degradation state at ith monitoring point t_i is $X(t_i)$. Intuitively, the RUL at t_i is equal to the FHT of a new stochastic process $\{G(l_i),\ l_i \geq 0\}$ crossing the threshold $\omega_{t_i} = \omega - X(t_i)$ by proper time shifts and threshold shifts, where the new stochastic process can be written as

$$\begin{aligned} G(l_i) &= X(l_i + t_i) - X(t_i) \\ &= G(0) + \int_0^{l_i} \mu'(\tau; \boldsymbol{\vartheta})\,\mathrm{d}\tau + \sigma_B B(l_i),\ \ l_i \geq 0 \end{aligned} \tag{10.7}$$

where $G(0) = 0$, $\int_{t_i}^{l_i+t_i} \mu(\tau; \boldsymbol{\vartheta})\,\mathrm{d}\tau = \int_0^{l_i} \mu'(\tau; \boldsymbol{\vartheta})\,\mathrm{d}\tau$ and $l_i = t - t_i$ denotes the realization of the RUL at the time t_i if t is the FHT of the degradation model $\{X(t),\ t \geq t_i\}$.

According to Remark 10.3 and Lemma 10.1, the PDF of the FHT of $G(l_i)$ is

$$\begin{aligned} f_{L_i}(l_i|\boldsymbol{\vartheta}, X(t_i), Y_{1:i}) \approx{}& \frac{1}{\sqrt{2\pi l_i}}\left[\frac{\omega - X(t_i) - \int_0^{l_i}\mu'(\tau; \boldsymbol{\vartheta})\,\mathrm{d}\tau}{\sigma_B l_i}\right. \\ &\left. + \frac{\mu'(l_i; \boldsymbol{\vartheta})}{\sigma_B}\right] \cdot \exp\left[-\frac{\left(\omega - X(t_i) - \int_0^{l_i}\mu'(\tau; \boldsymbol{\vartheta})\,\mathrm{d}\tau\right)^2}{2\sigma_B^2 l_i}\right], \end{aligned} \tag{10.8}$$

where $f_{L_i}(l_i|\boldsymbol{\vartheta}, X(t_i), Y_{1:i})$ denotes the approximate PDF of the FHT of $\{G(l_i), l_i \geq 0\}$ crossing the threshold ω_{t_i}.

However, the degradation state $X(t_i)$ at time t_i is unobservable and the accurate value is impossible to be known. Thus, we cannot directly use the degradation state, and instead we have to estimate the distribution of $X(t_i)$ at time t_i.

To identify the degradation state in the dynamic system, the state and measurement equations should be converted into the discrete time equations to facilitate the state estimation once the new observations are available at the CM point. For convenience, let $h(t;\boldsymbol{\vartheta})$ denote $\int_0^t \mu(\tau;\boldsymbol{\vartheta})\,\mathrm{d}\tau$. Then, we can obtain the transformed dynamic system equations at the discrete time point $t_k = k\Delta t, k = 1, 2, \ldots$ as

$$X_k = X_{k-1} + h(t_k;\boldsymbol{\vartheta}) - h(t_{k-1};\boldsymbol{\vartheta}) + \sigma_B\sqrt{\Delta t}\,\omega_k, \tag{10.9a}$$

$$Y_k = g(X_k;\boldsymbol{\xi}) + \sigma\upsilon_k, \tag{10.9b}$$

where Δt is the discretization step, $X_k = X(t_k)$ and $Y_k = Y(t_k)$ denote the state and the measurement at time t_k, respectively. $\{\omega_k\}_{k\geq 1}$ and $\{\upsilon_k\}_{k\geq 1}$ are independent and identically distributed (i.i.d.) noise sequences, respectively. Furthermore, we assume that $\omega_k \sim N(0,1)$ and $\upsilon_k \sim N(0,1)$. According to the established model (10.9), we utilize the EKF to estimate the hidden degradation state. First, we define $\hat{X}_{k|k} = \mathrm{E}(X_k|Y_{1:k})$ and $P_{k|k} = \mathrm{Var}(X_k|Y_{1:k})$ as the expectation and variance of X_k that is conditional on the measurement history available until the current point, respectively, where $Y_{1:k} \triangleq \{Y_1, Y_2, \ldots, Y_k\}$. We also define $\hat{X}_{k|k-1} = \mathrm{E}(X_k|Y_{1:k-1})$ and $P_{k|k-1} = \mathrm{Var}(X_k|Y_{1:k-1})$ as the one-step predicted expectation and variance, respectively. Then, to apply the EKF, $g(X_k;\boldsymbol{\xi})$ is linearized at $\hat{X}_{k|k-1}$:

$$g(X_k;\boldsymbol{\xi}) \approx g(\hat{X}_{k|k-1}) + g'_{k|k-1}(X_k - \hat{X}_{k|k-1}), \tag{10.10}$$

where $g'_{k|k-1}$ is the derivative of $g(X_k,\boldsymbol{\xi})$ at $X_k = \hat{X}_{k|k-1}$. Now we directly give the final formulation and omit the derivation process which can be found in [44, 45].

- **Extended Kalman filtering algorithm**

For $k = 1, 2, \ldots, i$, the filtering formulation can be summarized as

$$\begin{aligned}
\hat{X}_{k|k-1} &= \hat{X}_{k-1|k-1} + h(t_k;\boldsymbol{\vartheta}) - h(t_{k-1};\boldsymbol{\vartheta}),\\
\hat{X}_{k|k} &= \hat{X}_{k|k-1} + K(k)[Y_k - g(\hat{X}_{k|k-1})],\\
K(k) &= P_{k|k-1}\cdot g'_{k|k-1}\cdot[(g'_{k|k-1})^2 P_{k|k-1} + \sigma^2]^{-1},\\
P_{k|k-1} &= P_{k-1|k-1} + \sigma_B^2\Delta t,\\
P_{k|k} &= P_{k|k-1} - K(k)\cdot g'_{k|k-1}\cdot P_{k|k-1}.
\end{aligned}$$

Applying the EKF algorithm, after obtaining the PDF of $X(t_i)$ conditional on the measurement sequence $Y_{1:i}$ up to t_i, i.e., $X_i \sim N(\hat{X}_{i|i}, P_{i|i})$, the RUL distribution at time t_i can be derived with the estimated $X(t_i)$. In order to facilitate the derivation of the RUL distribution, we first give the following lemma.

Lemma 10.2 *If* $\rho \sim N(\mu,\sigma^2)$, *and* $\omega_1, \omega_2, \alpha, \beta \in \mathbb{R}$, $\gamma \in \mathbb{R}^{\dagger}$, *then the following holds*

$$\mathrm{E}_\rho\left[(\omega_1-\alpha\rho)\exp\left(-\frac{(\omega_2-\beta\rho)^2}{2\gamma}\right)\right]$$
$$=\sqrt{\frac{\gamma}{\beta^2\sigma^2+\gamma}}\left(\omega_1-\alpha\frac{\beta\sigma^2\omega_2+\mu\gamma}{\beta^2\sigma^2+\gamma}\right)\exp\left(-\frac{(\omega_2-\beta\mu)^2}{2(\beta^2\sigma^2+\gamma)}\right). \tag{10.11}$$

Proof If $\rho\sim N(\mu,\sigma^2)$, then we have

$$\mathrm{E}_\rho\left[(\omega_1-\alpha\rho)\cdot\exp\left(-\frac{(\omega_2-\beta\rho)^2}{2\gamma}\right)\right]=\omega_1 I_1-\alpha I_2,$$

where

$$I_1=\mathrm{E}_\rho\left[\exp\left(-\frac{(\omega_2-\beta\rho)^2}{2\gamma}\right)\right],\ I_2=\mathrm{E}_\rho\left[\rho\exp\left(-\frac{(\omega_2-\beta\rho)^2}{2\gamma}\right)\right].$$

So, we can derive I_1 as follows:

$$\begin{aligned}I_1&=\frac{1}{\sqrt{2\pi\sigma^2}}\int_{-\infty}^{\infty}\exp\left[-\frac{(\omega_2-\beta\rho)^2}{2\gamma}\right]\exp\left[-\frac{(\rho-\mu)^2}{2\sigma^2}\right]\mathrm{d}\rho\\&=\frac{1}{\sqrt{2\pi\sigma^2}}\int_{-\infty}^{\infty}\exp\left[-\frac{(\omega_2-\beta\rho)^2}{2\gamma}-\frac{(\rho-\mu)^2}{2\sigma^2}\right]\mathrm{d}\rho\\&=\frac{1}{\sqrt{2\pi\sigma^2}}\exp\left[-\frac{\sigma^2\omega_2^2+\gamma\mu^2}{2\sigma^2\gamma}\right]\int_{-\infty}^{\infty}\exp\left[-\frac{\rho^2-2\varphi\rho}{\psi}\right]\mathrm{d}\rho\\&=\frac{\sqrt{\psi\pi}}{\sqrt{2\pi\sigma^2}}\exp\left[-\frac{\sigma^2\omega_2^2+\gamma\mu^2}{2\sigma^2\gamma}\right]\exp\left(\frac{\varphi^2}{\psi}\right)\\&=\sqrt{\frac{\gamma}{\beta^2\sigma^2+\gamma}}\exp\left[-\frac{(\omega_2-\beta\mu)^2}{2(\beta^2\sigma^2+\gamma)}\right]\end{aligned}$$

with $\varphi=\dfrac{\sigma^2\beta\omega_2+\gamma\mu}{\beta^2\sigma^2+\gamma},\ \psi=\dfrac{2\sigma^2\gamma}{\beta^2\sigma^2+\gamma}$.

Similarly, I_2 can be formulated as follows:

$$\begin{aligned}I_2&=\frac{1}{\sqrt{2\pi\sigma^2}}\int_{-\infty}^{\infty}\rho\exp\left[-\frac{(\omega_2-\beta\rho)^2}{2\gamma}\right]\exp\left[-\frac{(\rho-\mu)^2}{2\sigma^2}\right]\mathrm{d}\rho\\&=\frac{1}{\sqrt{2\pi\sigma^2}}\exp\left[-\frac{\sigma^2\omega_2^2+\gamma\mu^2}{2\sigma^2\gamma}\right]\\&\quad\cdot\exp\left(\frac{\varphi^2}{\psi}\right)\int_{-\infty}^{\infty}\rho\exp\left[-\frac{(\rho-\varphi)^2}{\psi}\right]\mathrm{d}\rho\\&\overset{(\rho-\varphi)/\sqrt{\psi}=\phi}{=\!=\!=}\frac{1}{\sqrt{2\pi\sigma^2}}\exp\left[-\frac{\sigma^2\omega_2^2+\gamma\mu^2}{2\sigma^2\gamma}+\frac{\varphi^2}{\psi}\right]\end{aligned}$$

$$
\begin{aligned}
&\cdot \int_{-\infty}^{\infty} (\varphi + \phi\sqrt{\psi}) \exp(-\phi^2)\,\mathrm{d}\phi \\
&= \frac{\sqrt{\psi}}{\sqrt{2\pi\sigma^2}} \exp\left[-\frac{\sigma^2\omega_2^2 + \gamma\mu^2}{2\sigma^2\gamma} + \frac{\varphi^2}{\psi}\right] \varphi\sqrt{\pi} \\
&= \varphi \cdot \sqrt{\frac{\gamma}{\beta^2\sigma^2 + \gamma}} \exp\left[-\frac{(\omega_2 - \beta\mu)^2}{2(\beta^2\sigma^2 + \gamma)}\right] \\
&= \varphi \cdot I_1.
\end{aligned}
$$

Finally, we can obtain

$$
\begin{aligned}
&\mathrm{E}_\rho\left[(\omega_1 - \alpha\rho) \cdot \exp\left(-\frac{(\omega_2 - \beta\rho)^2}{2\gamma}\right)\right] \\
&= (\omega_1 - \alpha\varphi) I_1 \\
&= \sqrt{\frac{\gamma}{\beta^2\sigma^2 + \gamma}} \left(\omega_1 - \alpha\frac{\beta\sigma^2\omega_2 + \mu\gamma}{\beta^2\sigma^2 + \gamma}\right) \exp\left[-\frac{(\omega_2 - \beta\mu)^2}{2(\beta^2\sigma^2 + \gamma)}\right].
\end{aligned}
$$

This completes the proof.

Based on Lemma 10.2, we give the following theorem.

Theorem 10.1 *Under the same conditions as Lemma 10.1, if the degradation process $X(t)$ is hidden, then the PDF of the RUL can be formulated at time t_i with the estimated $X(t_i)$ conditional on the measurement sequence $\{Y_1, Y_2, \ldots, Y_i\}$ as*

$$
\begin{aligned}
f_{L_i}(l_i|\boldsymbol{\theta}, Y_{1:i}) \approx &\frac{1}{\sqrt{2\pi l_i^2 (P_{i|i} + \sigma_B^2 l_i)}} \Bigg[\omega - \lambda(l_i; \boldsymbol{\vartheta}) + l_i\mu(l_i + t_i; \boldsymbol{\vartheta}) \\
&- \frac{P_{i|i}(\omega - \lambda(l_i; \boldsymbol{\vartheta})) + \hat{X}_{i|i}\sigma_B^2 l_i}{P_{i|i} + \sigma_B^2 l_i}\Bigg] \exp\left(-\frac{(\omega - \lambda(l_i; \boldsymbol{\vartheta}) - \hat{X}_{i|i})^2}{2(P_{i|i} + \sigma_B^2 l_i)}\right)
\end{aligned}
\tag{10.12}
$$

where $\lambda(l_i; \boldsymbol{\vartheta}) = h(l_i + t_i; \boldsymbol{\vartheta}) - h(t_i; \boldsymbol{\vartheta})$, $\boldsymbol{\theta}$ is the unknown parameter vector of the state-space model, $\hat{X}_{i|i}$ and $P_{i|i}$ are the estimated expectation and variance of $X(t_i)$, respectively.

Proof It is easy to verify that $\{G(l_i),\ l_i \geq 0\}$ satisfies all the conditions of Lemma 10.1. So we directly derive

$$
\mu'(l_i; \boldsymbol{\vartheta}) = \mu(l_i + t_i; \boldsymbol{\vartheta}),
$$

$$
S_B(l_i) = \frac{1}{\sigma_B}[\omega_{t_i} - (h(l_i + t_i; \boldsymbol{\vartheta}) - h(t_i; \boldsymbol{\vartheta}))].
$$

Substituting the above equations into (10.8), we have

$$\begin{aligned} & f_{L_i}(l_i|\boldsymbol{\theta}, X(t_i), Y_{1:i}) = \\ & \frac{1}{\sqrt{2\pi\sigma_B^2 l_i^3}}\left[\omega - \lambda(l_i; \boldsymbol{\vartheta}) + l_i\mu(l_i + t_i; \boldsymbol{\vartheta}) - X(t_i)\right] \\ & \cdot \exp\left[-\frac{(\omega - \lambda(l_i; \boldsymbol{\vartheta}) - X(t_i))^2}{2\sigma_B^2 l_i}\right], \end{aligned} \tag{10.13}$$

where $\lambda(l_i; \boldsymbol{\vartheta}) = h(l_i + t_i; \boldsymbol{\vartheta}) - h(t_i; \boldsymbol{\vartheta})$.

Suppose the evaluated mean and variance of $X(t_i)$ are $\hat{X}_{i|i}$ and $P_{i|i}$ at t_i, respectively. Then we can obtain $X(t_i) \sim N(\hat{X}_{i|i}, P_{i|i})$. Let $p(x_i|Y_{1:i})$ denote the PDF of $X(t_i)$ conditional on the available observations $Y_{1:i}$. Using the law of total probability, we can obtain

$$\begin{aligned} & f_{L_i}(l_i|\boldsymbol{\theta}, Y_{1:i}) \\ = & \int_{-\infty}^{\infty} f_{L_i}(l_i|\boldsymbol{\theta}, X(t_i), Y_{1:i}) p(x_i|Y_{1:i})\, \mathrm{d}x_i \\ = & \mathrm{E}_{X(t_i)|Y_{1:i}}\left\{f_{L_i}(l_i|\boldsymbol{\theta}, X(t_i), Y_{1:i})\right\} \\ = & \frac{1}{\sqrt{2\pi\sigma_B^2 l_i^3}} \mathrm{E}_{X(t_i)|Y_{1:i}}\left\{\left[\omega - \lambda(l_i; \boldsymbol{\vartheta}) + l_i\mu(l_i + t_i; \boldsymbol{\vartheta}) - X(t_i)\right]\right. \\ & \left.\cdot \exp\left[-\frac{(\omega - \lambda(l_i; \boldsymbol{\vartheta}) - X(t_i))^2}{2\sigma_B^2 l_i}\right]\right\}. \end{aligned} \tag{10.14}$$

According to Lemma 10.2, let $\alpha = 1, \beta = 1, \gamma = \sigma_B^2 l_i, \omega_1 = \omega - \lambda(l_i; \boldsymbol{\vartheta}) + l_i\mu(l_i + t_i; \boldsymbol{\vartheta}), \omega_2 = \omega - \lambda(l_i; \boldsymbol{\vartheta})$, and then we obtain (10.12). This completes the proof of Theorem 10.1.

10.2.3 Comparative Discussions

We summarized the detailed comparisons of our results with the results obtained under the monotonic assumption as specified in Eq. (10.4) of the original version for reference.

In literature, many current researches define the lifetime T as we define in Eq. (10.3) of the original version, i.e. $T = \inf\{t : X(t) \geq \omega \mid X(0) < \omega\}$, but they obtained the lifetime distribution by

$$F(t) = \Pr(T \leq t) = \Pr(X(t) \geq \omega). \tag{10.15}$$

The last equation is achieved by the monotone assumption of degradation process $X(t)$. This implies that the process can only hit the threshold once and cannot go back. However, for BM-based models, the degradation path is not monotonic so the simple model cannot be used, or if used can only be considered as a crude approximation. In our chapter, we do not assume that degradation process is monotonic and we just say the probability of hitting the threshold before t is small enough. Our hitting time is not the FHT but close. This allows the degradation process be non-monotonic and is consistent with the characteristics of diffusion process.

In fact, there are two advantages to utilize our assumption compared with the monotonic assumption. First, in the deriving the lifetime distribution, we actually used the process information before t. Look at Eq. (10.9) in this supplemental material,

$$b(t) = \lim_{s \to t} (t - s)^{-1} \cdot \mathrm{E}_{W(s)|W(t)} \left[I(s, W)(S(s) - W(s)) | W(t) = S(t) \right].$$

Because there is no transformation in time scale and so the time scale in the $X(t)$ and the transformed process is the same. Therefore, in Eq. (10.9), using the information before t is equivalent to use the process information of $X(t)$ before t. However, applying $F(t) = \Pr(X(t) \geq \omega)$ to approximate the lifetime, only the degradation $X(t)$ at time t is used. In this sense, our model can utilize more information. Second, the obtained PDF of the lifetime T of our model can cover the FHT distribution of the linear model and Brownian motion with zero drift. Now, for comparison, we give the derivation of lifetime distribution under the monotone assumption. According to Eq. (10.15), the cumulative distribution function (CDF) is

$$\Pr(T \leq t) = \Pr(X(t) \geq \omega) = 1 - \Phi\left(\omega; h(t; \boldsymbol{\vartheta}), \sqrt{\sigma_B^2 t}\right) \tag{10.16}$$

where $h(t; \boldsymbol{\vartheta}) = \int_0^t \mu(\tau; \boldsymbol{\vartheta})\, d\tau$, and the PDF of the lifetime is

$$\begin{aligned} f_T(t) &= -\frac{d}{dt}\Phi\left(\omega; h(t; \boldsymbol{\vartheta}), \sqrt{\sigma_B^2 t}\right) \\ &= -\frac{d}{dt}\Phi\left(\frac{\omega - h(t; \boldsymbol{\vartheta})}{\sqrt{\sigma_B^2 t}}; 0, 1\right) \\ &= -\frac{1}{\sqrt{2\pi}}\exp\left[-\frac{(\omega - h(t; \boldsymbol{\vartheta}))^2}{2\sigma_B^2 t}\right] \cdot \frac{d}{dt}\frac{\omega - h(t; \boldsymbol{\vartheta})}{\sqrt{\sigma_B^2 t}}. \end{aligned} \tag{10.17}$$

Since

$$\frac{d}{dt}\frac{\omega - h(t; \boldsymbol{\vartheta})}{\sqrt{\sigma_B^2 t}} = \frac{1}{\sigma_B} \cdot \left[\frac{0.5h(t; \boldsymbol{\vartheta}) - 0.5\omega}{t^{3/2}} - \frac{\mu(t; \boldsymbol{\vartheta})}{\sqrt{t}}\right]$$

we have

$$f_T(t) = \frac{1}{\sigma_B\sqrt{2\pi}}\left[\frac{0.5\omega - 0.5h(t;\,\boldsymbol{\vartheta})}{t^{3/2}} - \frac{\mu(t;\,\boldsymbol{\vartheta})}{\sqrt{t}}\right]\cdot \exp\left[-\frac{(\omega - h(t;\,\boldsymbol{\vartheta}))^2}{2\sigma_B^2 t}\right]. \tag{10.18}$$

Take one of the three cases in case study: $\mu(t;\,\boldsymbol{\vartheta}) = abt^{b-1}$ for an example, under the monotone assumption, the PDF of the lifetime can be written as

$$f_T(t) = \frac{0.5\omega - at^b(0.5 - b)}{\sigma_B\sqrt{2\pi t^3}}\exp\left[-\frac{(\omega - at^b)^2}{2\sigma_B^2 t}\right]. \tag{10.19}$$

When $b = 1$, we have

$$f_T(t) = \frac{0.5\omega + 0.5at}{\sigma_B\sqrt{2\pi t^3}}\exp\left[-\frac{(\omega - at)^2}{2\sigma_B^2 t}\right], \tag{10.20}$$

and when $b = 0$, we have

$$f_T(t) = \frac{0.5\omega - 0.5a}{\sigma_B\sqrt{2\pi t^3}}\exp\left[-\frac{(\omega - a)^2}{2\sigma_B^2 t}\right]. \tag{10.21}$$

Clearly, these results cannot be consistent with the exact results for the linear model and BM with zero drift. However, under our assumption, when $b = 1$, from Eq. (10.35), the PDF of the lifetime can be formulated as

$$f_T(t) = \frac{\omega}{\sigma_B\sqrt{2\pi t^3}}\exp\left[-\frac{(\omega - at)^2}{2\sigma_B^2 t}\right], \tag{10.22}$$

which is exactly the PDF of the lifetime of the process $X(t) = at + \sigma_B B(t)$, known as the inverse Gaussian distribution; when $b = 0$, we have

$$f_T(t) = \frac{\omega - a}{\sigma_B\sqrt{2\pi t^3}}\exp\left[-\frac{(\omega - a)^2}{2\sigma_B^2 t}\right], \tag{10.23}$$

which is exactly the PDF of the lifetime of the process $X(t) = a + \sigma_B B(t)$.

Therefore, our method can cover the existing results for linear model or zero drift model. This is important since the exact results for the mentioned special cases exist and the nonlinear model should go back to the linear case if used for the linear case.

In order to show the usefulness of Theorem 10.1, we provide the specific results for several frequently used nonlinear functions in supplementary material. These results are obtained through applying Theorem 10.1 and will be applied in the case study for comparative studies.

10.3 Parameter Estimation

In the above section, we have already derived the PDF of the RUL at time t_i incorporating the estimated distribution of the degradation state $X(t_i)$ and the observation history up to t_i. Since the model parameters are unknown, we need to estimate them based on the available observations up to the current time. For convenience, we define the observable measurement and hidden degradation state sequences until the current time t_i as

$$Y_{1:i} \triangleq \{Y_1, Y_2, \ldots, Y_i\}, \quad X_{1:i} \triangleq \{X_1, X_2, \ldots, X_i\}. \tag{10.24}$$

Besides, we denote $\boldsymbol{\theta} = [\boldsymbol{\vartheta}, \boldsymbol{\xi}, \sigma_B, \sigma]$ as the unknown parameter vector. As a solution strategy, we employ the Maximum Likelihood Estimation (MLE) approach to calculate and update the estimation of the parameter $\boldsymbol{\theta}$ once the new observation data are available. In such case, the MLE with respect to the available measurements is computed as

$$\hat{\boldsymbol{\theta}}^{\mathrm{ML}} = \arg\max_{\boldsymbol{\theta}} L(Y_{1:i}|\boldsymbol{\theta}) = \arg\max_{\boldsymbol{\theta}} \log p(Y_{1:i}|\boldsymbol{\theta}) \tag{10.25}$$

where $L(Y_{1:i}|\boldsymbol{\theta})$ is the log-likelihood function and $p(Y_{1:i}|\boldsymbol{\theta})$ denotes the joint PDF of the observations $Y_{1:i}$ which is parameterized by $\boldsymbol{\theta}$. Then using the definition of the conditional probability, the log-likelihood function can be written as

$$\begin{aligned} L(Y_{1:i}|\boldsymbol{\theta}) &= \log\left(p(Y_1|\boldsymbol{\theta})\prod_{j=2}^{i} p(Y_j|Y_{1:j-1}, \boldsymbol{\theta})\right) \\ &= \sum_{j=2}^{i} \log p(Y_j|Y_{1:j-1}, \boldsymbol{\theta}) + \log p(Y_1|\boldsymbol{\theta}). \end{aligned} \tag{10.26}$$

The MLE of the unknown model parameters can be computed in many standard methods such as Newton's method or one of its related variants [45]. However, it is difficult to estimate the unknown parameter $\boldsymbol{\theta}$ because of the unobservable degradation states. In this chapter, the EM algorithm is used to estimate $\boldsymbol{\theta}$, which provides a framework to solve the MLE problem in the presence of the hidden states. By iteratively computing and maximizing the conditional expectation of log-likelihood function consisting on a complete data $(X_{1:i}, Y_{1:i})$, the EM algorithm can generate a sequence of parameter estimates which converge to the MLE of the parameters [45]. Let $\hat{\boldsymbol{\theta}}^{(j)}$ denote the estimation of the parameter $\boldsymbol{\theta}$ from the jth iteration of the EM algorithm and $\mathrm{E}_{\hat{\boldsymbol{\theta}}^{(j)}}\{\cdot|Y_{1:i}\}$ denote the conditional expectation operator with respect to a probability density function determined by $\hat{\boldsymbol{\theta}}^{(j)}$. Then, the conditional expectation of the complete data log-likelihood function can be written as

$$\begin{aligned}\mathscr{Q}(\boldsymbol{\theta}, \hat{\boldsymbol{\theta}}^{(j)}) &= \mathrm{E}_{\hat{\boldsymbol{\theta}}^{(j)}}\{[L(X_{1:i}, Y_{1:i}|\boldsymbol{\theta})]|Y_{1:i}\} \\ &= \mathrm{E}_{\hat{\boldsymbol{\theta}}^{(j)}}\{[\log p(X_{1:i}, Y_{1:i}|\boldsymbol{\theta})]|Y_{1:i}\}, \end{aligned} \tag{10.27}$$

where $L(X_{1:i}, Y_{1:i}|\boldsymbol{\theta}) = \log p(X_{1:i}, Y_{1:i}|\boldsymbol{\theta})$ is the complete data log-likelihood function. The more details can be found in [45, 46]. In sum, the parameter estimation procedure consists of the following two steps:

(1) ***E-step***

$$\text{Calculate:} \quad \mathscr{Q}(\boldsymbol{\theta}, \hat{\boldsymbol{\theta}}^{(j)}); \tag{10.28}$$

(2) ***M-step***

$$\text{Compute:} \quad \hat{\boldsymbol{\theta}}^{(j+1)} = \arg\max_{\boldsymbol{\theta}} \mathscr{Q}(\boldsymbol{\theta}, \hat{\boldsymbol{\theta}}^{(j)}). \tag{10.29}$$

The above steps are iterated with an initial guess $\hat{\boldsymbol{\theta}}^{(0)}$ until the criterion of convergence is satisfied.

For our model, according to the definition of multiplication formula of the conditional probability and the Markov property associated with the model (10.9), the joint log-likelihood function of both the measurement sequence $Y_{1:i}$ and the degradation state sequence $X_{1:i}$ until current time t_i can be formulated as

$$\begin{aligned}&L(X_{1:i}, Y_{1:i}|\boldsymbol{\theta}) \\ &= \log p(X_{1:i}, Y_{1:i}|\boldsymbol{\theta}) = \log\left[\prod_{k=1}^{i} p(X_k|X_{k-1};\boldsymbol{\theta})\prod_{k=1}^{i} p(Y_k|X_k;\boldsymbol{\theta})\right] \\ &= -i\log(2\pi\sqrt{\Delta t}) - i\log\sigma_B - i\log\sigma - \frac{1}{2\sigma^2}\sum_{k=1}^{i}\left[Y_k - g(X_k;\boldsymbol{\xi})\right]^2 \\ &\quad - \frac{1}{2\sigma_B^2\Delta t}\sum_{k=1}^{i}\left[X_k - X_{k-1} - (h(k;\boldsymbol{\vartheta}) - h(k-1;\boldsymbol{\vartheta}))\right]^2, \end{aligned} \tag{10.30}$$

where $h(k;\boldsymbol{\vartheta})$ denotes $h(t_k;\boldsymbol{\vartheta})$ in (10.9a) for convenience.

The next step is to compute the conditional expectation of the complete data log-likelihood function, i.e., $\mathscr{Q}(\boldsymbol{\theta}, \hat{\boldsymbol{\theta}}^{(j)}) = \mathrm{E}_{\hat{\boldsymbol{\theta}}^{(j)}}\{[\log p(X_{1:i}, Y_{1:i}|\boldsymbol{\theta})]|Y_{1:i}\}$. Specifically, we have

$$
\begin{aligned}
&\mathcal{Q}(\boldsymbol{\theta}, \hat{\boldsymbol{\theta}}^{(j)}) \propto \\
&- i \log \sigma_B - i \log \sigma - \frac{1}{2\sigma^2} \sum_{k=1}^{i} \mathrm{E}_{\hat{\boldsymbol{\theta}}^{(j)}} \left\{ [Y_k - g(X_k)]^2 \mid Y_{1:i} \right\} - \\
&\frac{1}{2\sigma_B^2 \Delta t} \sum_{k=1}^{i} \mathrm{E}_{\hat{\boldsymbol{\theta}}^{(j)}} \left\{ [X_k - X_{k-1} - (h(k;\boldsymbol{\vartheta}) - h(k-1;\boldsymbol{\vartheta}))]^2 \mid Y_{1:i} \right\}. \qquad (10.31)
\end{aligned}
$$

Note that the terms which are independent of $\boldsymbol{\theta}$ are neglected, since they do not affect the subsequent optimization problem.

To calculate $\mathcal{Q}(\boldsymbol{\theta}, \hat{\boldsymbol{\theta}}^{(j)})$, we must derive the conditional expectation of each term on the right-hand side of (10.31). With this in mind, for $k = 1, 2, \ldots, i$, we first define the following quantities:

$$
\begin{aligned}
\hat{X}_{k|i} &= \mathrm{E}_{\hat{\boldsymbol{\theta}}^{(j)}}(X_k | Y_{1:i}), \\
P_{k|i} &= \mathrm{E}_{\hat{\boldsymbol{\theta}}^{(j)}}(X_k^2 | Y_{1:i}) - \hat{X}_{k|i}^2, \\
P_{k,k-1|i} &= \mathrm{E}_{\hat{\boldsymbol{\theta}}^{(j)}}(X_k X_{k-1} | Y_{1:i}) - \hat{X}_{k|i} \hat{X}_{k-1|i}.
\end{aligned}
$$

According to the above-defined quantities, after some algebraic manipulations, we have

$$
\begin{aligned}
&\mathcal{Q}(\boldsymbol{\theta}, \hat{\boldsymbol{\theta}}^{(j)}) \propto \\
&- i \log \sigma_B - i \log \sigma - \frac{1}{2\sigma_B^2 \Delta t} \sum_{k=1}^{i} \Big[A_k - 2(h(k;\boldsymbol{\vartheta}) - h(k-1;\boldsymbol{\vartheta})) B_k \\
&+ (h(k;\boldsymbol{\vartheta}) - h(k-1;\boldsymbol{\vartheta}))^2 \Big] - \frac{1}{2\sigma^2} \sum_{k=1}^{i} \Big[Y_k^2 + \left(g(\hat{X}_{k|k-1}) \right)^2 \\
&+ (g'_{k|k-1})^2 C_k - 2 Y_k \left(g(\hat{X}_{k|k-1}) + g'_{k|k-1} D_k \right) \\
&+ 2 g(\hat{X}_{k|k-1}) g'_{k|k-1} D_k \Big] \qquad (10.32)
\end{aligned}
$$

where $g'_{k|k-1}$ is the derivative of $g(X_k, \boldsymbol{\xi})$ at $X_k = \hat{X}_{k|k-1}$ defined in (10.10), and,

$$
\begin{aligned}
A_k &= P_{k|i} + \hat{X}_{k|i}^2 + P_{k-1|i} + \hat{X}_{k-1|i}^2 - 2 P_{k,k-1|i} - 2 \hat{X}_{k|i} \hat{X}_{k-1|i}, \\
B_k &= \hat{X}_{k|i} - \hat{X}_{k-1|i}, \\
C_k &= P_{k|i} + \hat{X}_{k|i}^2 + \hat{X}_{k|k-1}^2 - 2 \hat{X}_{k|i} \hat{X}_{k|k-1}, \\
D_k &= \hat{X}_{k|i} - \hat{X}_{k|k-1}.
\end{aligned}
$$

Obviously, computing $\mathscr{Q}(\boldsymbol{\theta}, \hat{\boldsymbol{\theta}}^{(j)})$ requires to evaluate $\hat{X}_{k|i}$, $\hat{X}_{k-1|i}$, $P_{k|i}$, $P_{k-1|i}$ and $P_{k,k-1|i}$ with respect to the estimated parameter $\hat{\boldsymbol{\theta}}^{(j)}$ at the jth iteration. These quantities can be obtained by the Extended Kalman Smoother (EKS). EKS includes two parts: one part is forward recursion, i.e., fitering, and the other is backward recursion, i.e., smoothing. As same as the EKF, the detail about the EKS can be found in [44, 45]. We only give the final formulations.

(1) **Forward filtering**

The forward iteration by the EKF algorithm has been given in the above section.

(2) **Backward smoothing**

For $k = i, i-1, \ldots, 1$, the backward recursion can be summarized as follows:

$$\begin{aligned} J(k-1) &= P_{k-1|k-1} \cdot (P_{k|k-1})^{-1}, \\ \hat{X}_{k-1|i} &= \hat{X}_{k-1|k-1} + J(k-1)(\hat{X}_{k|i} - \hat{X}_{k|k-1}), \\ P_{k-1|i} &= P_{k-1|k-1} + J^2(k-1)(P_{k|i} - P_{k|k-1}) \end{aligned}$$

with the initial state $\hat{X}_{i|i}$ and $P_{i|i}$ derived from the forward filtering. Finally, the covariance $P_{k,k-1|i}$ can also be computed for $k = i-1, i-2, \ldots, 1$

$$P_{k,k-1|i} = P_{k|k} J(k-1) + J(k)J(k-1)(P_{k+1,k|i} - P_{k|k})$$

where the initial condition is

$$P_{i,i-1|i} = \left[1 - K(i) \cdot \left.\frac{\partial g}{\partial x}\right|_{x=\hat{X}_{i|i}}\right] P_{i-1|i-1}.$$

After deriving $\mathscr{Q}(\boldsymbol{\theta}, \hat{\boldsymbol{\theta}}^{(j)})$, the unknown parameter vector $\hat{\boldsymbol{\theta}}^{(j+1)}$ at the $(j+1)$th iteration can be obtained through maximizing $\mathscr{Q}(\boldsymbol{\theta}, \hat{\boldsymbol{\theta}}^{(j)})$ with respect to $\boldsymbol{\theta}$. If the unknown parameters in the state equation are independent of the parameters in the observation equation, for reducing the complexity of parameter estimation algorithm, $\mathscr{Q}(\boldsymbol{\theta}, \hat{\boldsymbol{\theta}}^{(j)})$ can be divided into two parts. One part only contains the parameter vector in the state equation $\boldsymbol{\theta}_1 = [\boldsymbol{\vartheta}, \sigma_B]$ and can be written as

$$\begin{aligned} \mathscr{Q}_1(\boldsymbol{\theta}_1, \hat{\boldsymbol{\theta}}_1^{(j)}) \propto -\frac{1}{2\sigma_B^2 \Delta t} \sum_{k=1}^{i} \Big[A_k - 2(h(k; \boldsymbol{\vartheta}) - h(k-1; \boldsymbol{\vartheta}))B_k + \\ (h(k; \boldsymbol{\vartheta}) - h(k-1; \boldsymbol{\vartheta}))^2 \Big] - i \log \sigma_B, \end{aligned} \tag{10.33}$$

and the other part only contains the parameter vector in the observation equation $\boldsymbol{\theta}_2 = [\boldsymbol{\xi}, \sigma]$, which can be formulated as

$$\mathcal{Q}_2(\boldsymbol{\theta}_2, \hat{\boldsymbol{\theta}}_2^{(j)}) \propto -i\log\sigma - \frac{1}{2\sigma^2}\sum_{k=1}^{i}\Big[Y_k^2 + \left(g(\hat{X}_{k|k-1})\right)^2 + (g'_{k|k-1})^2 C_k - 2Y_k\left(g(\hat{X}_{k|k-1}) + g'_{k|k-1}D_k\right) + 2g(\hat{X}_{k|k-1})g'_{k|k-1}D_k\Big]. \tag{10.34}$$

Therefore, we maximize $\mathcal{Q}(\boldsymbol{\theta}, \hat{\boldsymbol{\theta}}^{(j)})$ through maximizing $\mathcal{Q}_1(\boldsymbol{\theta}_1, \hat{\boldsymbol{\theta}}_1^{(j)})$ and $\mathcal{Q}_2(\boldsymbol{\theta}_2, \hat{\boldsymbol{\theta}}_2^{(j)})$, respectively. Clearly, this maximum policy is easier to realize since each of the two parts consists of less unknown parameters than maximizing $\mathcal{Q}(\boldsymbol{\theta}, \hat{\boldsymbol{\theta}}^{(j)})$ directly.

From the above derivations, we can observe that, once the new observation data are available, we first utilize the EM algorithm to evaluate the parameters. Then, we estimate and update the hidden degradation state at the current time by the EKF. Finally, we update the RUL distribution accordingly and realize real-time RUL estimation. Note that the EKS is just used in the derivation of parameter estimation, i.e., the estimated state at the current time t_i by the EKS is only used to estimate the parameters and cannot be taken as the final estimate of the state $X(t_i)$ at t_i. For practical applications, we need to determine the forms of nonlinear degradation function and observation equation. Misspecification of the parametric forms of these equations may reduce the accuracy of the estimated results. Thus, we use AIC and MSE as measures to choose the model with the best fitness results in the case study.

10.4 Illustrative Examples

In the following, we provide the illustrative results corresponding to several frequently adopted models in degradation modeling practices for reference.

10.4.1 The Derivation of the RUL for Three Cases

We consider three cases for the drift part: $\mu(t;\boldsymbol{\vartheta}) = abt^{b-1}$, $\mu(t;\boldsymbol{\vartheta}) = ab\exp(bt)$ and $\mu(t;\boldsymbol{\vartheta}) = abt^{b-1} + cd\cdot\exp(d\cdot t)$, corresponding to Case 1 (C_1), Case 2 (C_2) and Case 3 (C_3), respectively.

For Case 1, the hidden degradation process can be written as

$$X(t) = X(0) + at^b + \sigma_B B(t),$$

from Lemma 10.1, we have $S_B(t) = (\omega - at^b)/\sigma_B$. Then substituting $\mu(t;\boldsymbol{\vartheta})$ and $S_B(t)$ into Eq. (10.5) of Lemma 10.1, we can obtain the PDF of the FHT for Case 1

$$f_T(t \mid \mathrm{C}_1) = \frac{\omega - at^b(1-b)}{\sigma_B\sqrt{2\pi t^3}} \exp\left[-\frac{(\omega - at^b)^2}{2\sigma_B^2 t}\right]. \tag{10.35}$$

Suppose that the degradation state at time t_i is $X(t_i)$, we can derive $G(l_i) = a\left((l_i + t_i)^b - t_i^b\right) + \sigma_B B(l_i)$. According to Theorem 10.1, we formulate the PDF of the RUL at time t_i for Case 1 as

$$\begin{aligned} f_{L_i|\mathrm{C}_1}(l_i|\boldsymbol{\theta}, Y_{1:i}) = & \frac{1}{\sqrt{2\pi l_i^2 (P_{i|i} + \sigma_B^2 l_i)}} \left[\omega - a\lambda'(l_i) + abl_i(l_i + t_i)^{b-1}\right. \\ & \left. - \frac{P_{i|i}(\omega - a\lambda'(l_i)) + \hat{X}_{i|i}\sigma_B^2 l_i}{P_{i|i} + \sigma_B^2 l_i}\right] \\ & \cdot \exp\left(-\frac{(\omega - a\lambda'(l_i) - \hat{X}_{i|i})^2}{2(P_{i|i} + \sigma_B^2 l_i)}\right) \end{aligned} \tag{10.36}$$

where $\lambda'(l_i) = (l_i + t_i)^b - t_i^b$.

For Case 2, we have

$$X(t) = X(0) + a\exp(bt) - a + \sigma_B B(t),$$

similar to Case 1, we can derive $S_B(t) = (\omega - a\exp(bt) + a)/\sigma_B$, and then the PDF of the FHT can be formulated as

$$f_T(t \mid \mathrm{C}_2) = \frac{\omega - a(\exp(bt) - bt\exp(bt) - 1)}{\sigma_B\sqrt{2\pi t^3}} \cdot \exp\left[-\frac{(\omega - a\exp(bt) + a)^2}{2\sigma_B^2 t}\right]. \tag{10.37}$$

Similarly, from Theorem 10.1, the PDF of the RUL for Case 2 can be written as

$$\begin{aligned} f_{L_i|\mathrm{C}_2}(l_i|\boldsymbol{\theta}, Y_{1:i}) = & \frac{1}{\sqrt{2\pi l_i^2 (P_{i|i} + \sigma_B^2 l_i)}} \left[\omega - a\gamma'(l_i) + abl_i\exp(b(l_i + t_i))\right. \\ & \left. - \frac{P_{i|i}(\omega - a\gamma'(l_i)) + \hat{X}_{i|i}\sigma_B^2 l_i}{P_{i|i} + \sigma_B^2 l_i}\right] \\ & \cdot \exp\left(-\frac{(\omega - a\gamma'(l_i) - \hat{X}_{i|i})^2}{2(P_{i|i} + \sigma_B^2 l_i)}\right) \end{aligned} \tag{10.38}$$

where $\gamma'(l_i) = \exp(b(l_i + t_i)) - \exp(bt_i)$.

For Case 3, the degradation process is modeled as

$$X(t) = X(0) + at^b + c\exp(d \cdot t) - c + \sigma_B B(t),$$

and the PDF of the FHT can be written as

$$f_T(t \mid \mathrm{C}_3) = \frac{\omega - at^b(1-b) - c(\exp(d \cdot t) - d \cdot t\exp(d \cdot t) - 1)}{\sigma_B\sqrt{2\pi t^3}} \cdot \exp\left[-\frac{(\omega - at^b - c\exp(d \cdot t) + c)^2}{2\sigma_B^2 t}\right]. \tag{10.39}$$

Similar to Case 1 and Case 2, the PDF of the RUL for Case 3 can be formulated as

$$f_{L_i|\mathrm{C}_2}(l_i|\boldsymbol{\theta}, Y_{1:i}) = \frac{1}{\sqrt{2\pi l_i^2(P_{i|i} + \sigma_B^2 l_i)}}\left[\omega - \eta'(l_i) + abl_i(l_i + t_i)^{b-1} + cd \cdot l_i\exp(d \cdot (l_i + t_i)) - \frac{P_{i|i}(\omega - \eta'(l_i)) + \hat{X}_{i|i}\sigma_B^2 l_i}{P_{i|i} + \sigma_B^2 l_i}\right] \cdot \exp\left(-\frac{(\omega - \eta'(l_i) - \hat{X}_{i|i})^2}{2(P_{i|i} + \sigma_B^2 l_i)}\right) \tag{10.40}$$

where $\eta'(l_i) = a(l_i + t_i)^b - at_i^b + c\exp(d \cdot (l_i + t_i)) - c\exp(d \cdot t_i)$.

From Eq. (10.35), we can see that $f_T(t \mid \mathrm{C}_1)$ is reduced to the inverse Gaussian distribution exactly if $b = 1$. This is as expected since any properly developed nonlinear model should cover a linear model as a special case. However, if we use $\Pr(T \le t) = \Pr(X(t) \ge \omega)$ to obtain the FHT distribution for C_1 as an approximation under the monotonic assumption, it can be shown without difficulty that the obtained results cannot reduce to the linear case exactly.

10.4.2 *The Derivation of Parameter Estimation Algorithm for Three Cases*

In the simulation study, we considered the case where $\mu(t; \boldsymbol{\vartheta}) = abt^{b-1}$ and $g(X_k; \boldsymbol{\xi}) = \beta_0 + \beta_1\exp(X_k)$. Then the state-space model is

$$X_k = X_{k-1} + a(k\Delta t)^b - a((k-1)\Delta t)^b + \sigma_B\sqrt{\Delta t}\,\omega_k, \tag{10.41a}$$

$$Y_k = \beta_0 + \beta_1\exp(X_k) + \sigma\upsilon_k \tag{10.41b}$$

Thus, we give the derivation of parameter estimation algorithm for this case firstly. In Sect. 10.3, we have derived the general parameter estimation algorithm for our model. Now we give the conditional expectation of the complete data log-likelihood function directly for this case

$$
\begin{aligned}
&\mathcal{Q}(\boldsymbol{\theta}, \hat{\boldsymbol{\theta}}^{(j)}) \propto \\
&-i\log\sigma_B - \frac{1}{2\sigma_B^2\Delta t}\sum_{k=1}^{i}\Big[A_k - 2a\big((k\Delta t)^b - ((k-1)\Delta t)^b\big)B_k + a^2\big((k\Delta t)^b - ((k-1)\Delta t)^b\big)^2\Big] \\
&-i\log\sigma - \frac{1}{2\sigma^2}\sum_{k=1}^{i}\Big[Y_k^2 + \beta_0^2 + \beta_1^2\exp(2\hat{X}_{k|k-1})C_k - 2Y_k(\beta_0 + \beta_1\exp(\hat{X}_{k|k-1})(D_k+1)) \\
&+ 2\beta_1\exp(\hat{X}_{k|k-1})(\beta_0 + \beta_1\exp(\hat{X}_{k|k-1}))D_k + \beta_1^2\exp(2\hat{X}_{k|k-1}) + 2\beta_0\beta_1\exp(\hat{X}_{k|k-1})\Big],
\end{aligned} \tag{10.42}
$$

where A_k, B_k, C_k and D_k are the same as the general case.

After deriving $\mathcal{Q}(\boldsymbol{\theta}, \hat{\boldsymbol{\theta}}^{(j)})$, the unknown parameter vector $\hat{\boldsymbol{\theta}}^{(j+1)}$ at the $(j+1)$th iteration can be obtained through maximizing $\mathcal{Q}(\boldsymbol{\theta}, \hat{\boldsymbol{\theta}}^{(j)})$ with respect to $\boldsymbol{\theta}$. However, it is difficult to obtain the accurate estimation of $\hat{\boldsymbol{\theta}}^{(j+1)}$ since it consists of six unknown parameters. Now let $\boldsymbol{\theta}_1 = [a, b, \sigma_B]$ and $\boldsymbol{\theta}_2 = [\beta_0, \beta_1, \sigma]$. Then, we decompose $\mathcal{Q}(\boldsymbol{\theta}, \hat{\boldsymbol{\theta}}^{(j)})$ into two parts. The first part is

$$
\begin{aligned}
\mathcal{Q}_1(\boldsymbol{\theta}_1, \hat{\boldsymbol{\theta}}_1^{(j)}) \propto &-\frac{1}{2\sigma_B^2\Delta t}\sum_{k=1}^{i}\Big[A_k + a^2\left((k\Delta t)^b - ((k-1)\Delta t)^b\right)^2 \\
&- 2a\big((k\Delta t)^b - ((k-1)\Delta t)^b\big)B_k\Big] - i\log\sigma_B,
\end{aligned} \tag{10.43}
$$

and the second part is

$$
\begin{aligned}
\mathcal{Q}_2(\boldsymbol{\theta}_2, \hat{\boldsymbol{\theta}}_2^{(j)}) \propto &-\frac{1}{2\sigma^2}\sum_{k=1}^{i}\Big[Y_k^2 + \beta_0^2 + \beta_1^2\exp(2\hat{X}_{k|k-1}) + 2\beta_0\beta_1\exp(\hat{X}_{k|k-1}) \\
&+ \beta_1^2\exp(2\hat{X}_{k|k-1})C_k + 2\beta_1\exp(\hat{X}_{k|k-1})(\beta_0 + \beta_1\exp(\hat{X}_{k|k-1}))D_k \\
&- 2Y_k(\beta_0 + \beta_1\exp(\hat{X}_{k|k-1})(D_k+1))\Big] - i\log\sigma.
\end{aligned} \tag{10.44}
$$

From the above two formulations, we can find that $\mathcal{Q}_1(\boldsymbol{\theta}_1, \hat{\boldsymbol{\theta}}_1^{(j)})$ and $\mathcal{Q}_2(\boldsymbol{\theta}_2, \hat{\boldsymbol{\theta}}_2^{(j)})$ do not affect each other if maximizing these two parts. Therefore, we maximize $\mathcal{Q}(\boldsymbol{\theta}, \hat{\boldsymbol{\theta}}^{(j)})$ through maximizing $\mathcal{Q}_1(\boldsymbol{\theta}_1, \hat{\boldsymbol{\theta}}_1^{(j)})$ and $\mathcal{Q}_2(\boldsymbol{\theta}_2, \hat{\boldsymbol{\theta}}_2^{(j)})$, respectively. Clearly, this maximum policy is easier to realize since each of the two parts consists of less unknown parameters than maximizing $\mathcal{Q}(\boldsymbol{\theta}, \hat{\boldsymbol{\theta}}^{(j)})$ directly. For notational convenience, we omit the superscript $(j+1)$ indicating that the parameters are obtained at $(j+1)$th iteration in the following derivation. Now we compute $\hat{\boldsymbol{\theta}}_1$ and $\hat{\boldsymbol{\theta}}_2$ through the technique of profile log-likelihood function as follows.

First, we calculate the partial derivatives of $\mathcal{Q}_1(\boldsymbol{\theta}_1, \hat{\boldsymbol{\theta}}_1^{(j)})$ with respect to a and σ_B for specific value of b, and let these two partial derivatives equal to zero, we can obtain

$$\hat{a} = \frac{\sum_{k=1}^{i} \left((k\Delta t)^b - ((k-1)\Delta t)^b\right) B_k}{\sum_{k=1}^{i} \left((k\Delta t)^b - ((k-1)\Delta t)^b\right)^2} \tag{10.45}$$

and

$$\hat{\sigma}_B^2 = \left\{ \sum_{k=1}^{i} \left[A_k - 2a\left((k\Delta t)^b - ((k-1)\Delta t)^b\right) B_k + a^2 \left((k\Delta t)^b - ((k-1)\Delta t)^b\right)^2 \right] \right\} \Big/ (i \cdot \Delta t). \tag{10.46}$$

Similarly, let

$$\begin{aligned} Z_k =& Y_k^2 + \beta_0^2 + \beta_1^2 \exp(2\hat{X}_{k|k-1}) + 2\beta_0\beta_1 \exp(\hat{X}_{k|k-1}) + \beta_1^2 \exp(2\hat{X}_{k|k-1})C_k \\ &+ 2\beta_1 \exp(\hat{X}_{k|k-1})(\beta_0 + \beta_1 \exp(\hat{X}_{k|k-1}))D_k - 2Y_k(\beta_0 + \beta_1 \exp(\hat{X}_{k|k-1})(D_k + 1)). \end{aligned}$$

We compute the partial derivatives of $\mathscr{Q}_2(\boldsymbol{\theta}_2, \hat{\boldsymbol{\theta}}_2^{(j)})$ with respect to σ and β_0 for specific value of β_1, and set these two partial derivatives equal to zero, we have

$$\hat{\sigma}^2 = \frac{\sum_{k=1}^{i} Z_k}{i} \tag{10.47}$$

and

$$\hat{\beta}_0 = \frac{\sum_{k=1}^{i} Y_k - \sum_{k=1}^{i} [\beta_1 \exp(\hat{X}_{k|k-1}) + \beta_1 \exp(\hat{X}_{k|k-1})D_k]}{i}. \tag{10.48}$$

Substituting (10.45) and (10.46) into (10.43), the profile log-likelihood function can be obtained for b. Then, maximizing the profile log-likelihood function with respect to b through one-dimension search, the estimate of b can be derived. Under the estimated b, the estimates of a and σ_B can be obtained from (10.45) and (10.46). Same to the above calculating procedure, the estimate of $\boldsymbol{\theta}_2 = [\beta_0, \beta_1, \sigma]$ can also be obtained. Then we iterate the E-step and M-step until the convergence is achieved.

Now we give the derivation of parameter estimation algorithm for the other two forms in the case study.

In the case study, we consider one case where $\mu(t; \boldsymbol{\vartheta}) = ab\exp(bt)$. Then the state-space model is

$$X_k = X_{k-1} + a\exp((b \cdot k\Delta t)) - a\exp(b \cdot ((k-1)\Delta t)) + \sigma_B\sqrt{\Delta t}\,\omega_k, \tag{10.49a}$$

$$Y_k = X_k + \sigma\upsilon_k \tag{10.49b}$$

According to the general form, we have

$$\mathcal{Q}(\boldsymbol{\theta}, \hat{\boldsymbol{\theta}}^{(j)}) \propto -i \log \sigma_B - i \log \sigma - \frac{1}{2\sigma^2} \sum_{k=1}^{i} \left(Y_k^2 + P_{k|i} + \hat{X}_{k|i}^2 - 2Y_k \hat{X}_{k|i} \right)$$
$$- \frac{1}{2\sigma_B^2 \Delta t} \sum_{k=1}^{i} \Big[A_k - 2a\big(\exp(b(k\Delta t)) - \exp(b((k-1)\Delta t))\big) B_k$$
$$+ a^2 \big(\exp(b(k\Delta t)) - \exp(b((k-1)\Delta t))\big)^2 \Big]. \tag{10.50}$$

As the same as Case 1, we decompose $\mathcal{Q}(\boldsymbol{\theta}, \hat{\boldsymbol{\theta}}^{(j)})$ into two parts. The first part is

$$\mathcal{Q}_1(\boldsymbol{\theta}_1, \hat{\boldsymbol{\theta}}_1^{(j)}) \propto - \frac{1}{2\sigma_B^2 \Delta t} \sum_{k=1}^{i} \Big[A_k - 2a\big(\exp(b(k\Delta t)) - \exp(b((k-1)\Delta t))\big) B_k$$
$$+ a^2 \big(\exp(b(k\Delta t)) - \exp(b((k-1)\Delta t))\big)^2 \Big] - i \log \sigma_B, \tag{10.51}$$

and the second part is

$$\mathcal{Q}_2(\boldsymbol{\theta}_1, \hat{\boldsymbol{\theta}}_1^{(j)}) \propto -i \log \sigma - \frac{1}{2\sigma^2} \sum_{k=1}^{i} \left(Y_k^2 + P_{k|i} + \hat{X}_{k|i}^2 - 2Y_k \hat{X}_{k|i} \right). \tag{10.52}$$

Similarly, we can obtain

$$\hat{a} = \frac{\sum_{k=1}^{i} (\exp(b(k\Delta t)) - \exp(b((k-1)\Delta t)) B_k}{\sum_{k=1}^{i} (\exp(b(k\Delta t)) - \exp(b((k-1)\Delta t))^2}, \tag{10.53}$$

$$\hat{\sigma}_B^2 = \Bigg\{ \sum_{k=1}^{i} \Big[A_k - 2a\big(\exp(b(k\Delta t)) - \exp(b((k-1)\Delta t))\big) B_k$$
$$+ a^2 \big(\exp(b(k\Delta t)) - \exp(b((k-1)\Delta t))\big)^2 \Big] \Bigg\} \Big/ (i \cdot \Delta t) \tag{10.54}$$

and

$$\hat{\sigma}^2 = \frac{\sum_{k=1}^{i} \left(Y_k^2 + P_{k|i} + \hat{X}_{k|i}^2 - 2Y_k \hat{X}_{k|i} \right)}{i}. \tag{10.55}$$

The subsequent estimation procedure is the same as the first case.

We consider the third case where $\mu(t; \boldsymbol{\vartheta}) = abt^{b-1} + cd \cdot \exp(d \cdot t)$. Then the state-space model is

$$X_k = X_{k-1} + a(k\Delta t)^b - a((k-1)\Delta t)^b + c\exp(d \cdot k\Delta t) - c\exp(d \cdot ((k-1)\Delta t)) + \sigma_B\sqrt{\Delta t}\,\omega_k, \tag{10.56a}$$

$$Y_k = X_k + \sigma \upsilon_k \tag{10.56b}$$

Now we give the decomposed $\mathcal{Q}(\boldsymbol{\theta}, \hat{\boldsymbol{\theta}}^{(j)})$ directly as

$$\mathcal{Q}_1(\boldsymbol{\theta}_1, \hat{\boldsymbol{\theta}}_1^{(j)}) \propto -\frac{1}{2\sigma_B^2 \Delta t}\sum_{k=1}^{i}\left[A_k - 2aH_kB_k - 2cJ_kB_k + (aH_k + cJ_k)^2\right] - i\log\sigma_B, \tag{10.57}$$

and

$$\mathcal{Q}_2(\boldsymbol{\theta}_1, \hat{\boldsymbol{\theta}}_1^{(j)}) \propto -i\log\sigma - \frac{1}{2\sigma^2}\sum_{k=1}^{i}\left(Y_k^2 + P_{k|i} + \hat{X}_{k|i}^2 - 2Y_k\hat{X}_{k|i}\right), \tag{10.58}$$

where $H_k = (k\Delta t)^b - ((k-1)\Delta t)^b$, $J_k = \exp(d \cdot k\Delta t) - \exp(d \cdot (k-1)\Delta t)$. So we have

$$\hat{a} = \frac{\sum_{k=1}^{i}(H_kB_k - cH_kJ_k)}{\sum_{k=1}^{i}H_k^2}, \tag{10.59}$$

$$\hat{\sigma}_B^2 = \frac{\sum_{k=1}^{i}\left[A_k - 2aH_kB_k - 2cJ_kB_k + (aH_k + cJ_k)^2\right]}{i \cdot \Delta t}, \tag{10.60}$$

and

$$\hat{\sigma}^2 = \frac{\sum_{k=1}^{i}\left(Y_k^2 + P_{k|i} + \hat{X}_{k|i}^2 - 2Y_k\hat{X}_{k|i}\right)}{i}. \tag{10.61}$$

The subsequent estimation procedure is the same as the previous two cases.

10.5 Simulation Study

In this section, we provide a numerical simulation to verify the effectiveness of the proposed method. For illustration purpose, we consider the hidden degradation process represented by $X(t) = at^b + \sigma_B B(t)$ and the observation process expressed by $Y_k = \beta_0 + \beta_1\exp(X_k) + \sigma\upsilon_k$ to generate the simulation data. Then we use our approach to model these data and estimate the RUL, which is termed as Model 1.

The reason to use the exponential equation as the measurement equation is based on our observation that this kind of functional form is widely used in degradation modeling practice [6, 32, 34] and the other areas [47, 48]. For comparison, we also consider the linear drift-based BM as the hidden degradation process to achieve the RUL estimation, which is termed as Model 2. Specifically, the following state-space model is used to simulate the data.

$$X_k = X_{k-1} + a(k\Delta t)^b - a((k-1)\Delta t)^b + \sigma_B\sqrt{\Delta t}\,\omega_k, \tag{10.62a}$$

$$Y_k = \beta_0 + \beta_1 \exp(X_k) + \sigma \upsilon_k \tag{10.62b}$$

We set the parameters $a = 0.1, b = 2, \sigma_B = 0.4, \beta_0 = 1, \beta_1 = 0.01$ and $\sigma = 0.005$, and the sampling interval $\Delta t = 0.1$ s. When we set the threshold $\omega = 10$, the FHT for a particular sampling path is approximated to be 9.3 s after 93 sampling. Namely, a realization of the FHT for our simulation is 9.3 s. In the following, we utilize the simulation data corresponding to this realization of the FHT to illustrate our approach and suppose that the lifetime is 9.3 s in order to compare the mean of the estimated RUL with the actual RUL at each sampling time point. Then, at the current time, we use all up-to-date data to estimate the unknown parameters by EM algorithm, and when the new observation is available, the parameters can be re-estimated based on the new observed measurement. The total time of processing all 93 simulation data is approximately 0.9976 s in the MATLAB with the computer 2.4 GHz Intel Core 2 Duo and 2 GB memory. This implies that the mean time of processing each observation data is about 0.017 s, but the sampling interval in simulation is set as 0.1 s. This demonstrates the computational speed is very fast in the context of real-time CM. The evolving path of the estimated parameter vector $\boldsymbol{\theta}$ is illustrated in Fig. 10.1a–f.

From Fig. 10.1a–f, we can observe that the estimates of the unknown parameters eventually converge to their true values, which proves that our parameter estimation method presented in Sect. 10.3 can work well. In addition, the estimates of the parameters except σ_B can approach the true values as quickly as the available data accumulated. After obtaining the estimated and updated parameters at each sampling time, the EKF can be used to evaluate the hidden degradation state. For comparison, the actual degradation path and the estimated expectations of the degradation states using Model 1 and Model 2 are shown in Fig. 10.2, respectively.

From Fig. 10.2, we can find that both of these two models can track the actual degradation state well. Especially at the beginning stage, the Model 2 even outperforms the Model 1. It seems that the estimated results have almost no difference. For clarity, we also show the errors of the estimated expectations with the actual state, using Model 1 and Model 2, respectively. Instead, we show the absolute errors in Fig. 10.3.

As shown in Fig. 10.3, if there are only few sampling data at the early stage of the degradation, the linear model has a better tracking performance than our model before the 38th CM point. This is not uncommon since the available data are few at the beginning stage of the degradation. Compared with the linear model, our model needs to estimate additional parameters. However, with the sampling data accumulated, the

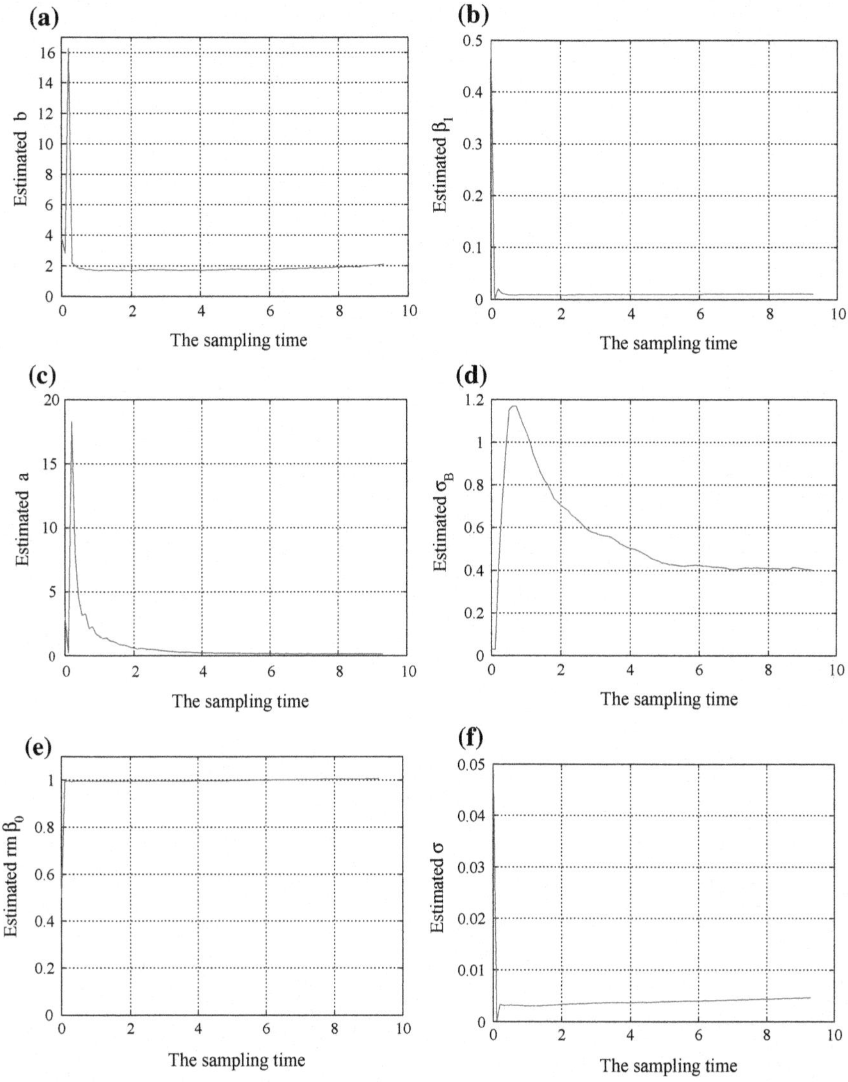

Fig. 10.1 Estimations of the parameters

error between the predicted path and the actual path becomes smaller and smaller. Particularly, after about the 38th CM point, Model 1 can track the degradation path very well, which is better than Model 2. Even so, the difference of the estimated RULs from two models is major, as shown in Figs. 10.4 and 10.5.

Obviously, with respect to the RUL estimation, the advantages of Model 1 are much greater than those of Model 2. First, the PDF of the RUL obtained by Model 1 is much more compact. This indicates that the uncertainty of the estimated RUL is

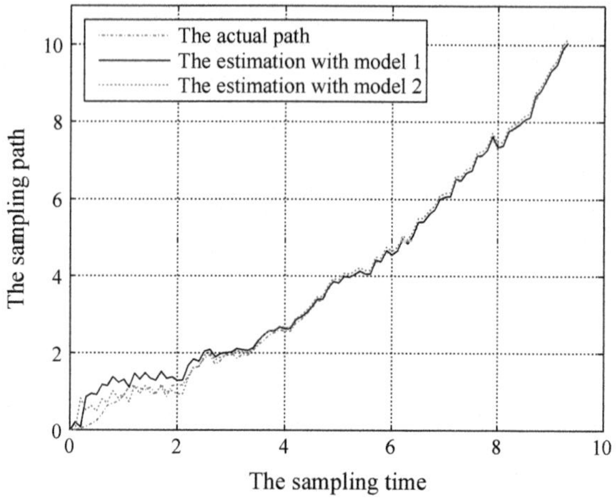

Fig. 10.2 Comparison of the state estimation results

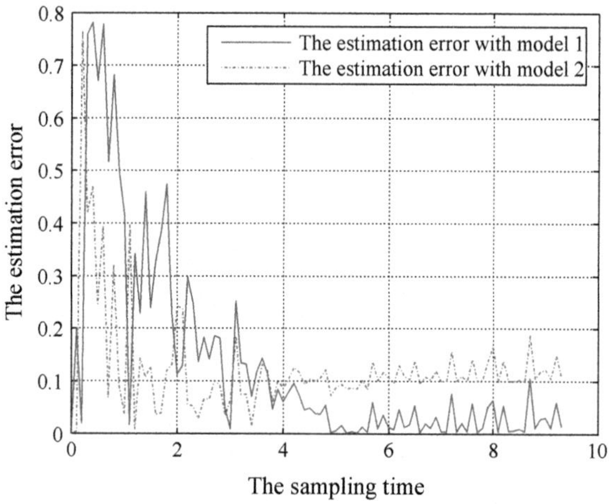

Fig. 10.3 Comparative results of the estimated errors

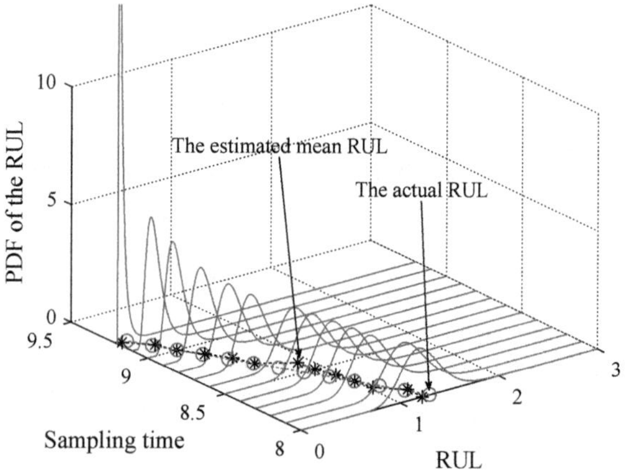

Fig. 10.4 Estimated results with Model 1

Fig. 10.5 Estimated results with Model 2

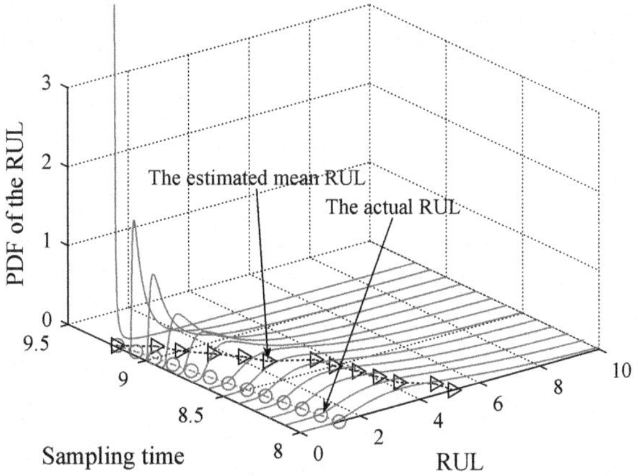

much smaller than that obtained by Model 2. Second, the mean RULs under Model 1 are more accurate, however, using the linear model to estimate the RUL leads to greater errors and the estimated RUL is far from the actual RUL. This shows that our method is effective for the case that the degradation process is hidden and has nonlinear characteristics. Note that from Figs. 10.2 and 10.3, all models can track the degradation path pretty well as the data are accumulated but the linear model fails to give a good estimate of the RUL distribution. In contrast, the proposed method can achieve accurate estimation and reduce the uncertainty in the estimated RUL, as illustrated by Figs. 10.4 and 10.5. This provides the evidence for the necessity of investigating the hidden and nonlinear degradation process and the application of the proposed method in prognostics and health management (PHM), since the uncertainty reduction is one of the important aspects for PHM. In addition, this simulation tells us that depending on the common statistical practice of fitting curves to data may lead to an inadequate model or even nonsense results for RUL estimation. This is true for estimating RUL since we care about not only the prediction accuracy for degradation state but also reducing the uncertainty of the estimated RUL. Fortunately, it seems that our developed method can balance such requirements. Actually, this phenomena is very consistent with the intuition that the longer the forecasting step is, the worse the prediction is. Namely, the long-term prediction will incur greater inaccuracy. For our case, the prediction for the degradation state is just one-step ahead prediction, but the RUL estimation is a long-term prediction up to the failure. Therefore, when the linear model is applied to nonlinear data at the current time, it will look at the future evolution of degradation path with a constant rate and thus ignore the time-dependent degradation rate for nonlinear process. As a result, greater inaccuracy in RUL estimation is caused. In contrast, our general model can take the time-varying degradation rate into account and thus leads to better RUL estimation in nonlinear case even for long-term prediction, since the data used in the simulation are nonlinear in nature. For verifying the above analysis, we compare the long-term

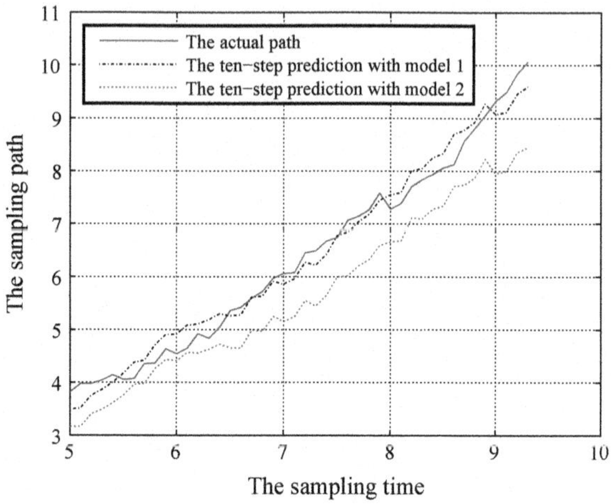

Fig. 10.6 Comparison of the ten-step prediction of the state

prediction of the state by these two models. The results indicate that the long-term predictions by Model 1 significantly outperform those of Model 2.

To make the comparison more comprehensive in the simulation study, we also consider the following two scenarios. First, we compare the long-term predictions of the state using Model 1 and Model 2. For illustrative purpose, we show the ten-step prediction of the state after observing until 4 s. Specifically, at time t_i, $40 \leq i \leq 83$ (since we have 93 observations in this simulation study), we used EKF to estimate the state $X(i)$ at time t_i, i.e., $\hat{X}_{i|i}$. Then, we predict the state $X(i+10)$ by $\hat{X}_{i|i}$ and obtain the ten-step prediction $\hat{X}_{i+10|i}$. When the new observation is available, the state estimation and ten-prediction can be updated accordingly. The result is shown in Fig. 10.6.

From Fig. 10.6, we can see that these two models have a large difference in long-term predictions. Moreover, the prediction accuracy of Model 1 outperforms Model 2 apparently. This phenomena indicates that the long-term predictions by linear model will incur greater inaccuracy than our model when the degradation process is nonlinear. Thus, as a long-term prediction up to failure, the RUL estimation by these two models will have a large difference as discussed in Sect. 10.4 of the original chapter.

Second, for illustrating that the linear model is a special case of our model in the linear case, we consider the hidden degradation process represented by the linear drifted-based BM $X(t) = at + \sigma_B B(t)$ and the observation process expressed by $Y_k = \beta_0 + \beta_1 \exp(X_k) + \sigma v_k$ to generate the simulation data. Then we compare the RUL estimation of Model 1 and Model 2 using these data, respectively. A realization of the FHT for this particular sampling path is approximated to be 4.5 s after 45 samplings. In the following, we utilize the simulation data corresponding to this realization of the FHT to illustrate our approach and suppose that the lifetime is 4.5 s in order to compare the mean of the estimated RUL with the actual RUL at each sampling time point. The evolving path of the estimated parameter b in our model is illustrated in Fig. 10.7.

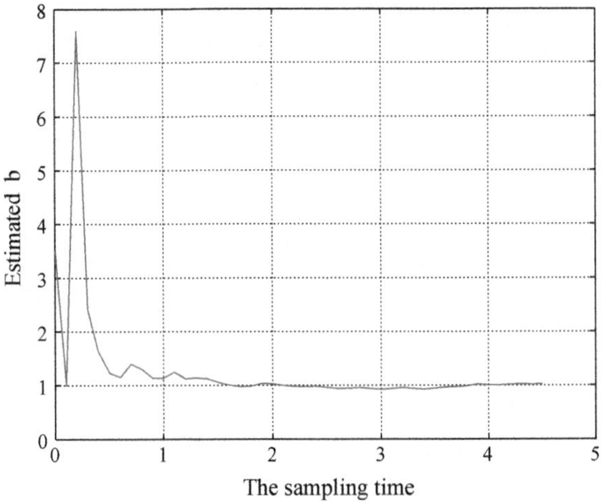

Fig. 10.7 Estimate of the parameter b of model 1

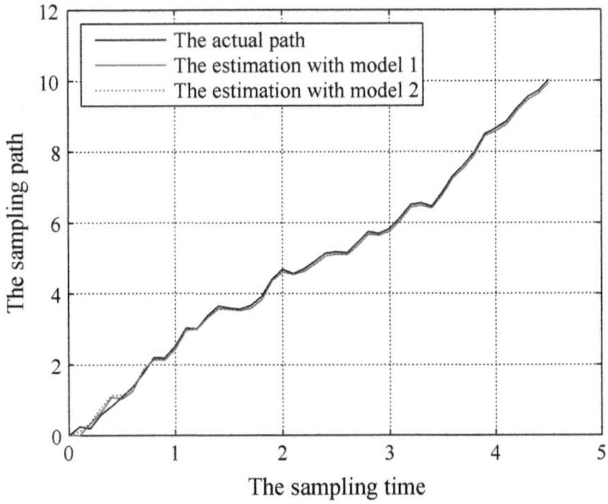

Fig. 10.8 Comparison of the state estimation results

From Fig. 10.2, it can be observed that as the number of observations increases, the estimated b approaches 1 in our model. This shows that our model can reduce to the linear model if the data are linear. As the same as the nonlinear scenario, we show the comparison of the state estimation in Figs. 10.8 and 10.9.

From Figs. 10.8 and 10.9, we can find that Model 1 and Model 2 can track the actual degradation state very well and the estimated results have almost no difference. Additionally, we illustrate the five-step prediction of the state by these two model after observing until 1.5 s, respectively, as shown in Fig. 10.10.

Only from Fig. 10.10, we can observe that overall, the performance of long-term prediction of the state by Model 2 displays a marginally better than that of Model 1. For further comparison, we show the RUL estimation by Model 1 and Model 2 in Figs. 10.11 and 10.12, respectively.

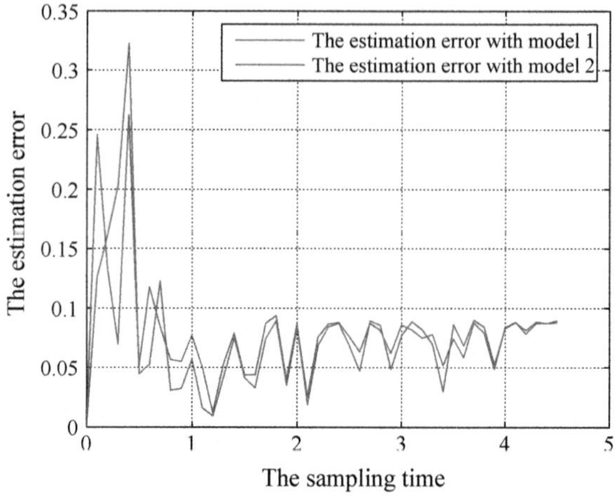

Fig. 10.9 Comparative results of the estimated errors

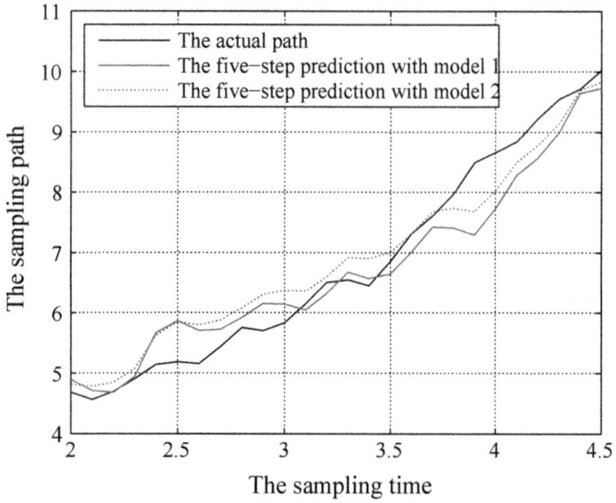

Fig. 10.10 Comparison of the five-step prediction of the state

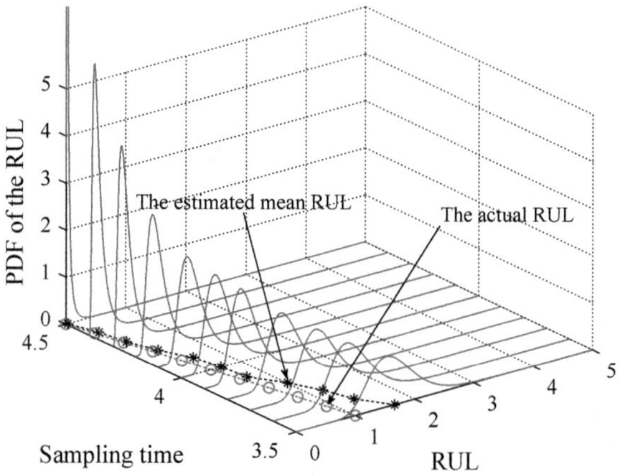

Fig. 10.11 Estimated results with model 1

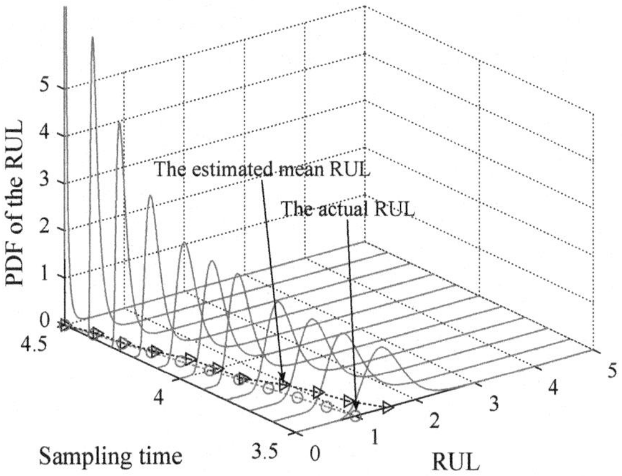

Fig. 10.12 Estimated results with model 2

From Figs. 10.11 and 10.12, we can find that there are marginal differences in RUL estimation between these two models and Model 2 performed a little better than Model 1. In a word, from the above simulations, we can observe that our model is more general and it can also estimate the RUL well in the linear case.

10.6 Case Study

Lithium-ion battery is a critical component of many electronic equipments and complex systems and many faults of these systems are caused by the failure of the Lithium-ion battery. The accurate RUL estimation of the battery can improve the system reliability and reduce the failure risk. Thus, there have been many studies regarding the RUL estimation of the Lithium-ion battery [49–53]. To demonstrate the effectiveness of our method, a case study was conducted on a NASA battery set [54]. The data can be found in the website http://ti.arc.nasa.gov/tech/dash/pcoe/prognostic-data-repository/.

10.6.1 The Data and State-Space-Based Degradation Model

The use of the Lithium-ion battery is a process of charging and discharging repeatedly. The battery capacity will degenerate with the increasing of the number of charge and discharge. The cycle life of the battery is defined as the number of times a battery can be recharged before its capacity has faded beyond acceptable limits (20∼30% of the rated capacity) [51]. Thus the battery capacity can be used as the actual degradation state. The data have been collected from a custom built battery prognostics testbed at the NASA Ames Prognostics Center of Excellence (PCoE).

In this testing, the Lithium-ion batteries were run through 3 different operational profiles including charge, discharge and electrochemical impedance spectrometry (EIS). Repeated charge and discharge cycles led to accelerated aging of the batteries. The experiments were stopped when the batteries reached the end-of-life (EOL) criteria of 30% fade in rated capacity. The capacities and the other inner parameters of these tested batteries were recorded in each cycle. Based on the data set, Saha and Geobel have conducted many important and excellent works [49–52]. They constructed state-space model and utilized filter technique to estimate the RUL of the battery. In these studies, the measured capacity which contained noise was used as the observed measurement, and the actual capacity was used as the hidden degradation state. Similar to the work of Saha and Geobel, in this chapter, we establish the state-space model as follows

$$X_k = X_{k-1} + h(t_k; \boldsymbol{\vartheta}) - h(t_{k-1}; \boldsymbol{\vartheta}) + \sigma_B \sqrt{\Delta t}\, \omega_k, \tag{10.63a}$$

$$Y_k = X_k + \sigma \upsilon_k \tag{10.63b}$$

where the unknown model parameter vector is $\boldsymbol{\theta} = [\boldsymbol{\vartheta}, \sigma_B, \sigma]$.

In practice, the hidden degradation process can be often fitted by our proposed model with different nonlinear forms. Thus, the selection of a suitable function $h(t; \boldsymbol{\vartheta})$ is essential when applying our method for a particular application scenario. To compare the fitness of different nonlinear forms, both the Akaike information criterion (AIC) [55] and total mean squared error (TMSE) [6] about the actual RUL obtained at each observation point are used as measures for comparative studies. the AIC considers both the log-likelihood and the number of parameters estimated, and provides a way of choosing the best fitness model. the AIC is calculated by

$$\text{AIC} = 2\left(k - \ln L(\hat{\boldsymbol{\theta}})\right) \tag{10.64}$$

where k is the number of model parameters, $\hat{\boldsymbol{\theta}}$ is the vector of estimated parameters and $L(\hat{\boldsymbol{\theta}})$ is the maximized likelihood function value.

The mean squared error (MSE) at each observation point is defined as

$$\text{MSE}_i = \int_0^\infty (\tilde{l}_i - l_i)^2 f_{L_i}(l_i|\boldsymbol{\theta}, Y_{1:i})\text{d}l_i, \tag{10.65}$$

where $\tilde{l}_i$ is the actual RUL at t_i and $f_{L_i}(l_i|\boldsymbol{\theta}, Y_{1:i})$ is the according PDF of the RUL estimated by (10.12).

Therefore, the TMSE can be defined as the sum of the MSE at each observation point over the whole life. Namely, if there are N observations, the TMSE can be formulated as

$$\mathrm{TMSE} = \sum_{i=1}^{N} \int_0^{\infty} (\tilde{l}_i - l_i)^2 f_{L_i}(l_i|\boldsymbol{\theta}, Y_{1:i}) \mathrm{d}l_i, \tag{10.66}$$

In both criteria, the smallest AIC and TMSE values correspond to the best fitness result, and thus we can choose the form with the smallest AIC and TMSE as the nonlinear degradation model for prognostics.

10.6.2 Results and Discussions

We utilize the data of No. 6 battery to verify the effectiveness of our method, which includes 168 sets of the CM data. Note that our method is derived based on the case that the degradation process has an increasing trend. However, the capacity of the battery has a decreasing trend over time. Thus, we take the reciprocals of the data as the observations and the critical threshold can be changed accordingly. In order to estimate and update the RUL of the No. 6 battery at each cycle accurately, we need to estimate and update the model parameters precisely. However, different initial values of parameters have an important impact on the parameter estimation. In this chapter, we take the data of the other batteries as the historical data to train the model, and then we can obtain the trained parameters which are taken as the initial values of parameters for the prognosis of No. 6 battery. The results of the model selection and final parameter estimates are given in Table 10.1.

For comparison, we also give the results of the linear model. Furthermore, we also show the total time of parameter estimation and updating. From Table 10.1, we can see that when $h(t;\boldsymbol{\vartheta}) = at^b + c\exp(d\cdot t) - c$, there is the smallest AIC and TMSE.

Table 10.1 The final estimated parameters and model selection results

	$h(t;\boldsymbol{\vartheta})$			
	$at^b + c\exp(d\cdot t) - c$	at^b	$a\exp(bt) - a$	at
a	0.2039	0.4839	1.0707e-8	0.005
b	0.5203	0.0242	0.089	1
c	0.0012	–	–	–
d	0.0002	–	–	–
σ_B	0.0465	0.0083	0.0389	0.0386
σ	0.0041	0.0026	0.0043	0.0043
$\ln L(\boldsymbol{\theta})$	515.653	493.4631	418.2522	410.6047
AIC	−1019.3	−978.9262	−828.5044	−815.2094
TMSE$\times 10^5$	3.5966	3.9083	4.9876	5.5932
Time of parameter updating (s)	3.4726	1.8769	1.7324	0.3172

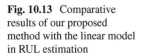

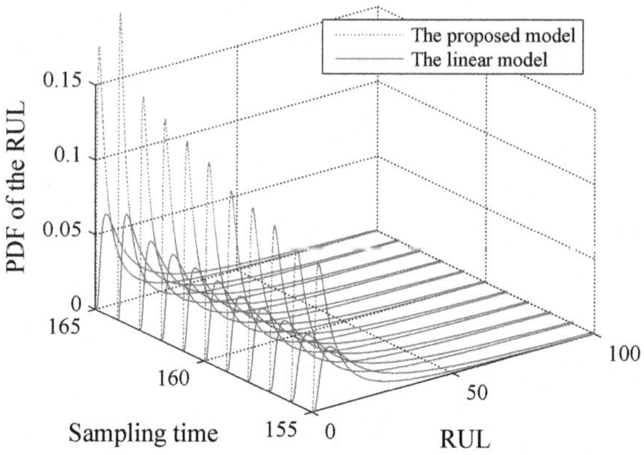

Fig. 10.13 Comparative results of our proposed method with the linear model in RUL estimation

This nonlinear form is most appropriate to fit the data of the battery. Thus, we utilize this form to model the degradation process. Note that the other nonlinear forms for this particular application scenario can be chosen. However, for illustrative purpose, we just choose these three nonlinear forms to show the feasibility and effectiveness of our method. As shown in Table 10.1, the estimated values of b in the three nonlinear models clearly confirm the nonlinear characteristics. All kinds of nonlinear models outperform the linear model in terms of $\ln L(\boldsymbol{\theta})$, AIC and TMSE. Since the linear model contains less parameters, the parameter updating of the linear model consumes the least time. However, for the updating speed, it has almost no difference with these three nonlinear models. The purpose of giving the total updating time is only showing the computational speed of our parameter estimation algorithm can satisfy the real-time requirement. For comparison, the PDFs of the RULs corresponding to the selected nonlinear model and the linear model from the 155th monitoring time point to the 165th monitoring time point are shown in Fig. 10.13.

It can be observed from Fig. 10.13 that our method outperforms the linear model in the respect of the results of the RUL estimation. In particularly, the PDF of the RUL is more dispersed when using the linear model, which reflects a greater uncertainty in the RUL estimation.

To further compare the performance of our method with the linear model, we show the MSE about the RUL at each monitoring time point associated with these two models in Fig. 10.14. Note that we only plot the results from the 60th monitoring time point to the 168th point since the changes of the parameter estimation will be smaller and smaller after the 60th monitoring time point.

As shown in Fig. 10.14, the MSE of the RUL obtained by our method maintains a lower level and approaches a small MSE value more quickly than the linear model. In addition, our method has a smoother curve of MSE compared with the linear model. All these results indicate that our method has a better RUL estimation than the linear model.

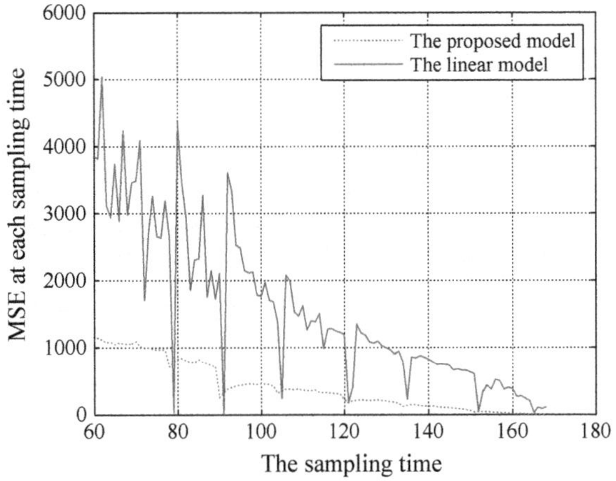

Fig. 10.14 Comparative results of our proposed method with the linear model in MSE of the estimated RUL

References

1. Jardine AKS, Lin D, Banjevic D (2006) A review on machinery diagnostics and prognostics implementing condition-based maintenance. Mech Syst Signal Process 20(7):1483–1510
2. Wang W (2007) A two-stage prognosis model in condition based maintenance. Eur J Op Res 182(3):1177–1187
3. You MY, Li L, Meng G, Ni J (2010) Cost-effective updated sequential predictive maintenance policy for continuously monitored degrading systems. IEEE Trans Autom Sci Eng 7(2):257–265
4. Camci F, Chinnam RB (2010) Health-state estimation and prognostics in machining processes. IEEE Trans Autom Sci Eng 7(3):581–597
5. Si XS, Wang W, Hu CH, Zhou DH (2011) Remaining useful life estimation-a review on the statistical data driven approaches. Eur J Op Res 213(1):1–14
6. Carr MJ, Wang W (2011) An approximate algorithm for prognostic modelling using condition monitoring information. Eur J Op Res 211(1):90–96
7. Wang W (2006) A prognosis model for wear prediction based on oil-based monitoring. J Op Res Soc 58(7):887–893
8. Gebraeel NZ, Lawley MA, Li R, Ryan JK (2005) Residual-life distributions from component degradation signals: a Bayesian approach. IIE Trans 37(6):543–557
9. Gebraeel NZ (2006) Sensory-updated residual life distributions for components with exponential degradation patterns. IEEE Trans Autom Sci Eng 3(4):382–393
10. Gebraeel NZ, Elwany A, Pan J (2009) Residual life predictions in the absence of prior degradation knowledge. IEEE Trans Reliab 58(1):106–117
11. Doksum KA, Hóyland A (1992) Models for variable-stress accelerated life testing experiments based on Wiener processes and the inverse Gaussian distribution. Technometrics 34(1):74–82
12. Wang X (2010) Wiener processes with random effects for degradation data. J Multivar Anal 101(2):340–351
13. Whitmore GA, Schenkelberg F (1997) Modelling accelerated degradation data using Wiener diffusion with a time scale transformation. Lifetime Data Anal 3(1):27–45
14. Lu CJ, Meeker WQ (1993) Using degradation measures to estimate a time-to-failure distribution. Technometrics 35(2):161–174
15. Si XS, Wang W, Hu CH, Pecht MG (2012) Remaining useful life estimation based on a nonlinear diffusion degradation process. IEEE Trans Reliab 61(1):50–67

16. Peng CY, Tseng ST (2009) Mis-specification analysis of linear degradation models. IEEE Trans Reliab 58(3):444–455
17. Si XS, Hu CH, Wang W, Chen MY (2011) An adaptive and nonlinear drift-based Wiener process for remaining useful life estimation. In: Prognostics and health management conference, vol 2011. Shenzhen, China, IEEE Xplore, pp 1–5
18. Zhou YF, Sun Y, Mathew J, Wolff R, Ma L (2011) Latent degradation indicators estimation and prediction: a Monte Carlo approach. Mech Syst Signal Process 25(1):222–236
19. Zhao JM, Feng TL (2011) Remaining useful life prediction based on nonlinear state space model. In: Prognostics and health management conference, vol 2011. Shenzhen, China, IEEE Xplore, pp 1–5
20. Li Z, Kapur KC (2010) Models and customer-centric system performance measures using fuzzy reliability. In: Proceedings of the 2010 IEEE international conference on systems man and cybernetics. Istanbul, Turkey
21. Li Z, Kapur KC (2011) Continuous-state reliability measures based on fuzzy sets. IIE Trans 44(11):1033–1044
22. Lu CJ, Park J, Yang Q (1997) Statistical inference of a time-to-failure distribution derived from linear degradation data. Technometrics 39(4):391–400
23. Robinson ME, Crowder MJ (2000) Bayesian methods for a growth-curve degradation model with repeated measures. Lifetime Data Anal 6(4):357–374
24. Kass RE, Raftery AE (1995) Bayes factors. J Am Stat Assoc 90(430):773–795
25. Pandey MD, Yuan XX, Van Noortwijk JM (2009) The influence of temporal uncertainty of deterioration on life-cycle management of structures. Struct Infrastruct Eng 5(2):145–156
26. Xu ZG, Ji YD, Zhou DH (2008) Real-time reliability prediction for a dynamic system based on the hidden degradation process identification. IEEE Trans Reliab 57(2):230–242
27. Christer AH, Wang W, Sharp JM (1997) A state space condition monitoring model for furnace erosion prediction and replacement. Eur J Op Res 101(1):1–14
28. Batzel TD, Swanson DC (2009) Prognostic health management of aircraft power generators. IEEE Trans Aerosp Electron Syst 45(2):473–483
29. Gašperin M, Juričić D, Boškoski P (2010) Model-based prognostics of gear health using stochastic dynamical models. Mech Syst Signal Process 25(2):537–548
30. Sun JZ, Zuo HF, Wang W, Pecht MG (2012) Application of a state space modeling technique to system prognostics based on a health index for condition-based maintenance. Mech Syst Signal Process 28:585–596
31. Zhou YF, Ma L, Sun Y, Mathew J (2008) Latent degradation indicator estimation using condition monitoring information. In: Proceedings of the WCEAM-IMS. Beijing, China, pp 1967–1980
32. Cadini F, Zio E, Avram D (2009) Monte Carlo-based filtering for fatigue crack growth estimation. Probab Eng Mech 24(3):367–373
33. Orchard ME, Vachtsevanos GJ (2009) A particle-filtering approach for on-line fault diagnosis and failure prognosis. Tran Inst Meas Control 31(3/4):221–246
34. Zio E, Peloni G (2011) Particle filtering prognostic estimation of the remaining useful life of nonlinear components. Reliab Eng Syst Saf 96(3):403–409
35. Barker CT, Newby MJ (2009) Optimal non-periodic inspection for a multivariate degradation model. Reliab Eng Syst Saf 94(1):33–43
36. Liao CM, Tseng ST (2006) Optimal design for step-stress accelerated degradation tests. IEEE Trans Reliab 55(1):59–66
37. Padgett WJ, Tomlinson MA (2004) Inference from accelerated degradation and failure data based on Gaussian process models. Lifetime Data Anal 10(2):191–206
38. Kloeden P, Platen E (1995) Numerical Solution of Stochastic Differential Equations. Springer, New York
39. Giorgio M, Guida M, Pulcini G (2010) A state-dependent wear model with an application to marine engine cylinder liners. Technometrics 52(2):172–187
40. Guida M, Pulcini G (2011) A continuous-state Markov model for age-and state-dependent degradation processes. Struct Saf 33(6):354–366

41. Giorgio M, Guida M, Pulcini G (2011) An age-and state-dependent Markov model for degradation processes. IIE Trans 43(9):621–632
42. Teresa Lam C, Yeh RH (1994) Optimal replacement policies for multistate deteriorating systems. Naval Res Logist (NRL) 41(3):303–315
43. Lee MLT, Whitmore GA (2006) Threshold regression for survival analysis: modeling event times by a stochastic process reaching a boundary. Stat Sci 21(4):501–513
44. Borkar VS, Ghosh MK, Rangarajan G (2010) Application of nonlinear filtering to credit risk. Op Res Lett 38(6):527–532
45. Schön TB (2009) An explanation of the expectation maximization algorithm. Division of Automatic Control, Linköping University, Linköping, Sweden, Tech. Rep. LITH-ISY-R-2915
46. Dempster A, Laird N, Rubin D (1977) Maximum likelihood from incomplete data via the EM algorithm. J R Stat Soc Ser B 39(1):1–38
47. Jin JH, Shi JJ (1999) State space modeling of sheet metal assembly for dimensional control. J Manuf Sci Eng 121(4):756–763
48. Zeng L, Zhou S (2007) Inferring the interactions in complex manufacturing processes using graphical models. Technometrics 49(4):373–381
49. Saha B, Poll S, Goebel K, Christophersen J (2007) An integrated approach to battery health monitoring using bayesian regression, classification and state estimation. In: Proceedings of IEEE Autotestcon2007. IEEE
50. Saha B, Goebel K, Christophersen J (2009) Comparison of prognostic algorithms for estimating remaining useful life of batteries. Trans Inst Meas Control 31(3):293–308
51. Saha B, Goebel K (2009) Modeling Li-ion battery capacity depletion in a particle filtering framework. In: Proceedings of the annual conference of the prognostics and health management society. San Diego, CA
52. Saha B, Goebel K (2009) Prognostics methods for battery health monitoring using a Bayesian framework. IEEE Trans Instrument Meas 58(2):291–296
53. Liu J, Wang W, Ma F, Yang YB, Yang CS (2012) A data-model-fusion prognostic framework for dynamic system state forecasting. Eng Appl Artif Intell 25(4):814–823
54. Saha B, Goebel K (2007) Battery Data Set, NASA Ames Prognostics Data Repository. http://ti.arc.nasa.gov/project/prognostic-data-repository/, NASA Ames, Moffett Field, CA
55. Akaike H (1974) A new look at the statistical model identification. IEEE Trans Autom Control 19(6):716–722

Chapter 11
Prognostics for Nonlinear Degrading Systems with Three-Source Variability

11.1 Introduction

Thanks to the rapid development of information and sensing technologies, the degradation signals of a system can be obtained relatively easily using CM techniques, and the past decade has witnessed an increasingly growing research interest on the RUL estimate of systems based on the sensed degradation signals [1, 2]. However, it is well known that degradation occurs in a stochastic way for a number of engineering systems or components, such as the fatigue crack length, gyros, and battery systems [3–7]. Therefore, it is difficult to estimate the RUL of stochastic degrading systems with certainty, and attention is usually paid to the estimation of the PDF of the RUL by modeling the sensed degradation signals.

In reality, there are three main sources of variability which affect the uncertainty of the RUL, including the temporal variability, the unit-to-unit variability, and the measurement variability [8–12]. First, the temporal variability results from the inherent stochasticity of a degradation process over time, which is the reason for using a stochastic process to model the degradation. Second, the unit-to-unit variability is referred to as the heterogeneity of the systems with the same kind, which results in different degradation paths of the same kind of units. Last, but not least, the measurement variability results from imperfect measurements in practice. For example, measurement error is inevitable in the measured data, such as noise, disturbance, etc. [2]. In this case, the underlying degradation state can only be partially reflected by the observed degradation signals.

There are three main sources of variability affecting the measured degradation signals and the associated RUL estimate. Under such circumstances, it is expected that fully taking into account the temporal variability, unit-to-unit variability and measurement variability can decrease the uncertainty resulting from these sources simultaneously and improve the accuracy of the RUL estimate. In the existing literature, there have been various prognostic models to describe such variability in the degradation model and RUL estimate, such as random coefficient regression models [13, 14], Gamma processes [15–17], inverse Gaussian processes [18–20],

X.-S. Si et al., *Data-Driven Remaining Useful Life Prognosis Techniques*,
Springer Series in Reliability Engineering, DOI 10.1007/978-3-662-54030-5_11

and Wiener processes [21–23]. More detailed discussions can be found in a review [24]. In these studies, degradation models driven by Wiener processes are particularly attractive because they have favorable mathematical properties and can model nonmonotonic degradation signals. Thus, we focus mainly on the models based on Wiener processes. In the model framework of a Wiener process, several prognostic models have been developed. For example, in [25], a Wiener process with a deterministic drift parameter was used to estimate the RUL, and the drift parameter was adaptively updated based on the Kalman filtering technique. However, the authors only considered the temporal variability in the RUL estimate and omitted the unit-to-unit variability and measurement variability. Further, in [26], the authors considered the unit-to-unit variability on the RUL estimate by modeling the uncertainty of the drift parameter, but they ignored the presence of the measurement variability.

It is observed that the above-mentioned works partially considered the temporal variability, unit-to-unit variability, and measurement variability, and the research on the RUL estimate using degradation models with three-source variability was very limited. Recently, the work in [3] characterized the three-source variability simultaneously to incorporate the effect of them into the RUL estimate, and a random effect parameter was adaptively estimated by the Kalman filtering technique. However, they only adopted the BM with a linear drift to model the degradation process. This was also the case for the above-mentioned studies, and most of them considered linear degradation processes for the RUL estimate, including [3, 27, 28].

In practice, the degradation nonlinearity exists extensively in complicated degrading systems, and the linear model cannot track the dynamics of such degradation processes effectively. In the current literature, some nonlinear processes can be approximated to be nearly linear by some kind of transformations on the degradation data, such as a log-transformation [29, 30] or a timescale transformation [31, 32]. It is worth noting that these transformations were limited to cases where such transformations existed, but few nonlinear processes can be transformed in these ways. At the same time, there was an implicit assumption used in the above transformations that the random part of the transformed process was still BM, which may not always be the case. In addition, the particle filter or a Monte Carlo simulation was used to estimate the RUL of nonlinear degradation processes [33, 34]. While a simulation can only evaluate the PDF of the RUL numerically, its analytical form is preferred for health management.

To handle the above problem, the work in [4] formulated an analytical approximation of the PDF of the RUL based on a well-known time-space transformation. However, they only considered the temporal variability and unit-to-unit variability for the PDF of the RUL. There are reported works considering the measurement variability using nonlinear state-space models in [34–37]. However, the other two sources of variability were not considered simultaneously together in these works. For example, in [34], the authors applied a nonlinear state-space model and the particle filter technique to estimate the hidden fatigue crack growth process. Similar ideas can also be found in [35, 36]. Recently, the work in [37] considered the temporal variability and measurement variability to estimate the PDF of the RUL by a nonlinear state-space-based prognostic model and the extended Kalman filtering technique, but the unit-to-unit variability was omitted for the PDF of the RUL.

In this chapter, a general nonlinear degradation model is presented to characterize the three-source variability and the degradation nonlinearity simultaneously. By constructing a state-space model and applying the Kalman filtering technique, we derive the analytical form of the PDF of the RUL with three-source variability and the degradation nonlinearity approximately, which can be real-time updated with the available observations. The presented work is distinguished from existing results such as [4, 29, 31–37] in the following major aspects: (1) we explicitly consider three-source variability simultaneously in modeling nonlinear degradation progress and derive the analytical form of the PDF of the RUL in this case, which can be updated with the newly observed degradation data; (2) for the nonlinear degradation model, the random effect parameters and the hidden degradation states can be real-time estimated by the constructed state-space model and the Kalman filtering technique; and (3) using Kalman filtering, the effects of the degradation nonlinearity and the three-source variability are propagated into the PDF of the RUL. It is worth noting that the constructed state-space model is initialized through the MLE of the parameters based on heterogeneous degradation observations of multiple units. Thus, the linkage between current and past data is established. Finally, the proposed approach is demonstrated by a numerical example and a practical case study. The results verify that the proposed approach improves the modeling fitting and the accuracy of the RUL estimate.

The remainder of the chapter is organized as follows. Section 11.2 gives a description of the nonlinear degradation model with three-source variability for prognostics. In Sect. 11.3, we present the method of the RUL estimate with three-source variability. Section 11.4 provides experimental results for demonstration.

11.2 Nonlinear Prognostic Model Description

To describe the temporal variability, let $\{X(t), t \geq 0\}$ denote the stochastic process describing the nonlinear degradation progression over the operating time t. Furthermore, to characterize the temporal variability and the unit-to-unit variability, $\{X(t), t \geq 0\}$ is modeled by a diffusion process as

$$X(t) = X(0) + f(t; \theta_1)^T \theta_2 + \sigma_B B(t) \tag{11.1}$$

where the degradation process $\{X(t), t \geq 0\}$ is driven by a standard BM $\{B(t), t \geq 0\}$ with a drift term $f(t; \theta_1)^T \theta_2$. Here, $f(t; \theta_1)$ is a n-dimensional vector whose elements can be some fundamental functions θ_1 and θ_2 are parameter vectors, and particularly, there is $\theta_2 \in R^n$.

It is noted that Eq. (11.1) is a general form that can include some existing models as special cases. As an illustrative example, we consider a hybrid deteriorating model presented in [40], which was used to describe the gyros' degradation process as

$X(t) = \lambda t + \alpha \int_0^t \beta \gamma^{\beta-1} d\gamma + \sigma_B B(t)$. In this case, if we let $X(0) = 0$ $f(t; \theta_1)^T = [t \int_0^t \beta \gamma^{\beta-1} d\gamma], \theta_1 = [1 \beta]$, and $\theta_2 = [\lambda \alpha]^T$, then the model in [40] can be considered as a special case of Eq. (11.1).

Now, we summarize the main settings for Eq. (11.1). $\{B(t), t \geq 0\}$ is used to characterize the temporal variability in the degradation process; θ_2 is a random parameter vector, which is used to represent the individual variation; θ_1 and σ_B are fixed parameters representing the degradation features that are common to all systems in the population. For simplicity, we assume that θ_2 and $B(t)$ are s-independent and θ_2 follows a multivariable normal distribution, i.e., $\theta_2 \sim MVN(\mu_{\theta_2}, \Sigma_{\theta_2})$. Similar assumptions can be found in the existing degradation model literature, e.g., [3, 9, 24].

Remark 11.1 Equation (11.2) is a more general expression than the linear degradation model [38], nonlinear degradation model [39], and hybrid degradation model [40]. Particularly, when $f(t; \theta_1)^T \theta_2$ is a one-dimensional linear or nonlinear function, Eq. (1) is a linear or nonlinear model, for example, if $f(t; \theta_1)^T = t$, Eq. (11.1) becomes the conventional linear model in [41]. Furthermore, when $f(t; \theta_1)^T \theta_2$ is a two-dimensional vector containing linear and nonlinear functions, Eq. (11.1) can describe the hybrid degradation model given in [40].

In addition, to model the effect of the measurement variability, the relationship between the underlying degradation state and the observable, but uncertain, measurements at time t is described by the measurement process $\{Y(t), t \geq 0\}$ as

$$Y(t) = X(t) + \varepsilon, \tag{11.2}$$

where ε is the random measurement error, assumed to be independent and identically distributed ($i.i.d.$) with $\varepsilon \sim N(0, \sigma_\varepsilon^2)$. It is further assumed that ε, θ_2, and $B(t)$ are mutually s-independent. It is worth noting that Eq. (11.2) has been widely used in degradation model studies such as [3, 10, 22]. Of course, other observation models can also be used but will require complicated computations like the extended Kalman filter or the particle filter rather than the Kalman filter used in this chapter.

Similar to other degradation modeling works [4, 24, 42], we adopt the concept of the first hitting time (FHT) to define the lifetime. Namely, if the degradation process $\{X(t), t \geq 0\}$is equal to or beyond a preset failure threshold level, it is declared that system failure occurs. Based on the FHT concept, the lifetime T of a system can be defined as

$$T = \inf \{t : X(t) \geq w | X(0) < w\}, \tag{11.3}$$

where w is the preset failure threshold level.

Now, let us focus on the main objective of how to estimate and update the PDF of the RUL of an individual system in service with the newly obtained degradation signals. Suppose the degradation process is discretely monitored at time $0 = t_0 < t_1 < ... < t_k$, and let $Y_k = Y(t_k)$ denote the degradation measurement at time t_k. The set of the degradation measurements up to t_k is represented by $Y_{1:k} = \{Y_1, Y_2, ..., Y_k\}$, and the corresponding set of the degradation states up to t_k is represented by

$X_{1:k} = \{X_1, X_2, ..., X_k\}$, where $X_k = X(t_k)$. From Eq. (11.2), we further express the discrete measurement at t_k as $Y_k = X_k + \varepsilon_k$, where the measurement errors ε_k are assumed to be $i.i.d.$ realizations of ε

Therefore, the RUL L_k of a system at time t_k can be defined as

$$L_k = \inf\{l_k > 0 : X(l_k + t_k) \geq w\}, \tag{11.4}$$

with the conditional PDF $f_{L_k|Y_{1:k}}(l_k|\, Y_{1:k})$

In the following sections, the primary goal is to derive the conditional PDF $f_{L_k|Y_{1:k}}(l_k|\, Y_{1:k})$ of the RUL based on $Y_{1:k}$.

11.3 RUL Estimate Method with Three-Source Variability

To obtain the RUL estimate with three-source variability for model (11.1), we first summarize, by a synoptic, the main steps of the presented RUL estimate method with three-source variability as follows.

Step 1: estimate the RUL only with the temporal variability.

Step 2: estimate the RUL with the temporal variability and the unit-to-unit variability based on step 1.

Step 3: estimate the RUL with the temporal variability and the measurement variability based on step 1.

Step 4: estimate the RUL with three-source variability simultaneously based on step 3.

Step 5: estimate the unknown parameters for the degradation models based on the history data.

In the following, we elaborate the details of the presented method step by step.

11.3.1 RUL Estimate Only with the Temporal Variability

In this case, the degradation process is described in Eq. (11.1), and the degradation observations can be observed directly. Inspired by the work in [4], we use a time-space transformation on the model to obtain an analytical approximation of the distribution of the FHT in a closed form. Then, the following Lemma can be established.

Lemma 11.1 *For the degradation process $X(t)$ given by (11.2), if $f'(t;\theta_1)^T$ is a continuous function of time t with $t \in [0, \infty)$, then the PDF of the FHT of $X(t)$ crossing a constant boundary w can be approximated with an explicit form as follows:*

$$f_T(t|\theta_2) \cong \frac{w - f(t;\theta_1)^T \theta_2 + tf'(t;\theta_1)^T \theta_2}{\sigma_B \sqrt{2\pi t^3}} \times \exp\left[-\frac{(w - f(t;\theta_1)^T \theta_2)^2}{2t\sigma_B^2}\right]. \tag{11.5}$$

The proof of Lemma 11.1 and the practical meaning of Lemma 11.1 can be found in [4].

Based on the Lemma 11.1, the RUL could be estimated at the kth monitoring point t_k from the starting time as the following theorem.

Theorem 11.1 *If the unknown parameters are fixed and the current degradation state X_k can be observed, i.e., there are no random effects, the PDF of the RUL can be formulated at t_k by incorporating the real-time degradation state X_k as*

$$f_{L_k|\theta_2,X_k}(l_k|\theta_2,X_k) = \frac{w_k - f^*(l_k,\theta_1)^T\theta_2 - l_k[f^*(l_k,\theta_1)^T]'\theta_2}{\sigma_B\sqrt{2\pi l_k^3}} \times \exp\left[-\frac{(w_k - f^*(l_k,\theta_1)^T\theta_2)^2}{2l_k\sigma_B^2}\right], \tag{11.6}$$

where $f^*(l_k;\theta_1)^T = f(l_k + t_k;\theta_1)^T - f(t_k;\theta_1)^T$, $w_k = w - X_k$

Proof Once we observe $X(t_k)$at t_k, for $t \geq t_k$, the degradation process can be written as $X(t) = X(t_k) + \left[f(t;\theta_1)^T - f(t_k;\theta_1)^T\right]\theta_2 + \sigma_B B(t - t_k)$. In such a case, the residual $t - t_k$corresponds to the realization of the RUL at time t_k if t is the FHT of $\{X(t), t \geq t_k\}$. Having this in mind, we take the transformation $l_k = t - t_k$with $l_k \geq 0$ and the process$\{X(t), t \geq t_k\}$ can be transformed into

$$X(l_k + t_k) - X(t_k) = \left[f(l_k + t_k;\theta_1)^T - f(t_k;\theta_1)^T\right]\theta_2 + \sigma_B B(l_k), \text{ with } l_k \geq 0. \tag{11.7}$$

As a result, the RUL at time t_k is equal to the FHT of the process $\{Z(l_k), l_k \geq 0\}$, crossing the threshold $w_k = w - X(t_k)$, where $Z(l_k) = X(l_k + t_k) - X(t_k)$ and $Z(0) = 0$. That is to say,

$$Z(l_k) = \left[f(l_k + t_k;\theta_1)^T - f(t_k;\theta_1)^T\right]\theta_2 + \sigma_B B(l_k). \tag{11.8}$$

Based on the Lemma 11.1, we have $f^*(l_k;\theta_1)^T = f(l_k + t_k;\theta_1)^T - f(t_k;\theta_1)^T$, and we can obtain the PDF of the RUL as summarized in Eq. (11.6).

The proof is completed.

11.3.2 RUL Estimate with the Temporal Variability and the Unit-to-Unit Variability

On the basis of the temporal variability, we further consider the random effect of θ_2, which characterizes the unit-to-unit variability. As such, the PDF of the lifetime can be calculated based on the law of total probability as

$$f_T(t) = \int_{-\infty}^{+\infty} f_{T|\theta_2}(t|\theta_2)p(\theta_2)d\theta_2 = E_{\theta_2}[f_{T|\theta_2}(t|\theta_2)], \tag{11.9}$$

where $p(\theta_2)$ is the PDF of θ_2, and $E_{\theta_2}[\cdot]$ is the expectation operator with respect to θ_2

To calculate the integral in Eq. (11.9) explicitly, we derive the following lemma.

Lemma 11.2 *If $\rho \sim MVN(\mu, \Sigma)$, $w_1, w_2 \in R\gamma \in R^+$, $a, b \in R^n$, n is the number of the dimensions of ρ and I is a n-dimension identity matrix, then the following holds:*

$$E_\rho\left[(w_1 - a^T\rho)\exp\left(-\frac{(w_2 - b^T\rho)^2}{2\gamma}\right)\right] = \sqrt{\frac{\gamma^n}{|bb^T\Sigma + \gamma I|}} \times \left(w_1 - \frac{w_2 a^T\Sigma b + \gamma a^T\mu}{\gamma + b^T\Sigma b}\right) \times \exp\left[-\frac{(w_2 - b^T\mu)^2}{2(\gamma + b^T\Sigma b)}\right]. \tag{11.10}$$

Proof We first obtain

$$\begin{aligned} &E_\rho\left[(w_1 - a^T\rho)\exp\left(-\frac{(w_2 - b^T\rho)^2}{2\gamma}\right)\right] \\ &= w_1 E_\rho\left[\exp\left(-\frac{(w_2 - b^T\rho)^2}{2\gamma}\right)\right] - E_\rho\left[a^T\rho\exp\left(-\frac{(w_2 - b^T\rho)^2}{2\gamma}\right)\right] \\ &= w_1 I_1 - I_2, \end{aligned}$$

where I_1 and I_2 can be formulated separately as follows:

$$I_1 = \frac{1}{\sqrt{|\Sigma|}\sqrt{(2\pi)^n}}\exp\left[-\frac{w_2^2 + \gamma\mu^T\Sigma^{-1}\mu}{2\gamma}\right]\int_{-\infty}^{+\infty}\cdots\int_{-\infty}^{+\infty}\exp\left(-\frac{\rho^T M\rho - 2q^T\rho}{2\gamma}\right)d\rho_1\cdots d\rho_n,$$

with $M = bb^T + \gamma\Sigma^{-1}q^T = w_2 b^T + \gamma\mu^T\Sigma^{-1}$.

Using the integral formulation

$$\int_{-\infty}^{+\infty}\exp\left(-\frac{\omega z^2 - 2\psi z}{\varphi}\right)dz = \sqrt{\frac{\varphi\pi}{\omega}}\exp\left[\frac{\psi^2}{\omega\varphi}\right].$$

We can compute I_1 as

$$I_1 = \sqrt{\frac{\gamma^n}{|\Sigma||M|}}\exp\left[-\frac{w_2^2 + \gamma\mu^T\Sigma^{-1}\mu - q^T M^{-1}q}{2\gamma}\right].$$

In a similar way, I_2 can be obtained as

$$I_2 = \frac{1}{\sqrt{|\Sigma|}\sqrt{(2\pi)^n}} \exp\left[-\frac{w_2^2 + \gamma\mu^T\Sigma^{-1}\mu}{2\gamma}\right] \int_{-\infty}^{+\infty} \cdots \int_{-\infty}^{+\infty} a^T\rho \exp\left(-\frac{\rho^T M\rho - 2q^T\rho}{2\gamma}\right) d\rho_1 \cdots d\rho_n.$$

Using the integral formulation

$$\int_{-\infty}^{+\infty} z \exp\left(-\frac{\omega z^2 - 2\psi z}{\varphi}\right) dz = \frac{\psi}{\omega}\sqrt{\frac{\varphi\pi}{\omega}} \exp\left[\frac{\psi^2}{\omega\varphi}\right].$$

We can compute I_2 as

$$I_2 = a^T M^{-1} q \sqrt{\frac{\gamma^n}{|\Sigma||M|}} \exp\left[-\frac{w_2^2 + \gamma\mu^T\Sigma^{-1}\mu - q^T M^{-1} q}{2\gamma}\right].$$

Thus, the formulation of the expectation can be obtained as

$$E_\rho\left[(w_1 - a^T\rho)\exp\left(-\frac{(w_2 - b^T\rho)^2}{2\gamma}\right)\right] = \sqrt{\frac{\gamma^n}{|\Sigma||M|}} \left(w_1 - a^T M^{-1} q\right) \exp\left[-\frac{w_2^2 + \gamma\mu^T\Sigma^{-1}\mu - q^T M^{-1} q}{2\gamma}\right].$$

Furthermore, simplifying the above formulation by matrix operation, we have

$$E_\rho\left[(w_1 - a^T\rho)\exp\left(-\frac{(w_2 - b^T\rho)^2}{2\gamma}\right)\right] = \sqrt{\frac{\gamma^n}{|bb^T\Sigma + \gamma I|}} \times \left(w_1 - a^T M^{-1} q\right) \times \exp\left[\frac{w_2 - b^T\mu^2}{2(\gamma + b^T\Sigma b)}\right].$$

Then, we obtain

$$E_\rho\left[(w_1 - a^T\rho)\exp\left(-\frac{(w_2 - b^T\rho)^2}{2\gamma}\right)\right] = \sqrt{\frac{\gamma^n}{|bb^T\Sigma + \gamma I|}} \times \left(w_1 - \frac{w_2 a^T\Sigma b + \gamma a^T\mu}{\gamma + b^T\Sigma b}\right) \times \exp\left[\frac{w_2 - b^T\mu^2}{2(\gamma + b^T\Sigma b)}\right].$$

This completes the proof of Lemma 11.2.

Similar to Theorem 11.1, the RUL estimate considering the temporal variability and the unit-to-unit variability is summarized as follows.

Theorem 11.2 *For the degradation process in Eq. (11.1) and the definition of the RUL in Eq. (11.4), given the current degradation state* X_k *and* $\theta_2 \sim MVN(\mu_{\theta_2}, \Sigma_{\theta_2})$, *the RUL estimate at time t_k can be formulated as*

$$f_{l_k|X_k}(l_k|X_k) \cong \frac{1}{\sigma_B\sqrt{2\pi l_k^3}}\sqrt{\frac{\gamma^n}{|bb^T\Sigma_{\theta_2}+\gamma I_n|}}\left[w_k - \frac{w_k a^T\Sigma_{\theta_2}b+\gamma a^T\mu_{\theta_2}}{\gamma+b^T\Sigma_{\theta_2}b}\right] \times \exp\left[-\frac{(w_k-b^T\mu_{\theta_2})^2}{2(\gamma+b^T\Sigma_{\theta_2}b)}\right], \tag{11.11}$$

where $a^T = f^*(l_k,\theta_1)^T - t[f^*(l_k,\boldsymbol{\theta}_1)^T]'$, $b^T = f^*(l_k,\theta_1)^T$, $w_k = w - X_k$ *and* $\gamma = \sigma_B^2 l_k$.

Proof Using the law of total probability, Theorem 11.1 and Lemma 11.2, we obtain

$$\begin{aligned}
f_{l_k|X_k}(l_k|X_k) &= \int_{-\infty}^{+\infty} f_{L_k|\theta_2,X_k}(l_k|\,\theta_2, X_k)p(\theta_2)d\theta_2 \\
&= E_{\theta_2|X_k}\left\{f_{L_k|\theta_2,X_k}(l_k|\,\theta_2, X_k)\right\} \\
&\cong \frac{E_{\theta_2|X_k}\left\{\left[w_k - f^*(l_k,\theta_1)^T\theta_2 - l_k[f^*(l_k,\theta_1)^T]'\theta_2\right]\exp\left[-\frac{(w_k-f^*(l_k,\theta_1)^T\theta_2)^2}{2l_k\sigma_B^2}\right]\right\}}{\sigma_B\sqrt{2\pi l_k^3}} \\
&= \frac{1}{\sigma_B\sqrt{2\pi l_k^3}}\sqrt{\frac{\gamma^n}{|bb^T\Sigma_{\theta_2}+\gamma I_n|}}\left(w_k - \frac{w_k a^T\Sigma_{\theta_2}b+\gamma a^T\mu_{\boldsymbol{\theta}_2}}{\gamma+b^T\Sigma_{\theta_2}b}\right)\exp\left[-\frac{(w_k-b^T\mu_{\theta_2})^2}{2(\gamma+b^T\Sigma_{\theta_2}b)}\right],
\end{aligned} \tag{11.12}$$

where $a^T = f^*(l_k,\theta_1)^T - t[f^*(l_k,\boldsymbol{\theta}_1)^T]'$, $b^T = f^*(l_k,\theta_1)^T$, $w_k = w - X_k$ and $\gamma = \sigma_B^2 l_k$.

This completes the proof of Theorem 11.2.

The above results are derived under the assumption that the current degradation state can be observed directly and exactly. However, this directly observed case is rather difficult in reality for the appearance of the measurement variability. Therefore, to reduce the effect of the measurement variability, the unobservable degradation state should be estimated from the degradation measurements. The case is considered in the following part.

11.3.3 RUL Estimate with the Temporal Variability and the Measurement Variability

Because only uncertain measurements $Y_{1:k}$ up to the current time t_k are available and the degradation state X_k cannot be directly used, we have to estimate the distribution of X_k at time t_k to account for the impact of the measurement variability on the RUL estimate. To identify the hidden degradation state in the dynamic system, we convert the state and measurement equations into discrete time equations to facilitate real-time state estimation. Specifically, once the new observations are available at the discrete time point t_k, $k = 1, 2, \ldots$, we have

$$\begin{cases} X_k = X_{k-1} + f(t_k; \theta_1)^T \theta_2 - f(t_{k-1}; \theta_1)^T \theta_2 + \nu_k \\ Y_k = X_k + \varepsilon_k \end{cases} \tag{11.13}$$

where $\nu_k = \sigma_B \left[B(t_k) - B(t_{k-1}) \right]$ and ε_k is the realization of ε at t_k. From the model settings in Sect. 11.2, we know that $\{\nu_k\}_{k \geq 1}$ and $\{\varepsilon_k\}_{k \geq 1}$ are $i.i.d.$ noise sequences, respectively. We further have $\nu_k \sim N\left(0, \sigma_B^2 (t_k - t_{k-1})\right)$ and $\varepsilon_k \sim N(0, \sigma_\varepsilon^2)$

According to Eq. (11.13), we utilize the Kalman filter to estimate the hidden degradation state. To do so, define $\hat{X}_{k|k} = E(X_k | Y_{1:k}, \theta_2)$ and $P_{k|k} = var(X_k | Y_{1:k}, \theta_2)$ as the conditional expectation and variance of X_k, respectively. Further, we define $\hat{X}_{k|k-1} = E(X_k | Y_{1:k-1}, \theta_2)$ and $P_{k|k-1} = var(X_k | Y_{1:k-1}, \theta_2)$ as the one-step-ahead predicted expectation and variance, respectively. Therefore, at time t_k, according to Eq. (11.13), the Kalman filter for the degradation state estimation can be summarized as follows:

State estimation:

$$\begin{aligned} \hat{X}_{k|k-1} &= \hat{X}_{k-1|k-1} + f(t_k; \theta_1)^T \theta_2 - f(t_{k-1}; \theta_1)^T \theta_2, \\ \hat{X}_{k|k} &= \hat{X}_{k|k-1} + K(k)(Y_k - \hat{X}_{k|k-1}), \\ K(k) &= P_{k|k-1}(P_{k|k-1} + \gamma^2)^{-1}, \\ P_{k|k-1} &= P_{k-1|k-1} + \sigma_B^2 (t_k - t_{k-1}). \end{aligned}$$

Updating variance:

$$P_{k|k} = (1 - K(k)) P_{k|k-1},$$

where the initial values are set as $\hat{X}_{0|0} = 0$ and $P_{0|0} = 0$ based on the model setting $X_0 = 0$.

Under the Kalman filtering algorithm, the posterior estimate of the X_k conditional on the measurement sequence $Y_{1:k}$ up to t_k is Gaussian and analytically tractable, i.e., $X_k | \theta_2, Y_{1:k} \sim N(\hat{X}_{k|k}, P_{k|k})$. Therefore, considering the estimation uncertainty, the RUL estimate is calculated by

$$\begin{aligned} f_{L_k|\theta_2, Y_{1:k}}(l_k | \theta_2, Y_{1:k}) &= \int_{-\infty}^{+\infty} f_{L_k|\theta_2, X_k, Y_{1:k}}(l_k | \theta_2, X_k, Y_{1:k}) p(X_k | \theta_2, Y_{1:k}) dX_k \\ &= E_{X_k|\theta_2, Y_{1:k}} \left[f_{L_k|\theta_2, X_k, Y_{1:k}}(l_k | \theta_2, X_k, Y_{1:k}) \right] \\ &= E_{X_k|\theta_2, Y_{1:k}} \left[f_{L_k|\theta_2, X_k}(l_k | \theta_2, X_k) \right], \end{aligned}$$

where $p(X_k | \theta_2, Y_{1:k})$ is the conditional PDF of $X_k | \theta_2, Y_{1:k}$, with mean $\hat{X}_{k|k}$ and variance $P_{k|k}$.

Based on Lemma 11.2, the RUL estimate with the temporal variability and measurement variability can be derived as the following theorem.

Theorem 11.3 *For the diffusion process in Eqs. (11.1) and (11.4), given θ_2 and uncertain measurements $Y_{1:k}$ up to the current time t_k, the following result for the RUL estimate at time t_k holds:*

$$f_{L_k|\theta_2,Y_{1:k}}(l_k|\theta_2,Y_{1:k}) \cong \frac{1}{\sqrt{2\pi l_k^2\gamma^*}}\left(w^* - l_k[f^*(l_k,\theta_1)^T]'\theta_2 - \frac{P_{k|k}w^* + \sigma_B^2 l_k\hat{X}_{k|k}}{\gamma^*}\right) \times \exp\left(-\frac{\left(\hat{X}_{k|k} - w^*\right)^2}{2\gamma^*}\right), \quad (11.14)$$

where $\gamma^* = P_{k|k} + \sigma_B^2 l_k$, $w^* = w - f^*(l_k,\theta_1)^T\theta_2$, *and* $f^*(l_k;\theta_1)^T = f(l_k + t_k;\theta_1)^T - f(t_k;\theta_1)^T$

Proof Using Theorem 11.1 and Lemma 11.2, we obtain

$$\begin{aligned}
f_{L_k|\theta_2,Y_{1:k}}(l_k|\theta_2,Y_{1:k}) &= E_{X_k|\theta_2,Y_{1:k}}\left[f_{L_k|\theta_2,X_k,Y_{1:k}}(l_k|\theta_2,X_k,Y_{1:k})\right] \\
&= E_{X_k|\theta_2,Y_{1:k}}\left[f_{L_k|\theta_2,X_k}(l_k|\theta_2,X_k)\right] \\
&\cong \frac{E_{X_k|\theta_2,Y_{1:k}}\left\{\left(w_k - f^*(l_k,\theta_1)^T\theta_2 - l_k[f^*(l_k,\theta_1)^T]'\theta_2\right)\times\exp\left[-\frac{(w_k - f^*(l_k,\theta_1)^T\theta_2)^2}{2l_k\sigma_B^2}\right]\right\}}{\sigma_B\sqrt{2\pi l_k^3}} \\
&= \frac{E_{X_k|\theta_2,Y_{1:k}}\left\{\left(w - f^*(l_k,\theta_1)^T\theta_2 - l_k[f^*(l_k,\theta_1)^T]'\theta_2 - X_k\right)\times\exp\left[-\frac{(w - f^*(l_k,\theta_1)^T\theta_2 - X_k)^2}{2l_k\sigma_B^2}\right]\right\}}{\sigma_B\sqrt{2\pi l_k^3}} \\
&= \frac{\left(w - f^*(l_k,\theta_1)^T\theta_2 - l_k[f^*(l_k,\theta_1)^T]'\theta_2 - \frac{P_{k|k}(w - f^*(l_k,\theta_1)^T\theta_2) + \sigma_B^2 l_k\hat{X}_{k|k}}{P_{k|k} + \sigma_B^2 l_k}\right)}{\sqrt{2\pi l_k^2(P_{k|k} + \sigma_B^2 l_k)}} \\
&\quad\times\exp\left[-\frac{\left(w - f^*(l_k,\theta_1)^T\theta_2 - \hat{X}_{k|k}\right)^2}{2(P_{k|k} + \sigma_B^2 l_k)}\right] \\
&= \frac{1}{\sqrt{2\pi l_k^2\gamma^*}}\left(w^* - l_k[f^*(l_k,\theta_1)^T]'\theta_2 - \frac{P_{k|k}w^* + \sigma_B^2 l_k\hat{X}_{k|k}}{\gamma^*}\right)\times\exp\left(-\frac{\left(w^* - \hat{X}_{k|k}\right)^2}{2\gamma^*}\right),
\end{aligned}$$

where $\gamma^* = P_{k|k} + \sigma_B^2 l_k$, $w^* = w - f^*(l_k,\theta_1)^T\theta_2$, $f^*(l_k;\theta_1)^T = f(l_k + t_k;\theta_1)^T - f(t_k;\theta_1)^T$

This completes the proof of Theorem 11.3.

Theorem 11.3 incorporates the estimation uncertainty to the PDF of the RUL because the measurement variability is considered in the derived the PDF of the

RUL. However, the parameter θ_2 is assumed to be given, and the random effect from θ_2 is not considered in this case. In addition, the parameter θ_2 is not real-time updated as the newly obtained measurements $Y_{1:k}$. In the following part, the random effect and the updating mechanism for $\theta_2$are considered for the RUL estimate.

11.3.4 RUL Estimate with Three-Source Variability

To estimate the PDF of the RUL with three-source variability based on the Step 3, we consider an updating procedure for the parameter $\theta_2$by making $\theta_{2,k} = \theta_{2,k-1}$, where $\theta_{2,0} \sim MVN(\mu_{\theta_2}, \Sigma_{\theta_2})$ is the initial distribution. The posterior distribution of θ_2 can be calculated based on the measurements up to t_k. Furthermore, based on Eq. (11.13), the degradation equation with three-source variability can be reconstructed as the following state-space model:

$$\begin{cases} X_k = X_{k-1} + f(t_k;\theta_1)^T \theta_{2,k-1} - f(t_{k-1};\theta_1)^T \theta_{2,k-1} + \upsilon_k \\ \theta_{2,k} = \theta_{2,k-1} \\ Y_k = X_k + \varepsilon_k \end{cases}, \tag{11.15}$$

where $\{\upsilon_k\}_{k\geq 1}$ and $\{\varepsilon_k\}_{k\geq 1}$ are *i.i.d.* noise sequences, i.e., $\upsilon_k \sim N\left(0, \sigma_B^2(t_k - t_{k-1})\right)$ and $\varepsilon_k \sim N(0, \sigma_\varepsilon^2)$, respectively.

In Eq. (11.15), the hidden degradation state and random parameter θ_2 need to be estimated from the uncertain measurements $Y_{1:k}$ up to the current time t_k. We use the Kalman filter to estimate the hidden degradation state and random parameter and further formulate Eq. (11.15) as

$$\begin{cases} Z_k = A_k Z_{k-1} + \eta_k \\ Y_k = C Z_k + \varepsilon_k \end{cases}, \tag{11.16}$$

where $Z_k \in R^{(n+1)\times 1}$, $\eta_k \in R^{(n+1)\times 1}$, $A_k \in R^{(n+1)\times(n+1)}$, $C \in R^{1\times(n+1)}$, $\eta_k \sim MVN(0, Q_k)$, and $Q_k \in R^{(n+1)\times(n+1)}$, with

$$Z_k = \begin{bmatrix} X_k \\ \theta_{2,k} \end{bmatrix}, \eta_k = \begin{bmatrix} \upsilon_k \\ 0 \end{bmatrix}, A_k = \begin{bmatrix} 1 & f(t_k;\theta_1)^T - f(t_{k-1};\theta_1)^T \\ 0 & 1 \end{bmatrix},$$
$$C = [1, 0], Q_k = \begin{bmatrix} \sigma_B^2(t_k - t_{k-1}) & 0 \\ 0 & 0 \end{bmatrix}$$

respectively.

Similarly, the expectation and variance of Z_k are defined as

$$\hat{Z}_{k|k} = \begin{bmatrix} \hat{X}_{k|k} \\ \hat{\theta}_{2,k|k} \end{bmatrix} = E(Z_k | Y_{1:k}), \tag{11.17}$$

$$P_{k|k}=\begin{bmatrix}\kappa_{X,k} & \kappa_{c,k}^{T}\\ \kappa_{c,k} & \kappa_{\theta_2,k}\end{bmatrix}=cov(Z_k|\,Y_{1:k}),\tag{11.18}$$

where

$\hat{X}_{k|k}=E(X_k|\,Y_{1:k}),\hat{\theta}_{2,k|k}=E(\theta_{2,k}\big|\,Y_{1:k}),\kappa_{X,k}=var(X_k|\,Y_{1:k}),\kappa_{\theta_2,k}=var(\theta_{2,k}\big|\,Y_{1:k}),$
$\kappa_{c,k}=\left[cov(X_k,\theta_{2,1,k}\big|\,Y_{1:k}),cov(X_k,\theta_{2,2,k}\big|\,Y_{1:k})...cov(X_k,\theta_{2,n,k}\big|\,Y_{1:k})\right]^T.$

As such, we further define the one-step-ahead predicted expectation and variance, respectively, as

$$\hat{Z}_{k|k-1}=\begin{bmatrix}\hat{X}_{k|k-1}\\ \hat{\theta}_{2,k|k-1}\end{bmatrix}=E(Z_k|\,Y_{1:k-1})\tag{11.19}$$

$$P_{k|k-1}=\begin{bmatrix}\kappa_{X,k|k-1} & \kappa_{c,k|k-1}^{T}\\ \kappa_{c,k|k-1} & \kappa_{\theta_2,k|k-1}\end{bmatrix}=cov(Z_k|\,Y_{1:k-1}).\tag{11.20}$$

Under the above settings and definitions, once the measurement is available at t_k, the hidden degradation state and random parameter, i.e., Z_k, can be estimated by the Kalman filtering algorithm as follows:

State estimation:

$$\begin{aligned}
\hat{Z}_{k|k-1}&=A_k\hat{Z}_{k-1|k-1},\\
\hat{Z}_{k|k}&=\hat{Z}_{k|k-1}+K(k)(Y_k-C\hat{Z}_{k|k-1}),\\
K(k)&=P_{k|k-1}C^T[CP_{k|k-1}C^T+\sigma_\varepsilon^2]^{-1},\\
P_{k|k-1}&=A_kP_{k-1|k-1}A_k^T+Q_k.
\end{aligned}$$

Updating variance:

$$P_{k|k}=P_{k|k-1}-K(k)CP_{k|k-1},$$

where the initial values of the states are specified as

$$\hat{Z}_{0|0}=\begin{bmatrix}0\\ \mu_{\theta_2}\end{bmatrix},\quad P_{0|0}=\begin{bmatrix}0 & 0\\ 0 & \Sigma_{\theta_2}\end{bmatrix},$$

where μ_{θ_2} and Σ_{θ_2} can be obtained by the maximum likelihood estimation discussed in Sect. 11.4.

According to the Gaussian nature of the Kalman filter, the PDF of Z_k in Eq. (11.16), conditional on $Y_{1:k}$ is Gaussian distributed with $Z_k\sim N(\hat{Z}_{k|k},P_{k|k})$. Based on the properties of the multivariate Gaussian distribution, we have

$$\theta_{2,k}\big|\,Y_{1:k}\sim MVN(\hat{\theta}_{2,k|k},\kappa_{\theta_2,k}),\tag{11.21}$$

$$X_k | Y_{1:k} \sim N(\hat{X}_{k|k}, \kappa_{X,k}), \tag{11.22}$$

$$X_k | \theta_{2,k}, Y_{1:k} \sim N\left(\mu_{X_k|\theta_{2,k}}, \sigma^2_{X_k|\theta_{2,k}}\right) \tag{11.23}$$

with

$$\mu_{X_k|\theta_{2,k}} = \hat{X}_{k|k} + \kappa^T_{c,k}\kappa^{-1}_{\theta_2,k}(\theta_{2,k} - \hat{\theta}_{2,k|k}), \tag{11.24}$$

$$\sigma^2_{X_k|\theta_{2,k}} = \kappa_{X,k} - \kappa^T_{c,k}\kappa^{-1}_{\theta_2,k}\kappa_{c,k}. \tag{11.25}$$

Now, we can derive $f_{L_k|Y_{1:k}}(l_k | Y_{1:k})$ at t_k with three-source variability. As such, based on the law of total probability, we can obtain

$$\begin{aligned} f_{L_k|Y_{1:k}}(l_k | Y_{1:k}) &= \int_{-\infty}^{+\infty} f_{L_k|Z_k,Y_{1:k}}(l_k | \mathbf{Z}_k, Y_{1:k}) p(\mathbf{Z}_k | Y_{1:k}) d\mathbf{Z}_k \\ &= E_{\theta_{2,k}|Y_{1:k}}\left[E_{X_k|\theta_{2,k},Y_{1:k}}[f_{L_k|\theta_{2,k},X_k,Y_{1:k}}(l_k | \theta_{2,k}, X_k, Y_{1:k})]\right]. \end{aligned} \tag{11.26}$$

Based on Lemmas 11.1 and 11.2 and Theorem 11.3, we can estimate the PDF of the RUL using the degradation model with three-source variability.

Theorem 11.4 *For the diffusion process Eqs. (11.1) and (11.4), given the uncertain measurements $Y_{1:k}$ up to current time t_k, the following result for the RUL L_k at time t_k holds:*

$$f_{L_k|Y_{1:k}}(l_k | Y_{1:k}) \cong \sqrt{\frac{C_k^{n-3}}{2\pi |B_k B_k^T \kappa_{\theta_2,k} + C_k I_n|}} \left(\omega_{1,k} - \frac{\omega_{2,k} A_k \kappa_{\theta_2,k} B_k + C_k A_k \hat{\theta}_{2,k}}{C_k + B_k^T \kappa_{\theta_2,k} B_k}\right) \tag{11.27}$$

$$\times \exp\left(-\frac{\left((w - \hat{X}_{k|k} - f^*(l_k, \theta_1)^T \hat{\theta}_{2,k})\right)^2}{2(C_k + B_k^T \kappa_{\theta_2,k} B_k)}\right), \tag{11.28}$$

where $f^*(l_k; \theta_1)^T$, $w_{1,k}$, $w_{2,k}$ A_k, B_k, and C_k are specified as follows:

$$\begin{aligned} & f^*(l_k; \theta_1)^T = f(l_k + t_k; \theta_1)^T - f(t_k; \theta_1)^T \\ & w_{1,k} = (w - \hat{X}_{k|k} + \kappa^T_{c,k}\kappa^{-1}_{\theta_2,k}\hat{\theta}_{2,k|k})\sigma^2_B \\ & w_{2,k} = w - \hat{X}_{k|k} + \kappa^T_{c,k}\kappa^{-1}_{\theta_2,k}\hat{\theta}_{2,k|k} \\ & A_k = \kappa^T_{c,k}\kappa^{-1}_{\theta_2,k}\sigma^2_B + f^*(l_k, \theta_1)^T\sigma^2_B + l_k[f^*(l_k, \theta_1)^T]'\sigma^2_B - [f^*(l_k, \theta_1)^T]'\sigma^2_{X_k|\theta_2,k} \\ & B_k = \left(f^*(l_k, \theta_1)^T + \kappa^T_{c,k}\kappa^{-1}_{\theta_2,k}\right)^T, C_k = \sigma^2_{X_k|\theta_2,k} + \sigma^2_B l_k. \end{aligned}$$

Proof To derive the result for $f_{L_k|Y_{1:k}}(l_k | Y_{1:k})$, we first have the following result:

$$f_{L_k|\theta_{2,k},X_k,Y_{1:k}}(l_k|\,\theta_{2,k},X_k,Y_{1:k}) = f_{L_k|\theta_{2,k},X_k}(l_k|\,\theta_{2,k},X_k)$$
$$= \frac{w_k - f^*(l_k,\theta_1)^T\theta_{2,k} - l_k[f^*(l_k,\theta_1)^T]'\theta_{2,k}}{\sigma_B\sqrt{2\pi l_k^3}}\exp\left[-\frac{(w_k - f^*(l_k,\theta_1)^T\theta_{2,k})^2}{2l_k\sigma_B^2}\right]. \tag{11.29}$$

In this case, because $X_k|\,\theta_{2,k}, Y_{1:k}\sim N\left(\mu_{X_k|\theta_{2,k}}, \sigma^2_{X_k|\theta_{2,k}}\right)$, we obtain the following from Theorem 11.3:

$$E_{X_k|\theta_{2,k},Y_{1:k}}[f_{L_k|\theta_{2,k},X_k,Y_{1:k}}(l_k|\,\theta_{2,k},X_k,Y_{1:k})]$$
$$\cong \frac{1}{\sqrt{2\pi l_k^2 C_k}}\left(w^* - l_k[f^*(l_k,\theta_1)^T]'\theta_{2,k} - \frac{\sigma^2_{X_k|\theta_{2,k}}w^* + \sigma_B^2 l_k\mu_{X_k|\theta_{2,k}}}{C_k}\right)\exp\left(-\frac{\left(\mu_{X_k|\theta_{2,k}} - w^*\right)^2}{2C_k}\right), \tag{11.30}$$

where $\mu_{X_k|\theta_{2,k}}$ and $\sigma^2_{X_k|\theta_{2,k}}$ are specified in Eqs. (11.24) and (11.25), and $\mu_{X_k|\theta_{2,k}}$ is a function of $\theta_{2,k}$.

From Eq. (11.26) and $\theta_{2,k}\big|\,Y_{1:k}\sim MVN(\hat{\theta}_{2,k|k}, \kappa_{\theta_{2,k}})$, we know

$$f_{L_k|Y_{1:k}}(l_k|\,Y_{1:k}) = E_{\theta_{2,k}|Y_{1:k}}\left[E_{X_k|\theta_{2,k},Y_{1:k}}[f_{L_k|\theta_{2,k},X_k,Y_{1:k}}(l_k|\,\theta_{2,k},X_k,Y_{1:k})]\right]$$
$$\cong E_{\theta_{2,k}|Y_{1:k}}\left[\frac{1}{\sqrt{2\pi l_k^2\gamma^*}}\left(w^* - l_k[f^*(l_k,\theta_1)^T]'\theta_{2,k} - \frac{\sigma^2_{X_k|\theta_{2,k}}w^* + \sigma_B^2 l_k\mu_{X_k|\theta_{2,k}}}{C_k}\right)\right.$$
$$\left.\times\exp\left(-\frac{\left(\mu_{X_k|\theta_{2,k}} - w^*\right)^2}{2C_k}\right)\right]$$
$$= E_{\theta_{2,k}|Y_{1:k}}\left[\frac{1}{\sqrt{2\pi l_k^2 C_k}}\left(w^* - l_k[f^*(l_k,\theta_1)^T]'\theta_{2,k} - \frac{\sigma^2_{X_k|\theta_{2,k}}w^* + \sigma_B^2 l_k(\hat{X}_{k|k} + \kappa_{c,k}^T\kappa_{\theta_{2,k}}^{-1}(\theta_{2,k} - \hat{\theta}_{2,k|k}))}{C_k}\right)\right.,$$
$$\left.\times\exp\left(-\frac{\left((\hat{X}_{k|k} + \kappa_{c,k}^T\kappa_{\theta_{2,k}}^{-1}(\theta_{2,k} - \hat{\theta}_{2,k|k})) - w^*\right)^2}{2C_k}\right)\right]$$
$$= \sqrt{\frac{C_k^{n-3}}{2\pi|B_kB_k^T\kappa_{\theta_{2,k}}^{+}C_kI_n|}}\left(\omega_{1,k} - \frac{\omega_{2,k}A_k\kappa_{\theta_{2,k}}B_k + C_kA_k\hat{\theta}_{2,k}}{C_k + B_k^T\kappa_{\theta_{2,k}}B_k}\right)\times\exp\left(-\frac{\left((w - \hat{X}_{k|k} - f^*(l_k,\theta_1)^T\hat{\theta}_{2,k})\right)^2}{2(C_k + B_k^T\kappa_{\theta_{2,k}}B_k)}\right)$$

where the last "=" is implied by Lemma 11.2, and $f^*(l_k;\theta_1)^T$, w^*, $w_{1,k}$, $w_{2,k}$ A_k, B_k, and C_k are specified as follows:

$$f^*(l_k;\theta_1)^T = f\,(l_k + t_k;\theta_1)^T - f\,(t_k;\theta_1)^T$$
$$w^* = w - f^*(l_k,\theta_1)^T\theta_{2,k}$$
$$w_{1,k} = (w - \hat{X}_{k|k} + \kappa_{c,k}^T\kappa_{\theta_{2,k}}^{-1}\hat{\theta}_{2,k|k})\sigma_B^2$$
$$w_{2,k} = w - \hat{X}_{k|k} + \kappa_{c,k}^T\kappa_{\theta_{2,k}}^{-1}\hat{\theta}_{2,k|k}$$
$$A_k = \kappa_{c,k}^T\kappa_{\theta_{2,k}}^{-1}\sigma_B^2 + f^*(l_k,\theta_1)^T\sigma_B^2 + l_k[f^*(l_k,\theta_1)^T]'\sigma_B^2 - [f^*(l_k,\theta_1)^T]'\sigma^2_{X_k|\theta_{2,k}}$$
$$B_k = \left(f^*(l_k,\theta_1)^T + \kappa_{c,k}^T\kappa_{\theta_{2,k}}^{-1}\right)^T,\ C_k = \sigma^2_{X_k|\theta_{2,k}} + \sigma_B^2 l_k.$$

This completes the proof of Theorem 11.4.

In the above case, as a new degradation measurement is observed, we can calculate the estimation of $Z_k$conditional on $Y_{1:k}$ by Eq. (11.16) with $Z_k \sim N(\hat{Z}_{k|k}, P_{k|k})$. Therefore, we can estimate the RUL of this dynamic system adaptively by utilizing Theorem 11.4. For the underlying stochastic degradation process, different from Theorems 11.2 and 11.3, Theorem 11.4 incorporates the uncertainties in estimating the degradation state and random effect part into the PDF of the RUL, $f_{L_k|Y_{1:k}}(l_k|\, Y_{1:k})$. At the same time, once the new observation is available, the parameters $\hat{\theta}_{2,k|k}$ and $\kappa_{\theta_2,k}$in $f_{L_k|Y_{1:k}}(l_k|\, Y_{1:k})$ can be real-time updated by the Kalman filter.

For the relationship between Theorem 11.4 and the previous results about the RUL estimate, we have the following remark.

Remark 11.2 It is not difficult to verify from the derivation process of Theorem 11.4 that the results in Theorem 11.4 can reduce to the RUL estimate results in [3] by simplifying the drift part as $f^*(l_k; \theta_1)^T = l_k$. In addition, by selecting the parameters $\sigma_\varepsilon^2$and Σ_{θ_2}, the results in Theorem 11.4 can reduce to the RUL estimate results, which only consider one- or two-source variability.

11.3.5 Parameter Estimation

In addition to the real-time estimated parameters in Eq. (11.13), there are several unknown initial parameters, including μ_{θ_2}, Σ_{θ_2}, θ_1, σ_B^2, and σ_ε^2, which need to be estimated based on the historical data. These parameters can be estimated using the MLE approach, and the detailed method can be found in [3, 4].

11.4 Experimental Studies

The purpose of this section is to demonstrate the presented modeling framework. To compare the fitness of the models, both the AIC [43] and TMSE [44] are employed. The AIC is used to balance the log-likelihood with the number of the estimated parameters to avoid the problem of over-parameterization. The AIC is calculated by

$$AIC = 2(p - \max \ell), \tag{11.31}$$

where p is the number of estimated model parameters and $\max \ell$ is the maximized likelihood.

The other useful measure of goodness of fit, MSE, can assess the fit to the data directly, defined as [44]

$$MSE_k = E\left[(L_k - \tilde{L}_k)^2\right], \tag{11.32}$$

where $\tilde{L}_k$ denotes the actual RUL obtained at t_k, and the expectation is calculated based on the PDF of the RUL through the available data. The TMSE is the sum of the MSE at each CM point over the whole life cycle. In other words, if there are m observations, $TMSE = \sum_{k=1}^{m} MSE_k$.

Based on the above criteria, we can choose the model with the smallest AIC and MSE values, which correspond to the best fitting accuracy.

To show the superiority considering the three-source variability, we consider the following three competing cases for comparative studies.

(1) Case 1: $\Sigma_{\theta_2} = 0 \quad \sigma_\varepsilon = 0$. In this case, the random effect and the measurement uncertainty are ignored, and the result in Theorem 1 with only the temporal variability is used for the RUL estimate.
(2) Case 2: $\sigma_\varepsilon = 0$. In this case, only the measurement uncertainty is ignored, and the result in Theorem 2 with the temporal variability and the random effect is used for the RUL estimate.
(3) Case 3: $\Sigma_{\theta_2} = 0$ In this case, only the random effect is ignored, and the result in Theorem 3 with the temporal variability and the measurement uncertainty is used for the RUL estimate.

It is noted that Case 1 corresponds to the case only taking one-source variability. Case 2 and Case 3 correspond to the cases only considering two-source variability. By contrast, we term the model with three-source case as Case 4. In the following part, we compare the model fitting for the above four cases through a numerical simulation study and a practical case for fatigue crack data [4].

11.4.1 Simulation Study

Here, we first provide a numerical simulation. The purpose of this numerical example is to illustrate the effectiveness of the presented method by comparing it with the results of Cases 1–3. Another purpose is to verify the usefulness of the proposed parameter estimation algorithm.

According to Eqs. (11.1) and (11.2), the degradation data of N units are generated with the following state-space model:

$$\begin{cases} X_k = X_{k-1} + a(k\Delta t)^b - a[(k-1)\Delta t]^b + v_k \\ Y_k = X_k + \varepsilon_k \end{cases}, \tag{11.33}$$

where $a \sim N(\mu_a, \sigma_a^2)$, $v_k \sim N\left(0, \sigma_B^2(t_k - t_{k-1})\right)$, and $\varepsilon_k \sim N(0, \sigma_\varepsilon^2)$. The parameters are set as $\mu_a = 0.2$, $\sigma_a = 0.03$, $b = 2$, $\sigma_B = 0.05$, and $\sigma_\varepsilon = 0.3$, and the sampling interval is $\Delta t = 0.1$. Each unit is sampled m times, and the failure threshold is specified as $w = 13$. Here, we simulate data with $N = 101$ and $m = 100$. The former 100 units are used to identify the unknown parameters, and the last unit is used as a testing sample to compare the results of the RUL estimate for different cases. In the following, we compare the four cases of model fitting and the RUL estimate.

Table 11.1 Comparison of the four degradation models with the simulation data

	μ_a	σ_a	b	σ_B	σ_ε	log-LF	AIC
Case 1	0.2038	–	2.0030	1.3471	–	−5648.9928	11303.9856
Case 2	0.1985	0.0181	2.0138	1.3474	–	−5647.0081	11302.0162
Case 3	0.2054	–	1.9999	0.4456	0.27989	−3922.7418	7853.4836
Case 4	0.2104	0.0312	1.9901	0.3056	0.29291	−3646.8234	7303.6448

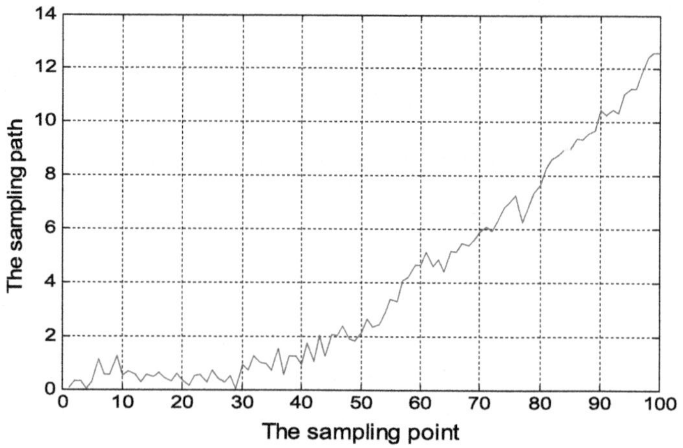

Fig. 11.1 The degradation path of the testing sample

Based on the parameter estimation method presented in Sect. 4, we estimate the parameters for the four cases. For comparison, the parameter estimates, the associated values of log-LF, and AIC are shown in Table 11.1.

As shown in Table 11.1, the estimated results of the parameters are close to the values used for generating the simulation data. Therefore, the parameter estimation method in this chapter is effective. In addition, the model in Case 1 with one-source variability has the least log-LF and the greatest AIC. Comparatively, the models in Case 2 and Case 3 with two-source variability have a greater log-LF and a lower AIC than Case 1, but the greatest log-LF and the least AIC are obtained by the model of Case 4 with three-source variability. These observations indicate that considering three-source variability in degradation modeling is necessary and can effectively improve the model fitting in term of the AIC.

To further illustrate the effectiveness of the developed model and method in RUL estimate, we use the testing sample, which can be observed in Fig. 11.1 from the simulation data, to show the MSEs of the RUL under four different cases in the lifecycle.

The MSEs can be calculated based on the definition of the MSE and the RUL estimate. Here, the evolving path of the MSEs of Cases 1–3 against Case 4 at each sampling point is shown in Fig. 11.2.

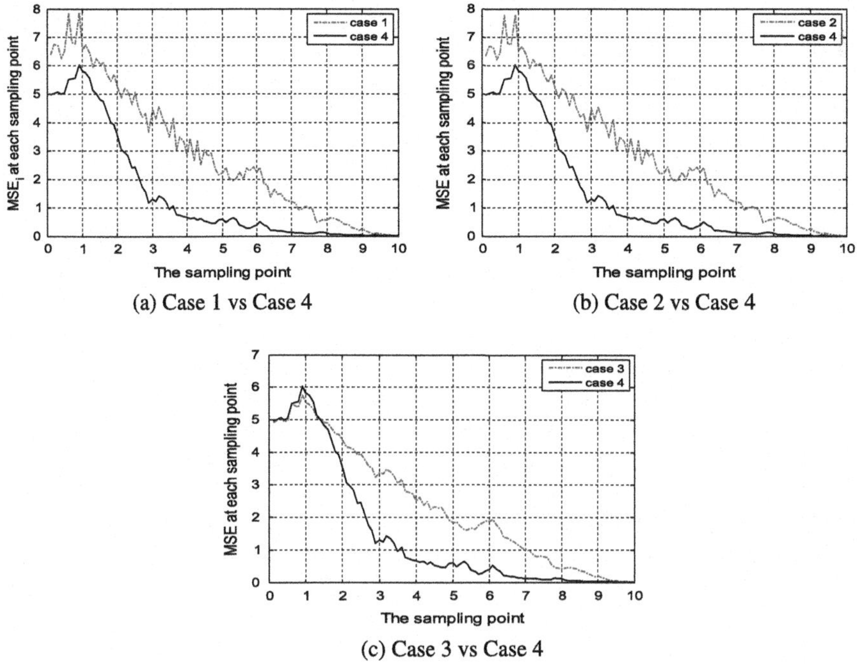

Fig. 11.2 Comparisons for MSE of the RUL estimate for the examining sample

As shown in Fig. 11.2, the MSEs of the RUL estimate with three-source variability maintain a relatively low level compared to the other three cases. In Cases 1–3, the MSEs decrease slowly and suffer some large fluctuations, which result from the uncertain degradation signals. These three cases do not fully consider the uncertainties in the degradation signals, i.e., taking either only one-source or two-source variability. By contrast, in Case 4, the MSEs decrease quickly because it considers all three-source variability simultaneously.

Furthermore, we note that Case 3 behaves better than Cases 1 and 2, though the random effect is not considered in Fig. 11.2. The reason is that the observation history of the degradation signals up to the current sampling point is fully used to estimate the RUL based on Kalman filter, as implied by Theorem 11.3. In addition, the MSEs in Case 3 have a better performance than Case 4 before the 15th sampling point, which is not uncommon because there are few available data at the initial stage of the degradation. Clearly, as the degradation data are accumulated, the results in Case 4 are better than those in Case 3.

Finally, Case 1 and Case 2 have larger MSEs in all of CM points because the observation data contain measurement errors and relatively large variance, which can be observed in Table 11.1. Accordingly, the values of the TMSEs for Cases 1–4 are 290.2352, 288.9250, 234.0842, and 146.0553, respectively. It is not difficult to find that these results are consistent with the above discussions, and the results further demonstrate that degradation modeling with three-source variability has a better RUL estimate than the other three cases and can improve the estimation accuracy.

To further verify the effectiveness of the developed method in practical application, we estimate the RUL of the fatigue crack length in 2017-T4 aluminum alloy [4]. Accordingly, the results are compared with the method presented in [4], which mainly considers the temporality or both the temporal variability and individual variability.

11.4.2 Case Study

The original degradation data representing the crack length propagation contain four test specimens of 2017-T4 aluminum alloy, as given in Chap. 7. For each unit, the fatigue crack length is recorded per 0.1 million cycles from 1.5 through 2.4 million cycles. During testing, ten crack levels were recorded for each unit. The data have been shown graphically in Fig. 7.2. Here, the failed threshold level is set as 5.6 mm. The main parameter estimation results are summarized in Table 11.2.

As expected, Table 11.2 shows that the model of Case 4 with three-source variability outperforms Cases 1–3 in terms of log-LF, the AIC, and TMSE. Specifically, the estimated results show that the results of Case 4 have better fit than those in Cases 1–3 in terms of both the AIC and TMSE. This case study demonstrates the better model fitting performance of the presented model with three-source variability over considering one- or two-source variability. However, the estimated results in Table 11.2 deserve some comments. It is observed that the estimated values of σ_a are much lower than the values of μ_a, σ_B, and σ_ε, which may indicate that the influence brought by the random effects (represented by σ_a) can be neglected. On the other hand, according to Eq. (11.1) and the measurement model in Eq. (11.2), the variance of the degradation quantity at a discrete time t_k is $var\left(X\left(t_k\right)\right) = \sigma_a^2 t_k^{2b} + \sigma_B^2 t_k$. Thus, $\sigma_a$has more impact on the degradation uncertainty as t increases. In this sense, the influence of the random effects cannot be simply neglected. On the other hand, using the parameters in Table 11.2 (Case 4), at $t_k = 2$ million cycles, we have $\sigma_a^2 t_k^{2b} = 3.04367e - 003$ and $\sigma_B^2 t_k = 2.58e - 004$. This computed result indicates that the effect of the random effects is greater than the temporal variability. Thus, we cannot neglect the effect of σ_a, particularly when the system's age t is large.

To further illustrate the effectiveness of the developed model and the method in the RUL estimate, we use the degradation data of the third unit to show the PDF of

Table 11.2 Comparisons of the four degradation models with the fatigue crack growth data

	μ_a	σ_a	b	σ_B	σ_ε	log-LF	AIC	TMSE
Case 1	3.5654e−004	–	11.0746	1.9819	–	−43.5131	93.0262	0.0572
Case 2	3.9477e−005	8.9347e−006	13.4820	1.8977	–	−43.1125	94.2250	0.0518
Case 3	4.9223e−005	–	13.3145	0.534649	0.490743	−37.4882	82.9764	0.0259
Case 4	4.9e−003	1.9582e−004	8.1382	0.011358	0.51211	−28.9682	67.9364	0.0063

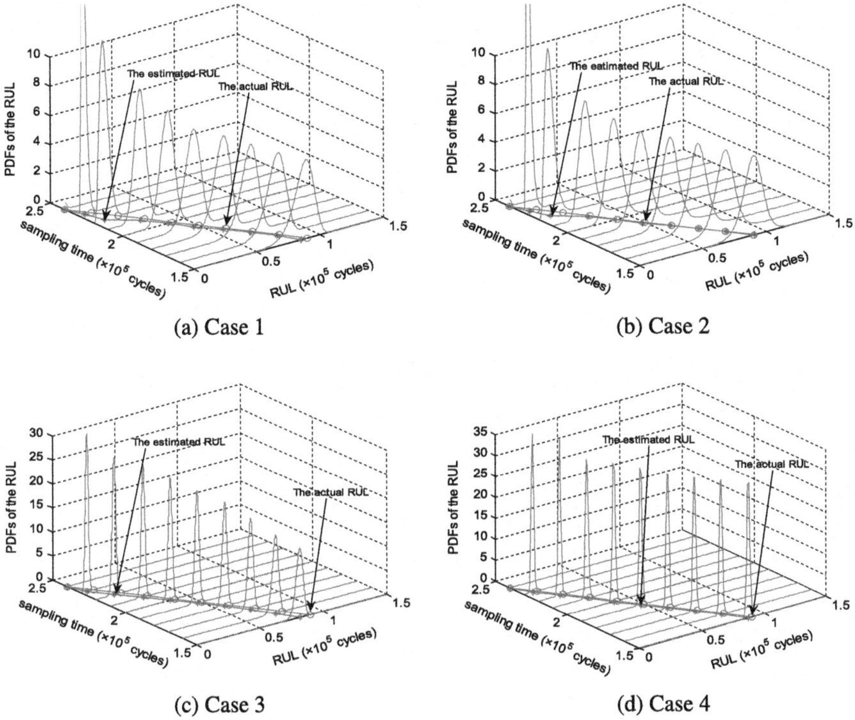

Fig. 11.3 Comparative results of the RUL estimate for Cases 1–4 with the fatigue crack data

the RUL estimate of Cases 1–4. Correspondingly, the actual RULs and the estimated mean RULs are shown graphically in Fig. 11.3 for comparison.

As observed in Fig. 11.3, the PDF of the RUL is more compact around the actual RUL. This denotes that the uncertainty of the RUL estimate is much smaller than other cases. On the other hand, the mean RUL is more accurate in Case 4 than other cases. In addition, at the last four sampling times, the results in Case 4 and Case 3 outperform the results in Case 1 and Case 2. The reason behind this observation is elaborated as follows. It is observed that the degradation rate of the crack length increases from 2.1 million cycles, but Case 1 and Case 2 cannot track such changes in time. In contrast, Case 3 can update the estimated state of the crack length by the Karman filter, and Case 4 can adaptively adjust the drift coefficient a by the updating mechanism. Similarly, the updating process is shown in Fig. 11.4.

As shown in Fig. 11.4, the posterior mean $\mu_{a,k|k}$ of a is changed as the obtained degradation data because the drift coefficient a is changing as the degradation accumulates, which is different from the simulation case in the previous section. In addition, the standard deviation $\sigma_{a,k}$ is still decreasing as the update process develops.

For further comparisons, we compute the corresponding MSEs of the RUL estimates for the four cases, as illustrated in Fig. 11.5. As in Fig. 11.2, we can obtain

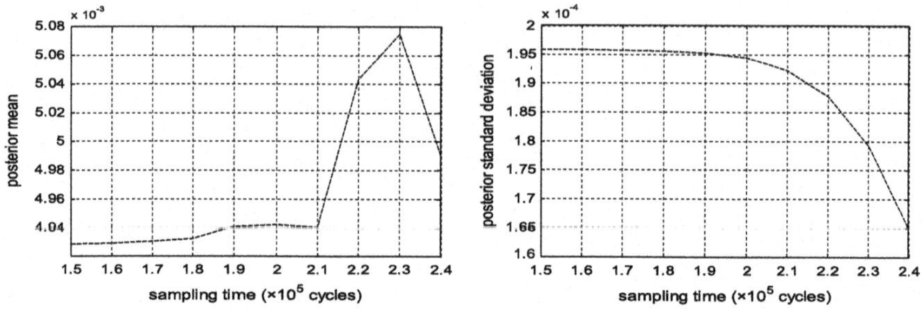

(a) $\mu_{a,k|k}$ (b) $\sigma_{a,k}$

Fig. 11.4 Updates of $\mu_{a,k|k}$ and $\sigma_{a,k}$ using the fatigue crack data

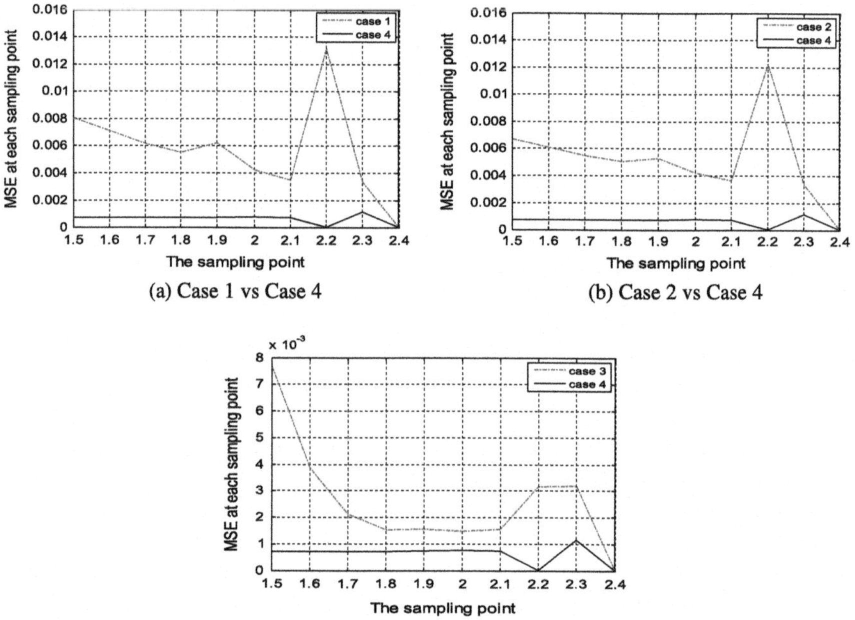

(a) Case 1 vs Case 4 (b) Case 2 vs Case 4 (c) Case 3 vs Case 4

Fig. 11.5 Comparisons for the MSE of the RUL estimate for the fatigue crack data

similar observations from Fig. 11.5, and the associated TMSEs can be found in Table 11.2. Together with these comparative results, we can conclude that the results of the RUL estimate with three-source variability are much better than the other cases in practical applications.

References

1. Zio E, Maio FD (2010) A data-driven fuzzy approach for predicting the remaining useful life in dynamic failure scenarios of a nuclear system. Reliab Eng Syst Saf 95(1):49–57
2. Huynh KT, Barros A, Bérenguer C (2012) Maintenance decision-making for systems operating under indirect condition monitoring: value of online information and impact of measurement uncertainty. IEEE Trans Reliab 61(2):410–425
3. Si XS, Wang W, Hu CH, Zhou DH (2014) Estimating remaining useful life with three-source variability in degradation modeling. IEEE Trans Reliab 63(1):167–190
4. Si XS, Wang W, Hu CH, Zhou DH, Pecht M (2012) Remaining useful life estimation based on a nonlinear diffusion degradation process. IEEE Trans Reliab 61(1):50–67
5. Vasan ASS, Long B, Pecht M (2013) Diagnostics and prognostics method for analog electronic circuits. IEEE Trans Ind Electron 60(11):5277–5291
6. Orchard ME, Hevia-Koch P, Zhang B, Tang L (2013) Risk measures for particle-filtering-based state-of-charge prognosis in lithium-ion batteries. IEEE Trans Ind Electron 60(11):5260–5269
7. Shahriari M, Farrokhi M (2013) Online state-of-health estimation of VRLA batteries using state of charge. IEEE Trans Ind Electron 60(1):191–202
8. Whitmore GA (1995) Estimating degradation by a Wiener diffusion process subject to measurement error. Lifetime Data Anal 1(3):307–319
9. Meeker WQ, Escobar LA (1998) Statistical methods for reliability data. Wiley, New York
10. Peng CY, Tseng ST (2009) Mis-specification analysis of linear degradation models. IEEE Trans Reliab 58(3):444–455
11. Ye ZS, Shen Y, Xie M (2012) Degradation-based burn-in with preventive maintenance. Eur J Oper Res 221(2):360–367
12. Zio E, Maio FD, Tong JJ (2010) Safety margins confidence estimation for a passive residual heat removal system. Reliab Eng Syst Saf 95(8):828–836
13. Chen N, Tsui KL (2013) Condition monitoring and residual life prediction using degradation signals: revisited. IIE Trans 45(9):939–952
14. Park JI, Bae SJ (2010) Direct prediction methods on lifetime distribution of organic light-emitting diodes from accelerated degradation tests. IEEE Trans Reliab 59(1):74–90
15. Pandey MD, Yuan XX, Van Noortwijk JM (2009) The influence of temporal uncertainty of deterioration on life-cycle management of structures. Struct Infrastruct Eng Maint Manag Lifecycle Des Perform 5(2):145–156
16. Liao HT, Elsayed EA, Chan LY (2006) Maintenance of continuously monitored degrading systems. Eur J Oper Res 175(2):821–835
17. Ye ZS, Hong YL, Xie YM (2013) How do heterogeneities in operating environments affect field failure predictions and test planning. Ann Appl Stat 7(4):2249–2271
18. Chen N, Ye ZS, Xiang YS, Zhang LM (2015) Condition-based maintenance using the inverse Gaussian degradation model. Eur J Oper Res 243(1):190–199
19. Ye ZS, Chen N (2013) The inverse Gaussian process as a degradation model. Technometrics 56(3):309–320
20. Ye ZS (2013) On the conditional increments of degradation processes. Stat Probab Lett 83(11):2531–2536
21. Ye ZS, Chen N, Shen Y (2015) A new class of wiener process models for degradation analysis. Reliab Eng Syst Saf 139:58–67. http://dx.doi.org/10.1016/j.ress.2015.02.005
22. Ye ZS, Wang Y, Tsui KL, Pecht M (2013) Degradation data analysis using Wiener processes with measurement errors. IEEE Trans Reliab 62(4):772–780
23. Ye ZS, Xie M (2015) Stochastic modelling and analysis of degradation for highly reliable products. Appl Stoch Models Bus Ind 31(1):16–36
24. Si XS, Wang W, Hu CH, Zhou DH (2011) Remaining useful life estimation—a review on the statistical data driven approaches. Eur J Oper Res 213(1):1–14
25. Wang W, Carr M, Xu W, Kobbacy AKH (2011) A model for residual life prediction based on Brownian motion with an adaptive drift. Microelectron Reliab 51(2):285–293

26. Si XS, Wang W, Chen MY, Hu CH, Zhou DH (2013) A Wiener process-based degradation model with a recursive filter algorithm for remaining useful life estimation. Mech Syst Signal Process 35(1–2):219–237
27. Christer AH, Wang W, Sharp JM (1997) A state space condition monitoring model for furnace erosion prediction and replacement. Eur J Oper Res 101(1):1–14
28. Batzel TD, Swanson DC (2009) Prognostic health management of aircraft power generators. IEEE Trans Aerosp Electron Syst 45(2):473–483
29. Gebraeel N, Lawley MA, Li R, Ryan JK (2005) Residual-life distributions from component degradation signals: a Bayesian approach. IIE Trans 37(6):543–557
30. Park C, Padgett WJ (2005) Accelerated degradation models for failure based on geometric Brownian motion and gamma processes. Lifetime Data Anal 11(4):511–527
31. Wang X (2010) Wiener processes with random effects for degradation data. J Multivar Anal 101(2):340–351
32. Whitmore GA, Schenkelberg F (1997) Modeling accelerated degradation data using Wiener diffusion with a time scale transformation. Lifetime Data Anal 3(1):27–45
33. Zio E, Peloni G (2011) Particle filtering prognostic estimation of the remaining useful life of nonlinear components. Reliab Eng Syst Saf 96(3):403–409
34. Cadini F, Zio E, Avram D (2009) Monte Carlo-based filtering for fatigue crack growth estimation. Probab Eng Mech 24(3):367–373
35. Zhou YF, Sun Y, Mathew J, Wolff R, Ma L (2011) Latent degradation indicators estimation and prediction: a Monte Carlo approach. Mech Syst Signal Process 25(1):222–236
36. Zhao JM, Feng TL (2011) Remaining useful life prediction based on nonlinear state space model, In: 2011 Prognostics and Health Management Conference, IEEE Xplore, Shenzhen, China, pp 1–5
37. Feng L, Wang HL, Si XS, Zou HX (2013) A state-space-based prognostic model for hidden and age-dependent nonlinear degradation process. IEEE Trans Autom Sci Eng 10(4):1072–1086
38. Si XS, Wang W, Chen MY, Hu CH, Zhou DH (2013) A degradation path-dependent approach for remaining useful life estimation with an exact and closed-form solution. Eur J Oper Res 226(1):53–66
39. Nardo ED, Nobile A, Pirozzi E, Ricciardi LM (2001) A computational approach to first passage-time problems for Gauss-Markov processes. Adv Appl Probab 33(2):453–482
40. Wang ZQ, Wang W, Hu CH, Si XS (2014) An additive Wiener process-based prognostic model for hybrid deteriorating systems. IEEE Trans Reliab 63(1):208–222
41. Kharoufeh JP, Mixon DG (2009) On a Markov-modulated shock and wear process. Nav Res Logist 56(6):563–576
42. Lee MLT, Whitmore GA (2006) Threshold regression for survival analysis: modeling event times by a stochastic process reaching a boundary. Stat Sci 21(4):501–513
43. Akaike H (1974) A new look at the statistical model identification. IEEE Trans Autom Control 19(6):716–722
44. Carr MJ, Wang W (2011) An approximate algorithm for prognostic modelling using condition monitoring information. Eur J Oper Res 211(1):90–96

Chapter 12
RSL Prediction Approach for Systems with Operation State Switches

12.1 Introduction

Predicting the residual life is of significant importance in proactive maintenance, and prognostics and health management of systems [1–3]. Many highly critical systems in military and aerospace fields, like missiles, rockets, and their associated systems, are required of long-term storage before used [4–6]. For such systems, storage is an essential part of their lifecycles and the operating time of such systems is usually very short compared with the time in storage. Therefore, the investigation of the residual storage life (RSL) prediction is of significant importance in that it can help to plan efficient monitoring policy to extend the system's life.

Thanks to recent advances in sensor and condition monitoring (CM) technologies, the last decades have witnessed a great deal of efforts made to predict the residual life of systems using sensor signals [7]. In literature, sensor signals used in prognostics mainly include health monitoring data of the deterioration process of a system over time, since most failures of assets arise from a degradation mechanism and there are characteristics that deteriorate over time. Examples include gyroscopes in inertial navigation platforms of rockets, batteries, motor drives, and capacitors in analog electronic circuits [8–11]. However, current works mainly focus on predicting the residual life of continuous operating systems based on their stochastic degradation signals and the scale for the predicted residual life is therefore the continuous operating time [7, 12, 13]. There is neither the storage concept nor the operation state switches for these systems.

To our knowledge, the reports about the RSL prediction for systems with switches between the working state and storage state are very limited. Because it is often deemed that the system's performance is unchanged when it is in storage or non-operating conditions. However, this does not signify no degradation in the system performance. For some systems not used frequently, such as weapon systems, inertial devices equipped in rockets, and so on, their performance may decrease gradually when they are in storage, due to the changes of external or environmental factors, such as temperature, humidity, and man-made interference [14]. In practice, such

X.-S. Si et al., *Data-Driven Remaining Useful Life Prognosis Techniques*,
Springer Series in Reliability Engineering, DOI 10.1007/978-3-662-54030-5_12

critical systems are always in the storage condition during their lifecycles unless some missions are required to run such systems. Besides, to ensure high precision and high stability once such critical systems are put into battlefield or military exercises, even after long-term storage, routine CM is also required to put these systems into operating after storing such systems for several months, so as to be familiar with their performance characteristics and health conditions.

Generally, the operating of such systems tends to significantly speed up the degradation of systems compared with the case in storage. Take the inertial device (e.g., gyroscopes, which own electromechanical structures) used in rockets for example, the lifetime of such device is about 200 h in continuous operating condition due to the wear of the gyroscope's electric motor [15]. However, its lifetime in the storage condition could be several years because of no wear. In addition, the observation from the degradation data of gyroscopes indicates that the values of gyroscopes' drift before and after storage exhibit an increasing trend, though such increase is not significant in a short-term storage period as compared with the quick degradation in the operating period. But the effect of long-term storage cannot be neglected. This phenomenon encourages us to consider the prediction of the RSL to be essential since the degradation also occurs in the storage condition, even though with a low degradation rate. In addition, the operating of such system is basically driven by the mission or routine CM; and therefore the associated duration of the system's working is frequently random. As a result, the transitions between the working state and storage state are random.

Together with the above discussions, there are two main characteristics of the concerned systems in this chapter: (1) the degradation rate depends on the system's operation state; and (2) the system's operation process experiences a number of state transitions between the working state and storage state. Accordingly, two challenging issues for a success of predicting the RSL of such system are how to establish the linkage between the degradation process and the system operation state, and how to formulate the probabilistic law of the number of state transitions and their transition times, so as to incorporate the future possible transitions into the RSL prediction. Unfortunately, to the best of our knowledge, these issues have not been fully investigated despite their potential in practical applications, and the purpose of this chapter is therefore to shorten such a gap by providing a degradation modeling framework for RSL prediction.

The main contributions of this chapter are summarized as follows. (1) Using the monitored degradation data in the working state, we present a novel degradation model to account for the dependency of the degradation process on the system's operation states, where a two-state homogeneous continuous-time Markov chain (CTMC) is used to approximate the switching process between the working state and storage state. The reason selecting the CTMC is twofold. On the one hand, there are two states in the system's operation process and both their transitions and sojourn times are random. In this case, the CMTC provides a natural tool to describe such random state transition mechanism [16, 17]. On the other hand, this kind of stochastic processes affords a great deal of modeling flexibility and appealing properties including the explicit law of the state transition in the future time span, which facilitates the computation. (2) Under Bayesian paradigm, the posterior probabilistic law of the number of state transitions and their transition times are derived and the posterior estimates of the parameters in the presented model can be derived and updated by the CM data. This makes the most recently estimated values accurately reflect the current reality of the system. (3) The formulation for the RSL prediction is established, which explicitly accounts for the effect of state switches on the RSL prediction and incorporates the probabilistic law of the number of state transitions and their transition times in the future. To be solvable, a numerical simulation algorithm is provided to calculate the distribution of the predicted RSL. Finally, we demonstrate the proposed approach by a case study of the inertial platform.

The remainder parts are organized as follows. Section 12.2 provides the literature review related to the RSL prediction. The problem description is given in Sect. 12.3. In Sect. 12.4, we formulate the distribution of state transitions between the working state and storage state. Section 12.5 models the system's degradation process to predict the RSL and gives the simulation-based solution algorithm. In Sect. 12.6, we provide a case study for demonstration.

12.2 Literature Review

Traditionally, if the past failure data from either fields or experiments are available, the storage lifetime can be estimated from the failure data of systems using likelihood-based inference methods [4, 5]. However, for expensive and highly reliable assets, failure data are scarce or limited, or the time for data acquisition is very consuming, even in the accelerated condition. Another feature of this kind of methods is to make predictions for a population of systems, rather than a specific system in service as the CM data for this individual system are not considered. These limitations restrict the applications of failure data based storage lifetime estimation methods.

In practice, most failures of assets arise from a degradation mechanism and there are characteristics that deteriorate over time, such as the drift of gyroscopes in inertial navigation platforms of rockets, batteries, motor drives, and capacitors in analog electronic circuits [8–11, 18, 19]. Therefore, the monitored degradation data of these characteristics can be used as a feasible alternative for life estimation task (see [7] for a comprehensive review). However, the works surveyed in [7] mainly focus on estimating the residual life of continuous operating systems. There is neither the storage concept nor the operation state switches for these systems. Very few efforts have been made to address the degradation process which involves switching points [20, 21]. Note that the common shared by these relevant works is that the switching points are assumed to be deterministic and the possible state transitions in future time span are not considered in the life prediction. The most recent work in [22] developed a residual life prediction approach for the system with a single but random switching point. The random switching point is estimated within a Bayesian framework. However, once the degradation process goes into the second stage, there are no operation state switches as if the system operation state is fixed. Thus, these surveyed methods are not suitable to predict the RSL since the aforementioned characteristics of our concerned systems cannot be sufficiently considered by these methods.

The other line of research possibly related to the concerned problem tries to predict the system's residual life by considering the environmental effect on the system's degradation, where the environments could be either time-varying or random [23]. However, the basic assumption in these works is that the future environments are unchanged. From these recent works, we can observe that they all focus on continuous operating systems and the possible environmental changes in the future are not appropriately considered, but the operation state switches are the primary feature having to be modeled for reasonable RSL predictions. Thus, it is desired to develop a new method to predict the RSL by considering the switches between the system's operation states.

12.3 Problem Description for RSL Estimation

Let the stochastic process $\{X(t), t \geq 0\}$ denote the underlying degradation process of the system. Without loss of generality, we simply assume the starting reading $x(0)$ for the degradation process to be fixed. During the system's operation process, the system can occupy one of two possible states: the working state and the storage state. Moreover, the transitions and state sojourn times are random. In the working state, the system may suffer additional load and environmental patterns. Therefore, it is reasonable to assume that the system will degrade in a higher degradation rate than the case in the storage state. The system's lifetime is defined as the first passage time (FPT) that the degradation process $\{X(t), t \geq 0\}$ hits a failure threshold w, which can be determined by expert knowledge or well-accepted industrial standards (e.g., ISO 2372 and ISO 10816 for vibration level) [24].

From the above general description, we know that at each time t, the system can only occupy one of the operation states in a state set $\Psi = \{1, 2\}$, where "1" denotes the working state and "2" the storage state. Let $\nu : [0, \infty) \to \Psi$ be a deterministic and piecewise constant function so that $\nu(t)$ represents the system operation state at time t. Without loss of generality, we assume $\nu(0) = 1$. Under this setting, the system's operation states are sequential but the duration of each state is random. Further, we define a degradation rate function $\lambda : \Psi \to \Re^+$ so that $\lambda(\nu(t))$ represents the system's degradation rate at time t. For example, if the system is at the kth operation state at time t, then $\nu(t) = k$ and $\lambda(k)$ is the degradation rate of the kth operation state, where $k = 1, 2$. For notation brevity, we simplify $\lambda(k)$ as λ_k throughout the remaining parts.

Under the above settings, the evolution of the monitored degradation process over time is described by a Wiener process-based degradation model with piecewise constant drift coefficients as follows:

$$X(t) = x(0) + \int_0^t \lambda\,(\nu(u))\mathrm{d}u + \sigma B(t). \tag{12.1}$$

where $\sigma\,(\sigma > 0)$ is the diffusion parameter representing the common character shared by both operation states, and $\{B(t), t \geq 0\}$ is a standard Brownian motion (BM) process, with $\sigma B(t) \sim N(0, \sigma^2 t)$ for $t \geq 0$. This term characterizes the degradation effect that cannot be attributed to the system operation process.

The main objective here is to estimate and update the RSL distribution of the system based on the discretely observed information up to the current time. Here, the observed information up to the current time includes twofold: the degradation data and the information regarding the transition times and transition numbers of the system's operation states. Suppose the degradation in the working state is discretely monitored at times $0 = t_0 < t_1 < \cdots < t_i$ and let $x_k = X(t_k)$ denote the degradation observation at time t_k, where $k = 0, 1, \ldots, i$. The set of the degradation observations up to the current time t_i is represented by $\boldsymbol{X}_i = \{x_0, x_1, x_2, \ldots, x_i\}$. In addition, we can actually observe the transition times of the system's operation states up to t_i. Therefore, the set of the transition times and the transition numbers of the system's operation states up to t_i is denoted as $\boldsymbol{C}_i = \{u_j, n(t_i) : j = 1, 2, \ldots, n(t_i)\}$, where u_j is the jth state transition time before t_i and $n(t_i)$ is the cumulative number of state transitions up to t_i. Under these settings, we know that the system's operation process will maintain the state $\nu(u_{j-1})$ over the interval $[u_{j-1}, u_j)$, $j = 1, 2, \ldots, n(t_i)$, $u_0 = 0$.

Based on the above settings, the storage lifetime of the system can be defined using the concept of the FPT as [24]

$$T = \inf\{t > 0 : X(t) \geq w\}. \tag{12.2}$$

As a result, we define the RSL S_i of the system at time t_i as

$$S_i = \inf\{s_i > 0 : X(s_i + t_i) \geq w\}. \tag{12.3}$$

with conditional cumulative distribution function (CDF)

$$\Pr(S_i \leq s_i | \boldsymbol{X}_i, \boldsymbol{C}_i) = \Pr\left(\sup_{s_i > 0} X(t_i + s_i) \geq w \,\middle|\, \boldsymbol{X}_i, \boldsymbol{C}_i\right), \tag{12.4}$$

where $(\boldsymbol{X}_i, \boldsymbol{C}_i)$ is the information available up to t_i.

In the remaining parts, we will present specific stochastic models and updating procedures to compute $\Pr(S_i \leq s_i | \boldsymbol{X}_i, \boldsymbol{C}_i)$ for the RSL prediction.

12.4 Model Formulation for Transitions Between the Operating State and Storage State

12.4.1 Randomly Varying System Operation Process

In this modeling aspect, we utilize a two state CTMC to model $\{\nu(t), t \geq 0\}$, with the state set $\Psi = \{1, 2\}$, i.e., $\nu(t) = 1$ or $\nu(t) = 2$. The transition rate matrix of this CTMC is represented as $\mathbf{Q} = \begin{bmatrix} -\gamma & \gamma \\ \upsilon & -\upsilon \end{bmatrix}$, where $1/\gamma$ and $1/\upsilon$ are the expected sojourn times of the working state and storage state, respectively. For notation convenience, we let $\boldsymbol{\kappa} = (\gamma, \upsilon)$ be the parameter set in $\{\nu(t), t \geq 0\}$.

According to (12.3), in order to predict $\Pr(S_i \leq s_i | \boldsymbol{X}_i, \boldsymbol{C}_i)$, it is necessary to formulate the probabilistic law of the system's operation state transition times and the number of the transitions over a future time span. Toward this end, suppose that the system's current operation state $\nu(t_i)$ is known, and define the transition times over a future time span $(t_i, s + t_i)$ as $\left\{\tau_{0,i}, \tau_{1,i}, \tau_{2,i}, \ldots, \tau_{N_i(s),i}\right\}$ and the according number of transitions in $(t_i, s + t_i)$ as $N_i(s)$, respectively, where $\tau_{n,i}, n = 0, 1, 2, \ldots, N_i(s)$ is the time of the nth transition with $t_i \leq \tau_{n,i} \leq t_i + s$ and $N_i(s) = 0, 1, 2, \ldots$. Here, it is worth noting that $\tau_{0,i}$ is used to represent the event of no state transition and thus is equal to the current time t_i. Further, given $\boldsymbol{\kappa}$, we use $f_i\left(\tau_{0,i}, \tau_{1,i}, \tau_{2,i}, \ldots, \tau_{N_i(s),i}, N_i(s) \middle| \boldsymbol{\kappa}\right)$ to represent the joint PDF of $\left\{\tau_{0,i}, \tau_{1,i}, \tau_{2,i}, \ldots, \tau_{N_i(s),i}\right\}$ and $N_i(s)$ at t_i. Then, given $\nu(t_i)$ and $\boldsymbol{\kappa}$, $f_i\left(\tau_{0,i}, \tau_{1,i}, \tau_{2,i}, \ldots, \tau_{N_i(s),i}, N_i(s) \middle| \boldsymbol{\kappa}\right)$ can be determined by the following theorem.

Theorem 12.1 *For a two-state CTMC with the transition rate matrix* $\mathbf{Q}$*, given the current system operation state* $\nu(t_i)$ *and* $\boldsymbol{\kappa}$*,* $f_i\left(\tau_{0,i}, \tau_{1,i}, \tau_{2,i}, \ldots, \tau_{N_i(s),i}, N_i(s) \middle| \boldsymbol{\kappa}\right)$ *can be formulated as follows.*

(1) If $\nu(t_i) = 1$*, then* $f_i\left(\tau_{0,i}, \tau_{1,i}, \tau_{2,i}, \ldots, \tau_{N_i(s),i}, N_i(s) \middle| \boldsymbol{\kappa}\right)$ *can be formulated as*

$$f_i\left(\tau_{0,i}, \tau_{1,i}, \tau_{2,i}, \ldots, \tau_{L,i}, N_i(s) = L \middle| \boldsymbol{\kappa}\right) = \begin{cases} (\gamma\, \upsilon)^n\, e^{\sum_{k=0}^{L} (\gamma - \upsilon)(-1)^k \left(\tau_{k,i} - t_i\right)} e^{-\gamma s}, & L = 2n,\ n = 0, 1, 2, \ldots \\ \gamma\, (\gamma\, \upsilon)^n\, e^{\sum_{k=1}^{L} (\gamma - \upsilon)(-1)^k \left(\tau_{k,i} - t_i\right)} e^{-\upsilon s}, & L = 2n + 1,\ n = 0, 1, 2, \ldots \end{cases} \tag{12.5}$$

(2) If $\nu(t_i) = 2$*, then* $f_i\left(\tau_{0,i}, \tau_{1,i}, \tau_{2,i}, \ldots, \tau_{N_i(s),i}, N_i(s) \middle| \boldsymbol{\kappa}\right)$ *can be formulated as*

$$f_i(\tau_{0,i}, \tau_{1,i}, \tau_{2,i}, \ldots, \tau_{L,i}, N_i(s) = L \big| \kappa) =$$
$$\begin{cases} (\gamma\, \upsilon)^n\, e^{\sum\limits_{k=0}^{L} (\upsilon-\gamma)(-1)^k \left(\tau_{k,i}-t_i\right)} e^{-\upsilon s}, \; L = 2n, \; n = 0, 1, 2, \ldots \\ \upsilon\, (\gamma\, \upsilon)^n\, e^{\sum\limits_{k=1}^{L} (\upsilon-\gamma)(-1)^k \left(\tau_{k,i}-t_i\right)} e^{-\gamma s}, \; L = 2n+1, \; n = 0, 1, 2, \ldots \end{cases} \tag{12.6}$$

where $\tau_{0,i} = t_i$.

Proof Suppose $\nu(t_i) = 1$ and the case for $\nu(t_i) = 2$ can be proved in the same way. Let $D_{k,i}$ denote the time length from the current time t_i to the kth transition time over $(t_i, s + t_i)$, e.g., $D_{0,i} = \tau_{0,i} - t_i = 0$, where $k = 0, 1, 2, \ldots, N_i(s)$. Then, if there is no state transition in $(t_i, s + t_i)$, we have $N_i(s) = 0$. In this case, we have

$$\begin{aligned} &\Pr(N_i(s) = 0 | \, \kappa) = e^{-\gamma s}, \\ &f_i(\tau_{0,i}, N_i(s) = 0 \big| \kappa) = e^{-\gamma s} \delta(\tau_{0,i}), \end{aligned} \tag{12.7}$$

where $\delta(\tau_{0,i}) = 1$ when $\tau_{0,i} = t_i$; otherwise, $\delta(\tau_{0,i}) = 0$.

If $N_i(s) = 1$, we have

$$\begin{aligned} &\Pr(\tau_{1,i} - t_i < D_{1,i} \le \tau_{1,i} - t_i + \Delta_1, \, N_i(s) = 1 | \, \kappa) \\ &= \Pr(N_i(\tau_{1,i} - t_i) = 0, \, N_i(\tau_{1,i} - t_i + \Delta_1) - N_i(\tau_{1,i} - t_i) = 1, \\ &N_i(s) - N_i(\tau_{1,i} - t_i + \Delta_1) = 0 \big| \kappa) \\ &= e^{-\gamma(\tau_{1,i}-t_i)} \cdot \gamma\, \Delta_1 e^{-\gamma \Delta_1} \cdot e^{-\upsilon(s-\tau_{1,i}+t_i-\Delta_1)} \\ &= \gamma\, \Delta_1 e^{-(\gamma-\upsilon)(\tau_{1,i}-t_i)} \cdot e^{-(\gamma-\upsilon)\Delta_1} \cdot e^{-\upsilon s}. \end{aligned} \tag{12.8}$$

Thus, the according PDF is

$$\begin{aligned} &f_i(\tau_{0,i}, \tau_{1,i}, N_i(s) = 1 \big| \kappa) \\ &= \lim_{\Delta_1 \to 0} \frac{\Pr(\tau_{1,i} - t_i < D_{1,i} \le \tau_{1,i} - t_i + \Delta_1, \, N_i(s) = 1 | \, \kappa)}{\Delta_1} \\ &= \lim_{\Delta_1 \to 0} \frac{\gamma\, \Delta_1 e^{-(\gamma-\upsilon)(\tau_{1,i}-t_i)} \cdot e^{-(\gamma-\upsilon)\Delta_1} \cdot e^{-\upsilon s}}{\Delta_1} \\ &= \gamma\, e^{-(\gamma-\upsilon)(\tau_{1,i}-t_i)} \cdot e^{-\upsilon s}. \end{aligned} \tag{12.9}$$

Following the same procedure and the principle of mathematical induction, we can obtain the following results: when $L = 2n, \; n = 0, 1, 2, \ldots$,

$$\begin{aligned} &f_i(\tau_{0,i}, \tau_{1,i}, \tau_{2,i}, \ldots, \tau_{L,i}, N_i(s) = L \big| \kappa) \\ &= (\gamma\, \upsilon)^n\, e^{\sum\limits_{k=0}^{L} (\gamma-\upsilon)(-1)^k \left(\tau_{k,i}-t_i\right)} e^{-\gamma s}, \end{aligned}$$

and when $L = 2n + 1,\ n = 0, 1, 2, \ldots,$

$$f_i(\tau_{0,i}, \tau_{1,i}, \tau_{2,i}, \ldots, \tau_{L,i}, N_i(s) = L \mid \boldsymbol{\kappa})$$
$$= \gamma\ (\gamma\ \upsilon)^n\ e^{\sum_{k=1}^{L} (\gamma - \upsilon)(-1)^k \left(\tau_{k,i} - t_i\right)} e^{-\upsilon s}.$$

This completes the proof.

For simplifying notation, we denote the future operation state transitions from current time t_i to a fixed future time span s and the according number of state transitions by a set $\boldsymbol{F}_i(s) = \left\{\tau_{j,i}, N_i(s) : \tau_{j,i} \in (t_i, s + t_i], j = 0, 1, \ldots, N_i(s)\right\}$. This information will be used in deriving the predicted RSL distribution since the degradation path and the RSL prediction are dependent on the future operation state transitions and the number of state transitions.

12.4.2 *Bayesian Estimation for Parameters in the System's Operation Process*

Here we first describe a Bayesian framework for updating the parameters in $\{\nu(t), t \geq 0\}$, i.e., $\boldsymbol{\kappa}$. Namely, we assume that the parameters in the transition rate matrix are random with prior distributions and then the Bayesian rule is used to obtain the posterior estimates of these parameters once $\boldsymbol{C}_i$ is available.

Let the prior distribution of $\boldsymbol{\kappa}$ be $p(\boldsymbol{\kappa})$. At the current time t_i, the observation set $\boldsymbol{C}_i = \left\{u_j, n(t_i) : j = 1, 2, \ldots, n(t_i)\right\}$ associated with $\{\nu(t), t \geq 0\}$ is available, where mainly contains two kinds of information: the number of the system's operation state transitions between the working state "1" and the storage state "2" in the interval $[0, t_i]$, and the total dwelling times in "1" and "2" on this interval. Let $n_{1\to2}(t_i)$ be the number of the system's operation state transitions from "1" to "2" and $n_{2\to1}(t_i)$ be the number from "2" to "1" during $[0, t_i]$, respectively; then $n(t_i) = n_{1\to2}(t_i) + n_{2\to1}(t_i)$. Further, let $d_1(t_i)$ and $d_2(t_i)$ be the total dwelling times in the states "1" and "2" during $[0, t_i]$, respectively; then $t_i = d_1(t_i) + d_2(t_i)$. Once obtaining $n_{1\to2}(t_i), n_{2\to1}(t_i), d_1(t_i)$ and $d_2(t_i)$, the likelihood function of $\boldsymbol{\kappa}$ for the observed $\boldsymbol{C}_i$ can be formulated by the Markov property as

$$L(\boldsymbol{\kappa}) = \gamma^{n_{1\to2}(t_i)} \exp\left(-\gamma\, d_1(t_i)\right) \cdot \upsilon^{n_{2\to1}(t_i)} \exp\left(-\upsilon d_2(t_i)\right) \tag{12.10}$$

To achieve Bayesian updating and consider the fact of $\gamma > 0$ and $\upsilon > 0$, we assume that the elements in $\boldsymbol{\kappa}$ have independent gamma prior distributions as $\gamma \sim \text{gama}(\alpha_1, \beta_1)$ and $\upsilon \sim \text{gama}(\alpha_2, \beta_2)$ where $\alpha_k, \beta_k, k = 1, 2$ are the shape and scale parameters for γ and υ, respectively. Thus we have $p(\boldsymbol{\kappa}) = p(\gamma) \cdot p(\upsilon)$, where the PDFs of γ and υ are formulated as

$$\begin{aligned} p(\gamma) &= \frac{\gamma^{\alpha_1-1}\exp(-\gamma/\beta_1)}{\Gamma(\alpha_1)\beta_1^{\alpha_1}} \text{ for } \gamma > 0 \\ p(\upsilon) &= \frac{\upsilon^{\alpha_2-1}\exp(-\upsilon/\beta_2)}{\Gamma(\alpha_2)\beta_2^{\alpha_2}} \text{ for } \upsilon > 0. \end{aligned} \tag{12.11}$$

By applying the Bayesian formula, the posterior distribution of $\boldsymbol{\kappa}$ is

$$f_{\kappa}(\boldsymbol{\kappa} | \boldsymbol{C}_i) \propto L(\boldsymbol{\kappa}) \times p(\boldsymbol{\kappa}). \tag{12.12}$$

Based on some mathematical manipulations, we establish the following theorem for the posterior estimate of $\boldsymbol{\kappa}$.

Theorem 12.2 *Consider that* $\{\nu(t), t \geq 0\}$ *is modeled by a CTMC process with the transition rate matrix* $\mathbf{Q}$ *and the observations regarding* $\{\nu(t), t \geq 0\}$ *up to* t_i *is* $\boldsymbol{C}_i$*, from which* $n_{1\to2}(t_i)$*,* $n_{2\to1}(t_i)$*, and* $d_1(t_i)$ *and* $d_2(t_i)$ *are obtained. If the prior distributions for each element of* $\boldsymbol{\kappa}$ *are gamma as specified in (12.8), then the posterior distributions for each element of* $\boldsymbol{\kappa}$ *are still gamma with the updated shape and scale parameters* $(\alpha_{k,i}, \beta_{k,i})$*,* $k = 1, 2$*, as follows:*

$$\begin{aligned} \alpha_{1,i} &= \alpha_1 + n_{1\to2}(t_i), \beta_{1,i} = \left(\beta_1^{-1} + d_1(t_i)\right)^{-1} \\ \alpha_{2,i} &= \alpha_2 + n_{2\to1}(t_i), \beta_{2,i} = \left(\beta_2^{-1} + d_2(t_i)\right)^{-1}. \end{aligned} \tag{12.13}$$

Proof To derive the posterior estimate of $\boldsymbol{\kappa}$, we first obtain $n_{1\to2}(t_i)$, $n_{2\to1}(t_i)$, and $d_1(t_i)$ and $d_2(t_i)$ from $\boldsymbol{C}_i$. According to (12.10)–(12.12), we have

$$\begin{aligned} f_{\kappa}(\boldsymbol{\kappa} | \boldsymbol{C}_i) &\propto L(\boldsymbol{\kappa}) \times p(\boldsymbol{\kappa}) \\ &\propto \left(\gamma^{n_{1\to2}(t_i)+\alpha_1-1} \exp\left(-\gamma\left(d_1(t_i) + \beta_1^{-1}\right)\right)\right) \\ &\cdot \left(\upsilon^{n_{2\to1}(t_i)+\alpha_2-1} \exp\left(-\upsilon\left(d_2(t_i) + \beta_2^{-1}\right)\right)\right). \end{aligned} \tag{12.14}$$

By comparing the form in the above equation with the form of $p(\boldsymbol{\kappa}) = p(\gamma) \cdot p(\upsilon)$, we learn that the unique solution to the posterior PDF $f_{\kappa}(\boldsymbol{\kappa} | \boldsymbol{C}_i)$ satisfying the condition of a probability measure is

$$\begin{aligned} f_{\kappa}(\boldsymbol{\kappa} | \boldsymbol{C}_i) &= \frac{\gamma^{\alpha_{1,i}-1}\exp(-\gamma/\beta_{1,i})}{\Gamma(\alpha_{1,i})\beta_{1,i}^{\alpha_{1,i}}} \cdot \frac{\upsilon^{\alpha_{2,i}-1}\exp(-\upsilon/\beta_{2,i})}{\Gamma(\alpha_{2,i})\beta_{2,i}^{\alpha_{2,i}}} \\ &= f_{\gamma}(\gamma | \boldsymbol{C}_i) \cdot f_{\upsilon}(\upsilon | \boldsymbol{C}_i), \end{aligned} \tag{12.15}$$

where $(\alpha_{k,i}, \beta_{k,i})$, $k = 1, 2$ are determined by (12.13).

This completes the proof.

As such, we have

$$f_{\kappa}(\boldsymbol{\kappa} | \boldsymbol{C}_i) = \frac{\gamma^{\alpha_{1,i}-1}\exp(-\gamma/\beta_{1,i})}{\Gamma(\alpha_{1,i})\beta_{1,i}^{\alpha_{1,i}}} \cdot \frac{\upsilon^{\alpha_{2,i}-1}\exp(-\upsilon/\beta_{2,i})}{\Gamma(\alpha_{2,i})\beta_{2,i}^{\alpha_{2,i}}}. \tag{12.16}$$

As a result, $f_i(\tau_{0,i}, \tau_{1,i}, \tau_{2,i}, \ldots, \tau_{N_i(s),i}, N_i(s)|\, \boldsymbol{C}_i)$ can also be updated by $\boldsymbol{C}_i$ as

$$f_i(\tau_{0,i}, \tau_{1,i}, \tau_{2,i}, \ldots, \tau_{N_i(s),i}, N_i(s)|\, \boldsymbol{C}_i) = \int_0^{+\infty} f_i(\tau_{0,i}, \tau_{1,i}, \tau_{2,i}, \ldots, \tau_{N_i(s),i}, N_i(s)|\, \kappa) f_\kappa(\kappa|\, \boldsymbol{C}_i)\mathrm{d}\kappa. \quad (12.17)$$

This updated information will be used to derive the RSL distribution which considers the future state transitions and the according number of transitions in the next section.

12.5 Model Formulation of the System Degradation Process to Predict the RSL

This section focuses on formulating the predicted RSL distribution of the system based on the information $(\boldsymbol{X}_i, \boldsymbol{C}_i)$ available up to t_i.

12.5.1 Predicting the RSL Conditional on the Model Parameters and Fixed System Operation Process

We first reformulate (12.1) into two parts: deterministic and stochastic terms as

$$X(t) = \eta(t) + \sigma B(t), \quad (12.18)$$

where $\eta(t) = x(0) + \int_0^t \lambda(\nu(u))\mathrm{d}u$ is the deterministic part and $\sigma B(t)$ is the stochastic part.

From our formulation of $\{X(t), t \geq 0\}$ in Sect. 12.3, we learn that $\eta(t)$ is a piecewise linear function whose line segments are defined by the system's operation state transition times $\{\tau_{j,0}\}$ at the initial time $t_0 = 0$, where $\tau_{j,0} \in \boldsymbol{F}_0(t)$. Namely, if the system's operation process is in the kth state during $[\tau_{j-1,0}, \tau_{j,0})$, then the degradation rate over this interval is λ_k and $\int_{\tau_{j-1,0}}^t \lambda(\nu(u))\mathrm{d}u = \lambda_k(t-\tau_{j-1,0})$ for $t \in [\tau_{j-1,0}, \tau_{j,0})$ and $k = 1, 2$. After the jth transition at $\tau_{j,0}$ to the operation state $m(m \neq k)$, the new degradation rate becomes λ_m, and so on.

In order to derive the RSL distribution, we first have a look at the distribution of $X(t)$. Then the probability that the degradation is below w at time t is

$$\begin{aligned}
&\Pr(X(t) \leq w|\boldsymbol{X}_0, \boldsymbol{C}_0, \boldsymbol{\theta}, \boldsymbol{F}_0(t)) \\
&= \Pr(\sigma B(t) < w - \eta(t)|\, \boldsymbol{X}_0, \boldsymbol{C}_0, \boldsymbol{\theta}, \boldsymbol{F}_0(t)) \\
&= \Pr\left(B(t) < \frac{w - \eta(t)}{\sigma} \middle| \boldsymbol{X}_0, \boldsymbol{C}_0, \boldsymbol{\theta}, \boldsymbol{F}_0(t)\right),
\end{aligned} \quad (12.19)$$

where $\boldsymbol{\theta}$ is the parameter vector associated with $\{X(t), t \geq 0\}$ and $\{\nu(t), t \geq 0\}$, consisting of $\boldsymbol{\kappa}$ and $\lambda_1, \lambda_2, \sigma^2$, which will be estimated from the available data. For notation simplification, we denote $\boldsymbol{\theta} = (\boldsymbol{\kappa}, \boldsymbol{\vartheta})$, where $\boldsymbol{\vartheta} = (\lambda_1, \lambda_2, \sigma^2)$. As such, the prediction for the storage lifetime T at $t_0 = 0$ can be written as

$$\begin{aligned}\Pr(T \leq t|\boldsymbol{X}_i, \boldsymbol{C}_i) &= \Pr\left(\sup_{t>0} X(t) \geq w \middle| \boldsymbol{X}_0, \boldsymbol{C}_0, \boldsymbol{\theta}, \boldsymbol{F}_0(t)\right) \\ &= \Pr\left(\sup_{t>0}\left(B(t) - \frac{w-\eta(t)}{\sigma}\right) \geq 0 \middle| \boldsymbol{X}_0, \boldsymbol{C}_0, \boldsymbol{\theta}, \boldsymbol{F}_0(t)\right). \end{aligned} \tag{12.20}$$

To further simplify notation, let $b(t) = \frac{w-\eta(t)}{\sigma}$. Thus, predicting the storage lifetime can be achieved by deriving the probability that $\{B(t), t \geq 0\}$ crosses the boundary $b(t)$. The method of this derivation is based on the idea that the event that $\{B(t), t \geq 0\}$ does not cross $b(t)$ over the interval $[0, U_0]$ (where $U_0 = \tau_{N_0(t),0}$ is the $N_0(t)$th transition time) may be split into $N_0(t)$ conditional events that $\{B(t), t \geq 0\}$ does not cross $b(t)$ over the interval $[\tau_{j-1,0}, \tau_{j,0})$ given that $\{B(t), t \geq 0\}$ does not cross $b(t)$ over the interval $[\tau_{j-2,0}, \tau_{j-1,0})$, where $\tau_{j,0}$ is the jth transition time. Let $b_j = b(\tau_{j,0}) = \frac{w-\eta(\tau_{j,0})}{\sigma}, j = 0, 1, \ldots, N_0(t)$, and further denote $N_0(t)$ as the number of state transitions on the interval $[0, U_0]$. Then we have $[0, U_0] = \bigcup_{j=1}^{N_0(t)} [\tau_{j-1,0}, \tau_{j,0})$. After U_0, no transition occurs.

Based on the above settings, we have the following lemma to derive the distribution of the lifetime T associated with $\{X(t), t \geq 0\}$ given in (12.1).

Lemma 12.1 *Let* $0 = \tau_{0,0} < \tau_{1,0} < \cdots < \tau_{N_0(t),0} = U_0$ *denote* $N_0(t)$ *transition times of the system's operation states and denote* $\boldsymbol{b} = (b_0, b_1, b_2, \ldots, b_{N_0(t)})'$. *The probability distribution of the lifetime* T *associated with* $\{X(t), t \geq 0\}$ *for the piecewise linear boundary* $b(t)$ *within* $[\tau_{j-1,0}, \tau_{j,0}), j = 1, \ldots, N_0(t)$ *is given as*

$$\begin{aligned}&\Pr(T \leq t|\boldsymbol{X}_0, \boldsymbol{C}_0, \boldsymbol{\theta}, \boldsymbol{F}_0(t)) = 1 - \\ &E\left[g\left(B(\tau_{1,0}), B(\tau_{2,0}), \ldots, B(\tau_{N_0(t),0}); \boldsymbol{b}\right) \middle| \boldsymbol{X}_0, \boldsymbol{C}_0, \boldsymbol{\theta}, \boldsymbol{F}_0(t)\right],\end{aligned} \tag{12.21}$$

where the function $g(\cdot)$ *is formulated as*

$$\begin{aligned}&g\left(z_1, z_2, \ldots, z_{N_0(t)}; b\right) = \\ &\prod_{j=1}^{N_0(t)} \boldsymbol{I}(z_j < b_j)\left(1 - \exp\left[-\frac{2(b_{j-1} - z_{j-1})(b_j - z_j)}{\tau_{j,0} - \tau_{j-1,0}}\right]\right)\end{aligned} \tag{12.22}$$

and $\boldsymbol{I}(\cdot)$ *is the indicator function.*

The proof of Lemma 12.1 can be achieved similarly by the method in [25] with some minor changes and thus is omitted here. In the following, we focus on predicting the RSL of the system at t_i ($t_i < U_i$), where $U_i = \tau_{N_i(s_i)}$. Suppose the degradation observations have been obtained at times $t_0, t_1, \ldots, t_i$ as $\boldsymbol{X}_i$ and let $\{\nu(t) : t_i \leq t \leq U_i\}$ denote the future system operation profile from t_i to U_i. Now the probability distribution of the RSL S_i at t_i given that the degradation process has not crossed the threshold w up to t_i can be obtained by the crossing probability of $\{X(t), t \geq 0\}$ to the threshold w. Specifically, we have the following theorem regarding the probability distribution of the RSL S_i at time t_i.

Theorem 12.3 *Let* $b_i(t) = \frac{1}{\sigma}(w - x_i - \int_{t_i}^{t} \lambda(v(u))du)$ *for* $t \in (t_i, t_i + s_i]$ *and* $A_i(s_i) = \{j : \tau_{j,i} \in [t_i, t_i + s_i]\}$. *The conditional probability distribution of the RSL* S_i *at time* t_i *can be formulated as*

$$\Pr(S_i \leq s_i | \boldsymbol{X}_i, \boldsymbol{C}_i, \boldsymbol{\theta}, \boldsymbol{F}_i(s_i)) = 1 - E\left[\prod_{j \in A_i(s_i)} \boldsymbol{I}(W_i(\tau_{j,i}) < b_i(\tau_{j,i})) \times \left(1 - \Lambda_{i,j}\right) \middle| \boldsymbol{X}_i, \boldsymbol{C}_i, \boldsymbol{\theta}, \boldsymbol{F}_i(s_i)\right] \quad (12.23)$$

with

$$\Lambda_{i,j} = \exp\left[-\frac{2\left(b_i(\tau_{j-1,i}) - W_i(\tau_{j-1,i})\right)\left(b_i(\tau_{j,i}) - W_i(\tau_{j,i})\right)}{\tau_{j,i} - \tau_{j-1,i}}\right],$$

where $W_i(\tau_{j,i}) = B(\tau_{j,i}) - B(t_i)$, *and* $|A_i(s_i)| = N_i(s_i) + 1$ *denotes the number of elements in* $A_i(s_i)$.

Proof Conditionally on $\boldsymbol{X}_i$, the degradation process can be formulated for each $t > t_i$ as

$$X(t) = x_i + \int_{t_i}^{t} \lambda(v(u))du + \sigma\left(B(t) - B(t_i)\right). \quad (12.24)$$

Then, by utilizing the notation of the RSL realization, s_i, the degradation process can be revised for each $s_i > 0$ as

$$X(t_i + s_i) = x_i + \int_{t_i}^{t_i + s_i} \lambda(v(u))du + \sigma\left(B(t_i + s_i) - B(t_i)\right). \quad (12.25)$$

It can be easily verified that $\{B(t_i + s_i) - B(t_i), s_i \geq 0\}$ is still a standard BM over time s_i [26]. Therefore, the RSL at time t_i corresponds to the FPT of $\{B(t_i + s_i) - B(t_i), s_i \geq 0\}$ crossing the boundary $b_i(t_i + s_i)$, denoted by $b_i(t_i + s_i) = \frac{1}{\sigma}(w - x_i - \int_{t_i}^{t_i+s_i} \lambda(v(u))du)$. Applying Lemma 12.1, we can directly complete the proof.

Noted that, to evaluate the expectation operator in (12.23), we need the joint PDF of $W_i(\tau_{j,i})$, $j \in A_i(s_i)$. For example, since $|A_i(s_i)| = N_i(s_i) + 1$, we can represent $A_i(s_i)$ as $A_i(s_i) = \{0, 1, 2, \ldots, N_i(s_i)\}$, and then by independent increments properties of the standard BM, the joint PDF of $W_i(\tau_{j,i})$, $j = 0, 1, 2, \ldots, N_i(s_i)$ can be formulated as

$$
\begin{aligned}
& f\left(W_i(\tau_{j,i}) = z_j,\, j \in A_i(s_i)\right) \\
& = \frac{1}{\sqrt{2\pi(\tau_{1,i} - t_i)}} \exp\left[-\frac{z_1^2}{2(\tau_{1,i} - t_i)}\right] \\
& \times \prod_{j=2}^{N_i(s_i)} \frac{1}{\sqrt{2\pi(\tau_{j,i} - \tau_{j-1,i})}} \exp\left[-\frac{(z_j - z_{j-1})^2}{2(\tau_{j,i} - \tau_{j-1,i})}\right].
\end{aligned} \tag{12.26}
$$

Remark 12.1 When σ is varying with the states, as the case of the degradation rate function $\lambda\,(\nu(t))$, we can define a function $\sigma : \Psi \to \Re^+$ so that $\sigma\,(\nu(t))$ represents the diffusion parameter at time t. For example, if the system is at the kth operation state at time t, then $\nu(t) = k$ and $\sigma\,(k) = \sigma_k$ is the diffusion parameter of the kth operation state, where $k = 1, 2$. In this case, (12.19) can be revised as

$$
\begin{aligned}
& \Pr\left(X(t) \le w | \boldsymbol{X}_0, \boldsymbol{C}_0, \boldsymbol{\theta}, \boldsymbol{F}_0(t)\right) \\
& = \Pr\left(B(t) < \frac{w - \eta(t)}{\sigma\,(\nu(t))} \middle| \boldsymbol{X}_0, \boldsymbol{C}_0, \boldsymbol{\theta}, \boldsymbol{F}_0(t)\right).
\end{aligned}
$$

Then, we can define $b(t)$ as $b(t) = \frac{w-\eta(t)}{\sigma(\nu(t))}$. Thus, predicting the storage lifetime can be achieved by deriving the probability that $\{B(t), t \ge 0\}$ crosses the boundary $b(t)$. According to the above setting, we know that $\sigma\,(\nu(t))$ is a piecewise constant function. As a result, $b(t) = \frac{w-\eta(t)}{\sigma(\nu(t))}$ in this case is still a piecewise linear function. Therefore, similar results to Lemma 12.1 and Theorem 12.3 can be directly established in this case by making some minor changes. ■

In the following, we focus on estimating the parameters in $\{X(t), t \ge 0\}$, based on $(\boldsymbol{X}_i, \boldsymbol{C}_i)$.

12.5.2 Bayesian Estimation for Parameters in the Degradation Process

In many applications, historical database and empirical knowledge might be available. By combining both historical and real-time data, we are able to estimate the unknown parameters in the presented model within a Bayesian framework. Recent studies about Bayesian-based state and parameter updating can be found in [19, 27, 28]. In this chapter, we develop a Bayesian approach to update the parameters of the degradation model using prior information estimated from historical data and real-time degradation observations $(\boldsymbol{X}_i, \boldsymbol{C}_i)$ available up to t_i.

The updating for the presented model, according to (12.18), requires to update the parameters λ_1, λ_2, and σ^2. For notational brevity, we denote them as a parameter vector $\boldsymbol{\vartheta}$. Accordingly, the prior distribution of $\boldsymbol{\vartheta}$ is represented as $p(\boldsymbol{\vartheta}) = p(\lambda_1, \lambda_2, \sigma^2)$. Regarding how to determine the prior distribution, similar method can be used based on the historical database by referring to [22]. After obtaining $(\boldsymbol{X}_i, \boldsymbol{C}_i)$, the posterior distribution of the model parameters can be calculated under the Bayesian framework as

$$f_{\vartheta}(\boldsymbol{\vartheta}|\,\boldsymbol{X}_i, \boldsymbol{C}_i) = f_{\vartheta}(\boldsymbol{\vartheta}|\,\boldsymbol{X}_i, \boldsymbol{C}_i) \propto p(\boldsymbol{\vartheta}) \times f_{X_i}(\boldsymbol{X}_i|\,\boldsymbol{C}_i, \boldsymbol{\vartheta}), \tag{12.27}$$

where $f_{X_i}(\boldsymbol{X}_i|\,\boldsymbol{C}_i, \boldsymbol{\vartheta})$ is the joint PDF of the degradation observations $\boldsymbol{X}_i$ up to t_i, given $\boldsymbol{C}_i$ and $\boldsymbol{\vartheta}$.

For an illustration, we assume the prior marginal distributions of $\lambda_1, \lambda_2, \sigma^2$ are mutually independent following $\lambda_1 \sim N(\mu_1, \sigma_1^2)$, $\lambda_2 \sim N(\mu_2, \sigma_2^2)$ and $\sigma^2 \sim \log N(\mu_3, \sigma_3^2)$, respectively, where the log-normal distribution is used to ensure $\sigma^2 > 0$. The parameters in these prior distributions can be determined by the historical database or empirical knowledge. In order to obtain the posterior estimates of $\boldsymbol{\vartheta}$, we need to derive the likelihood function $f_{X_i}(\boldsymbol{X}_i|\,\boldsymbol{C}_i, \boldsymbol{\vartheta})$. To do so, we consider the increments $X(t_k) - X(t_{k-1}), k = 1, 2, \ldots, i$. From the degradation model, we have

$$X(t_k) - X(t_{k-1}) = \int_{t_{k-1}}^{t_k} \lambda\,(v(u))\mathrm{d}u + \sigma\,(B(t_k) - B(t_{k-1}))\,, \tag{12.28}$$

where $k = 1, 2, \ldots, i$.

Thanks to the independent increments property of the standard BM, we know $\sigma\,(B(t_k) - B(t_{k-1})) \sim N(0, \sigma^2(t_k - t_{k-1}))$ for $k = 1, 2, \ldots, i$, and they are independent of each other. As a result, given $\boldsymbol{C}_i$, we can easily formulate $f_{X_i}(\boldsymbol{X}_i|\,\boldsymbol{C}_i, \boldsymbol{\vartheta})$ as

$$f_{X_i}(\boldsymbol{X}_i|\,\boldsymbol{C}_i, \boldsymbol{\vartheta}) = \prod\nolimits_{k=1}^{i} \phi_k(x_k - x_{k-1}), \tag{12.29}$$

where $\phi_k(x_k - x_{k-1})$ is the PDF of a normal random variable with mean $\int_{t_{k-1}}^{t_k} \lambda\,(v(u))\,du$ and variance $\sigma^2(t_k - t_{k-1})$.

Then, the posterior distribution of $\boldsymbol{\vartheta}$ is given by

$$f_{\vartheta}(\boldsymbol{\vartheta}|\,\boldsymbol{X}_i, \boldsymbol{C}_i) = \frac{p(\boldsymbol{\vartheta}) \times \prod_{k=1}^{i} \phi_k(x_k - x_{k-1})}{\int_{\vartheta} p(\boldsymbol{\vartheta}) \times \prod_{k=1}^{i} \phi_k(x_k - x_{k-1})\mathrm{d}\boldsymbol{\vartheta}}. \tag{12.30}$$

From the above procedure, the posterior distribution of $\boldsymbol{\vartheta}$ can be updated in line with the arrivals of the degradation observations when the system is in the working state.

12.5.3 RSL Prediction Considering the Future Transitions and Updated Parameters

Based on the result in Theorem 12.3 and the posterior estimates of the parameters, the predicted conditional RSL distribution at time t_i can be formulated by the law of total probability as (12.31)

$$\begin{aligned}\Pr(S_i \le s_i|\boldsymbol{X}_i, \boldsymbol{C}_i) &= E_{\theta,\boldsymbol{F}_i(s_i)|\boldsymbol{X}_i,\boldsymbol{C}_i}\left[\Pr(S_i \le s_i|\boldsymbol{X}_i, \boldsymbol{C}_i, \boldsymbol{\theta}, \boldsymbol{F}_i(s_i))\right] \\ &\qquad\qquad (\Pr(S_i \le s_i|\boldsymbol{X}_i, \boldsymbol{C}_i, \boldsymbol{\theta}, \boldsymbol{F}_i(s_i))\cdot \\ &= \int_{\theta,\tau_j\in\boldsymbol{F}_i(s_i),j\in A_i(s_i)} f_{\boldsymbol{F}_i}(\boldsymbol{F}_i(s_i)|\boldsymbol{X}_i, \boldsymbol{C}_i, \boldsymbol{\theta}) f_{\boldsymbol{\theta}}(\boldsymbol{\theta}|\,\boldsymbol{X}_i, \boldsymbol{C}_i))\mathrm{d}\boldsymbol{\theta}\mathrm{d}\tau_j,\end{aligned} \tag{12.31}$$

where $f_{\boldsymbol{F}_i}(\boldsymbol{F}_i(s_i)|\boldsymbol{X}_i, \boldsymbol{C}_i, \boldsymbol{\theta})$ is the joint PDF of the future operation state transition times and the number of transitions from time t_i to $t_i + s_i$; $f_{\boldsymbol{\theta}}(\boldsymbol{\theta}|\,\boldsymbol{X}_i, \boldsymbol{C}_i)$ is the posterior estimate of the model parameters; $\Pr(S_i \le s_i|\boldsymbol{X}_i, \boldsymbol{C}_i, \boldsymbol{\theta}, \boldsymbol{F}_i(s_i))$ is the RSL distribution given the parameters and future operation profile, determined by Theorem 12.3.

Based on the model development in Sects. 12.4.2 and 12.5.2, we have

$$f_{\boldsymbol{\theta}}(\boldsymbol{\theta}|\,\boldsymbol{X}_i, \boldsymbol{C}_i) = f_{\boldsymbol{\kappa}}(\boldsymbol{\kappa}|\,\boldsymbol{C}_i)\cdot f_{\boldsymbol{\vartheta}}(\boldsymbol{\vartheta}|\,\boldsymbol{X}_i, \boldsymbol{C}_i), \tag{12.32}$$

and

$$\begin{aligned}f_{\boldsymbol{F}_i}(\boldsymbol{F}_i(s_i)|\boldsymbol{X}_i, \boldsymbol{C}_i, \boldsymbol{\theta}) &= f_{\boldsymbol{F}_i}(\boldsymbol{F}_i(s_i)|\boldsymbol{\kappa}) \\ &= f_i(\tau_{0,i}, \tau_{1,i}, \tau_{2,i}, \ldots, \tau_{N_i(s_i)}, N_i(s_i)\big|\,\boldsymbol{\kappa}),\end{aligned} \tag{12.33}$$

and one of the specifications of $f_{\boldsymbol{F}_i}(\boldsymbol{F}_i(s_i)|\boldsymbol{X}_i, \boldsymbol{C}_i, \boldsymbol{\theta})$ is

$$f_{\boldsymbol{F}_i}(\boldsymbol{F}_i(s_i)|\boldsymbol{X}_i, \boldsymbol{C}_i, \boldsymbol{\theta}) = f_i(\tau_{0,i}, \tau_{1,i}, \tau_{2,i}, \ldots, \tau_{L,i}, N_i(s_i) = L\big|\,\boldsymbol{\kappa}), \tag{12.34}$$

where $f_i(\tau_{0,i}, \tau_{1,i}, \tau_{2,i}, \ldots, \tau_{L,i}, N_i(s_i) = L\big|\,\boldsymbol{\kappa})$, $f_{\boldsymbol{\kappa}}(\boldsymbol{\kappa}|\,\boldsymbol{C}_i)$, and $f_{\boldsymbol{\vartheta}}(\boldsymbol{\vartheta}|\,\boldsymbol{X}_i, \boldsymbol{C}_i)$ are determined by Theorems 12.1 and 12.2 (or (12.16)), and (12.30), respectively.

Therefore, (12.31) can be further written as

$$\begin{aligned}&\Pr(S_i \le s_i|\boldsymbol{X}_i, \boldsymbol{C}_i) = \\ &\sum_{L=0}^{\infty}\int_{\boldsymbol{\theta},\tau_{j,i}\in\boldsymbol{F}_i(s_i)}\left[\begin{array}{l}\Pr(S_i \le s_i|\boldsymbol{X}_i, \boldsymbol{C}_i, \boldsymbol{\theta}, \tau_{0,i}, \tau_{1,i}, \\ \tau_{2,i}, \ldots, \tau_{L,i}, N_i(s_i) = L)\times \\ f_i(\tau_{0,i}, \tau_{1,i}, \tau_{2,i}, \ldots, \tau_{L,i}, N_i(s_i) = L\big|\,\boldsymbol{\kappa})\times \\ f_{\kappa}(\boldsymbol{\kappa}|\,\boldsymbol{C}_i) f_{\vartheta}(\vartheta|\,\boldsymbol{X}_i, \boldsymbol{C}_i)\mathrm{d}\boldsymbol{\theta}\mathrm{d}\tau_{j,i}\end{array}\right]\end{aligned} \tag{12.35}$$

From (12.35), we can also observe the dependency of the predicted RSL on the degradation observations and the system's operation progression history up to t_i. However, despite the simple and straightforward formulations, it is not easy to directly and analytically compute the right-hand side of (12.35) because the computation involves multi-dimensional integration. As such, we need a simulation-based solution method to tackle this problem. Targeting this objective, we employ a Monte Carlo simulation algorithm to calculate $\Pr(S_i \le s_i|\boldsymbol{X}_i, \boldsymbol{C}_i)$ at time t_i for every given s_i.

Algorithm 12.1 (*Simulation-based solution algorithm*)

Step 1: Based on $(\boldsymbol{X}_i, \boldsymbol{C}_i)$, calculate the posterior distributions $f_{\boldsymbol{\kappa}}(\boldsymbol{\kappa}|\,\boldsymbol{C}_i)$ and $f_{\boldsymbol{\vartheta}}(\boldsymbol{\vartheta}|\,\boldsymbol{X}_i, \boldsymbol{C}_i)$ of $\boldsymbol{\kappa}$ and $\boldsymbol{\vartheta}$ by (12.16) and (12.30), respectively.

Step 2: Select a sufficiently large number (denoted by M_1) to simulate the realizations of $\boldsymbol{\theta}$ from $f_{\boldsymbol{\theta}}(\boldsymbol{\theta} | \boldsymbol{X}_i, \boldsymbol{C}_i)$. These realizations of $\boldsymbol{\theta}$ are expressed by $\boldsymbol{\theta}_r = (\boldsymbol{\kappa}_r, \boldsymbol{\vartheta}_r), r = 1, \ldots, M_1$, where $\boldsymbol{\kappa}_r$ is generated from $f_{\boldsymbol{\kappa}}(\boldsymbol{\kappa} | \boldsymbol{C}_i)$ and $\boldsymbol{\vartheta}_r$ from $f_{\boldsymbol{\vartheta}}(\boldsymbol{\vartheta} | \boldsymbol{X}_i, \boldsymbol{C}_i)$.

Step 3: Select the number of the future system's operation process profiles for each realization $\boldsymbol{\theta}_r$ of $\boldsymbol{\theta}$, $K_r, r = 1, \ldots, M_1$.

Step 4: For each $\boldsymbol{\theta}_r$ and K_r, given $v(t_i)$ and s_i, simulate system's operation process paths on the interval $[t_i, t_i + s_i]$, $\left\{v_{r,m}(u) : t_i < u \le t_i + s_i\right\}$ for $m = 1, 2, \ldots, K_r$ and obtain the future system's operation profile sets $\boldsymbol{F}_{i,r,m}(s_i) = \{\tau_{0,i,r,m}, \tau_{1,i,r,m}, \ldots, \tau_{N_{i,r,m}(s_i),i,r,m}, N_{i,r,m}(s_i)\}$ and $A_{i,r,m}(s_i) = \{0, 1, 2, \ldots N_{i,r,m}(s_i)\}$, where $r = 1, \ldots, M_1$. If $N_{i,r,m}(s_i) = 0$, then $\tau_{0,i,r,m} = t_i$, representing no transition.

Step 5: For each $\boldsymbol{F}_{i,r,m}(s_i)$, calculate $b_{i,r,m}(\tau_{j,i,r,m}) = x_i + \int_{t_i}^{\tau_{j,i,r,m}} \lambda\left(v(u)\right) du$, $j = 0, 1, \ldots, N_{i,r,m}(s_i)$, where $m = 1, 2, \ldots, K_r$ and $r = 1, \ldots, M_1$.

Step 6: Calculate $\Pr(S_i \le s_i | \boldsymbol{X}_i, \boldsymbol{C}_i, \boldsymbol{\theta}_r, \boldsymbol{F}_{i,r,m}(s_i))$ via Theorem 12.3 for each sampling path generating $\boldsymbol{F}_{i,r,m}(s_i)$, where $m = 1, 2, \ldots, K_r$. Specifically, we compute $\Pr(S_i \le s_i | \boldsymbol{X}_i, \boldsymbol{C}_i, \boldsymbol{\theta}_r, \boldsymbol{F}_{i,r,m}(s_i))$ by the following steps:

1. Select a sufficiently large number (e.g. M_2) of realizations for $N_{i,r,m}(s_i)$-dimensional BM process $\{B(\tau_{j,i,r,m}) - B(t_i), j = 1, \ldots, N_{i,r,m}(s_i)\}$.
2. For each $l = 1, 2, \ldots, M_2$, generate the lth realization $z^l_{j,i,r,m}$ of $B(\tau_{j,i,r,m}) - B(t_i)$, $j = 1, \ldots, N_{i,r,m}(s_i)$. This can be achieved by the simulation technique of a standard BM based on the independent increments property of BM.
3. By the law of large numbers, for sufficiently large M_2, the conditional RSL distribution can be calculated by

$$\Pr(S_i \le s_i | \boldsymbol{X}_i, \boldsymbol{C}_i, \boldsymbol{\theta}_r, \boldsymbol{F}_{i,r,m}(s_i)) \approx 1 - \frac{1}{M_2} \sum_{l=1}^{M_2} g^l_{i,r,m}, \tag{12.36}$$

where $g^l_{i,r,m}$ is given in (12.37):

$$g^l_{i,r,m} = \prod_{j \in A_{i,r,m}(s_i)} \boldsymbol{I}(z^l_{j,i,r,m} < b_{i,r,m}(\tau_{j,i,r,m})) \times \\ \left(1 - \exp\left[-\frac{2\left(b_{i,r,m}(\tau_{j-1,i,r,m}) - z^l_{j-1,i,r,m}\right)\left(b_{i,r,m}(\tau_{j,i,r,m}) - z^l_{j,i,r,m}\right)}{\tau_{j,i,r,m} - \tau_{j-1,i,r,m}}\right]\right). \tag{12.37}$$

Step 7: By the law of large numbers, for sufficiently large M_1 and K_r, the RSL distribution can be calculated by

$$\Pr(S_i \le s_i | \boldsymbol{X}_i, \boldsymbol{C}_i) \approx \\ \frac{1}{M_1} \sum_{r=1}^{M_1} \left(\frac{1}{K_r} \sum_{m=1}^{K_r} \Pr(S_i \le s_i | \boldsymbol{X}_i, \boldsymbol{C}_i, \boldsymbol{\theta}_r, \boldsymbol{F}_{i,r,m}(s_i))\right). \tag{12.38}$$

Based on the above algorithm, we can also obtain the conditional reliability, point estimate, and confidence interval (CI) of the predicted RSL at t_i. Specifically, the according conditional reliability at t_i can be computed as

$$R(s_i|\boldsymbol{X}_i, \boldsymbol{C}_i) = 1 - \Pr(S_i \leq s_i|\boldsymbol{X}_i, \boldsymbol{C}_i). \tag{12.39}$$

For the point estimate of the predicted RSL at t_i, we use the median of the RSL distribution computed at t_i, denoted by $\hat{s}_i$, which can be obtained by solving $\Pr(S_i \leq \hat{s}_i|\boldsymbol{X}_i, \boldsymbol{C}_i) = 0.5$. Here the median is chosen as an approximate value for the point estimate of the predicted RSL since the distribution of the RSL is often highly skewed and thus it is reasonable to use the median as a measure of central tendency rather than the mean. Another quantity of interest is the CI of the predicted RSL. For any constant $0 < \alpha < 1$, let $s_{\alpha,i}$ be the time satisfying

$$\Pr(S_i \leq s_{\alpha,i}|\boldsymbol{X}_i, \boldsymbol{C}_i) = 1 - \alpha. \tag{12.40}$$

Then we can employ (12.35) and the aforementioned simulation algorithm to calculate $s_{\alpha/2,i}$ and $s_{1-\alpha/2,i}$ numerically. Thereafter, the $100\%(1-\alpha)$ CI of the predicted RSL at t_i is represented as $(s_{1-\alpha/2,i}, s_{\alpha/2,i})$.

12.6 Case Study

In this section, we provide a practical case study for the inertial platform used in the inertial navigation systems (INS) of rockets to illustrate the application of the developed RSL prediction approach.

12.6.1 Background and Data Description

As a key system, an inertial platform, which is a electromechanical system, plays an important and irreplaceable role in the INS. Its health condition has a direct and dominant influence on the navigation precision. For an illustrative purpose, we provide an illustration of the structure of the inertial platform in Fig. 12.1, which consists of gyroscopes, motors, accelerometers, etc. The sensors fixed in an inertial platform include three gyroscopes and three accelerometers, which measure angular velocity and linear acceleration, respectively.

Most of the inertial platform's lifecycle is spent in the storage condition, unless some missions or routine CM are required to put it into the working state. Namely, the inertial platform will experience switches between the working state and storage state. For an illustrative purpose, we provide an illustration of a deformed bearing of the motor (see Fig. 12.2), which was obtained by scanning electron microscopy S-3700N with amplification factor 70. It can be found that the maximum length of the

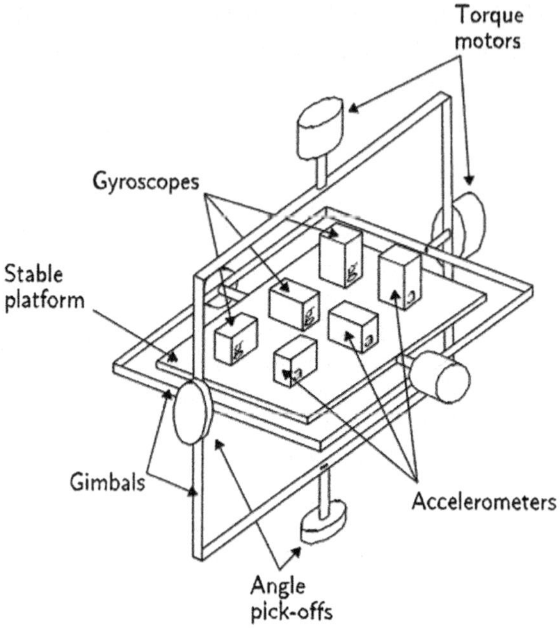

Fig. 12.1 Illustration of the structure of the inertial platform

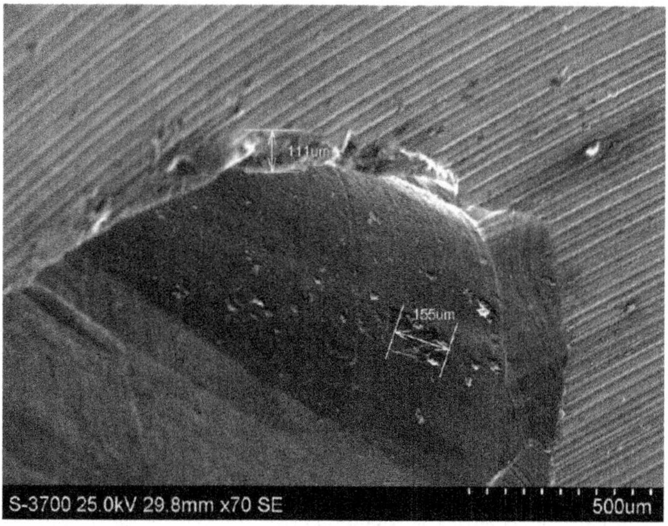

Fig. 12.2 Illustration of a deformed bearing

metal flake is 155 μm. Historical statistical data show that almost 80% of the failures of the INS result from the inertial platform due to the wear of its bearings, whose performance is reflected by the drift values measured by gyroscopes. It is therefore natural to utilize the collected drift data as the degradation signals to predict the RSL of the inertial platform.

The gyroscope fixed on the inertial platform is a electromechanical structure having two degrees of freedom from the driver and sense axis. Generally, the drift measurement along the sense axis, K_{SX}, plays a dominant role in the health assessment of the inertial platform. In our study, we use the CM data of K_{SX} to predict the RSL. It is noted here the selection of the failure threshold is an important problem in practice. In the case study of this chapter, the failure threshold w is determined according to the performance requirement of the INS. In the INS health monitoring practice, it is required that K_{SX} should not exceed 0.36°/h, i.e., $w = 0.36°/\text{h}$. This threshold is predetermined at the design stage and is strictly enforced in practice since an INS is a critical device used in a navigated rocket system.

The used data are collected from January 2004 to April 2008 for a real-life inertial platform armed in the INS, where a PC-based data acquisition system is used to acquire and store the drift data. The number of the CM data in this data set is 109, including the initial value $x(0) = 0$. In addition, this data set records the system's operation information during January 2004–April 2008 such as the times for each transition between the working state and storage state, and the number of total transitions. Namely, the system's operation information $\boldsymbol{C}_i$ required at each CM time in this chapter is readily available. In addition, the storage lifetime of this investigated inertial platform is known, since the drift value of the gyroscope has crossed the failure threshold and the recorded failure time corresponds to the storage lifetime, i.e., 37,200 h. Thus, the actual RSL at each CM time is known and used for subsequent validation. Here, a year is equal to 8760 h (a year = 365×24 h). For illustration, the partial system's operation information from January 2004 to January 2008 is summarized in Table 12.1.

The data of the monitored drift coefficient K_{SX} in the working state can also be obtained over the number of the CM. Note that, to protect sensitive proprietary information, we have had to change the timescale and make some manipulations on

Table 12.1 Partial system's operation information from January 2004 to April 2008

t_i (hours)	System operation state	Number of the CM
0–15	"1"	6
15–2880	"2"	0
2880–3000	"1"	8
3000–7320	"2"	0
7320–7330	"1"	4
7330–8740	"2"	0

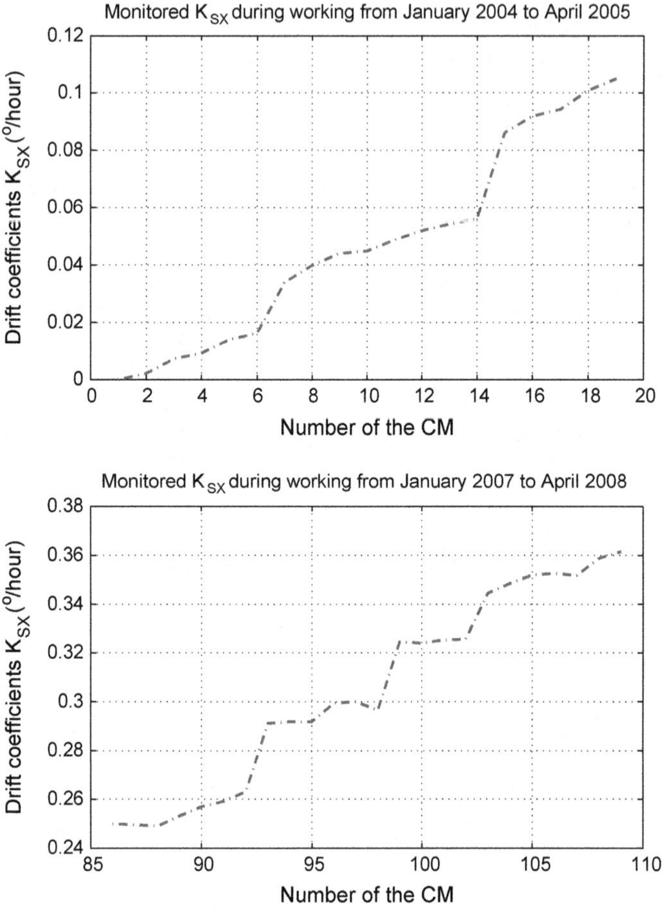

Fig. 12.3 The CM data of K_{SX} collected in the working state

the data but not lose the basic features in the data. Here we illustrate the CM data from January 2004 to April 2008 in Fig. 12.3.

From the data illustrated in Fig. 12.3, which corresponds to the system's operation process given in Table 12.1, it can be observed that the system will partially degrade during the storage period as the increments of the degradation before and after the storage are positive for our case of the inertial platform. This is consistent with the previous discussions. In the following, we utilize the data described above to illustrate the prediction results of the RSL.

12.6.2 Results and Discussions

In order to apply our proposed approach for the RSL prediction, we first need to specify the hyper-parameters in prior distributions of $\boldsymbol{\kappa}$ and $\boldsymbol{\vartheta}$. The hyper-parameters in prior distributions of $\boldsymbol{\kappa}$ and $\boldsymbol{\vartheta}$ can be determined by the history data or empirical knowledge. For example, the routine CM often lasts fifteen hours while the duration of each storage period is usually about three or four months. Based on the history data or engineering knowledge of the inertial platform, we empirically specify the hyper-parameters in prior distributions as follows: $\gamma \sim \text{gama}(13.6, 1.1)$, $\upsilon \sim \text{gama}(16, 180)$, $\lambda_1 \sim N(0.0013, 4 \times 10^{-8})$, $\lambda_2 \sim N(9 \times 10^{-6}, 1.4 \times 10^{-12})$, $\sigma^2 \sim \log N(-5.49, 1.05 \times 10^{-10})$. In such cases, we have $E[\gamma] = 14.96$ (about 15 h) and $E[\upsilon] = 2880$ (about 4 months).

Based on the specified prior parameters, we can use the CM data illustrated in Fig. 12.3 and Table 12.1 to predict the RSL distribution of the inertial platform, i.e., $\Pr(S_i \leq s_i | \boldsymbol{X}_i, \boldsymbol{C}_i)$. In performing the RSL predictions, the simulation algorithm in Sect. 12.5 is used, in which we set $M_1 = 1000$, $M_2 = 1000$, and $K_r = 500$ for each $r \in \{1, 2, \ldots, 1000\}$. The calculated values of $\Pr(S_i \leq s_i | \boldsymbol{X}_i, \boldsymbol{C}_i)$ at the 6th CM point and the 100th CM point by the simulation algorithm are illustrated in Fig. 12.4 as follows.

From Fig. 12.4, it is found that the presented method can achieve the RSL prediction and find the median of the RSL at each CM point by solving $\Pr(S_i \leq s_i | \boldsymbol{X}_i, \boldsymbol{C}_i) = 0.5$ numerically. To have a further look at the accuracy of the presented method in predicting the RSL, we calculate the point estimations by median and the CIs at several different CM points, as summarized in Table 12.2, where the relative error (RE) of the observed RSL and the predicted RSL are used for comparisons. Let $\tilde{S}_i$ denote the actual RSL at t_i, and let $\hat{s}_i$ be the predicted RSL by its median at t_i. Hence, the RE for the RSL at t_i is defined as $RE_i = \left(\left| \tilde{S}_i - \hat{s}_i \right| / \tilde{S}_i \right) \times 100$.

From Table 12.2, we have a conclusion that the accuracy of the predicted RSL by our presented approach will be improved with the accumulation of the CM data in the working state. Also, all REs between the actual RSL and the predicted RSL are below 15%. This further reflects the effectiveness of our method. By contrast, if we only consider the working state, then the predicted storage lifetime is about 280 h, which is very poor, since the long-term storage period is not taken into account. On the other hand, if we ignore the loss of the life in the working period, then the predicted storage lifetime will be more than seven years corresponding to 61,320 h, which is also unacceptable. In contrast, our presented method can generate reasonable results and has the potential application in predicting the RSL.

It is noted that, to implement the presented RSL prediction approach, the hyper-parameters in prior distributions should be first specified. In the above computational results, the hyper-parameters are empirically determined. To have a look at the robustness of the RSL prediction over hyper-parameters, we randomly generate 100 sets of hyper-parameters of prior distributions and then perform the RSL predictions at the 6th CM point, the 50th CM point and the 100th CM point with these hyper-

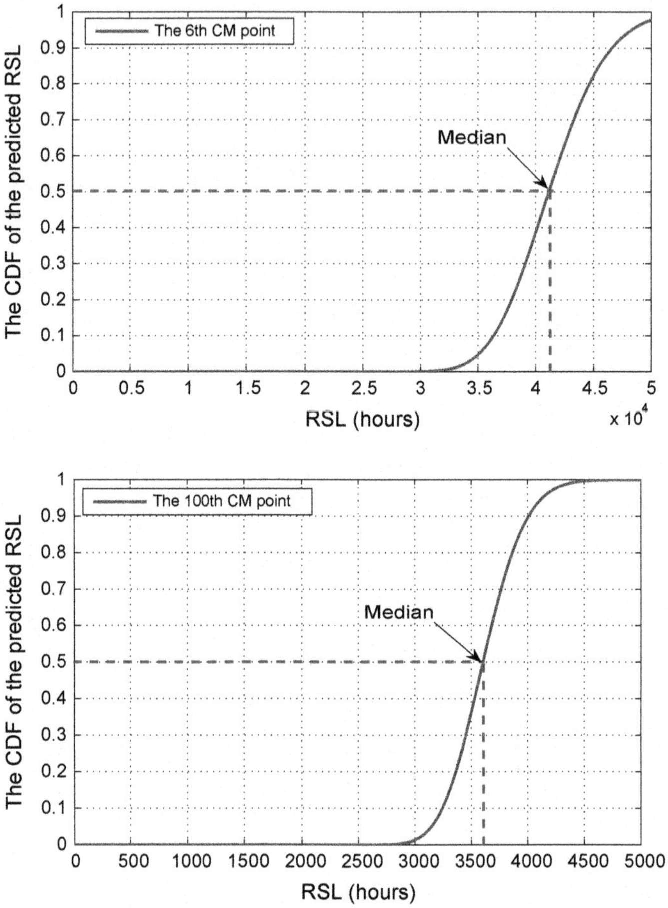

Fig. 12.4 Values of $\Pr(S_i \leq s_i | \boldsymbol{X}_i, \boldsymbol{C}_i)$ at the 6th CM point and the 100th CM point

Table 12.2 Estimated mean RSLs (hours) and corresponding REs

	Predicted RSL	RE (%)	95% CI
The 6th CM point	41,500	11.56	(34,051, 49,671)
The 50th CM point	18,026	8.45	(14,982, 21,456)
The 100th CM point	3625	4.52	(3064, 4240)

Table 12.3 Mean statistics of the RSL predictions with randomly selected hyper-parameters

	Predicted RSL	RE (%)	95% CI
The 6th CM point	45,299	21.82	(41,562, 52,639)
The 50th CM point	18,770	12.93	(15,108, 22,627)
The 100th CM point	3705	6.83	(3147, 4352)

parameters by the simulation algorithm. The mean statistics of the computational results are summarized in Table 12.3.

From the results in Table 12.3, we can find that, in contrast with the results in Table 12.3, the accuracy of the RSL predictions is affected by the selection of hyper-parameters, particularly at the early stage with less data. However, as the CM data are accumulated, the impact of hyper-parameters can be effectively reduced and the accuracy of the RSL predictions can be improved to the same level as that of Table 12.2. This indicates that the presented RSL prediction approach has certain robustness over hyper-parameters in the prior distributions and thus has application potentials.

References

1. Pecht M (2008) Prognostics and health management of electronics. Wiley, New Jersey
2. Ye ZS, Shen Y, Xie M (2012) Degradation-based burn-in with preventive maintenance. Eur J Oper Res 221(2):360–367
3. Lall P, Lowe R, Goebel K (2012) Prognostics health management of electronic systems under mechanical shock and vibration using Kalman filter models and metrics. IEEE Trans Ind Electron 59(11):4301–4314
4. Mclain K, Warren R (1990) Automated reliability life data analysis of missiles in storage and flight. In: Proceedings annals reliability and maintainability symposium, pp 490–493
5. Zhao M, Xie M, Zhang YT (1995) A study of a storage reliability estimation problem. Qual Reliab Eng Int 11(1):123–127
6. Luo K, Han LM, Fang HZ, Hou SM (2012) Research on storage life prediction method for strap-down inertial navigation system. In: Prognostics and system health management conference (PHM-2012 Beijing), MU3026, pp 1–6
7. Si XS, Wang W, Hu CH, Zhou DH (2011) Remaining useful life estimation-A review on the statistical data driven approaches. Eur J Oper Res 213(1):1–14
8. Soualhi A, Razik H, Clerc G, Doan D (2013) Prognosis of bearing failures using hidden Markov models and the adaptive neuro-fuzzy inference system. IEEE Trans Ind Electron. doi:10.1109/TIE.2013.2274415
9. Vasan ASS, Long B, Pecht M (2013) Diagnostics and prognostics method for analog electronic circuits. IEEE Trans Ind Electron 60(11):5277–5291
10. Shahriari M, Farrokhi M (2013) Online state-of-health estimation of VRLA batteries using state of charge. IEEE Trans Ind Electron 60(1):191–202
11. Strangas EG, Aviyente S, Neely JD, Zaidi SSH (2013) The effect of failure prognosis and mitigation on the reliability of permanent-magnet AC motor drives. IEEE Trans Ind Electron 60(8):3519–3528
12. Orchard ME, Hevia-Koch P, Zhang B, Tang L (2013) Risk measures for particle-filtering-based state-of-charge prognosis in lithium-ion batteries. IEEE Trans Ind Electron 60(11):5260–5269

13. Ye ZS, Wang Y, Tsui KL, Pecht M (2013) Degradation data analysis using Wiener processes with measurement errors. IEEE Trans Reliab. doi:10.1109/TR.2013.2284733
14. Perfetti G, Aubert T, Wildeboer WJ, Meesters GMH (2011) Influence of handling and storage conditions on morphological and mechanical properties of polymer-coated particles: characterization and modeling. Powder Technol 206(1):99–111
15. Si XS, Wang W, Chen MY, Hu CH, Zhou DH (2013) A Wiener-process-based degradation model with a recursive filter algorithm for remaining useful life estimation. Mech Syst Signal Process 35(1–2):219–237
16. Çinlar E, Ëzekici S (1987) Reliability of complex devices in random environments. Probab Eng Inf Sci 1(1):97–115
17. Hawkes AG, Cui LR, Zheng ZH (2012) Modeling the evolution of system reliability performance under alternative environments. IIE Trans 43(11):761–772
18. Ye ZS, Chen N (2013) The inverse Gaussian process as a degradation model. Technometrics. doi:10.1080/00401706.2013.830074
19. He W, Williard N, Osterman M, Pecht M (2011) Prognostics of lithium-ion batteries based on Dempster-Shafer theory and the Bayesian Monte Carlo method. J Power Sources 196(23):10314–10321
20. Bae SJ, Kvam PH (2006) A change-point analysis for modeling incomplete burn-in for light displays. IIE Trans 38:489–498
21. Feng J, Sun Q, Jin TD (2012) Storage life prediction for a high-performance capacitor using multi-phase Wiener degradation model. Commun Stat-Simul Comput 41(8):1317–1335
22. Chen N, Tsui KL (2013) Condition monitoring and residual life prediction using degradation signals: revisited. IIE Trans 45(9):939–952
23. Gebraeel NZ, Pan J (2008) Prognostic degradation models for computing and updating residual life distributions in a time-varying environment. IEEE Trans Reliab 57(4):539–550
24. Lee M-LT, Whitmore GA (2006) Threshold regression for survival analysis: modeling event times by a stochastic process reaching a boundary. Stat Sci 21(4):501–513
25. Wang L, Potzelberger K (1997) Boundary crossing probability for Brownian motion and general boundaries. J Appl Probab 34(1):54–65
26. Cox DR, Miller HD (1965) The theory of stochastic processes. Methuen and Company, London
27. Rahimi-Eichi H, Baronti F, Chow MY (2014) Online adaptive parameter identification and state-of-charge coestimation for lithium-polymer battery cells. IEEE Trans Ind Electron 61(4):2053–2061
28. Gholizadeh M, Salmasi FR (2014) Estimation of state of charge, unknown nonlinearities, and state of health of a lithium-ion battery based on a comprehensive unobservable model. IEEE Trans Ind Electron 61(3):1335–1344

Part IV
Applications of Prognostic Information

Chapter 13
Reliability Estimation Approach for PMS

13.1 Introduction

Many complex systems are designed to perform missions that consist of several phases in which the deterioration and configuration of systems may change from phase to phase. These systems are called phased-mission systems (PMSs) [1]. PMSs are formally defined to be the systems subject to multiple, consecutive, nonoverlapping phases of operation required to finish the final product or service [2]. A typical PMS is the on-board systems for the aided guide of aircraft, whose mission consists of takeoff, ascent, cruise, approach, and landing phases. For mission success, all phases must be completed without failure. Other PMSs include safety-critical systems (such as aerospace systems and weapon systems), and modern manufacturing processes (e.g., assembly, machining, semiconductor fabrication and pharmaceutical manufacturing) [3]. As an important measure for system design, operation and maintenance of PMSs [4, 5], reliability can be used to quantify the performance of PMSs. Accurate estimation of the reliability is very helpful for efficient maintenances and logistic supports of such systems, which actually lead to lifecycle cost reduction and the avoidance of catastrophic failures.

In the past decades, many methods have been developed to analyze the reliability of PMSs based on fault trees (FTs). Esary and Ziehms in [1] first introduced a FT-based method to transform a phased mission into an equivalent single phase mission, and analyzed the system's reliability. Burdick et al. in [6] and Veatch in [7] analyzed PMSs with non-repairable components. Alam and Al-Saggaf in [8] analyzed repairable systems with deterministic phase durations. Kim and Park in [9] considered systems with generally distributed phase durations. Dugan in [10] proposed a methodology which combined all phases into one model and used a standard Markov chain solution technique to calculate different reliability measures. Vaurio in [11] analyzed PMSs with repairable and non-repairable components using FTs.

Due to the dependencies across phases, FT analysis of PMSs is computationally expensive to calculate the reliability of a mission, if the system structure is complex, or if there are a high number of phases [12]. Several chapters tried to reduce the

X.-S. Si et al., *Data-Driven Remaining Useful Life Prognosis Techniques*,
Springer Series in Reliability Engineering, DOI 10.1007/978-3-662-54030-5_13

computational burden [13–16]. In [16], the phased algebra was used together with noncoherent mission FTs to directly evaluate the probability of failure in individual phases. However, FT-based methods are still unsuitable to analyze large systems, particularly those with noncoherent FTs. This led to the appearance of binary decision diagram (BDD) techniques [5, 12, 17].

Zang et al. in [17] developed a BDD-based algorithm for non-repairable systems with general failure distributions. Xing and Dugan in [18] analyzed a more general class of systems including PMSs with combinational phase requirement and imperfect coverage. Using BDD, Tang and Dugan in [19] analyzed PMSs with multimode failures, in which different failure modes had different failure rates and effects. Tang et al. in [20] analyzed the reliability of a PMS with common cause failures. This model was extended by Xing [21] to include imperfect coverage and multiple common causes. Wang and Trivedi in [22] proposed a hierarchical modeling approach based on BDD for PMSs with repairable components, where the deterioration process of each component was modeled by a finite state continuous-time Markov chain.

Due to the nature of the BDD, the cancelation of common components among phases can be combined with the BDD generation, without additional operations, and the SDP can be implicitly represented by the final PMS BDD. Zang et al. in [17] showed that BDD-based algorithm was more efficient than other algorithms in both computation time and storage space. To apply the technique to reliability analysis, FTs can be analyzed through the BDD method. However, for the BDD-based FT analysis of PMSs, it is important to generate the PMS BDD efficiently. For the BDD generation from a FT, the ordering of variables is critical since the size of a BDD (the number of nodes) heavily depends on the ordering. But the problem of computing the ordering that minimizes the size of the BDD is itself a NP-Complete problem [23]. Çekyay and Özekici in [24] analyzed the reliability of PMSs under a general setting by proposing different reliability definitions. In a recent study, Çekyay and Özekici in [25] analyzed MTTF and availability of PMSs under the maximal repair policy where the mission process was represented by a semi-Markov process with random sequence and durations of phases.

By the literature, current works strongly depend on the knowledge of the structure of PMSs to generate the FT or BDD and then to use logic rule to estimate the reliability of PMSs. However, in practice, the structure of the mission system is too complicated and the complete knowledge is not always available. This leads to a great difficulty to apply existing approaches for reliability estimation of a practical PMS. In addition, current approaches focus more on a population of common type of PMSs and there is no work directly establishing the linkage between the reliability and the historical data/real-time CM information of individual PMS in service. Most works only consider the static scenario in off-line manner. As far as we know, real-time reliability analysis is desirable in practice so that the reliability of a PMS is repeatedly updated during its operation to ensure the most recently calculated reliability value accurately reflects the current reality of the PMS. Finally, most works assume that the degradation of the PMS follows a finite state continuous/discrete-time Markov chain. This makes the lifetime estimation of the PMS depend only upon the current state but ignore the possible history information to date due to the memory-less nature of Markov chain. The focus of this chapter is to develop a reliability method for PMS when the structure of the PMS is not available but the monitored data can be collected.

In this chapter, we will use the available CM data to analyze the reliability of a PMS in service. We present a novel condition-based approach to estimate the reliability of PMSs using the CM information and the degradation data of such system under dynamic operating scenario. This chapter aims to estimate the reliability for an individual PMS in service but not for a population, which is totally different from most existing methods only considering the static scenario without using the real-time information. In order to establish a linkage among the reliability, the historical data, and real-time information of the individual PMS, we adopt a stochastic filtering model to model the phase duration and obtain the updated estimation of the mission time by Bayesian law. At the meantime, the lifetime of PMS is estimated by degradation data based on a Brownian motion (BM) with a time-varying drift coefficient depending on the mission phase, in which the lifetime distribution of PMS can be updated under Bayesian framework once new information is available. Unique to this work is the union of the CM data and degradation data of PMS to real-time estimate the mission reliability through the estimated distribution of the mission time in conjunction with the estimated lifetime distribution, which considers the dependency of the degradation rate of PMS on mission phase. We demonstrate the usefulness of the developed approach via a numerical example and a case study.

The main differences between our work and existing works are summarized in the following aspects. (1) We do not use a FT or a BDD to model PMSs while most existing works required the structural knowledge of the mission system in each phase. (2) We use the CM information and the degradation data to estimate the reliability of PMSs under dynamic operating scenario considering the dependency of the degradation rate of PMS on mission phase. However, most existing methods only consider the static scenario without using real-time information. (3) We consider the mission reliability estimation for an individual PMS in service while previous studies only considered the mission reliability for a population of the same type of PMSs. (4) The mission phase durations in our work can follow log-normal distributions estimated from the CM information while exponential distributions were used in existing works. (5) The lifetime of the mission system, the posterior probability of the system phase duration and the mission time depend on the history of observations. Hence the estimated mission reliability is also dependent on the history of observations and free of the structural constraints.

The remainder parts are organized as follows. In Sect. 13.2, we present the notation, the problem description, and the model assumptions. In Sect. 13.3, we formulate the mission time estimation from the CM information. Section 13.4 formulates the degradation data-based lifetime estimation for the mission system. Section 13.5 presents the formulations for the mission reliability estimation. In Sect. 13.6, we provide a numerical study and a case study to illustrate the effectiveness of the developed approach.

13.2 Assumptions and Problem Description

13.2.1 Problem Description

In general, there are two main definitions of the reliability of PMSs. One is the probability that the mission can be successfully accomplished before a given time under the condition that the system lifetime is longer than the total mission time. Another one is the probability that the mission can be successfully accomplished before the system fails, i.e., the mission system lifetime is longer than the total mission time. If we can model both the mission system lifetime and mission phase duration, the mission reliability is estimated straightforwardly through evaluating the probability of events corresponding to the above two definitions, respectively.

Specifically, we consider a PMS having N phases. Let X_n denote the duration of the nth phase, which is a random variable in $\Re^+ = [0, +\infty)$. Let a random variable T_M denote the total time of completing the mission. In principle, T_M can be represented as $T_M = \sum_{n=1}^{N} X_n$. If there are some conditional probability linkages among $X_n, n = 1, \ldots, N$, such as the PDF $p_{X_n|X_1,\ldots,X_{n-1}}(x_n|x_1, \ldots, x_{n-1})$ for $2 \leq n \leq N$, the PDF of the mission time T_M can be formulated. However, this mechanism is aimed for the population of mission systems with the same type. For a specific system in service, we need to estimate the remaining mission time at each phase using the history of the CM data to the current time t_i, denoted by $\Phi_{i,n}$, which is related to X_n. Then, we represent the estimated PDF of the mission time as $p_{T_M|\Phi_{i,n}}(t_m|\Phi_{i,n})$, which is relying on the CM information to date. In order to characterize the lifetime of the mission system, let a random variable T_d denote the lifetime of the mission system. To estimate the PDF of the lifetime of the mission system from the observed degradation data to t_i, denoted by Y_i, we use the degradation modeling technique and then the estimated PDF of T_d can be obtained, which is represented as $f_{T_d|Y_i,C_i,\Phi_{i,n}}(t|Y_i, C_i, \Phi_{i,n})$.

After obtaining the estimated $p_{T_M|\Phi_{i,n}}(t_m|\Phi_{i,n})$ and $f_{T_d|Y_i,C_i,\Phi_{i,n}}(t|Y_i, C_i, \Phi_{i,n})$, the main objective is to compute the above two kinds of the mission reliability. Here we summarize the general formulations for these two cases:

- To compute the probability that the mission can be successfully accomplished before a given time R under condition that the system lifetime T_d is longer than the total mission time T_M at each phase, i.e., $\Pr(T_M \leq R|\, T_d \geq T_M, Y_i, C_i, \Phi_{i,n})$.
- To compute the probability that the mission can be successfully accomplished before the system fails, i.e., $\Pr(T_d \geq T_M|\, Y_i, C_i, \Phi_{i,n})$.

For an intuitive vision, a schematic of degradation progression and three-phase mission process is illustrated in Fig. 13.1.

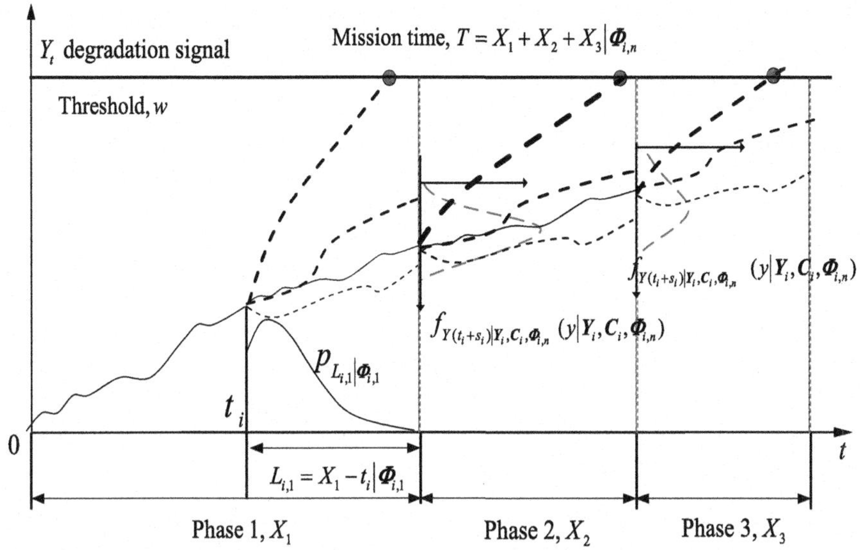

Fig. 13.1 Schematic of degradation progression and three-phase mission process (○—possible failure)

13.2.2 Assumptions

In this following, we summarize the main assumptions used in this chapter.

(1) No maintenance activities are involved during the process of carrying out a mission.

(2) The mission consists of a set of consecutive phases.

(3) For mission success, all phases must be completed.

(4) The phases of mission are sequential, i.e., the order of the mission phase is deterministic.

(5) The durations of the different phases are dependent.

(6) Duration of every phase is random following a log-normal distribution and the exact time of ending phase is unknown during the process of the operating phase, but the exact ending time is known once the phase is completed.

(7) The degradation process is dependent on the mission process and different mission phases correspond to different degradation rates.

(8) Failure resulting from degradation will lead to a mission failure.

(9) Given a particular phase and the observed history in this phase, the duration of the future phase is only dependent on the current and previous phases' duration, e.g.,

$$p_{X_{n+1}|X_1,\ldots,X_n,\Phi_{i,n}}(x_{n+1}|x_1,\ldots,x_n,\Phi_{i,n}) = p_{X_{n+1}|X_1,\ldots,X_n}(x_{n+1}|x_1,\ldots,x_n).$$

To clarify the above assumptions, we give the following remark.

Remark 13.1 Assumption 1 is commonly used in degradation practice. The motivation is that once the system is maintained such as replacement, the system renews. Then the degradation process restarts from the initial state. Assumptions 2–4 come from the definition of the PMS. Assumption 5 is easily understood because the phases of mission are sequential according to Assumption 4. The first part in Assumption 6 makes us focus on the random phase duration with a general distribution rather than the exponential distribution and the second part is practical since the sequence of the mission phase is deterministic as revealed by Assumption 4. Assumption 7 incorporates the effects of the mission process on the degradation process. Assumption 8 follows from the definition of the mission reliability. Assumption 9 is for model simplification, but is also practical. For example, at the first phase, we only observe the CM information $\Phi_{i,1}$ related to the duration of the first phase. Therefore, given X_1 and $\Phi_{i,1}$, it is reasonable to assume that the duration X_2 of the second phase is only dependent on X_1. Similarly, given X_1, X_2, and $\Phi_{i,1}$, the duration X_3 of the third phase is only dependent on X_1 and X_2, and so on.

Remark 13.2 Through the above assumptions, we can observe that, to solve the concerned problem, the PMS should have the following characteristics. First, there are two types of the CM information obtained from the observation process: the observed degradation data and the CM information related to the phase durations. The degradation data can reflect the health state of the PMS and are used to model the degradation process of the PMS for estimating its lifetime while the CM information is related to the duration of the phase and thus is used to estimate the probability distributions of durations of the mission phases. As such, the history of all samples collected to date can be used to calculate the posterior probability distribution of the system lifetime and the mission time. After obtaining the PDFs of the lifetime of the mission system and the distribution of the mission completion time, we can estimate the mission reliability at the current CM point.

13.3 Mission Process to Estimate the Mission Time

In this section, we consider a general multiphase mission process, as illustrated in Fig. 13.1. To do so, we mainly formulate the model for the mission process in the nth phase.

Considering that the exact duration of the phase is unknown in its operation, but one thing we do know is that over a monitoring interval of time, the duration is just an interval shorter at the end of the interval than at the beginning of the interval if nothing happened during that interval. In the meantime, we may observe an increasing or decreasing trend of the monitored CM information $\varphi_{i,n}$. Based on these observations, the problem can be formulated as follows with a simple and intuitive form, at least in principle. If we define $L_{i,n}$ as the remaining duration of the first phase at time t_i with the realization $l_{i,n}$, and the relationship between $L_{i,n}$ and $L_{i+1,n}$ can be described as $L_{i+1,n} = L_{i,n} - (t_{i+1} - t_i),\ if\ L_{i,n} > t_{i+1} - t_i$. Note that $L_{0,n}$ is actually the duration of the first phase. Furthermore, the duration of the mission time is always positive and thus we use the transformation $Z_{i,n} = \ln L_{i,n}$ with the realization $z_{i,n}$ to guarantee $L_{i,n} > 0$. In order to estimate $L_{i,n}$ from $\varphi_{i,n}$, we need to model the stochastic relationship between $l_{i,n}$ and $\varphi_{i,n}$. To do so, we use a concept called a floating scale parameter to model the relationship between $z_{i,n}$ and $\varphi_{i,n}$ which is modeled by a stochastic distribution to characterize their relation [26–28]. The basic idea is to make the mean parameter of $\varphi_{i,n}$be a function of $z_{i,n}$, which enables an updating mechanism of the mean parameter.

Together with the above description, the relationship among $L_{i,n}$, $L_{i+1,n}$, $Z_{i,n}$, and $\varphi_{i,n}$ can be described as follows according to [28]:

$$\begin{aligned} L_{i+1,n} &= \begin{cases} L_{i,n} - (t_{i+1} - t_i) \; if \; L_{i,n} > t_{i+1} - t_i \\ not \text{ defined} \qquad otherwise \end{cases}, \\ Z_{i,n} &= \ln L_{i,n} \\ \varphi_{i,n} &= g_n(z_{i,n}) + \eta_{i,n}, \end{aligned} , \tag{13.1}$$

where $g_n(z_{i,n})$ is a function to be determined, describing the relationship between the mission process and the CM data relative to the duration of the phase, and $\eta_{i,n}$ is the measurement error normally distributed as $\eta_{i,n} \sim N(0, \sigma_n^2)$.

Therefore, the key for estimating the remaining phase duration is to formulate the relationship between $l_{i+1,n}$ and the CM history $\Phi_{i,n}$. By the classical stochastic filtering theory, it can be shown that this relationship can be established recursively as follows

$$p_{L_{i+1,n}|\Phi_{i,n}}\left(l_{i+1,n}\middle|\, \Phi_{i,n}\right) = \frac{p\left(\varphi_{i,n}\middle|\, l_{i+1,n}\right) p_{L_{i,n}|\Phi_{i-1,n}}\left(l_{i+1,n} + t_{i+1} - t_i\middle|\, \Phi_{i-1,n}\right)}{\int_0^\infty p\left(\varphi_{i,n}\middle|\, l_{i+1,n}\right) p_{L_{i,n}|\Phi_{i-1,n}}\left(l_{i+1,n} + t_{i+1} - t_i\middle|\, \Phi_{i-1,n}\right) dl_{i+1,n}}. \tag{13.2}$$

It is noted that, in principle, any distribution with the positive support can be used to model $L_{0,n}$. However, in this case it is difficult to analytically evaluate (13.2) due to its general filtering nature and thus numerical methods have to be used, e.g., particle filter. In order to solve and formulate above equation explicitly, we apply a method using the Extended Kalman Filtering (EKF) technique based on the work in [28], in which the EKF technique was used to estimate the residual life. It is noted that using EKF instead of general particle filtering saves computational complexity, but comes at a cost of losing some generality, since the EKF is based on the Taylor expansion of functions of Gaussian random variables. Therefore, using the general filtering technique such as particle filtering might be a good alternative for complicated engineering applications, but this is not the focus of the current chapter.

As mentioned before, the duration of the mission time must be positive. As such, we define $L_{i,n}$ as a log-normal random variable and thus $Z_{i,n} = \ln L_{i,n}$ as the unknown state of the model (13.1) which is normally distributed. After obtaining the CM information about $\varphi_{i,n}$ at t_i, we use the EKF methodology to estimate/update the conditional PDF of $Z_{i,n}$ and further the remaining duration $L_{i,n}$ on the basis of the CM information to date, i.e., $\Phi_{i,n}$. We denote the updated and one-step predicted conditional PDF of $Z_{i,n}$ and $Z_{i+1,n}$as $Z_{i,n}\big|\,\Phi_{i,n} \sim N(z_{i|i,n}, P_{i|i,n})$, and $Z_{i+1,n}\big|\,\Phi_{i,n} \sim N(z_{i+1|i,n}, P_{i+1|i,n})$, respectively. In this chapter, $z_{i|i,n}$, $P_{i|i,n}$, $z_{i+1|i,n}$, and $P_{i+1|i,n}$ can be obtained by the EKF. The main results are summarized as the following proposition and its proof is placed in the supplementary material to save the space.

Proposition 13.1 *As for (13.1), the estimates of* $z_{i|i,n}$, $P_{i|i,n}$, $z_{i+1|i,n}$, *and* $P_{i+1|i,n}$ *can be obtained by the EKF as follows:*

$$z_{i|i,n} = z_{i|i-1,n} + K_{i,n}\left[\varphi_{i,n} - g_n(z_{i|i-1,n})\right], \tag{13.3}$$

$$K_{i,n} = \left[P_{i|i-1,n} g_n'(z_{i|i-1,n})\right]\left[g_n'(z_{i|i-1,n})^2 P_{i|i-1,n} + \sigma_n^2\right]^{-1}, \tag{13.4}$$

$$P_{i|i,n} = P_{i|i-1,n}\left(1 - K_{i,n} g_n'(z_{i|i-1,n})\right). \tag{13.5}$$

$$\begin{aligned} z_{i|i-1,n} = &\ln\left[e^{z_{i-1|i-1,n}+0.5P_{i-1|i-1,n}} - (t_i - t_{i-1})\right] - \\ &0.5\ln\left(1 + \frac{\left(e^{P_{i-1|i-1,n}}-1\right)e^{2z_{i-1|i-1,n}+P_{i-1|i-1,n}}}{\left[e^{z_{i-1|i-1,n}+0.5P_{i-1|i-1,n}} - (t_i - t_{i-1})\right]^2}\right) \end{aligned}, \tag{13.6}$$

and

$$P_{i|i-1,n} = P_{i-1|i-1,n}. \tag{13.7}$$

where $K_{i,n}$ *is the Kalman gain, and*

$$g_n'(z_{i|i-1,n}) = dg_n(z_{i,n})/dz_{i,n}\big|_{z_{i,n}=z_{i|i-1,n}}.$$

Proof Specifically, according to the principle of the EKF, the updating equation of the expectation of $Z_{i,n}$ can be formulated as

$$z_{i|i,n} = z_{i|i-1,n} + K_{i,n}\left[\varphi_{i,n} - g_n(z_{i|i-1,n})\right],$$

where $K_{i,n}$ is the Kalman gain function:

$$K_{i,n} = \left[P_{i|i-1,n} g_n'(z_{i|i-1,n})\right]\left[g_n'(z_{i|i-1,n})^2 P_{i|i-1,n} + \sigma_n^2\right]^{-1},$$

where $g_n'(z_{i|i-1,n}) = dg_n(z_{i,n})/dz_{i,n}\big|_{z_{i,n}=z_{i|i-1,n}}$.

Correspondingly, the updating equation for the estimation variance can be formulated as

$$P_{i|i,n} = P_{i|i-1,n}\left(1 - K_{i,n} g_n'(z_{i|i-1,n})\right).$$

When applying above EKF methodology, we need to initiate the algorithm at the start of the mission phase using $z_{0|0,n}$ and $P_{0|0,n}$, which can be estimated from historical data. In addition, in the above updating equations, it is required to calculate the one-step estimation for the expectation $z_{i|i-1,n}$ and variance $P_{i|i-1,n}$. In the following, we try to obtain these quantities.

Considering that $Z_{i,n}\big|\,\Phi_{i,n} \sim N(z_{i|i,n}, P_{i|i,n})$ and $Z_{i,n} = \ln L_{i,n}$, we get

$$E\left[L_{i,n}\big|\,\Phi_{i,n}\right] = e^{z_{i|i,n}+0.5P_{i|i,n}}, \tag{13.8}$$

and

$$var\left[L_{i,n}\big|\,\Phi_{i,n}\right] = \left(e^{P_{i|i,n}} - 1\right)e^{2z_{i|i,n}+P_{i|i,n}}. \tag{13.9}$$

Thus, we further have $L_{i,n}\big|\,\Phi_{i,n} \sim \log N(z_{i|i,n}, P_{i|i,n})$. Then, based on the first equation in Eq. (13.1), a one-step forecasting of the remaining mission phase duration from t_i to t_{i+1} is

$$E\left[L_{i+1,n}\big|\,\Phi_{i,n}\right] = \begin{cases} E\left[L_{i,n}\big|\,\Phi_{i,n}\right] - (t_{i+1} - t_i),\ if\,E\left[L_{i,n}\big|\,\Phi_{i,n}\right] > t_{i+1} - t_i \\ 0, \text{otherwise} \end{cases}.$$

Since the change in the established state equation is deterministic over (t_i, t_{i+1}), the variance about the mean estimate is given by

$$var\left[L_{i+1,n}\big|\,\Phi_{i,n}\right] = var\left[L_{i,n}\big|\,\Phi_{i,n}\right].$$

By reversing the relationships given in Eqs. (13.8) and (13.9), $E\left[L_{i+1,n}\big|\,\Phi_{i,n}\right]$ can be transformed into $z_{i+1|i,n}$ for the next CM time as

$$\begin{aligned} z_{i+1|i,n} &= \ln\left[E\left(L_{i+1,n}\big|\,\Phi_{i,n}\right)\right] - 0.5\ln\left(1 + \frac{var\left(L_{i+1,n}|\Phi_{i,n}\right)}{E\left(L_{i+1,n}|\Phi_{i,n}\right)^2}\right) \\ &= \ln\left[e^{z_{i|i,n}+0.5P_{i|i,n}} - (t_{i+1} - t_i)\right] - 0.5\ln\left(1 + \frac{\left(e^{P_{i|i,n}}-1\right)e^{2z_{i|i,n}+P_{i|i,n}}}{\left[e^{z_{i|i,n}+0.5P_{i|i,n}} - (t_{i+1}-t_i)\right]^2}\right), \end{aligned}$$

Without any random variation in the prediction of the state, we have

$$P_{i+1|i,n} = P_{i|i,n}.$$

Then, the expectation $z_{i|i-1,n}$ and variance $P_{i|i,n}$ can be straightforwardly computed as follows

$$z_{i|i,n} = \ln\left[e^{z_{i-1|i,n}+0.5P_{i-1|i,n}} - (t_i - t_{i-1})\right] - 0.5\ln\left(1 + \frac{\left(e^{P_{i-1|i-1,n}} - 1\right)e^{2z_{i-1|i,n}+P_{i-1|i,n}}}{\left[e^{z_{i-1|i-1,n}+0.5P_{i-1|i,n}} - (t_i - t_{i-1})\right]^2}\right),$$

and

$$P_{i|i-1,n} = P_{i-1|i,n}.$$

This completes the proof.

From the above proposition, upon obtaining the CM information $\varphi_{i,n}$ at t_i, the estimated PDF of the remaining duration of the nth phase, $p_{L_{i,n}|\Phi_{i,n}}\left(l_{i,n}\middle|\,\Phi_{i,n}\right)$, can be formulated as

$$p_{L_{i,n}|\Phi_{i,n}}\left(l_{i,n}\middle|\,\Phi_{i,n}\right) = \frac{1}{l_{i,n}\sqrt{2\pi P_{i|i,n}}}e^{-\left(2P_{i|i,n}\right)^{-1}\left(\ln l_{i,n}-z_{i|i,n}\right)^2}. \tag{13.10}$$

From the relationship between the duration of the current phase and its remaining duration, we have

$$X_n|\,\Phi_{i,n} = L_{i,n} + t_i - \sum_{j=1}^{n-1} x_j. \tag{13.11}$$

Then, we directly estimate the distribution of X_n at time t_i for the phased-mission system by:

$$p_{X_n|\Phi_{i,n}}\left(x_n|\,\Phi_{i,n}\right) = p_{L_{i,n}|\Phi_{i,n}}\left(x_n + \sum_{j=1}^{n-1} x_j - t_i \middle|\, \Phi_{i,n}\right), \tag{13.12}$$

where $p_{L_{i,n}|\Phi_{i,n}}\left(x_n + \sum_{j=1}^{n-1} x_j - t_i\middle|\,\Phi_{i,n}\right)$ can be calculated by (13.10).

Therefore, the duration of the next phase conditional on the data up to t_i and X_n can be computed according to

$$\begin{aligned} p_{X_{n+1}|\Phi_{i,n}}\left(x_{n+1}|\,\Phi_{i,n}\right) &= \int p_{X_n,X_{n+1}|\Phi_{i,n}}\left(x_n, x_{n+1}|\,\Phi_{i,n}\right)dx_n \\ &= \int p_{X_{n+1}|X_n,\Phi_{i,n}}\left(x_{n+1}|\,x_n, \Phi_{i,n}\right)p_{X_n|\Phi_{i,n}}\left(x_n|\,\Phi_{i,n}\right)dx_n \\ &= \int p_{X_{n+1}|X_n}\left(x_{n+1}|\,x_n\right)p_{X_n|\Phi_{i,n}}\left(x_n|\,\Phi_{i,n}\right)dx_n \\ &= \int p_{X_{n+1}|X_n}\left(x_{n+1}|\,x_n\right)p_{L_{i,n}|\Phi_{i,n}}\left(x_n + \sum_{j=1}^{n-1} x_j - t_i\middle|\,\Phi_{i,n}\right)dx_n \end{aligned} \tag{13.13}$$

Similarly, the duration of the remaining phases, conditional on the data up to t_i, can be obtained by

$$\begin{aligned} &p_{X_s|\Phi_{i,n}}\left(x_s|\,\Phi_{i,n}\right) \\ &= \int p_{X_n,X_{n+1},\ldots,X_s|\Phi_{i,n}}\left(x_n, x_{n+1}, \ldots, x_s|\,\Phi_{i,n}\right)dx_n dx_{n+1}\cdots dx_{s-1}, s = n+1, \ldots, N. \end{aligned} \tag{13.14}$$

According to the Assumption 9, we have

$$\begin{aligned} &p_{X_n,X_{n+1},\ldots,X_s|\Phi_{i,n}}\left(x_n, x_{n+1}, \ldots, x_s|\,\Phi_{i,n}\right) \\ &= p_{X_n|\Phi_{i,n}}(x_n|\,\Phi_{i,n})p_{X_{n+1}|X_n}(x_{n+1}|\,x_n)\cdots p_{X_s|X_n,X_{n+1},\ldots X_{s-1}}(x_s|\,x_n, x_{n+1}, \ldots, x_{s-1}) \end{aligned} \tag{13.15}$$

Based on the above results, the distribution of the mission time T_M conditional on the related CM information $\Phi_{i,n}$, denoted by $T_M | \Phi_{i,n} = \left(\sum_{k=1}^{N} X_k\right) \Big| \Phi_{i,n}$, can be calculated by (13.16). Then by differentiating $\Pr\left(T_M \leq t_m | \Phi_{i,n}\right)$ regarding to t_m, we have the PDF of the mission time conditional on $\Phi_{i,n}$ at t_i as (13.17).

$$
\begin{aligned}
&\Pr\left(T_M \leq t_m | \Phi_{i,n}\right) = \Pr\left(\sum_{j=1}^{n-1} x_j + \sum_{k=n}^{N} X_k \leq t_m \middle| \Phi_{i,n}\right) = \Pr\left(\sum_{k=n}^{N} X_k \leq t_m - \sum_{j=1}^{n-1} x_j \middle| \Phi_{i,n}\right) \\
&= \Pr\left(0 < X_n \leq t_m - \sum_{j=1}^{n-1} x_j, 0 < X_{n+1} \leq t_m - \sum_{j=1}^{n-1} x_j - X_n, \ldots, 0 < X_N \leq t_m \right. \\
&\qquad \left. - \sum_{j=1}^{n-1} x_j - \sum_{k=n}^{N-1} X_k \middle| \Phi_{i,n}\right) \\
&= \int_0^{t_m - \sum_{j=1}^{n-1} x_j} \int_0^{t_m - \sum_{j=1}^{n-1} x_j - x_n} \cdots \int_0^{t_m - \sum_{j=1}^{n-1} x_j - \sum_{k=n}^{N-1} x_k} p_{X_n, X_{n+1}, \ldots, X_N | \Phi_{i,n}} \\
&\qquad \times \left(x_n, x_{n+1}, \ldots, x_N | \Phi_{i,n}\right) dx_n dx_{n+1} \cdots dx_N \\
&= \int_0^{t_m - \sum_{j=1}^{n-1} x_j} p_{X_n | \Phi_{i,n}}\left(x_n | \Phi_{i,n}\right) \int_0^{t_m - \sum_{j=1}^{n-1} x_j - x_n} p_{X_{n+1} | X_n}\left(x_{n+1} | x_n\right) \cdots \Pr{}_{X_N | X_n, X_{n+1}, \ldots X_{N-1}} \\
&\qquad \times \left(t_m - \sum_{j=1}^{n-1} x_j - \sum_{k=n}^{N-1} x_k \middle| x_n, x_{n+1}, \ldots, x_{N-1}\right) dx_n dx_{n+1} \cdots dx_{N-1}
\end{aligned}
\tag{13.16}
$$

$$
\begin{aligned}
&p_{T_M | \Phi_{i,n}}\left(t_m | \Phi_{i,n}\right) \\
&= \int_0^{t_m - \sum_{j=1}^{n-1} x_j} p_{X_n | \Phi_{i,n}}\left(x_n | \Phi_{i,n}\right) \int_0^{t_m - \sum_{j=1}^{n-1} x_j - x_n} p_{X_{n+1} | X_n}\left(x_{n+1} | x_n\right) \cdots p_{X_N | X_n, X_{n+1}, \ldots X_{N-1}} \\
&\quad \times \left(t_m - \sum_{j=1}^{n-1} x_j - \sum_{k=n}^{N-1} x_k \middle| x_n, x_{n+1}, \ldots, x_{N-1}\right) dx_n dx_{n+1} \cdots dx_{N-1}.
\end{aligned}
\tag{13.17}
$$

From the above formulation, we can obtain the estimated mission time from the observed CM data. This will make the estimation of the mission reliability feasible. So far, we have completed the task of formulating the mission time distribution conditional on the related CM information. In the next section, we will focus on how to estimate the lifetime or the remaining useful life of the PMS from the observed degradation data.

13.4 System Degradation Process to Estimate the Lifetime

13.4.1 Model Description

Here we use a Wiener process to model the degradation process of the mission system over time, and to estimate the RUL of PMSs. It is noted that this model has been widely used to characterize the evolving path of the degradation process [29–39].

For $t \geq 0$, we let $Y(t)$ denote the degradation variable at time t. Without loss of generality, assume the starting reading $y(0)$ for the degradation process $\{Y(t), t \geq 0\}$ to be fixed. From the model formulation of the mission, we know that at each time t, the PMS can only occupy one of the phases in a phase set $\Psi = \{1, 2, \ldots, N\}$. Let $\nu : [0, \infty) \rightarrow \Psi$ be a deterministic and piecewise constant function so that $\nu(t)$ represents the mission phase at time t. Under this setting, the mission phases are sequential but the duration of each phase is random. Further, we define a degradation rate function $\lambda : \Psi \rightarrow \Re^+$ so that $\lambda(\nu(t))$ represents the degradation rate of the PMS at t. For example, if the PMS is at the nth phase at t, then $\nu(t) = n$ and $\lambda(n)$ is the degradation rate of the nth phase. It is noted that the rate of degradation is dependent on the mission phase. For notation convenience, we simplify $\lambda(n)$ as λ_n throughout the remaining parts.

Under the above descriptions, the evolution of the monitored degradation variable $Y(t)$ over time can be described using a Wiener process based degradation model with piecewise constant degradation rate as follows,

$$Y(t) = y(0) + \int_0^t \lambda(\nu(u))\mathrm{d}u + \sigma B(t). \tag{13.18}$$

where $\{B(t), t \geq 0\}$ is a standard Brownian motion (BM) process and $\sigma(\sigma > 0)$ is the diffusion parameter, with $\sigma B(t) \sim N(0, \sigma^2 t)$ for $t > 0$. This term characterizes degradation effects that cannot be attributed to the mission process.

Remark 13.3 As for the model (13.18), the difference made here from the other models in the literature is that the degradation process is dependent on the mission phase and thus is modeled as a multistage manner by $\lambda(\nu(t))$ It is noted that the model (13.18) is similar to the model with multiple change points such as [38] since the degradation rates of the PMS is dependent on the mission phases, and the degradation rate changes when the mission phase changes. The major difference of these two kinds of models is that a change-point detection procedure is needed in the model with multiple change points while the exact ending time of the mission phase is known once the phase is completed and thus there is no need to detect the change point for the proposed model. At the same time, during the process of the operating phase, we can estimate the phase duration by the CM data and the method has been given in Sect. 13.3. In contrast, estimating the time of the change point is not considered in the model with multiple change points.

Remark 13.4 In the model (13.18), it is assumed that the degradation process in all phases is linked to the same degradation characteristic. Actually, the health states of many systems can be reflected by one key characteristic such as capacity of battery systems and the vibration amplitude of mechanical systems. As a result, the main focus is placed on modeling the single degradation characteristic in degradation modeling field [31–34]. In this chapter, we also follow this line by considering the effect of the mission process on the degradation rate. On the other hand, for a complex system, there may be more than one degradation characteristics. In this case, it usually constructs a composite health index via fusion of multiple degradation characteristics. Then, the fused health index can be modeled by the method for one degradation characteristics (e.g., [40]). It can be observed that once the composite health index via fusion of multiple degradation characteristics is available, the presented model can also applied.

Remark 13.5 Considering that phased-mission systems are multistage systems in essence, we use $\lambda\left(\nu(t)\right)$ to model such multistage manner in (13.18). However, it is noted that (13.18) is a data-driven model in stochastic framework to fit the degradation data of the monitored health state of the system, rather than the physics-based model. In general, $\{Y\left(t\right), t \geq 0\}$ is a nonstationary process to model the degradation process with an increasing or decreasing trend.

In the above degradation modeling framework, the lifetime of PMS is defined as the first passage time (FPT) that the degradation process $\{Y(t), t \geq 0\}$ hits a fixed failure threshold w [37]. Such failure threshold can be determined by either engineering domain knowledge or accepted industrial standards. For example, the ISO 2372 and ISO 10816 are frequently adopted for defining acceptable vibration threshold levels. Based on this threshold, the RUL modeling principle is presented as follows. When degradation $Y(t)$ modeled by (13.18) reaches the threshold w, the PMS can be declared to be failed and thus there is no useful lifetime left. This FPT requirement may be considered to be restrictive to some cases since the degradation may go back after the first hit. However, for critical equipment, it is usually mandatory for putting this into practice and once the observed degradation is equal or above the set threshold level, the system must be stopped for inspecting.

The main objective here is to estimate and update the RUL distribution of PMS based on the discrete observations of the degradation process and the mission process over time. Suppose the degradation process is discretely monitored at time $0 = t_0 < t_1 < \cdots < t_i$ and let $y_k = y(t_k)$ denote the degradation observation at time $t_k, k = 0, 1, \ldots, i$. The set of the degradation observations up to t_i is represented by $Y_i = \{y_0, y_1, y_2, \ldots, y_i\}$. In addition, we can actually observe the transition times of the mission phases up to t_i. Therefore, we observe the set of the mission transition times $C_i = \left\{u_j : j = 1, 2, \ldots, n(t_i)\right\}$, where u_j is the jth mission phase transition time and $n(t_i)$ is the cumulative number of mission transitions up to t_i. Since the degradation path and the RUL estimation is dependent on the future mission transitions, we denote the future mission phase transitions from current time t_i to a fixed future time t by $F_i(t) = \left\{u_j : u_j \in (t_i, t]\right\}$. Under these settings, we know that the mission process will maintain the phase $\nu(u_{j-1})$ over the interval $[u_{j-1}, u_j)$,

$j = 1, 2 \ldots, n(t_i)$, $u_0 = 0$. Based on the above discussions, using the concept of the FPT, the lifetime of PMS can be defined as

$$T_d = \inf\{t > 0 : Y(t) \geq w\}. \tag{13.19}$$

As a result, we define the RUL S_i of PMS at time t_i as

$$S_i = \inf\{s_i > 0 : Y(s_i + t_i) \geq w\}. \tag{13.20}$$

with conditional cumulative distribution function (CDF)

$$\Pr(S_i \leq s_i | Y_i, C_i) = \Pr\left(\sup_{s_i > 0} Y(t_i + s_i) \geq w \middle| Y_i, C_i, \Phi_{i,n}\right), \tag{13.21}$$

where $\left(Y_i, C_i, \Phi_{i,n}\right)$ is the information available up to t_i.

13.4.2 Bayesian Updating of Model Parameters

On the basis of the information $\left(Y_i, C_i, \Phi_{i,n}\right)$ available up to t_i, we develop a Bayesian approach for updating the parameters of the degradation model using prior information estimated from historical data and real-time degradation observations obtained from a fielded PMS. In many applications, historical databases of the degradation data and mission phase transition times are available for estimating the prior information. However, even systems originating from the same source can exhibit widely varying reliability characteristics due to the heterogeneity in materials, operation environments, etc. By combining both historical and real-time data, we are able to account for these inherent differences.

The updating of the degradation model, according to its formulation in (13.18), requires updating parameters associated with the degradation model parameters λ and σ, as well as the parameters relevant to the mission process, denoted by γ. For notational brevity, we denote the parameter vector as $\theta = (\lambda, \sigma, \gamma)$. Denote the prior distribution of θ by $\pi(\theta) = \pi(\lambda, \sigma, \gamma)$. Regarding how to determine the prior distribution, similar method can be used based on the historical databases by referring to [38]. After obtaining the monitoring information $\left(Y_i, F_i, \Phi_{i,n}\right)$ available up to t_i, to achieve model updating under Bayesian framework, it is equivalent to compute the posterior distribution of the model parameters as follows:

$$\begin{aligned} f_\theta(\theta | Y_i, C_i, \Phi_{i,n}) &= f_\theta(\lambda, \sigma, \gamma | Y_i, C_i, \Phi_{i,n}) \\ &= \frac{\pi(\lambda, \sigma, \gamma) f_C(C_i, \Phi_{i,n} | \lambda, \sigma, \gamma) f_Y(Y_i | C_i, \Phi_{i,n}, \lambda, \sigma, \gamma)}{\int_\theta f_{C,Y}(Y_i, C_i, \Phi_{i,n} | \lambda, \sigma, \gamma) d\theta}, \\ &\propto \pi(\lambda, \sigma, \gamma) f_C(C_i, \Phi_{i,n} | \lambda, \sigma, \gamma) f_Y(Y_i | C_i, \Phi_{i,n}, \lambda, \sigma, \gamma) \end{aligned} \tag{13.22}$$

where $f_C(C_i, \Phi_{i,n} | \lambda, \sigma, \gamma) f_Y(Y_i | C_i, \Phi_{i,n}, \lambda, \sigma, \gamma)$ is the likelihood function associated with the degradation process and the mission process.

13.4.3 Estimating the RUL of PMS

In the following, we formulate the RUL estimation of the PMS based on the information $(Y_i, C_i, \Phi_{i,n})$ available up to t_i. This problem can be solved by computing the boundary crossing probability of a standard BM process. We elaborate this computing process in the following.

We first reformulate the degradation process into two parts: deterministic and stochastic terms as

$$Y(t) = \eta(t) + \sigma B(t), \tag{13.23}$$

where $\eta(t) = y(0) + \int_0^t \lambda(\nu(u))du$ is the deterministic part and $\sigma B(t)$ is the stochastic part. From the formulation of degradation process in Sect. 13.4.1, we learn that $\eta(t)$ is a piecewise linear function whose line segments are defined by the mission phase transition $\{u_n\}$. Namely, if the mission process is in the nth phase during $[u_{n-1}, u_n)$, then the degradation rate over this interval is λ_n and $\int_{u_{n-1}}^t \lambda(\nu(u))du = \lambda_n(t - u_{n-1})$ for $t \in [u_{n-1}, u_n)$. After the transition to the n+1th phase at u_n, the new degradation rate becomes λ_{n+1}, and so on.

According to the above model formulation, the probability that the degradation is below the threshold w at time t is

$$\begin{aligned} &\Pr\left(Y(t) \le w | Y_0, C_0, \theta, F_0(t), \Phi_{0,0}\right) = \Pr\left(\sigma B(t) < w - \eta(t) | Y_0, C_0, \theta, F_0(t), \Phi_{0,0}\right) \\ &= \Pr\left(B(t) < \frac{w-\eta(t)}{\sigma} \middle| Y_0, C_0, \theta, F_0(t), \Phi_{0,0}\right) \end{aligned}, \tag{13.24}$$

where θ is the parameter vector associated with the degradation process and the mission process, which will be estimated from the available data.

To further simplify notation, let $b(t) = \frac{w-\eta(t)}{\sigma}$. Thus, estimating the lifetime can be achieved by deriving the probability that $\{B(t), t \ge 0\}$ crosses the boundary $b(t)$. The method of this derivation is based on the idea that the event that $\{B(t), t \ge 0\}$ does not cross $b(t)$ over the interval $[0, U]$ (where $U = u_{N-1}$) can be split into n conditional events that $\{B(t), t \ge 0\}$ does not cross $b(t)$ over the interval $[u_j, u_{j+1})$ given that $\{B(t), t \ge 0\}$ does not cross $b(t)$ over the interval $[u_{j-1}, u_j)$, where u_j is the transition time from the jth phase to the j+1th phase. For notation simplicity, let $b_j = b(u_j) = \frac{w-\eta(u_j)}{\sigma}$, $j = 0, 1, \ldots, N-1$, and let $n(U) = N-1$ denote the number of phase transitions on the interval $[0, U]$. Then, we have $[0, U] = \bigcup_{j=1}^{N-1} [u_{j-1}, u_j)$. Based on these settings, we have the following theorem to derive the FPT associated with the degradation process $\{Y(t), t \ge 0\}$.

Theorem 13.1 *Let* $0 = u_0 < u_1 < \cdots < u_{N-1} = U$ *denote* $N-1$ *transition times of the mission phase and* $b = (b_1, b_2, \ldots, b_{N-1})'$. *The probability distribution of the*

FPT associated with $\{Y(t), t \geq 0\}$ *is given as*

$$\begin{aligned}&\Pr(T \leq t|Y_0, C_0, \theta, F_0(t), \Phi_{0,0})\\&= 1 - E\left[g\left(B(u_1), B(u_2), \ldots, B(u_{N-1}); b\right)|\, Y_0, C_0, \theta, F_0(t), \Phi_{0,0}\right],\end{aligned} \tag{13.25}$$

where the function $g(\cdot)$ *can be formulated as*

$$\begin{aligned}&g(z_1, z_2, \ldots, z_{N-1}; b)\\&= \prod_{j=1}^{N-1} I(z_j < d_j)\left(1 - \exp\left[-\tfrac{2(d_{j-1}-z_{j-1})(d_j-z_j)}{u_j-u_{j-1}}\right]\right),\end{aligned} \tag{13.26}$$

and $I(\cdot)$ *is the indicator function.*

Proof For a single linear boundary on the interval $[0, U]$ of the form $b(u) = c + au$, Siegmund in [41] proved that the (conditional) probability that a standard BM process does not cross the boundary in this interval is given by

$$\Pr\left(B(u) \leq c + au, u < U|B(U) = z\right) = 1 - \exp\left(-\frac{2c(aU + c - z)}{U}\right).$$

Considering the degradation model (13.16), we have

$$\begin{aligned}&\Pr\left(B(u) \leq b(u), u < U\right)\\&= \int_{-\infty}^{b_1} \Pr\left(B(u) \leq b(u), u < u_1|\, B(u_1) = z_1\right) \Pr\left(B(u) \leq b(u), u_1 < u \leq U|\, B(u_1) = z_1\right) \mathrm{d}\Pr_{u_1}(z_1),\end{aligned} \tag{13.27}$$

where

$$\mathrm{d}\Pr_u(z) = \frac{1}{\sqrt{2\pi u}} \exp\left(-\frac{z^2}{2u}\right) \mathrm{d}z,$$

and $\mathrm{d}\Pr_u(z)$ is the PDF of $B(u)$.

By the result of [41], we know that the first term in the integral part of Eq. (13.27) is

$$\Pr\left(B(u) \leq b(u), u < u_1|\, B(u_1) = z_1\right) = 1 - \exp\left(-\frac{2b_0(b_1 - z_1)}{u_1}\right).$$

It is further noted that, given $B(u_1) = z_1$, the process $\{B(u + u_1) - z_1\}$ is still a BM start from 0 and thus the second factor in the integral (B2) can be formulated as

$$\begin{aligned}&\Pr\left(B(u) \leq b(u), u_1 < u \leq U|\, B(u_1) = z_1\right) = \Pr\left(B(u) \leq b(u + u_1) - z_1, u \leq U - u_1\right)\\&= \int_{-\infty}^{b_2 - z_1} \left(1 - \exp\left(-\tfrac{2(b_1-z_1)(b_2-z_1-z_2)}{u_2-u_1}\right)\right)\\&\Pr\left(B(u) \leq b(u + u_1) - z_1, u_2 - u_1 < u \leq U - u_1|\, B(u_2 - u_1) = z_2\right) \mathrm{d}\Pr_{u_2-u_1}(z_2 - z_1)\\&= \int_{-\infty}^{b_2} \left(1 - \exp\left(-\tfrac{2(b_1-z_1)(b_2-z_2)}{u_2-u_1}\right)\right) \Pr\left(B(u) \leq b(u + u_2) - z_2, u \leq U - u_2\right) \mathrm{d}\Pr_{u_2-u_1}(z_2 - z_1)\end{aligned} \tag{13.28}$$

Applying the similar procedure, we can obtain

$$\begin{aligned}&\Pr\left(B(u)\le b(u+u_{n-1})-z_{n-1},u\le u_n-u_{n-1}\right)\\&=\int_{-\infty}^{b_n}\left(1-\exp\left(-\frac{2(b_{n-1}-z_{n-1})(b_n-z_n)}{u_n-u_{n-1}}\right)\right)\\&\Pr\left(B(u)\le b(u+u_n)-z_n,u\le U-u_n\right)\mathrm{d}\Pr_{u_n-u_{n-1}}(z_n-z_{n-1})\end{aligned}, \qquad (13.29)$$

where by the definition of $\{B(u), u \ge 0\}$,

$$\mathrm{d}\Pr_{u_n-u_{n-1}}(z_n-z_{n-1})=\frac{1}{\sqrt{2\pi(u_n-u_{n-1})}}\exp\left(-\frac{(z_n-z_{n-1})^2}{2(u_n-u_{n-1})}\right)\mathrm{d}z_n, n=1,\ldots,N-1.$$

Based on the result in Eq. (13.29), we finally have

$$\Pr\left(B(u)\le b(u),u<U\right)=\int_{(-\infty,b)}\prod_{j=1}^{N-1}\left(1-\exp\left[-\frac{2(d_{j-1}-z_{j-1})(d_j-z_j)}{u_j-u_{j-1}}\right]\right)f(z)\mathrm{d}z,$$

where the integral region is $(-\infty, b) = (-\infty, b_1) \times (-\infty, b_2) \times \cdots \times (-\infty, b_{N-1})$ and $f(z)$ is the PDF of $z = (z_1, z_2, \ldots, z_{N-1})'$. By the independent increments property of BM, we have

$$f(z)=\prod_{j=1}^{N-1}\frac{1}{\sqrt{2\pi(u_j-u_{j-1})}}\exp\left(-\frac{(z_j-z_{j-1})^2}{2(u_j-u_{j-1})}\right).$$

This completes the proof.

In the following, we focus on estimating the RUL of PMS at the current monitoring time t_i ($t_i < U$). Suppose the degradation observations have been obtained at times $0, t_1, \ldots, t_i$ and let $\{\nu(t) : t_i \le t \le U\}$ denote the future mission profile from the current time t_i to the time U. Now the probability distribution of the RUL S_i at time t_i given that the degradation process has not crossed the threshold w up to t_i, can be obtained by the crossing probability of $\{Y(t), t \ge 0\}$ to the threshold w. Specifically, we have the following theorem regarding the probability distribution of the RUL S_i at time t_i.

Theorem 13.2 *Let* $b_i(t) = \frac{1}{\sigma}(w - y_i - \int_{t_i}^{t} \lambda(\nu(u))du)$ *for* $t \in (t_i, U]$ *and* $A_i(t) = \{j : u_j \in [t_i, t]\}$. *The conditional probability distribution of the RUL* S_i *at time* t_i *is*

$$\begin{aligned}&\Pr(S_i\le s_i|Y_i,C_i,\Phi_{i,n},\theta,F_i(t_i+s_i))\\&=1-E\left[\prod_{j\in A_i(t_i+s_i)}\left[\begin{matrix}I(W_i(u_j)<b_i(u_j))\times\\\left(1-\exp\left[-\frac{2Z_{i,j}}{u_j-u_{j-1}}\right]\right)\end{matrix}\right]\middle| Y_i,C_i,\Phi_{i,n,\theta,F_i(t_i+s_i)}\right],\end{aligned} \qquad (13.30)$$

where $I(\cdot)$ *is the indicator function and* $W_i(u_j) = B(u_j) - B(t_i)$, $Z_{i,j} = \left(b_i(u_{j-1}) - W_i(u_{j-1})\right)\left(b_i(u_j) - W_i(u_j)\right)$.

Proof Conditionally on the degradation observations up to time t_i, the degradation process can be formulated for each $t \in (t_i, U]$ as

$$Y(t) = y_i + \int_{t_i}^{t} \lambda\left(\nu(u)\right)\mathrm{d}u + \sigma\left(B(t) - B(t_i)\right).$$

Then, by utilizing the notation of the RUL realization, s_i, the degradation process can be revised for each $t \in (t_i, U]$ as

$$Y(t_i + s_i) = y_i + \int_{t_i}^{t_i+s_i} \lambda\left(\nu(u)\right)\mathrm{d}u + \sigma\left(B(t_i + s_i) - B(t_i)\right).$$

It can be easily verified that $\{B(t_i+s_i) - B(t_i), s_i \geq 0\}$ is still a standard BM over time s_i. Therefore, the RUL at time t_i corresponds to the FPT of $\{B(t_i+s_i) - B(t_i), s_i \geq 0\}$ crossing the boundary $b_i(t) = b_i(t_i + s_i)$. Applying Theorem 13.2, we can directly complete the proof.

This completes the proof.

It is noted that, to evaluate the expectation operator in (13.30), we need the PDF of $W_i(u_j)$, $j \in A_i(t) = \left\{j : u_j \in [t_i, t_i + s_i]\right\}$. For example, we suppose $A_i(t_i + s_i]) = \{l, l+1, \ldots, l+m\}$, and then by the stationary and independent increments properties of the standard BM, the PDF of $W_i(u_j)$, $j = l, l+1, \ldots, l+m$ can be formulated as

$$f\left(W_i(u_j) = w_j, j \in A_i(t_i + s_i)\right) = \frac{1}{\sqrt{2\pi(u_l - t_i)}}\exp\left[-\frac{z_l^2}{2(u_l - t_i)}\right] \times \prod_{j=l+1}^{l+m} \frac{1}{\sqrt{2\pi(u_j - u_{j-1})}}\exp\left[-\frac{(w_j - w_{j-1})^2}{2(u_j - u_{j-1})}\right].$$

Based on the result in Theorem 13.2 and the posterior distribution of the parameters, the RUL distribution at time t_i can be formulated by the law of total probability as

$$\begin{aligned}&\Pr(S_i \leq s_i | Y_i, C_i, \Phi_{i,n})\\ &= \int_{\theta, u_j \in F_i(t_i+s_i)} \begin{bmatrix}\Pr(S_i \leq s_i | Y_i, C_i, \Phi_{i,n}, \theta, F_i(t_i + s_i))\cdot \\ f_{F_i}(F_i(t_i + s_i)|Y_i, C_i, \Phi_{i,n}, \theta) f_\theta(\theta|\, Y_i, C_i, \Phi_{i,n})\end{bmatrix} \mathrm{d}\theta \mathrm{d}u_j,\end{aligned} \tag{13.31}$$

where $f_{F_i}(F_i(t_i + s_i)|Y_i, C_i, \Phi_{i,n}, \theta)$ is the joint PDF of the remaining mission phase transition times from time t_i to $t_i + s_i$. This quantity can be evaluated by the joint PDF of the remaining current mission phase duration and subsequent mission phase durations. For example, if the current mission phase process occupies the nth phase, then the PDF of the remaining duration of the current phase is $p_{L_{i,n}|\Phi_{i,n}}\left(l_{i,n}\middle|\, \Phi_{i,n}\right)$ as formulated in (13.10). Then the joint PDF of the remaining current mission phase duration $L_{i,n}$, and subsequent mission phase durations $X_{n+1}, \ldots, X_{N-1}$ can be obtained by

$$
\begin{aligned}
& f_{F_i}(F_i(t_i+s_i)|Y_i, C_i, \Phi_{i,n}, \theta) \\
&= p_{L_{i,n}, X_{n+1},\ldots,X_{N-1}|\Phi_{i,n}}\left(l_{i,n}, x_{n+1}, \ldots, x_{N-1} \middle| \Phi_{i,n}\right) \\
&= p_{L_{i,n}|\Phi_{i,n}}\left(l_{i,n} \middle| \Phi_{i,n}\right) \cdot p_{X_{n+1}|X_n}\left(x_{n+1} \middle| t_i + l_{i,n} - \sum_{l=1}^{n-1} x_l\right) \cdots \\
& \cdot p_{X_{N-1}|X_n, X_{n+1},\ldots X_{s-1}}\left(x_{N-1} \middle| t_i + l_{i,n} - \sum_{l=1}^{n-1} x_l, x_{n+1}, \ldots, x_{s-1}\right)
\end{aligned}
\tag{13.32}
$$

Denote the future mission phase transition times by $u_n, u_{n+1}, \ldots, u_{N-1}$. Considering the fact that $u_n = t_i + L_{i,n}$, $u_{n+1} = t_i + L_{i,n} + X_{n+1}, \ldots u_{N-1} = t_i + L_{i,n} + \sum_{k=n+1}^{N-1} X_k$, the joint PDF of $u_n, u_{n+1}, \ldots, u_{N-1}$ can be achieved straightforwardly by the PDF of $L_{i,n}$ and $X_{n+1}, \ldots, X_{N-1}$ via the variable transformation technique.

From (13.31), we can also observe the dependency of the estimated lifetime of the system on the degradation observations and the mission progression history up to t_i. Unfortunately, despite the simple and straightforward formulations, it is not easy to directly and analytically compute the right-hand side of (13.31) because the computation involves multidimensional integration. As such, we need a simulation method to tackle this problem. In this chapter, we employ a Monte Carlo simulation algorithm to estimate the right-hand side of (13.31) at time t_i.

Algorithm 13.1 (*Monte Carlo algorithm for mission reliability estimation*)

Step 1: Select a sufficiently large number (denoted by M_1) to simulate the realizations of θ from $f_\theta(\theta| Y_i, C_i, \Phi_{i,n})$. These realizations of θ are expressed by $\theta_r, r = 1, \ldots, M_1$.

Step 2: Select the number of the future mission process profiles for each realization θ_r of θ, $K_r, r = 1, \ldots, M_1$.

Step 3: For each θ_r, given the current mission state $\nu(t_i) = n$, simulate the mission process paths on the interval $[t_i, U]$, $\{\nu_m(u) : t_i < u \le U\}$ for $m = 1, 2, \ldots, K_r$ and obtain the sets of the future mission transition times $F_i(t_i + s_i) = \{u_{n+1,m}, \ldots, u_{N-1,m}\}$ and $A_i(t_i + s_i) = \{n+1, \ldots, N-1\}$.

Step 4: Calculate $b_i^m(u_{j,m}) = \frac{1}{\sigma}\left(w - y_i - \int_{t_i}^{u_{j,m}} \lambda(\nu(u))du\right)$ for each sampling path, $m = 1, 2, \ldots, K_r$, $j = n+1, \ldots, N-1$.

Step 5: Calculate the RUL distribution for each sampling path, $\Pr_{m,r}(S_i \le s_i|Y_i, C_i, \Phi_{i,n}, \theta_r, F_i(t_i + s_i))$ via Theorem 2, $m = 1, 2, \ldots, K_r$. Specifically, we compute $\Pr_{m,r}(S_i \le s_i|Y_i, C_i, \Phi_{i,n}, \theta_r, F_i(t_i + s_i))$ by the following steps:

(1) Select a sufficiently large number (e.g., M_2) of realizations for $N - n - 1$-dimensional BM process $\{B(u_{j,m}) - B(t_i), j = n+1, \ldots, N-1\}$.

(2) For each $l = 1, 2, \ldots, M_2$, generate the lth realization z_j^l of $B(u_{j,m}) - B(t_i)$, $j = n+1, \ldots, N-1$. This can be achieved by the simulation technique of a standard BM based on the independent increments property of BM.

(3) By the law of large numbers, for sufficiently large M_2, the conditional RUL distribution can be estimated by

$$\begin{aligned}&\Pr_{m,r}(S_i \le s_i|Y_i, C_i, \Phi_{i,n}, \theta_r, F_i(t_i+s_i)) = 1 - \tfrac{1}{M_2}\textstyle\sum_{l=1}^{M_2} g_{i,l,m,r}\\&g_{i,l,m,r}\\&= \textstyle\prod_{j\in A_i(t_i+s_i)}\left[\begin{array}{l} I(z_j^l < b_i^m(u_{j,m}))\times \\ \left(1-\exp\left[-\frac{2\left(b_i(u_{j-1})-z_{j-1}^l\right)\left(b_i(u_j)-z_j^l\right)}{u_{j,m}-u_{j-1,m}}\right]\right)\end{array}\right]\end{aligned} \quad . \tag{13.33}$$

Step 6: By the law of large numbers, for sufficiently large M_1 and K_r, the RUL distribution can be estimated by

$$\begin{aligned}&\Pr(S_i \le s_i|Y_i, C_i, \Phi_{i,n})\\&= \tfrac{1}{M_1}\textstyle\sum_{r=1}^{M_1}\left(\tfrac{1}{K_r}\sum_{m=1}^{K_r}\Pr_{m,r}(S_i \le s_i|Y_i, C_i, \Phi_{i,n}, \theta_r, F_i(t_i+s_i))\right)\end{aligned} \cdot \tag{13.34}$$

13.5 Reliability Estimation for PMS

After obtaining the estimated lifetime $f_{T_d|Y_i,C_i}(t|Y_i, C_i)$ of the mission system from the observed degradation data and the mission time $p_{T_M|\Phi_{i,n}}(t_m|\,\Phi_{i,n})$ from the CM information related with the duration of the mission phase to date, we can estimate the reliability of the mission process according to two definitions of the mission reliability given in Sect. 13.2.1.

Then, according to two different definitions, the reliability of PMS at t_i under the nth phase can be respectively formulated as

$$\begin{aligned}\Pr(T_d \ge T_M|\,Y_i, C_i, \Phi_{i,n}) &= \textstyle\int_{t_m>0} p_{T_M|\Phi_{i,n}}(t_m|\,\Phi_{i,n})\left(\int_{t\ge t_m} f_{T_d|Y_i,C_i,\Phi_{i,n}}(t|\,Y_i, C_i, \Phi_{i,n})dt\right)dt_m\\&= \textstyle\int_{t_m>0} p_{T_M|\Phi_{i,n}}(t_m|\,\Phi_{i,n})\Pr(S_i \le t_m - t_i|Y_i, C_i, \Phi_{i,n})dt_m\end{aligned}, \tag{13.35}$$

and

$$\begin{aligned}&\Pr(T_M \le R|\,T_d \ge T_M, Y_i, C_i, \Phi_{i,n})\\&= \frac{\Pr(T_M \le R, T_d \ge T_M|Y_i, C_i, \Phi_{i,n})}{\Pr(T_d \ge T_M|Y_i, C_i, \Phi_{i,n})}\\&= \frac{\iint_{t\ge t_m, 0<t_m\le R} p_{T_m,T_d|Y_i,C_i,\Phi_{i,n}}(t_m, t|Y_i, C_i, \Phi_{i,n})dtdt_m}{\Pr(T_d \ge T_M|Y_i, C_i, \Phi_{i,n})}\\&= \frac{\int_{0<t_m\le R}\Pr(S_i \le t_m - t_i|Y_i, C_i, \Phi_{i,n})p_{T_M|\Phi_{i,n}}(t_m|\Phi_{i,n})dt_m}{\Pr(T_d \ge T_M|Y_i, C_i, \Phi_{i,n})}\end{aligned}, \tag{13.36}$$

where $f_{T_d|Y_i,C_i,\Phi_{i,n}}(t|Y_i, C_i, \Phi_{i,n}) = \mathrm{d}\Pr(S_i \le t - t_i|Y_i, C_i, \Phi_{i,n})/\mathrm{d}t$ and $p_{T_M|\Phi_{i,n}}$ $(t_m|\,\Phi_{i,n})$ have been modeled in Sects. 13.3 and 13.4.

To achieve the reliability estimation, (13.35) and (13.36) can be similarly evaluated by the simulation-based technique as the presented algorithm for computation of the RUL distribution.

From (13.35) and (13.36), we observe that the presented approach for mission reliability estimation establishes a linkage among the mission reliability, the historical data, and real-time information of the individual PMS. As such, the mission reliability can be obtained straightforwardly through the estimated distribution of the mission time from the CM information related with the mission phase in conjunction with the estimated lifetime distribution from the degradation data in a real-time way. Note that there are some unknown parameters in the PDFs $f_{T_d|Y_i,C_i,\Phi_{i,n}}(t|Y_i, C_i, \Phi_{i,n})$ and $p_{T_M|\Phi_{i,n}}(t_m|\Phi_{i,n})$. These parameters can be estimated based on the historical data by the maximum likelihood approach naturally and thus we do not specifically discuss this estimation issue to limit our scope.

13.6 Experimental Studies

13.6.1 Numerical Simulations

In this section, a numerical example is used to illustrate the effectiveness of the presented approach. Suppose that there is a multiphase mission process with three phases. The phase durations are log-normally distributed but correlated, which are estimated using the historical phase duration data. For an individual PMS conducting a particular mission process, there are some sensors to monitor the CM information related with the phase duration and the degradation data related with the lifetime of the PMS. The CM information, once obtained, is used to update the phase duration and the mission time, while the degradation data are used to estimate the lifetime of the PMS. Specifically, we consider the following relationships among X_1, X_2, and X_3:

$$X_1 \sim \log N(\mu_{x1}, \sigma_{x1}^2), \quad X_2|X_1 \sim \log N(\mu_{x2} + \ln x_1 - \mu_{x1}, \sigma_{x2}^2)$$

$$X_3|X_1, X_2 \sim \log N(\mu_{x3} + \ln x_1 + \ln x_2 - \mu_{x1} - \mu_{x2}, \sigma_{x3}^2)$$

where $\mu_{xj}, \sigma_{xj}^2, j{=}1, 2, 3$ are parameters of log-normal distributions. Actually, these distributions are respectively corresponding to the distributions of $L_{0,1}$, $L_{0,2}$, $L_{0,3}$, in the filtering models.

In the presented approach, it is required to determine the functional form of the CM information and the remaining phase duration, i.e., $g_n(z_{i,n})$, $n = 1, 2, 3$. In the simulation, we use the following functional forms of $g_n(z_{i,n})$,

$$\varphi_{i,n} = g_n(z_{i,n}) + \eta_{i,n}, \tag{13.37}$$

with $g_n(z_{i,n}) = a_n + b_n \exp(-z_{i,n})$ and $\eta_{i,n} \sim N(0, \sigma_n^2)$, $n = 1, 2, 3$.

Actually, the selection of a suitable function of $g_n(z_{i,n})$ is essential when applying the EKF for a particular application and the above is just an idea to model the relationship between $z_{i,n}$ and $\varphi_{i,n}$. Of course any other forms for (13.37) can be used but the appropriate forms are frequently case dependent and can be selected by comparing the model fitting according to some criterion such as Akaike information criterion. For example, in [26, 28], the exponential form was adopted by the motivation of the vibration-based monitoring data and AIC was used for the mode selection.

Table 13.1 Parameters used for simulation study

1st phase	$\mu_{x1} = 3, \sigma_{x1}^2 = 0.04$	$a_1 = 5, b_1 = 4, \sigma_1 = 0.3$
2nd phase	$\mu_{x2} = 2.5, \sigma_{x2}^2 = 0.06$	$a_2 = 2, b_2 = 3, \sigma_2 = 0.2$
3rd phase	$\mu_{x3} = 2.2, \sigma_{x3}^2 = 0.02$	$a_3 = 3.5, b_3 = 0.8, \sigma_3 = 0.3$
Degradation process	$\lambda_1 = 0.5, \lambda_2 = 0.2, \lambda_3 = 0.3, \sigma = 0.4$	

Therefore, similar approach can be used to select $g_n(z_{i,n})$, but it is not the focus of this chapter and instead we just use (13.37) for illustrative purpose.

In order to generate the degradation data to estimate the lifetime of the PMS, we use the following discrete equation

$$y_{i+1} = y_i + \int_{t_i}^{t_{i+1}} \lambda\left(\nu(u)\right)\mathrm{d}u + \sigma\left(B(t_{i+1}) - B(t_i)\right) \tag{13.38}$$

where $B(t_{i+1}-t_i) \sim N(0, t_{i+1}-t_i)$ and $\lambda\left(\nu(t)\right) = \lambda_{n-1}$ if $\sum_{k=0}^{n-1} X_k \leq t < \sum_{k=0}^{n} X_k$ with $X_0 = 0$ and $n = 1, 2, 3$. Given the above settings and the parameters, we can simulate the required data for our modeling and reliability estimation. Table 13.1 shows the parameters used for the data simulation.

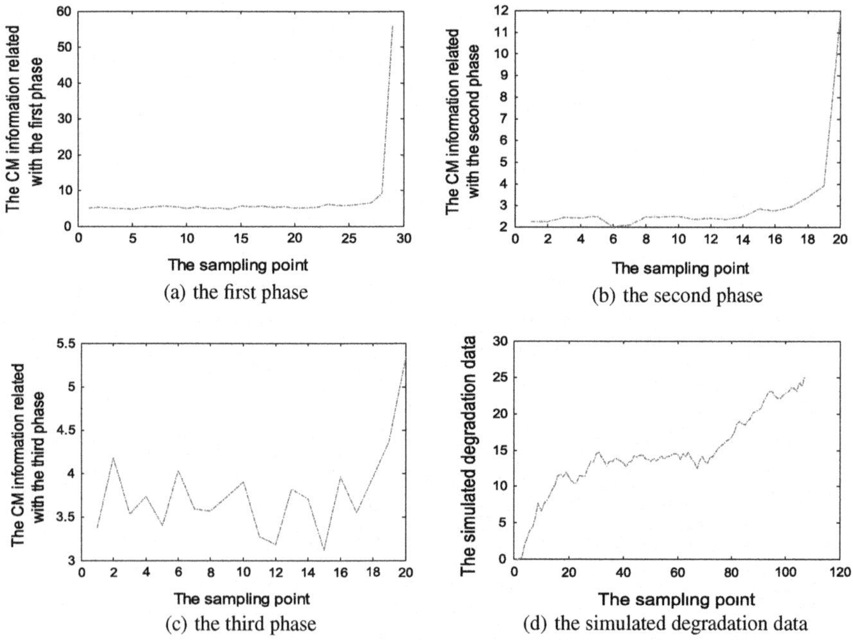

(a) the first phase
(b) the second phase
(c) the third phase
(d) the simulated degradation data

Fig. 13.2 The simulated CM information

Figure 13.2 shows the particular simulation data of the CM information related with the phase duration and the degradation process under the above model settings and parameters specifications.

Now we show the mission reliability results. When the mission process starts, we need to calculate the PDF of the remaining phase duration such as (13.10), and then update the PDF of the mission time (13.17) at each phase based on the CM data in this phase.

Figure 13.3 the estimated PDF of the remaining phase duration of each phase (a) the first phase, (b) the second phase, and (c) the third phase; (d) the estimated PDF of the RUL, at each sampling point.

Figure 13.3 illustrates the estimated PDF of the remaining phase duration of each phase and the estimated PDF of the RUL at each sampling point based on the method developed in Sect. 13.5. In estimating the RUL, we set the failure threshold as $w = 25$ and the numbers in the simulation algorithm are set as $M_1 = M_2 = K_r = 2000$ for calculate the RUL distribution. The obtained RUL distributions at different sampling times are illustrated in Fig. 13.3d. It can be found from Fig. 13.3 that both the developed filtering approach for remaining phase duration and RUL estimation approach for the PMS degradation can effectively update the associated PDFs using the CM information and degradation data to date. Based on these estimated phase

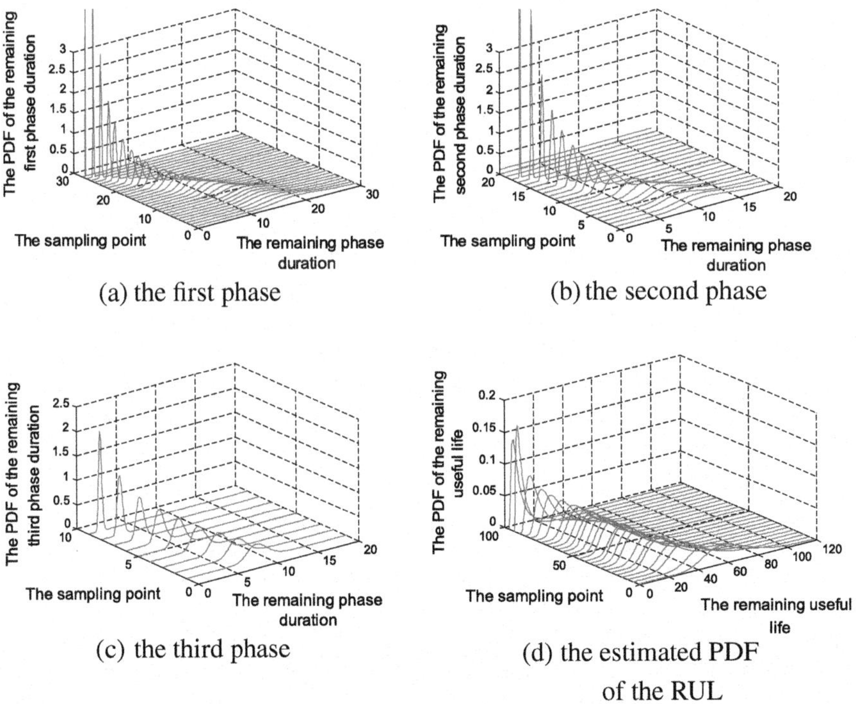

(a) the first phase

(b) the second phase

(c) the third phase

(d) the estimated PDF of the RUL

Fig. 13.3 the estimated PDF of the remaining phase duration of each phase at each sampling point

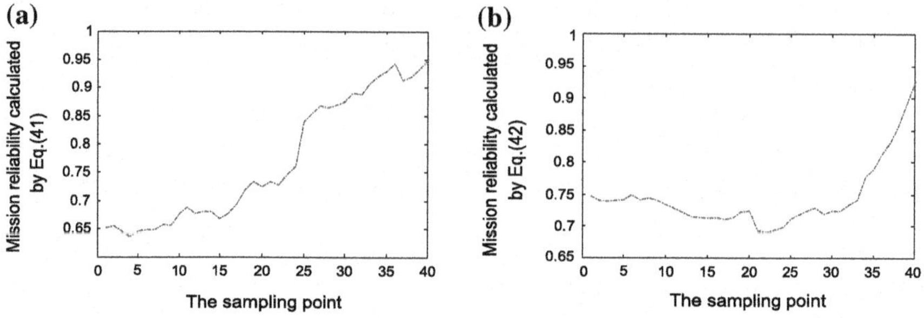

Fig. 13.4 The mission reliability calculated by **a** (13.35), **b** (13.36)

durations and RUL for PMS, we evaluate the mission reliability. In order to show the performance of the developed approach, we consider the following two cases:

- Case 1: The degradation quantity is subtle and thus the lifetime of mission system is long enough.

Since the degradation is subtle, the estimated RUL of the PMS is expected to be large. This is consistent with the intuition that for a newly installed PMS, its lifetime is naturally large enough and may be designed to own the ability of performing many missions. In order to simulate this case, we set a large failure threshold $w = 25$. The reason to do so is that a large threshold corresponds to a long lifetime of the PMS. First, we calculate the mission reliability at each sampling point according to the definition that the mission can be successfully accomplished before the system fails. The result is shown in Fig. 13.4a by evaluating (13.35).

As shown in Fig. 13.4, the success probability of PMSs will increase with the mission progressing. This can be well explained since with the mission process progressing the remaining mission time is less, but the reduced RUL of the PMS is not significant in contrast with its long lifetime in this case. Also, we can find that for the subtle degradation, the lifetime of the PMS will be large in stochastic sense. Therefore, the mission reliability will maintain in a relatively high level. Figure 13.4 reflects this fact. Accordingly, the probability that the mission can be successfully accomplished before a given time R under condition that the system lifetime is longer than the total mission time at each phase can be obtained by evaluating (13.36). For illustration, we set $R = 50$ and the result is shown in Fig. 13.4b.

It is not surprising that this kind of the mission reliability also has an increasing trend, as illustrated in Fig. 13.4b. The reason for this is similar to the above result because the lifetime of the PMS in this case is set to be long enough compared with the mission time. Therefore, the reduction of the RUL of the PMS is not significant as the mission progressing. In addition, we can observe that the mission reliability in the early phase is relatively low. This is resulted from the fact that this kind of reliability is a ratio, as seen from (13.36) and the denominator of (13.36) is relatively large in the early phase. However, as the mission progressing, the increase of the

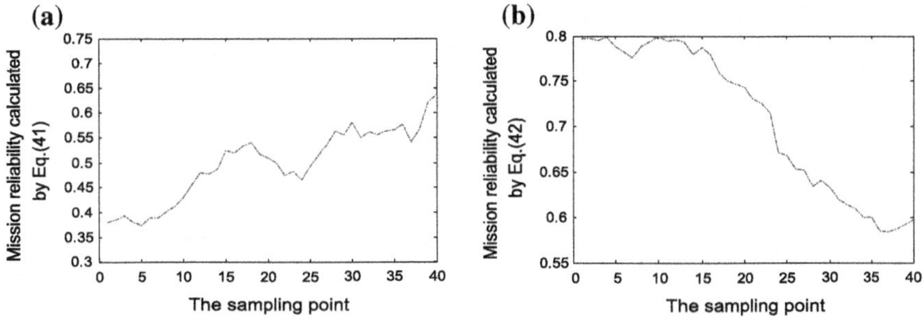

Fig. 13.5 The mission reliability calculated by **a** (13.35), **b** (13.36)

numerator is faster than the increase of the denominator. This leads to an increasing trend of this kind of mission reliability.

- Case 2: The degradation is dramatic and thus the lifetime of mission system is small.

Since the degradation is dramatic, the estimated RUL of the PMS is expected to be small. This is consistent with the intuition that for an aged PMS, its lifetime naturally approaches the end and thus there is high probability that the mission will fail. In order to simulate this case, we set a small failure threshold, such as $w = 15$. The reason is that a small threshold corresponds to a short lifetime of the PMS. Similar to Case 1, we first calculate the mission reliability which represents the probability that the mission can be successfully accomplished before the system fails. The result is illustrated in Fig. 13.5a.

Figure 13.5a shows that the mission reliability will be lower than the corresponding result in Case 1. Particularly, when the lifetime of mission system is small, the estimated mission reliability will fluctuate with the mission progressing to some extent though it still has certain increasing trend. These observations are largely resulted from the short lifetime of the PMS. Therefore, with the mission process evolving, the reduced RUL of the PMS is significant in contrast with the remaining mission time which is estimated from the CM information.

Accordingly, the probability that the mission can be successfully accomplished before a given time R under condition that the system lifetime is longer than the total mission time at each phase can be obtained by evaluating (13.36). As the same as the Case 1, the result is shown in Fig. 13.5b with the setting $R = 50$. It is interesting to note that the mission reliability for the success in the required time will experience a decreasing trend. This differs clearly from the previous results. Similar to the first type of the mission reliability, when the lifetime of the PMS is small, the estimated mission reliability of this type will fluctuate with the mission progressing. In this case, the mission reliability is a conditional probability as formulated in (13.36) and the denominator of this equation has an increasing trend as shown in Fig. 13.5a. As well as the result shown in Fig. 13.4, the numerator of (13.36) is a probability to characterize two events with the "AND" relationship: the event that the mission can be successfully accomplished before a given time R and the event that the system lifetime is longer than the total mission time. However, in this case, the lifetime of the PMS is small. Therefore, it is naturally expected that the increase of the numerator of (13.36) is not faster than that of the denominator. These observations finally result in the decreasing trend of the mission reliability of this type.

13.6.2 Case Study

In this part, we use the CM data of the inertial navigation system to illustrate the application of the developed method. As a key mission system of the control system in weapon systems and space systems, an inertial navigation system (INS) plays an important and irreplaceable role, and its operating state has a direct influence on navigation precision and the testing of the whole control system. In each control system testing, the INS will experience two phase: precision testing and function check. The success of these two phases will ensure the success of the testing of the whole control system. Thus, it is often desired to conduct the mission reliability estimation to mitigate the risk of the fail of the testing mission. However, the complicate structure of the INS makes the existing method such as BDD and FT difficult to apply. Thus, we use the presented method in this chapter to estimate the mission reliability.

In the practice, the sensors fixed in an INS include three gyros and three accelerometers, which measure angular velocity and linear acceleration, respectively. The gyro fixed on an inertial platform is a mechanical structure having two degrees of freedom from the driver and sense axis. When the inertial platform is operating, the wheels of the gyros rotate at very high speeds and can lead to rotation axis wear. In general, the drift data of gyros along the sense axis can be used as a performance indicator to evaluate the health condition of an INS and estimate its lifetime. Other data collected by the sensors can used to model the mission process. The detailed description of the INS can be referred to [30, 36].

For the used INS in certain space system, which is newly equipped systems, we conduct a control system testing mission. For the INS, the mission reliability is that the required mission time is that the mission can be successfully accomplished before the system fails. Namely, the second definition of the mission reliability is adopted. After the testing, the mission is completed successfully, and 99 points of the degradation data and the CM coefficients data are collected with regular CM intervals 1 h in field condition. The collected degradation data and the associated fitting cure used the presented method in this chapter are illustrated in Fig. 13.6, and the corresponding failure threshold is $0.15°/\text{h}$.

From Fig. 13.6, it can be observed that the presented degradation model can fit the practical degradation data well. In the following, we mainly give the estimation results of the mission reliability while the implementation details are similar to the numerical example and thus are omitted here due to the limited space. To implement Algorithm 13.1, we set $M_1 = M_2 = K_r = 1000$. The estimated mission reliability is shown in Fig. 13.7.

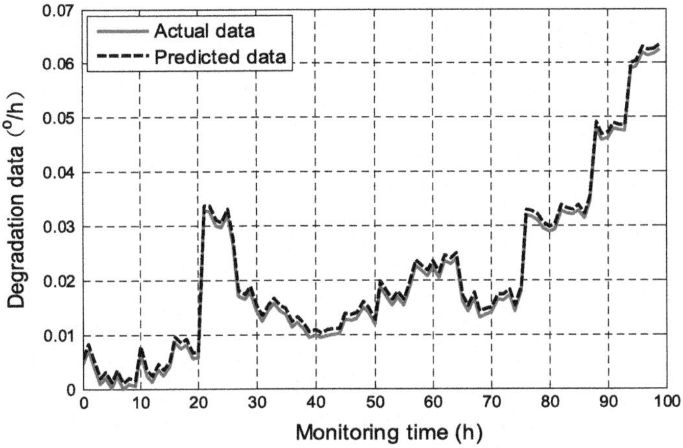

Fig. 13.6 The degradation data and the fitted degradation path

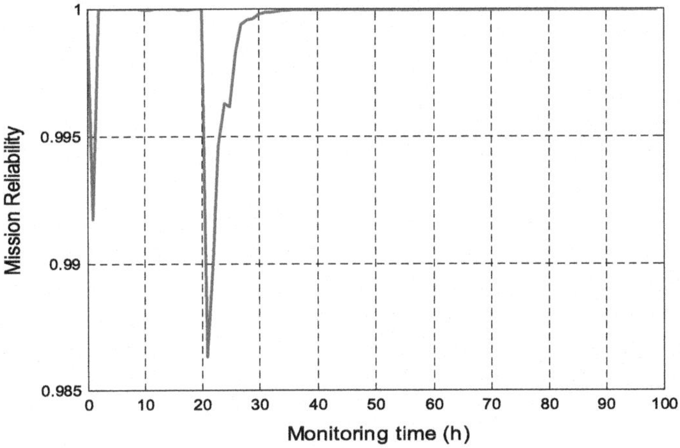

Fig. 13.7 The mission reliability

Figure 13.7 shows that the mission reliability of the INS during this testing process is high and more than 0.985 because this testing process is successes in fact. However, it can also be found that, during about the 20th CM time, the mission reliability decreases obviously. This observation can be easily verified by Fig. 13.6 since in about the 20th CM point, the mission process is switched to the second phase and there is a dramatic degradation trend. Thus, the mission reliability decreases. This observation is also consistent with the results of numerical example in Sect. 13.6.1. Through these results, we can find that the reliability estimation in the presented method can be repeatedly updated during its operation to ensure that the most recently calculated reliability value accurately reflects the current reality of the health state of the mission system.

Table 13.2 Computation time for estimating mission reliability

M_1	M_2	K_r	Computation time (s)
1000	1000	1000	84.6
2000	1000	1000	129.5
2000	2000	1000	135.6
3000	2000	3000	294.8

In this case study, it is interesting to compute the computation time of the whole algorithm to check whether the presented method can achieve online data processing for practical application. Therefore, we statistically compute the time needed to estimate the mission reliability in different cases of M_1, M_2, K_r. The main results are summarized in Table 13.2.

From the results in Table 13.2, it is found that the computation time increases with the increase of M_1, M_2, K_r. By comparing the computation time with the CM interval, it can be found that the computational time needed for the algorithm is much less than the CM interval (1 h) of the used INS. Thus, the online data processing can be achieved. It is also noted that, for the current application of the INS, the estimation performance for $M_1 = M_2 = K_r = 1000$ is actually satisfactory and in this case the computation time is less than 90 s. We believe that such computation time should be able to satisfy the demands of the updating frequency for other realistic applications.

In sum, the simulation study and case study illustrate the implementation of the developed approach and indicates that it can estimate the mission reliability of the PMS using the CM information related with the phase duration and the degradation data related with the PMS lifetime. Therefore, when the structure knowledge of the PMS is not available but the monitored data can be collected, the developed approach in this chapter could be potentially applied.

References

1. Esary J, Ziehms H (1975) reliability analysis of phased missions. Proceedings of the conference on reliability and fault tree analysis. SIAM, pp 213–236
2. Mo C (2009) New insights into the BDD-based reliability analysis of phased-mission systems. IEEE Trans Reliab 58(4):667–678
3. Shi J, Zhou SY (2009) Quality control and improvement for multistage systems: a survey. IIE Trans 41(9):744–753
4. Feyzioglu, Altmel L, Ozekici S (2008) Optimum component test plans for phased-mission systems. Eur J Oper Res 185(1):255–265
5. Reed S, Andrews JD, Dunnett SJ (2011) Improved efficiency in the analysis of phased mission systems with multiple failure mode components. IEEE Trans Reliab 60(1):70–79
6. Burdick G, Fussell J, Rasmuson D, Wilson J (1977) Phased mission analysis: A review of new developments and an application. IEEE Trans Reliab 2G:43–49
7. Veatch M (2006) Reliability of periodic, coherent, binary systems. IEEE Trans Reliab 35(5):504–507

8. Alam M, Al-Saggaf UM (1986) Quantitative reliability evaluation of repairable phased-mission systems using Markov approach. IEEE Trans Reliab R-35(4):498–503
9. Kim K, Park K (1994) Phased-mission system reliability under Markov environment. IEEE Trans Reliab 43(2):301–309
10. Dugan JB (1991) Automated analysis of phased-mission reliability. IEEE Trans Reliab 40(1):45–52
11. Vaurio J (2001) Fault tree analysis of phased-mission systems with repairable and non-repairable components. Reliab Eng Syst Saf 74(2):169–180
12. Shrestha, Xing LD, Dai YS (2011) Reliability analysis of multistate phased-mission systems with unordered and ordered states. IEEE Trans Syst Man Cybern-Part A: Syst Hum 41(4):625–636
13. Xue DZ, Wang XD (1989) A practical approach for phased mission analysis. Reliab Eng Syst Saf 25(4):333–347
14. Kohda T, Wada M, Inoue K (1994) A simple method for phased mission analysis. Reliab Eng Syst Saf 45(3):299–309
15. Somani K, Trivedi KS (1994) Boolean algebraic methods for phased-mission system analysis. In: Proceedings of sigmetrics, pp 98–107
16. La Band R, Andrews JD (2002) Phased mission modelling using fault tree analysis. Reliab Eng Syst Saf 78:45–56
17. Zang Z, Sun HR, Trivedi KS (1999) A BDD-based algorithm for reliability analysis of phased-mission systems. IEEE Trans Reliab 48(1):50–60
18. Xing LD, Dugan JB (2000) Analysis of general phased-mission system reliability, performance and sensitivity. IEEE Trans Reliab 51(2):199–211
19. Tang ZH, Dugan JB (2006) BDD-based reliability analysis of phased-mission systems with multimode failures. IEEE Trans Reliab 55(2):350–360
20. Tang ZH, Xu H, Dugan JB (2005) Reliability analysis of phased mission systems with common cause failures. In: Proceedings of annual reliability and maintainability symposium, pp 313–318
21. Xing L (2007) Reliability evaluation of phased-mission systems with imperfect fault coverage and common-cause failures. IEEE Trans Reliab 56(1):58–68
22. Wang D, Trivedi KS (2007) Reliability analysis of phased-mission system with independent component repairs. IEEE Trans Reliab 56(3):540–551
23. Bolling B, Wegener I (1996) Improving the variable ordering of OBDDs is NP-complete. IEEE Trans Comput C-35(8):677–691
24. Çekyay B, Özekici S (2008) Reliability of semi-Markov missions. Technical report, Koc University, Department of Industrial Engineering, Istanbul, Turkey
25. Çekyay B, Özekici S (2010) Mean time to failure and availability of semi-Markov missions with maximal repair. Eur J Oper Res 207(3):1442–1454
26. Wang W (2000) A model to predict the residual life of rolling element bearings given monitored condition information to date. IMA J Manag Math 13(1):3–16
27. Wang W, Zhang W (2005) A model to predict the residual life of aircraft engines based upon oil analysis data. Nav Res Logist 52(3):276–284
28. Carr MJ, Wang W (2011) An approximate algorithm for prognostic modelling using condition monitoring information. Eur J Oper Res 211(1):90–96
29. Si XS, Wang W, Hu CH, Zhou DH (2011) Remaining useful life estimation-a review on the statistical data driven approaches. Eur J Oper Res 213(1):1–14
30. Si XS, Wang W, Hu CH, Zhou DH (2014) Estimating remaining useful life with three-source variability in degradation modeling. IEEE Trans Reliab 63(1):167–190
31. Li H, Pan DH, Chen CLP (2014) Reliability modeling and life estimation using an expectation maximization based Wiener degradation model for momentum wheels. IEEE Trans Cybern. doi:10.1109/TCYB.2014.2341113
32. Ye ZS, Chen N, Shen Y, A new class of wiener process models for degradation analysis. Reliab Eng Syst Saf. http://dx.doi.org/10.1016/j.ress.2015.02.005

33. Ye ZS, Wang Y, Tsui KL, Pecht M (2013) Degradation data analysis using Wiener processes with measurement errors. IEEE Trans Reliab 62(4):772–780
34. Si XS, Zhou DH (2014) A generalized result for degradation model based reliability estimation. IEEE Trans Autom Sci Eng 11(2):632–637
35. He HF, Li J, Zhang QH, Sun GX (2014) A data-driven reliability estimation approach for phased-mission systems. Math Probl Eng. http://dx.doi.org/10.1155/2014/283740
36. Si XS, Chen MY, Wang W, Hu CH, Zhou DH (2013) Specifying measurement errors for required lifetime estimation performance. Eur J Oper Res 231(3):631–644
37. Lee M-LT, Whitmore GA (2006) Threshold regression for survival analysis: modeling event times by a stochastic process reaching a boundary. Stat Sci 21(4):501–513
38. Chen N, Tsui KL (2013) Condition monitoring and residual life prediction using degradation signals: revisited. IIE Trans 45(9):939–952
39. Liu K, Huang S (2015) Integration of data fusion methodology and degradation modeling process to improve prognostics. IEEE Trans Autom Sci Eng. doi:10.1109/TASE.2014.2349733
40. Ye ZS, Xie M (2015) Stochastic modelling and analysis of degradation for highly reliable products (with discussion). Appl Stoch Models Bus Ind 31(1):16–36
41. Wang L, Potzelberger K (1997) Boundary crossing probability for brownian motion and general boundaries. J Appl Probab 34(1):54–65

Chapter 14
A Real-Time Variable Cost-Based Maintenance Model

14.1 Introduction

With advances in condition monitoring technologies, the past decade has witnessed an increasingly growing research interest on various aspects of degradation modeling for prognostics from the observed signals by dedicated sensors [1–4]. This is (at least partly) caused by its importance in a variety of fields such as maintenance, inventory control, and public health surveillance and management. When the real-time monitoring information is involved in decision making, there are three key issues for a successful implementation of maintenance under prognostic information.

- How can we best model and predict degradation and remaining useful life (RUL) of the monitored equipment conditional on the observed information?
- How to measure the effect of the prognostic uncertainty on maintenance decisions?
- When should the maintenance action be done?

Although there are a great number of studies on these three key issues separately, it has long recognized that how to address these issues jointly in a general framework is challenging. Particularly, most of the preventive maintenance policies in the literature are evaluated with only the expected cost criteria. However, it is generally more critical to consider not only the expectation of the maintenance cost, but also the variability of the cost. For example, there are two maintenance policies, A and B. Under policy A, the maintenance costs $10,000 per month. Under policy B, there is an 80% probability of a zero cost and a 20% probability of a $50,000 cost for each month. For both policies, the long-run average of the costs is the same being $10,000 per month, if the maintenance budget allocation is $10,000 per month, which will always be affordable under policy A. However, under policy B, the large variability of the cost from one month to another leads to a high management risk under the anticipated budget allocation. As a result, policy A is preferable to policy B in terms of the budget allocation management because the cost of policy A has less cost variability. Figure 14.1 illustrates such uncertainty in maintenance costs resulted from the uncertainty of random failure.

X.-S. Si et al., *Data-Driven Remaining Useful Life Prognosis Techniques*,
Springer Series in Reliability Engineering, DOI 10.1007/978-3-662-54030-5_14

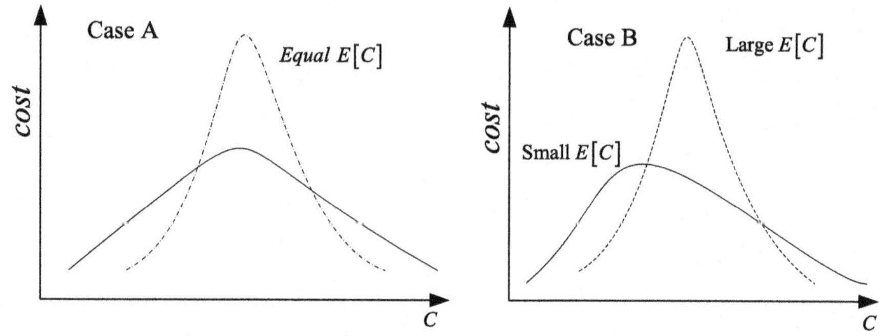

Fig. 14.1 Illustration of the uncertainty in maintenance costs

In the literature, Tapiero and Venezia first considered the variability in maintenance cost and presented a replacement policy based on such variability [5]. Motivated by Tapiero and Venezia [5], Chen and Jin [6], Gosavi [7], and Giri and Dohi [8] investigated many maintenance policies by minimizing the cost with the variability discount. However, all these works formulated the maintenance policy in the age-based replacement framework and thus the real-time condition monitoring information is not considered. Therefore, these policies aimed to schedule the maintenance activity for a population but not for an individual. However, considering real-time condition monitoring information is desirable so that the optimal replacement time is repeatedly updated during equipment operation to ensure the most recently calculated replacement time accurately reflects the current reality of the monitored equipment. Recently, Chen et al. in [9] considered a condition-based replacement model based on the variability of the maintenance cost in condition-based maintenance framework. However, this work did not incorporate the prognostic information. So far, no chapter considers the cost variability in the condition-based/predictive maintenance with the prognostic information about equipment's useful life.

In this chapter, we develop a new method to consider the effects of both the expectation of the maintenance cost and its variability. Given the prognostic information obtained from condition monitoring and variable maintenance cost, we obtain a maintenance decision which is different from that of the mean maintenance cost-based model under the same setting. The prognostic information is obtained from a degradation process modeled as an adaptive Wiener process in real time. One important proposition obtained shows that the decision from the proposed model is conservative as opposed to the case considering the mean cost only. We demonstrate the proposed method with a practical case study. The results indicate that our method can effectively mitigate the management risk, but with a small cost increase.

The remainder parts are organized as follows. In Sect. 14.2, we presented a Wiener-process-based degradation model for prognostics. Section 14.3 establishes the real-time variable cost-based maintenance model with the prognostic information involved. Section 14.4 provides a practical case study to demonstrate the developed approach.

14.2 Degradation Modeling for Prognostics

In this chapter, we consider applying a Wiener-process-based model with a recursive filter algorithm developed to obtain the prognostics information and then use this information to establish the replacement model.

14.2.1 Degradation Modeling

In general, a Wiener process $\{X(t), t \geq 0\}$ can be represented as

$$X(t) = \lambda t + \sigma B(t), \tag{14.1}$$

where λ is a defined drift coefficient, $\sigma > 0$ is a diffusion coefficient, and $B(t)$ is the standard Brownian motion. Without loss of generality, we assume $X(0) = 0$ in this chapter.

Since degradation is an inherent characteristic of operating plants, then for a stochastic model to be able to describe the degradation process, the model must be infinitely divisible [4]. This implies the model should be able to describe the degradation at any time and is independent of the previous sampling frequency and intervals. Here we give a formal definition of the property of infinite divisibility for a stochastic process, $\{X(t), t \geq 0\}$.

Definition 14.1 A stochastic process $\{X(t), t \geq 0\}$ is infinitely divisible if for a given fixed time t and any $n \in \mathbf{N}$, there exists a time sequence $\{t_i\}_{1 \leq i \leq n}$ with $0 \leq t_1 \leq t_2 \leq \ldots \leq t_n \leq t$ and independent, identically distributed random variables $X_{t_1}, X_{t_2-t_1}, \ldots, X_{t-t_n}$ where $X_{t_{i+1}-t_i} = X(t_{i+1}) - X(t_i)$ such that $X(t) = X_{t_1} + X_{t_2-t_1} + \cdots + X_{t-t_n}$.

We demonstrate in the following that a Wiener process is infinitely divisible and is a better choice for modeling a physical degradation process.

Proposition 14.1 *A Wiener process is infinitely divisible.*

Proof For an arbitrary time length t, and we divide it into n mutually exclusive intervals, $0 \leq t_1 \leq t_2 \leq \ldots \leq t_n \leq t_{n+1} = t$ where $n \in \mathbf{N}$. Define $X_{t_{i+1}-t_i} = X(t_{i+1}) - X(t_i)$. Then $X_{t_{i+1}-t_i}$ follows $X_{t_{i+1}-t_i} = \lambda(t_{i+1} - t_i) + \sigma B(t_{i+1} - t_i)$ from the property of the Wiener process. The total degradation is the summation of $X_{t_{i+1}-t_i}$, e.g., $X(t) = \lim|_{n\to\infty} \sum_{i=0}^{n} X_{t_{i+1}-t_i}$.

Since all $X_{t_{i+1}-t_i}$, $i = 0, 1, \ldots n$ are independent and follow $N\left(\lambda\,(t_{i+1} - t_i), \sigma^2 (t_{i+1} - t_i)\right)$ from the property of the Wiener process, it follows that $\sum_{i=0}^{n} (X(t_{i+1}) - X(t_i))$ is also normal with

$$E\left[\sum_{i=0}^{n} (X(t_{i+1}) - X(t_i))\right] = \sum_{i=0}^{n} E\left[(X(t_{i+1}) - X(t_i))\right] = \sum_{i=0}^{n} (\lambda\,(t_{i+1} - t_i)) = \lambda t_{n+1} = \lambda t, \tag{14.2}$$

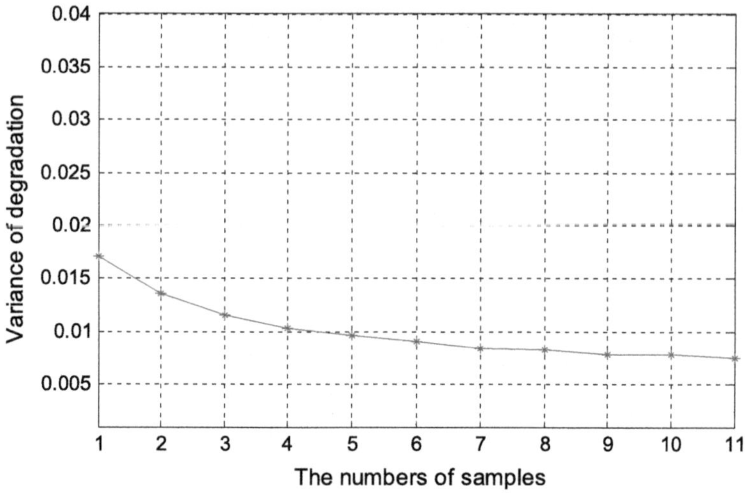

Fig. 14.2 The result for fatigue crack length data set

$$
\begin{aligned}
Var\left[\sum_{i=0}^{n}\left(X(t_{i+1}) - X(t_i)\right)\right] &= \sum_{i=0}^{n} Var\left[\left(X(t_{i+1}) - X(t_i)\right)\right] \\
&= \sum_{i=0}^{n}\left(\sigma^2\left(t_{i+1} - t_i\right)\right) = \sigma^2 t_{n+1} = \sigma^2 t, \qquad (14.3)
\end{aligned}
$$

which hold any $n \in \mathbf{N}$.

In order to further clarify the necessity of such infinite divisibility in modeling physical degradation process, we give an illustrating example as follows.

In the literature, there are many public degradation data sets from several devices of the same type, such as fatigue crack data and laser date in [10]. We also have two data sets of the LEDs at hand from the accelerating degradation tests at current 170 mA and 320 mA, respectively. These data really facilitate our illustration. Due to the limited space, we do not provide the data but the interested readers can obtain the used data on request. In order to show the relationship between the uncertainty during degradation process and the sampling numbers, we give a brief procedure as follows. We assume that there are M sampling paths, each path has N CM points, and the CM time points are same among different paths. The above-mentioned data sets are the case here. In simulation, we first fix the first and the last CM points and then randomly extract n CM point, where $n = 1, 2, \ldots, N - 2$. Once the random sampling is completed, we can obtain the variance of the degradation process during the first and the last CM points in statistical sense using the data from the extracted time points, the first and the last points. For each n, we repeat the random sampling process 1000 times and calculated the mean as the final variance. Figures 14.2, 14.3, 14.4, and 14.5 show the obtained results of the used data sets.

Clearly, we can observe from these figures that the sampling numbers have limited impact on the degradation characteristic, since our simulation is based on random

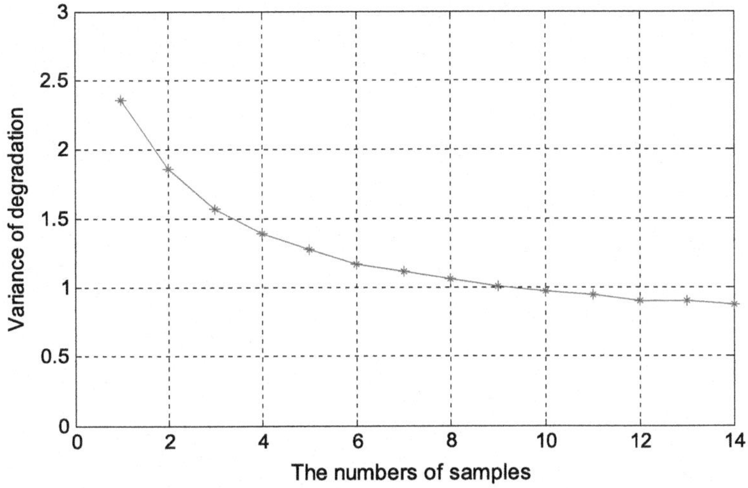

Fig. 14.3 The result for laser data set

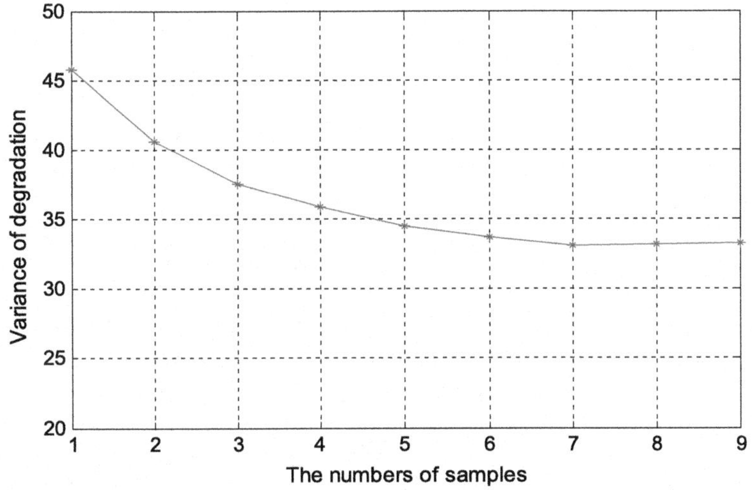

Fig. 14.4 The result for the LEDs data at 170 mA

sampling from the data sets. Therefore, we may reasonably infer that a physical degradation process should be free from the constraint of the sampling frequency and intervals.

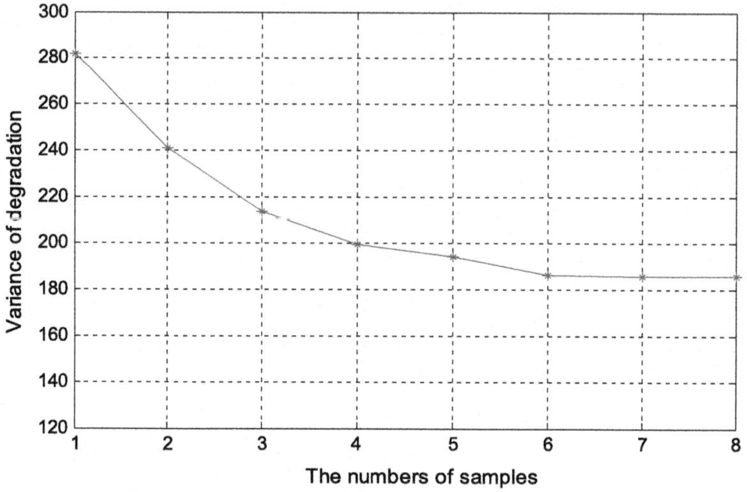

Fig. 14.5 The result for the LEDs data at 320 mA

14.2.2 RUL Estimation

Considering the potential for updating knowledge of the process when new degradation observations become available, for $t \geq t_i$, we model the degradation process over time since t_i as

$$X(t) = x_i + \lambda(t - t_i) + \sigma B(t - t_i). \tag{14.4}$$

To incorporate the history of the observations of the degradation, we consider an updating procedure for the drifting parameter λ by making λ evolving as $\lambda_i = \lambda_{i-1} + \eta$ over time, where $\eta \sim N(0, Q)$. As such, the drift parameter is evolving as a time-dependent variable with a random distribution, conditional on the observed data up to time t_i. The reason to make λ time-varying with the collecting data is that the degradation trend is dominated by the drift coefficient while the variance coefficient reflects the uncertainty in degradation process. In order to establish the linkage between the drift parameter and the observation history up to date, the degradation equation can be reconstructed and taken to be a *self-organizing state-space model* as

$$\begin{cases} \lambda_i = \lambda_{i-1} + \eta \\ x_i = x_{i-1} + \lambda_{i-1}(t_i - t_{i-1}) + \sigma\varepsilon_i \end{cases}, \tag{14.5}$$

where the error terms are distributed as $\eta \sim N(0, Q)$ and $\varepsilon_i \sim N(0, t_i - t_{i-1})$. The use of $t_i - t_{i-1}$ as the variance of ε_i is derived by the property of Brownian motion. In this chapter we assume that the initial drift λ_0 follows a normal distribution with the mean μ_0 and the variance P_0. As such, the drift parameter is considered as a latent "state" and can only be estimated from the historical information $\mathbf{X}_{1:i} = \{x_1, x_2, \ldots, x_i\}$ due to its unobservable nature. In the established state-space model (14.5), the state equation and the observation equation are linear and the associated noises are Gaussian. As such, the estimation of the drift parameter $\hat{\lambda}_i$ can be easily obtained by Kalman filtering once new observation x_i is available at t_i. The updated estimation of λ_i can be obtained from the following *Kalman filtering algorithm.*

Algorithm 14.1 (*Kalman filtering algorithm*)
Step 1: Initialize μ_0, P_0.
Step 2: State estimation at time t_i
$P_{i|i-1} = P_{i-1|i-1} + Q$
$K_i = (t_i - t_{i-1})^2 P_{i|i-1} + \sigma^2 (t_i - t_{i-1})$
$\hat{\lambda}_i = \hat{\lambda}_{i-1} + P_{i|i-1}(t_i - t_{i-1}) K_i^{-1} \left(x_i - x_{i-1} - \hat{\lambda}_{i-1}(t_i - t_{i-1}) \right)$
Step 3: Updating variance $P_{i|i} = P_{i|i-1} - P_{i|i-1}(t_i - t_{i-1})^2 K_i^{-1} P_{i|i-1}$.

Based on the threshold, the RUL modeling principle is presented as follows. When degradation $X(t)$ modeled as Eq. (14.4) reaches a preset critical level w, the plant can be declared to be failed and thus there has no useful lifetime left. From the first hitting time concept, the RUL L_i at t_i can be defined as

$$L_i = \inf \left\{ l_i : X(t_i + l_i) \geq w | \mathbf{X}_{1:i}, \forall 1 \leq j \leq i, x_j < w \right\}, \tag{14.6}$$

with the probability density function (PDF) $f_{L_i|\mathbf{X}_{1:i}}(l_i | \mathbf{X}_{1:i})$.

As a result of the law of total probability, the PDF of the RUL at time t_i can be formulated with the available degradation measurement $\mathbf{X}_{1:i}$ as

$$f_{L_i|\mathbf{X}_{1:i}}(l_i|\mathbf{X}_{1:i}) = \frac{w - x_i}{\sqrt{2\pi l_i^3 \left(P_{i|i} l_i + \sigma^2 \right)}} \exp \left(- \frac{\left(w - x_i - \hat{\lambda}_i l_i \right)^2}{2 l_i \left(P_{i|i} l_i + \sigma^2 \right)} \right), l_i > 0, \tag{14.7}$$

where $\hat{\lambda}_i$ and $P_{i|i}$ can be obtained from the previous filtering algorithm.

Once the new degradation observation is available and the parameter is updated, the RUL of this monitored asset can be computed using Eq. (14.7). In fact, the parameters in the degradation model and state-space model can be estimated simultaneously in conjunction with updating the drifting parameter.

14.3 Replacement Decision Modeling

In order to study a variable cost-based maintenance policy, the measure of the cost variability needs to be first defined. Several measures of variability for stochastic processes with rewards have been discussed and compared in [11]. In this chapter, the long-run variance measure recommended by Filar et al. [11] and Chen and Jin [6] is used. A discrete time scale is considered. Let t=1, 2,…denote the discrete time units, and C_t^π denote the cost spent at time unit t under maintenance policy π.

From [6, 11], the long-run average cost per unit time under maintenance policy π can be defined as

$$\varphi^\pi = \lim_{T\to\infty} \frac{1}{T} \sum_{t=1}^{T} C_t^\pi. \tag{14.8}$$

The *long-run variance* of the cost under maintenance policy π is defined as

$$V^\pi = \lim_{T\to\infty} \frac{1}{T} \sum_{t=1}^{T} \left[\left(C_t^\pi - \varphi^\pi \right)^2 \right]. \tag{14.9}$$

The *long-run mean square* cost under maintenance policy π is defined as

$$\psi^\pi = \lim_{T\to\infty} \frac{1}{T} \sum_{t=1}^{T} \left(C_t^\pi \right)^2. \tag{14.10}$$

Then the variable cost-based maintenance optimization problem can be formulated as

$$\min_{\pi} \left[(\varphi^\pi)^2 + \alpha V^\pi \right], \alpha \geq 0. \tag{14.11}$$

where π is the class of the considered maintenance polices, and α is the cost-variability-sensitive factor. Since $\varphi^\pi > 0$, the policies minimizing $(\varphi^\pi)^2$ must be the same as those minimizing φ^π. Therefore, this objective function is equivalent to the traditional cost policy when $\alpha = 0$.

In the above equation, α can be interpreted as the relative weight of variability. The intuitive meaning of $\alpha < 1$ is that the decision maker considers that the improvement in variability has less impact than the same degree of improvement in the mean cost. In other words, the cost variability is less significant than the mean cost. Similarly, $\alpha > 1$ means that the variability has a more significant impact than the mean cost on the decision making. We note that the most common intuition of the decision makers is to consider the mean cost to be more significant than the cost variability. Therefore, the decision makers should be cautious when they decide to use an α, with a value greater than one. When the cost variability is more important than the average cost for the decision makers, a choice of α with value greater than one might be reasonable.

Under prognostic information provided in Sect. 14.2, the above optimization problem can be formulated as Variance-Penalized Decision Problem. That is, in degradation observation x_i at the ith condition monitoring point t_i, denoted the state (i, x_i), the associated variable cost optimization problem can be represented as follows:

$$C^\pi(i, x_i) = \min_{t^* \in [t_i, \infty)} \left[(\varphi^\pi(i, x_i))^2 + \alpha V^\pi(i, x_i) \right], \alpha \in [0, \infty), \tag{14.12}$$

with

$$\varphi^{\pi}(i, x_i) = \frac{c_p + (c_f - c_p)\Pr(L_i < t - t_i \,|\mathbf{X}_{1:i})}{t_i + (t - t_i)(1 - \Pr(L_i < t - t_i \,|\mathbf{X}_{1:i})) + \int_{l_i=0}^{t-t_i} l_i f_{L_i}(l_i|\mathbf{X}_{1:i})dl_i}, \quad (14.13)$$

$$V_p = \frac{c_p^2 + \left(c_f^2 - c_p^2\right)\Pr(L_i < t - t_i \,|\mathbf{X}_{1:i}) - \left(c_p + (c_f - c_p)\Pr(L_i < t - t_i \,|\mathbf{X}_{1:i})\right)^2}{t_i + (t - t_i)(1 - \Pr(L_i < t - t_i \,|\mathbf{X}_{1:i})) + \int_0^{t-t_i} l_i f_{L_i}(l_i|\mathbf{X}_{1:i})dl_i}, \quad (14.14)$$

where $\Pr(L_i < t - t_i|\mathbf{X}_{1:i}) = \int_0^{t-t_i} f_{L_i|\mathbf{X}_{1:i}}(l_i|\mathbf{X}_{1:i})dl_i$ and $f_{L_i|\mathbf{X}_{1:i}}(l_i|\mathbf{X}_{1:i})$ can be obtained from Eq. (14.7), t is the decision variable representing the planned replacement time at the ith CM point, c_p is the mean cost of a preventive replacement, and c_f is the mean replacement cost associated with the failure; $\varphi^{\pi}(i, x_i)$ and $V^{\pi}(i, x_i)$ are the long-run average cost per unit time and *long-run variance* of the cost under maintenance policy π at the state (i, x_i), respectively.

Based on the above replacement decision modeling, we have the following main result.

Proposition 14.2 *Under the state (i, x_i), let $t_{\phi}^*(i, x_i) \equiv \inf A(\phi^{\pi}(i, x_i))$ where $A(\phi^{\pi}(i, x_i))$ is the set of all optimal solutions which minimize $\phi^{\pi}(i, x_i)$, and $t^*(i, x_i)$ be an optimal solution of the optimization problem $\min_{t^* \in [t_i, \infty)} \left[(\varphi^{\pi}(i, x_i))^2 + \alpha V^{\pi}(i, x_i)\right]$ with a fixed α. Then the followings hold:*

(1) if $\alpha > 0$, then $t^*(i, x_i) \le t_{\phi}^*(i, x_i)$. The inequality is strict when $t_{\phi}^*(i, x_i) < \infty$.
(2) let $\alpha_1 > \alpha_2 \ge 0$, $t_{\alpha_1}^*(i, x_i) \equiv \inf A(\alpha_1; i, x_i)$, and $t_{\alpha_2}^*(i, x_i) \equiv \inf A(\alpha_2; i, x_i)$, then $t_{\alpha_1}^*(i, x_i) \le t_{\alpha_2}^*(i, x_i)$, where $A(\alpha_i; i, x_i), i = 1, 2$ denote the sets of all optimal solutions which minimize $(\varphi^{\pi}(i, x_i))^2 + \alpha V^{\pi}(i, x_i)$ with $\alpha = \alpha_i$.

The optimal replacement interval under the variable cost-based maintenance policy tends to be shorter than that under the traditional mean cost maintenance policy. In other words, the variable cost optimal policy tends to be more conservative than the mean cost optimal policy. In addition, the larger the relative weight α, the shorter is the preventive replacement interval. This seems to be consistent with the intuition that the maintenance risk can be reduced through increasing the replacement frequency.

14.4 A Case Study

As a key device of the INS in weapon systems, an inertial platform plays an important and irreplaceable role [12]. Its operating state has a direct influence on navigation precision. The sensors fixed in an inertial platform mainly include three gyros and

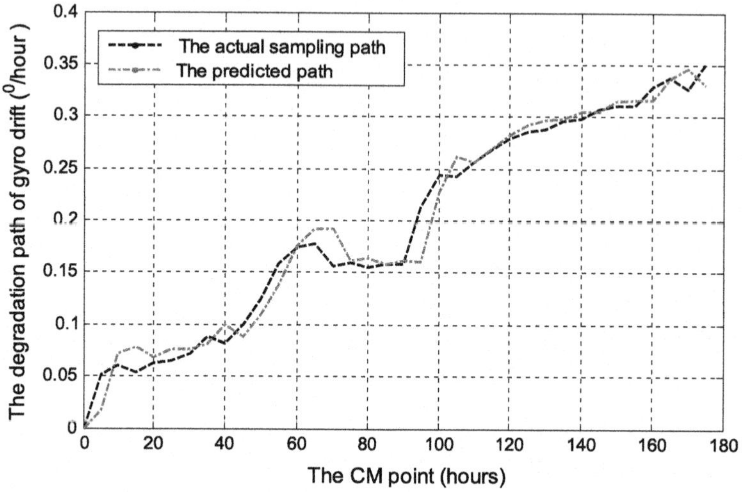

Fig. 14.6 The actual gyro's drift data and the predictions with adapted Brownian motion

three accelerometers, which measure angular velocity and linear acceleration, respectively. When the inertial platform is operating, the wheels of the gyros rotate at very high speeds and can lead to rotation axis wear and finally result in the gyros' drift. As the wear is accumulated, the drift degrades and finally leads to the failure of gyros. As such, the drift of gyros is often used as a performance indicator to evaluate the health condition of an inertial platform and scheduling maintenance activities. In our study, we take the drift degradation measurement along the sense axis for illustrative purposes, since this variable plays a dominant role in the assessment of gyro degradation. The obtained field data are illustrated in Fig. 14.6 with regular CM intervals 2.5 h.

In the practice of the INS health monitoring, it is usually required that the drift measurement along the sense axis should not exceed 0.37 (°/h). This threshold is predetermined at the design stage and is strictly enforced in the practice since an INS is a critical device used in a weapon system. Using the proposed method, the predictions of the gyro's drift and the distribution of the RUL can be obtained at each CM point. The predicted drifting path is shown in Fig. 14.6. Simultaneously, the PDF of the RUL can be obtained at each monitoring point.

In this case study, we set $c_f = 10000$ RMB and $c_p = 4000$ RMB. Figure 14.7 illustrates the expected cost per unit time against the associated time until replacement, i.e., $(\varphi^{\pi}(i, x_i))^2$, and the variable cost with $\alpha = 0.3$ and $\alpha = 0.05$, at three different CM points, the 50th, 59th, and 71st CM points, corresponding to the operating time 125, 147.5, and 177.5 h, respectively.

It can be seen from Fig. 14.7 that optimal replacement times (denoted by circles) under variable cost policy are always earlier than the corresponding mean cost policy, at different CM points. Additionally, it is noted that the optimal decision time is dynamic so that the decision is conditional on the CM history. To clearly have a

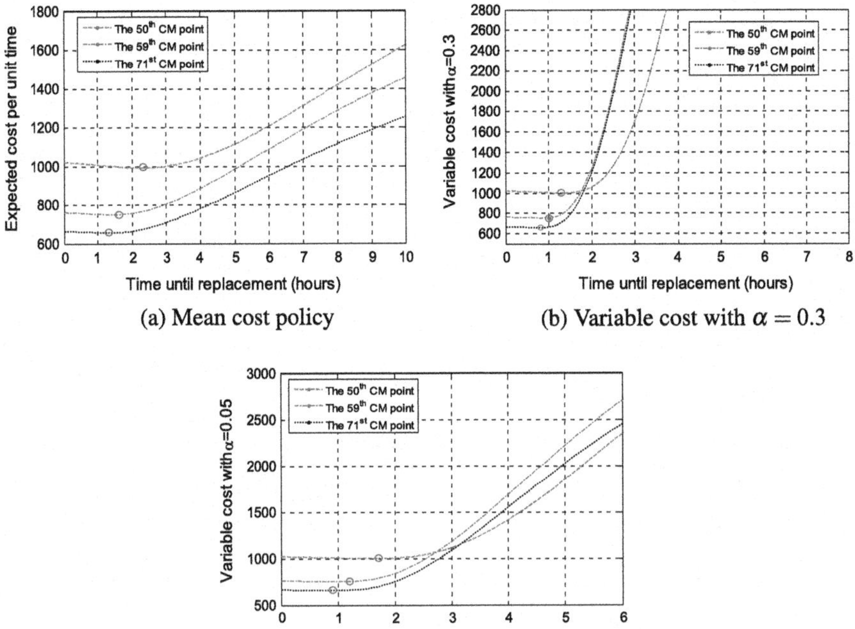

(a) Mean cost policy (b) Variable cost with $\alpha = 0.3$

(c) Variable cost with $\alpha = 0.05$

Fig. 14.7 Illustrations of the condition-based replacement decision at three different CM points

Table 14.1 Optimal cost associated with mean cost policy and variable cost policy

	Mean cost policy	Variable cost policy with $\alpha = 0.3$	Variable cost policy with $\alpha = 0.05$
50th point	748.59	753.21	751.6
59th point	656.41	660.03	658.85
71th point	993.23	1004.6	1000.6

look at the difference of the optimal maintenance cost between mean cost policy and variable cost, we summarize the minimized cost in the following Table 14.1.

The results in Table 14.1 indicate that our developed variable cost method has a small cost increase as opposed to the mean cost policy. However, our method can effectively mitigate the management risk by scheduling the replacement early.

References

1. Si XS, Wang W, Hu CH, Zhou DH (2011) Remaining useful life estimation- a review on the statistical data driven approaches. Eur J Oper Res 213(1):1–14
2. Wang W (2007) A two-stage prognosis model in condition based maintenance. Eur J Oper Res 182(3):1177–1187

3. Si XS, Hu CH, Wang W, Chen MY (2011) An adaptive and nonlinear drift-based Wiener process for remaining useful life estimation. In: 2011 prognostics and health management conference, PHM2011 Shenzhen, IEEE Xplore
4. Si XS, Wang W, Hu CH, Zhou DH, Pecht M (2012) Remaining useful life estimation based on a nonlinear diffusion degradation process. IEEE Trans Reliab 61(1):50–67
5. Tapiero CS, Venezia I (1979) A mean variance approach to the optimal machine maintenance and replacement problem. J Oper Res Soc 30(5):457–466
6. Chen Y, Jin JH (2003) Cost-variability-sensitive preventive maintenance considering management risk. IIE Trans 35(12):1091–1101
7. Gosavi A (2006) A risk-sensitive approach to total productive maintenance. Automatica 42:1321–1330
8. Giri BC, Dohi T (2010) Quantifying the risk in age and block replacement policies. J Oper Res Soc 61:1151–1158
9. Chen N, Chen Y, Li ZG, Zhou SY, Sievenpiper C (2011) Optimal variability sensitive condition-based maintenance with a Cox PH model. Int J Prod Res 49(7):2083–2100
10. Meeker WQ, Escobar LA (1998) Statistical methods for reliability data. Wiley, New York
11. Filar JA, Kallenberg LCM, Lee H-M (1989) Variance-penalized Markov decision processes. Math Oper Res 14(1):147–161
12. Si XS, Hu CH, Yang JB, Zhou ZJ (2011) A new prediction model based on belief rule base for system's behavior prediction. IEEE Trans Fuzzy Syst 19(4):456–471

Chapter 15
An Adaptive Spare Parts Demand Forecasting Method Based on Degradation Modeling

15.1 Introduction

System prognostics and health management (PHM) is a new health management methodology proposed for complex engineering systems to reduce maintenance costs, improve the system operating reliability and safety, and mitigate the failure risk [1, 2]. In PHM practice, proactive maintenance and spare parts ordering are two fundamental components for PHM of complex engineering systems. Spare parts need arises whenever a system fails or requires replacement and repair In some sectors, such as the aerospace and automotive industries, a wide range of spare parts are held in stock. Their management is therefore an important task [3].

It is pointed out by Wang and Syntetos in [4] that demand arising from system failure and preventive maintenance is scheduled and is stochastic with regard to the demand size but deterministic as far as the demand arrival is concerned. Demand arising from corrective maintenance, after a failure has occurred, is stochastic with regard to the time arrival but deterministic in quantity (being one in most cases). Such demand structures are typically intermittent in nature, meaning that demand arrives infrequently with time periods showing no demand at all. In this case, spare parts demand sizes may be highly variable leading to what is termed as 'lumpy' demand. Intermittent demand patterns are very difficult to model from a forecasting perspective because of the associated dual source of variation (demand arrivals, or correspondingly inter-demand intervals, and demand sizes) [4] There have been a number of considerable advancements in this area in the recent years, all of which though have been mainly focusing on coping, reactively, with the compound nature of the demand patterns under concern. However, no attempts have been made to characterize the sources of such demand patterns for the purpose of developing more effective and proactive mechanisms. Such an approach would require looking at the industrial failure processes that generate the relevant demand patterns.

Until now, many forecasting methods have been developed to predict spare parts demand. In these methods, if historical data of spare parts demand are available, time-series methods are used to forecast spare parts' requirements. Croston in [5] proposed

X.-S. Si et al., *Data-Driven Remaining Useful Life Prognosis Techniques*,
Springer Series in Reliability Engineering, DOI 10.1007/978-3-662-54030-5_15

a method that captures the compound nature of the relevant demand patterns. In particular, he suggested using Single Exponential Smoothing (SES) for separately forecasting the interval between demand incidences and the demand sizes. The ratio of the latter over the former may then be used in order to estimate the mean demand per time period. Based on this classical method, many extensions have been made to improve the accuracy of Croston's method in spare parts forecasting (e.g., [6–9]). Other bootstrapping methods for forecasting intermittent demands have also been discussed by [10, 11].

In maintenance related spare part forecasting studies, it is noted however that most researches have treated maintenance as an area of research on its own, and did not consider the impact of the availability of spare parts on the plant downtime and cost due to maintenance. There are some exceptions, in which inventory policies have been jointly considered with maintenance-related issues [4, 12] Attention has been paid to how equipment failures impact on the spare parts inventory policy. Overall, research in the area of forecasting for intermittent demand items has developed rapidly in recent years with new results implemented into software products because of their practical importance [13]. Nevertheless, all such studies share a common characteristic: they attempt to provide the best possible modeling of the underlying demand characteristics without questioning the demand generation process itself and the predictive results do not make use of the real-time condition monitoring (CM) degradation data of the system.

Together with the above discussions, it is concluded that spare parts demand arises from the maintenance activity. However, all maintenance actions are closely related with the system lifetime. Therefore, how to forecast spare parts demand via predicting the system lifetime is a novel idea which is different from traditional spare parts forecasting methods. Recently, Wang and Syntetos in [4] presented a novel idea to forecast spare parts demand considering the very sources of the demand generation process from maintenance. In [4], the maintenance process is modeled by the delay time model and the presented method was compared with a well known time-series method. However, the method in [4] is dependent on the statistical data of failure time and thus does not utilize the real-time operating state and CM data of the system.

The above observation leads to the main focus of this chapter to present an adaptive spare parts demand forecasting method based on degradation modeling of the CM data. Toward this end, a degradation modeling based method is proposed to adaptively forecast the demand of spare parts in this chapter. In the presented method, the system's lifetime is predicted in a probability form by modeling its degradation process as a Wiener process. To achieve adaptive lifetime estimation, the degradation rate parameter is recursively updated by Kalman filtering algorithm and then the lifetime is derived by considering the updated degradation rate parameter. Based on the lifetime distribution, the demand distribution of spare parts in a future time span can be forecasted by evaluating the convolution of the lifetime distribution. That is to say, spare parts forecasting is achieved by predicting the failure numbers of the system in a future time span. Finally, a case study for gyro's data is provided to illustrate the implementation of the proposed method.

The remainder of this chapter is structured as follows: In Sect. 15.2, the degradation modeling description is given. Section 15.3 provides the adaptive lifetime prediction method. In Sect. 15.4, we present an adaptive spare parts forecasting method. Section 15.5 provides a case study for illustration.

15.2 Degradation Modeling Description

In this section, we present the degradation modeling description of the system. In degradation modeling practice, Wiener processes have been extensively used to characterize the degradation processes of systems [2]. In physics, a Wiener process aims at modeling the movement of small particles in fluids and air with tiny fluctuations. A characteristic feature of this process in the context of reliability is that the plant's degradation can increase or decrease gradually and accumulatively over time. The tiny increase or decrease in degradation over a small time interval behaves similarly to the random walk of small particles in fluids and air. Therefore, this type of stochastic processes has been widely used to characterize the path of degradation processes where successive fluctuations in degradation can be observed, such as the degradation observations of rotating element bearings, LED lamps, self-regulating heating cables, laser generator, bridge beams, and gyros' drifting (e.g., [14–17]).

Specifically, a Wiener-process-based degradation model can be represented as

$$X(t) = \lambda t + \sigma B(t), \tag{15.1}$$

where λ is the drift coefficient, $\sigma > 0$ is the diffusion coefficient, and $B(t)$ is the standard Brownian motion representing the stochastic dynamics of the degradation process with $B(t) \sim N(0, t)$.

Based on the stochastic degradation process $\{X(t), t \geq 0\}$ represented by (15.1), the system lifetime can be defined by using the concept of the first hitting time (FHT) [18]. Based on the FHT, the principle of lifetime estimation is that the system failure is defined as the first time of $\{X(t), t \geq 0\}$ hits the failure threshold w. As such, the lifetime of the system is defined as

$$T = \inf\{t : X(t) \geq w | X(0) < w\} \tag{15.2}$$

According to the definition of the lifetime T in (15.2), the cumulative distribution function and the probability density function are denoted as $F_T(t)$ and $f_T(t)$, respectively. By the known results, the lifetime T follows an inverse Gaussian distribution, and the according CDF $F_T(t)$ and PDF $f_T(t)$ are as follows:

$$f_T(t) = \frac{w}{\sigma\sqrt{2\pi t^3}} \exp\left(-(w - \lambda t)^2/2\sigma^2 t\right) \tag{15.3}$$

$$F_T(t) = 1 - \Phi\left(\frac{w - \lambda t}{\sigma\sqrt{t}}\right) + \exp\left(\frac{2\lambda w}{\sigma^2}\right)\Phi\left(\frac{-w - \lambda t}{\sigma\sqrt{t}}\right). \tag{15.4}$$

In the system operating process, the degradation process is monitored at discrete times. As a result, the system lifetime can be estimated based on the monitoring data. Suppose the degradation process is discretely monitored at time $0 = t_0 < t_1 < \ldots < t_k$ and let $x_k = X(t_k)$ denote the degradation state at time t_k. The set of the degradation states up to t_k is represented by $X_{0:k} = \{x_0, x_1, x_2, \ldots, x_k\}$, where $t_0 = 0, k \geq 0$, and t_k is the current time.

Therefore, using the concept of the FHT, we define the remaining useful life (RUL) L_k of the system at time t_k as

$$L_k = \inf\{l_k > 0 : X(l_k + t_k) \geq w\}. \tag{15.5}$$

with PDF $f_{L_k|w,X_{0:k}}(l_k|\, w, X_{0:k})$ and CDF

$$F_{L_k|w,X_{0:k}}(l_k|\, w, X_{0:k}) = \Pr(L_k \leq l_k|w, X_{0:k}) = \Pr\left(\sup_{l_k>0} X(t_k + l_k) \geq w \,\middle|\, w, X_{0:k}\right), \tag{15.6}$$

where $X_{0:k}$ is the observed degradation states available up to t_k.

In the next sections, our primary goal is to adaptively estimate the conditional PDF $f_{L_k|w,X_{0:k}}(l_k|\, w, X_{0:k})$ and CDF $F_{L_k|w,X_{0:k}}(l_k|\, w, X_{0:k})$ based on $X_{0:k}$.

15.3 Adaptive Lifetime Estimation

In this chapter, we consider the updating mechanism for the lifetime estimation. For in service system at time t_k with the obtained degradation state x_k, we can use

$$X(t) = x_k + \lambda(t - t_k) + \sigma B(t - t_k),\ fort > t_k. \tag{15.7}$$

From (15.7), we directly have the PDF and CDF of the RUL at t_k as

$$f_{L_k|w,x_k}(l_k|\, w, x_k) = \frac{w - x_k}{\sigma\sqrt{2\pi l_k^3}} \exp\left(-\frac{(w - x_k - \lambda l_k)^2}{2\sigma^2 l_k}\right) \tag{15.8}$$

$$F_{L_k|w,x_k}(l_k|\, w, x_k) = 1 - \Phi\left(\frac{w - x_k - \lambda l_k}{\sigma\sqrt{l_k}}\right) + \exp\left(\frac{2\lambda(w - x_k)}{\sigma^2}\right)\Phi\left(\frac{-w + x_k - \lambda l_k}{\sigma\sqrt{l_k}}\right) \tag{15.9}$$

It is noted that (15.8) and (15.9) use only the current degradation data, but not its history before t_k. However, this is a Markov assumption. Ideally, the future FHT should depend on the path that the degradation has involved to date. Consequently, it is desired to utilize the degradation data to date for evaluating the RUL of the degraded system.

To incorporate the history of the observations, in this chapter the degradation rate parameter λ is modeled by a random walk model $\lambda_k = \lambda_{k-1} + \eta$ over time, where $\eta \sim N(0, Q)$. A similar idea can also be found in [19]. Motivated by the state space model, a self-organizing state-space model is constructed as

$$\lambda_k = \lambda_{k-1} + \eta, \tag{15.10}$$

$$x_k = x_{k-1} + \lambda_{k-1}(t_k - t_{k-1}) + \sigma \varepsilon_k, \tag{15.11}$$

where $t_0 = 0$, $x_0 = 0$, and $\varepsilon_k \sim N(0, t_k - t_{k-1})$. The use of $t_k - t_{k-1}$ as the variance of ε_k is required by the property of Brownian motion. Equation (15.10) is called the system equation, while (15.11) is the observation equation.

To adaptively update the degradation rate parameter λ based on $X_{0:k}$, it is assumed that λ_0 follows a normal distribution with mean a_0 and variance P_0. As such, the degradation rate parameter is considered as a hidden state and can be estimated from the observations up to t_k, denoted by $X_{0:k}$. Therefore, in (15.10), λ_k follows a distribution which can be estimated by a recursive filter once $X_{0:k}$ is available. We denote its mean by $\hat{\lambda}_k = E(\lambda_k | X_{0:k})$ and its variance by $P_{k|k} = var(\lambda_k | X_{0:k})$.

In stochastic filtering framework, recursion solution of $p(\lambda_k | X_{0:k})$ can be computed from $p(\lambda_{k-1} | X_{0:k-1})$ by Bayesian rule as follows:

$$p(\lambda_k | X_{0:k}) = \frac{\int p(\lambda_k | \lambda_{k-1}) p(x_k | \lambda_{k-1}, X_{0:k-1}) p(\lambda_{k-1} | X_{0:k-1}) d\lambda_{k-1}}{p(x_k | X_{0:k-1})}. \tag{15.12}$$

It has been well established that if (15.10) and (15.11) are used, (15.12) is Gaussian with mean $\hat{\lambda}_k$ and variance $P_{k|k}$ which can be computed by the Kalman filter [19]. The recursive estimations for $\hat{\lambda}_k$ and $P_{k|k}$ using Kalman filtering are summarized as Algorithm 15.1 in the following.

Algorithm 15.1 (*Kalman filtering algorithm*)

Step 1: Initialize $\hat{\lambda}_0 = a_0$, P_0.

Step 2: State estimation at time t_k

$$\begin{aligned} P_{k|k-1} &= P_{k-1|k-1} + Q \\ K_k &= (t_k - t_{k-1})^2 P_{k|k-1} + \sigma^2 (t_k - t_{k-1}) \\ \hat{\lambda}_k &= \hat{\lambda}_{k-1} + P_{k|k-1}(t_k - t_{k-1}) K_k^{-1} \left(x_k - x_{k-1} - \hat{\lambda}_{k-1}(t_k - t_{k-1}) \right) \end{aligned}.$$

Step 3: Updating variance $P_{k|k} = P_{k|k-1} - P_{k|k-1}(t_k - t_{k-1})^2 K_k^{-1} P_{k|k-1}$.

Based on (15.10)–(15.12), the PDF of λ_k conditional on $X_{0:k}$ is

$$p(\lambda_k | X_{0:k}) = \frac{1}{\sqrt{2\pi P_{k|k}}} \exp\left[-\left(\lambda_k - \hat{\lambda}_k\right)^2 \Big/ 2P_{k|k} \right], \tag{15.13}$$

where the dependence between λ_k and $X_{0:k}$ is contained in $\hat{\lambda}_k$ and $P_{k|k}$.

From (15.10) and (15.11), the PDF and CDF of the RUL at time t_k defined in (15.5) as follows [20]:

$$f_{L_k|w,\lambda_k,x_k}(l_k|\, w, \lambda_k, x_k) = \frac{w - x_k}{\sigma\sqrt{2\pi l_k^3}} \exp\left(-\frac{(w - x_k - \lambda l_k)^2}{2\sigma^2 l_k}\right) \tag{15.14}$$

$$F_{L_k|w,\lambda_k,x_k}(l_k|\, w, \lambda_k, x_k) = 1 - \Phi\left(\frac{w - x_k - \lambda l_k}{\sigma\sqrt{l_k}}\right) + \exp\left(\frac{2\lambda(w - x_k)}{\sigma^2}\right)\Phi\left(\frac{-w + x_k - \lambda l_k}{\sigma\sqrt{l_k}}\right) \tag{15.15}$$

As mentioned above, λ evolves as a random variable in (15.10) with a distribution, $p(\lambda_k|\, X_{0:k})$. By the total law of probability, the PDF and CDF of the updated RUL at t_k based on the updated PDF of λ_k can be obtained as

$$f_{L_k|w,X_{0:k}}(l_k|\, w, X_{0:k}) = \frac{w - x_k}{\sqrt{2\pi l_k^3\left(P_{k|k} l_k + \sigma^2\right)}} \exp\left(-\frac{\left(w - x_k - \hat{\lambda}_k l_k\right)^2}{2 l_k\left(P_{k|k} l_k + \sigma^2\right)}\right). \tag{15.16}$$

$$\begin{aligned} F_{L_k|w,X_{0:k}}(l_k|\, w, X_{0:k}) &= 1 - \Phi\left(\frac{w - x_k - \hat{\lambda}_k l_k}{\sqrt{P_{k|k} l_k^2 + \sigma^2 l_k}}\right) \\ &+ \exp\left(\frac{2\hat{\lambda}_k(w - x_k)}{\sigma^2} + \frac{2P_{k|k}(w - x_k)^2}{\sigma^4}\right)\Phi\left(-\frac{2P_{k|k}(w - x_k) l_k + \sigma^2\left(\hat{\lambda}_k l_k + w - x_k\right)}{\sigma^2\sqrt{P_{k|k} l_k^2 + \sigma^2 l_k}}\right). \end{aligned} \tag{15.17}$$

Comparing (15.17) with (15.15), we observe that the observation history and the variance of λ_k are involved in (15.17), which is also recursively updated. This RUL estimated results will be used to forecast the demand of spare parts. To do so, the forecasting results are related with the system operating state since the degradation state of the system is involved in (15.17).

15.4 Adaptively Forecasting Spare Parts Demand

As discussed previously, the system failure leads to the demand of spare parts [21–23]. Therefore, forecasting the spare parts demand can be achieved by predicting the failure number of the system in a future time span from the current time. We will elaborate on this issue in the following.

In this chapter, we assume that the system is immediately replaced once failure occurs. Further, assume that the degradation processes of the system after and before replacement are the same and the lifetimes of different systems of the same kind are independent. That is to say, the lifetimes and the system and its spare part are independent and identically distributed. For spare parts demand forecasting, let $D_k(\Delta)$ denote the demand of spare parts in a future time interval Δ from the current time t_k, i.e., failure number. Considering that the system lifetime is random, the demand of spare parts $D_k(\Delta)$ is also a random variable. Moreover, $D_k(\Delta)$ is discrete random variable. Namely, $D_k(\Delta) = n \ (n = 0, 1, \ldots)$ means that there are n failures of the system during the time interval $(t_k, t_k + \Delta]$. As a result, forecasting spare parts demand needs to evaluate the n-fold convolution of the RUL distribution.

By the convolution property of the inverse Gaussian distribution, the probability of $D_k(\Delta) = n$ at t_k can be formulated as

$$\Pr(D_k(\Delta) = n | X_{0:k}) = \Pr((L_k | nw, X_{0:k}) \leq \Delta, (L_k | (n+1)w, X_{0:k}) > \Delta) \tag{15.18}$$

where the event $\{(L_k | nw, X_{0:k}) \leq \Delta, (L_k | (n+1)w, X_{0:k}) > \Delta\}$ denotes that there are n failures during $(t_k, t_k + \Delta]$.

Further, we have

$$\Pr(D_k(\Delta) = n | X_{0:k}) = \Pr((L_k | nw, X_{0:k}) \leq \Delta) - \Pr((L_k | (n+1)w, X_{0:k}) \leq \Delta) \tag{15.19}$$

where $\Pr((L_k | nw, x_k) \leq \Delta)$ is probability that the first hitting time of $\{X(t), t \geq 0\}$ not greater than Δ while the initial degradation state is x_k and the according failure threshold is nw.

Specifically, according to the definition of the RUL in (15.5) and the formulation of the RUL distribution in (15.6), we directly have

$$\Pr((L_k | nw, x_k) \leq \Delta) = F_{L_k|nw, X_{0:k}}(\Delta | nw, X_{0:k}) \tag{15.20}$$

where $F_{L_k|nw, X_{0:k}}(\Delta | nw, X_{0:k})$ can be computed by (15.17) as

$$\begin{aligned} F_{L_k|nw, X_{0:k}}(\Delta | nw, X_{0:k}) &= 1 - \Phi\left(\frac{nw - x_k - \hat{\lambda}_k \Delta}{\sqrt{P_{k|k}\Delta^2 + \sigma^2 \Delta}}\right) \\ &+ \exp\left(\frac{2\hat{\lambda}_k (nw - x_k)}{\sigma^2} + \frac{2P_{k|k}(nw - x_k)^2}{\sigma^4}\right) \Phi\left(-\frac{2P_{k|k}(nw - x_k)\Delta + \sigma^2\left(\hat{\lambda}_k \Delta + nw - x_k\right)}{\sigma^2 \sqrt{P_{k|k}\Delta^2 + \sigma^2 \Delta}}\right) \end{aligned} \tag{15.21}$$

Therefore, the probability of the spare parts demand $D_k(\Delta) = n$ can be calculated as follows:

$$\Pr(D_k(\Delta) = n | X_{0:k}) = F_{L_k|nw, X_{0:k}}(\Delta | nw, X_{0:k}) - F_{L_k|(n+1)w, X_{0:k}}(\Delta | nw, X_{0:k}) \tag{15.22}$$

From (15.22), it is found that the probability distribution of the predicted spare parts demand $D_k(\Delta)$ is closely related with the system operating state $X_{0:k}$. Thus, such forecasting result reflects the actual operating state of the system in service and thus meets the practical desire of the system health management.

Based on (15.22), the expectation and variance of the forecasting spare parts demand $D_k(\Delta)$ during $(t_k, t_k + \Delta]$ can be respectively formulated as

$$E\left[D_k(\Delta)\,|X_{0:k}\right] = \sum_{n=0}^{\infty}(n \cdot \Pr(D_k(\Delta) = n|\, X_{0:k})) \tag{15.23}$$

$$var\left[D_k(\Delta)|\, X_{0:k}\right] = \sum_{n=0}^{\infty}\left(n^2 \cdot \Pr(D_k(\Delta) = n|\, X_{0:k})\right) - \left(E\left[D_k(\Delta)|\, X_{0:k}\right]\right)^2 \tag{15.24}$$

Together with the results in (15.22)–(15.24), we have formulated the probability distribution, the mean and variance of spare parts demand. These results can be used in PHM related decision makings, such as inventory control and joint spare parts ordering and replacement.

15.5 Adaptive Parameter Estimation

In (15.17) and (15.22)–(15.24), the model parameters, including a_0, P_0, Q and σ^2, should be estimated. Denote $\boldsymbol{\theta} = [a_0, P_0, Q, \sigma^2]^T$ as a parameter vector. In this chapter, we use the expectation maximization (EM) algorithm to estimate $\boldsymbol{\theta}$, since we treat λ_k as a hidden variable which is given in (15.10). For notation simplicity, let $\Upsilon_k = \{\lambda_0, \lambda_1, \ldots, \lambda_k\}$. Then, we have

$$L_k(\boldsymbol{\theta}) = p(X_{0:k}\,|\boldsymbol{\theta}) = \ell_k(\boldsymbol{\theta}) - \log p(\Upsilon_k|\, X_{0:k}, \boldsymbol{\theta}), \tag{15.25}$$

where

$$\ell_k(\boldsymbol{\theta}) = \log p(X_{0:k}, \Upsilon_k|\, \boldsymbol{\theta}). \tag{15.26}$$

As a result, the EM algorithm can be used for estimating $\boldsymbol{\theta}$ at t_k, denoted by $\hat{\boldsymbol{\theta}}_k$ according to the following two steps:

- ***E-step***: Calculate

$$\ell(\boldsymbol{\theta}|\, \hat{\boldsymbol{\theta}}_k^{(i)}) = \mathrm{E}_{\Upsilon_k|X_{0:k}, \hat{\boldsymbol{\theta}}_k^{(i)}}\left\{\ell_k(\boldsymbol{\theta})\right\}, \tag{15.27}$$

where $\hat{\boldsymbol{\theta}}_k^{(k)} = [a_{0,k}^{(i)}, P_{0,k}^{(i)}, Q_k^{(i)}, (\sigma_k^2)^{(i)}]^T$ denotes the estimated parameters in the ith step conditional on $X_{0:k}$.

- ***M-step***: Calculate

$$\hat{\boldsymbol{\theta}}_k^{(i+1)} = \arg\max_{\boldsymbol{\theta}}\left\{\mathrm{E}_{\Upsilon_k|X_{0:k}, \hat{\boldsymbol{\theta}}_k^{(i)}}\left\{\ell_k(\boldsymbol{\theta})\right\}\right\}. \tag{15.28}$$

Then, we iterate the E-step and M-step until a criterion of convergence is satisfied. The details of the algorithm can be referred to [19].

15.6 Case Study

In this section, we provide a practical case study for gyros in an inertial navigation system (INS) to illustrate the application of the presented forecasting method. As a key device of the INS, an inertial platform plays an important and irreplaceable role in the INS. The sensors fixed in an inertial platform include three gyros and three accelerometers. The gyro fixed on an inertial platform is a mechanical structure having two degrees of freedom from the driver and sense axis. When the inertial platform is operating, the wheels of the gyros rotate at very high speeds and can lead to rotation axis wear. As the wear is accumulated, the bearings on the gyros' electric motor will become deformed and such deformation can lead to the drift of the gyros. The increasing drift finally results in the failure of gyros and then the inertial platform. As such, the drift of gyros is often used as a performance indicator to evaluate the health condition of an inertial platform and to schedule maintenance activities.

In this study, we assume that CM values of drift coefficients reflect the performance of the inertial platform, and the larger the drift coefficients monitored are, the worse the performance is. Therefore, according to the CM data and technical index of the inertial platform, failure prediction can be implemented by modeling the drift coefficients. Generally, the drift degradation measurement along the sense axis, K_{SX}, plays a dominant role in the assessment of gyro degradation. In our study, we take the CM data of K_{SX} as the degradation signals and use them for predicting the spare parts demand of the gyro. For the monitored INS, 73 points of drift coefficients data were collected with regular CM intervals 2.5 h in field condition. The collected data are illustrated in Fig. 15.1.

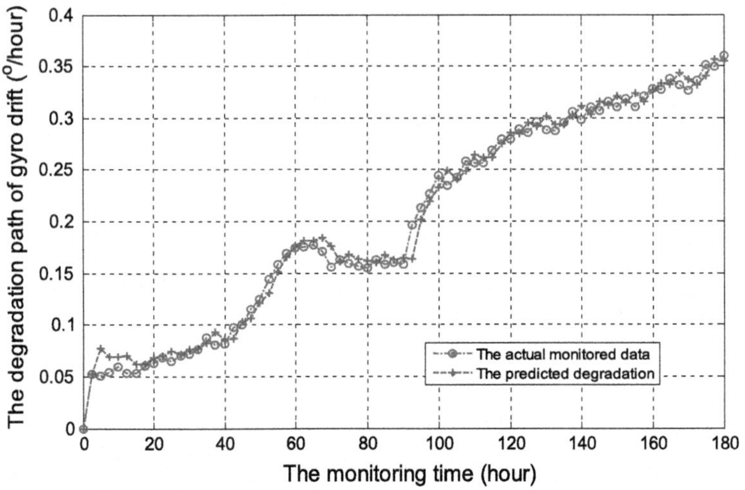

Fig. 15.1 The actual gyro's drift data and the predictions of the presented model

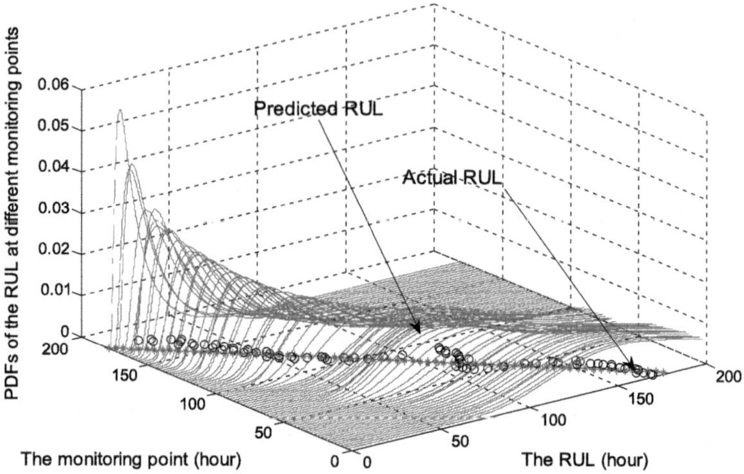

Fig. 15.2 Illustration of the PDF of the RUL at six different CM points

In the practice of the INS health monitoring, it is usually required that the drift measurement along the sense axis should not exceed $0.37(^{\circ}/\mathrm{h})$. Using our model, the predictions of the gyro's drift and the distribution of the RUL can be obtained at each CM point. The one step predicted drifting path is illustrated in Fig. 15.1 to show the fitness of the model to the gyro's degradation data. Clearly, the predicted results match with the actual data well. This demonstrates that our developed model can model the gyro's drift degradation data effectively. Figure 15.2 illustrates the PDFs of the RUL at different CM points.

As shown in Fig. 15.2, the actual RUL falls within the range of the estimated PDF of the RUL at each CM point and further the estimated PDF of the RUL becomes sharper as the degradation data are accumulated. This implies that the uncertainty of the estimated RUL is reduced since more data are utilized during estimating the model parameters. Additionally, it can be observed that the estimated mean RUL and the actual RUL match each other well.

Based on the above estimated RUL results, provided that the predictive interval Δ is given, the probability distribution of the spare parts demand can be obtained by (15.22). For example, when $\Delta = 60\,(\mathrm{h})$ and $\Delta = 300\,(\mathrm{h})$, the probability distributions of the spare parts demand at different CM times are summarized in Tables 15.1 and 15.2, respectively.

In Tables 15.1 and 15.2, for other values of n, the probability is 0 and thus results are omitted. From the results in Tables 15.1 and 15.2, it can be observed that the demand of the spare parts increases over time, and the demand will experience a monotone increasing trend with the predictive interval Δ. To further have a look at the forecasting results of the presented method, we show the progressions of the results of (15.23) and (15.24) under $\Delta = 300\,\mathrm{h}$ as the system operates. The calculated results are illustrated in Figs. 15.3 and 15.4.

Table 15.1 Probability distributions of the spare parts demand at different CM times with $\Delta = 60\,\text{h}$

	t_k	$\Delta = 60\,\text{h}$		
		$n=0$	$n=1$	$n=2$
$\Pr(D_k(\Delta) = n \mid X_{0:k})$	25 h	0.9510	0.490	0.00
	125 h	0.1238	0.8762	0.00
	180 h	0.0030	0.9970	0.00

Table 15.2 Probability distributions of the spare parts demand at different CM times with $\Delta = 300\,\text{h}$

	t_k	$\Delta = 300\,\text{h}$				
		$n=0$	$n=1$	$n=2$	$n=3$	$n=4$
$\Pr(D_k(\Delta) = n \mid X_{0:k})$	25 h	0.0005	0.0916	0.6312	0.2705	0.0062
	125 h	0.0000	0.231	0.7419	0.2347	0.003
	180 h	0.0000	0.288	0.8376	0.1336	0.0029

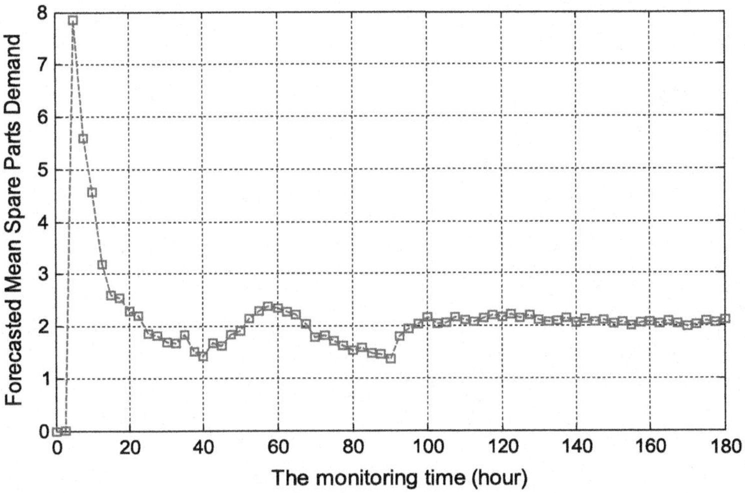

Fig. 15.3 Mean of the forecasting spare parts demand

The results in Figs. 15.3 and 15.4 indicate that the demand of the spare parts can be adaptively updated as the operating state of the system changes. In particular, the more the system approaches the failure, the more the variance of the spare parts demand reduces. This arises from the fact that the presented demand forecasting method can make full use of the CM data up to the current time and thus the estimated failure time of the system can accurately reflect the actual state of the operating system. Another observation is that the spare parts demand in the early stage of the system operation is small. The implication of this result is to suggest the manager not to prepare the spare parts too early so as to reduce the management costs of the system.

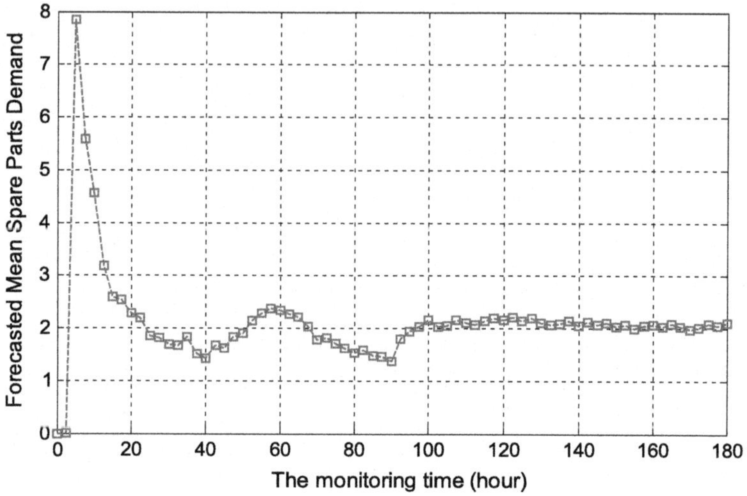

Fig. 15.4 Variance of the forecasting spare parts demand

References

1. Pecht M (2008) Prognostics and health management of electronics. Wiley, New Jersey
2. Si XS, Wang W, Hu CH, Zhou DH (2011) Remaining useful life estimation-a review on the statistical data driven approaches. Eur J Oper Res 213:1–14
3. Boylan JE, Syntetos AA (2008) Forecasting for inventory management of service parts. In: Murthy DNP, Kobbacy KAH (eds) Complex system maintenance handbook. Springer, London, pp 479–508
4. Wang W, Syntetos A (2011) Spare parts demand: linking forecasting to equipment maintenance. Transp Res Part E 47:1194–1209
5. Croston JD (1972) Forecasting and stock control for intermittent demand. Oper Res Quart 23:289–303
6. Syntetos AA, Boylan JE (2001) On the bias of intermittent demand estimates. Intern J Prod Econ 71:457–466
7. Shale EA, Boylan JE, Johnston FR (2006) Forecasting for intermittent demand: the estimation of an unbiased average. J Oper Res Soc 57:588–592
8. Teunter R, Sani B (2009) On the bias of Croston's forecasting method. Eur J Oper Res 194:177–183
9. Gutierrez RS, Solis AO, Mukhopadhyay S (2008) Lumpy demand forecasting using neural networks. Intern J Prod Econ 111:409–420
10. Porras EM, Dekker R (2008) An inventory control system for spare parts at a refinery: an empirical comparison of different reorder point methods. Eur J Oper Res 184:101–132
11. Teunter R, Duncan L (2009) Forecasting intermittent demand: a comparative study. J Oper Res Soc 60:321–329
12. de Smidt-Destombes KS, van der Heijden MC, van Harten A (2009) Joint optimisation of spare part inventory, maintenance frequency and repair capacity for k-out-of-N systems. Intern J Prod Econ 118:260–268
13. Fildes R, Nikolopoulos K, Crone S, Syntetos AA (2008) Forecasting and operational research: a review. J Oper Res Soc 59:1150–1172
14. Si XS, Wang W, Hu CH, Zhou DH, Pecht M (2012) Remaining useful life estimation based on a nonlinear diffusion degradation process. IEEE Trans Reliab 61(1):50–67

15. Si XS, Hu CH, Kong XY, Zhou DH (2014) A residual storage life prediction approach for systems with operation state switches. IEEE Trans Ind Electron 61(11):6304–6315
16. Si XS, Chen MY, Wang W, Hu CH, Zhou DH (2013) Specifying measurement errors for required lifetime estimation performance. Eur J Oper Res 231(3):631–644
17. Si XS, Wang W, Chen MY, Hu CH, Zhou DH (2013) A degradation path-dependent approach for remaining useful life estimation with an exact and closed-form solution. Eur J Oper Res 226(1):53–66
18. Lee M-LT, Whitmore GA (2006) Threshold regression for survival analysis: modeling event times by a stochastic process reaching a boundary. Stat Sci 21:501–513
19. Si XS, Wang W, Hu CH, Chen MY, Zhou DH (2013) A Wiener process-based degradation model with a recursive filter algorithm for remaining useful life estimation. Mech Syst Sign Process 35(1–2):219–237
20. Si XS, Wang W, Hu CH, Zhou DH (2014) Estimating remaining useful life with three-source variability in degradation modelling. IEEE Trans Reliab 63(1):167–190
21. Li R, Ryan JK (2011) A Bayesian inventory model using real-time condition monitoring information. Prod Oper Manag 20:754–771
22. Syntetos AA, Babai MZ, Altay N (2012) On the demand distributions of spare parts. Intern J Prod Res 50(8):2101–2117
23. Van Horenbeek A, Buré J, Cattrysse D, Pintelon L, Vansteenwegen P (2012) Joint maintenance and inventory optimization systems: a review. Intern J Prod Econ 143(2):499–508. doi:10.1016/j.ijpe.2012.04.001

Chapter 16
Variable Cost-Based Maintenance and Inventory Model

16.1 Introduction

The traditional maintenance and spare parts inventory decision models mainly rely on using population-specific reliability distribution, which cannot reflect the different degradation characteristics of single in-service equipment. With advances in condition monitoring (CM) technologies, many researchers advocated using the real-time prognostic information for individual equipment to optimize the subsequent maintenance and inventory decisions [1–4]. Carr and Wang in [5] utilized the semi-stochastic filtering approach to estimate the remaining useful life (RUL) of the equipment, and then, proposed a cost-based maintenance model with the prognostic information. Based on the prognostic model developed by [6], Kaiser and Gebraeel in [7] proposed a predictive maintenance policy and realized the real-time maintenance decision. However, they only considered the replacement model. Along the line of the work by Armstrong and Atkins [8], Elwany et al. in [9] considered the replacement and spare parts inventory decisions with the prognostic information simultaneously. Specifically, they assumed an equipment with room to store only one spare part, and focused on a sequential decision-making process where the optimal replacement time was first evaluated followed by the optimal ordering time. Once the optimal replacement time was computed, then it was used to decide when to order the spare part. Under predictive maintenance framework, the optimal replacement time and inventory ordering time can be real-time updated. Note that, the work in [9] just used the long-run average cost per unit time as the objective function of the replacement decision. In fact, most of the preventive maintenance policies in the literature are evaluated with the expected cost criteria only. However, it is generally more critical to consider not only the expectation of the maintenance cost, but also the variability of the cost. Furthermore, for the sequential replacement and inventory policy, the cost variability has a direct influence on the subsequent inventory decision and will result in an ineffective ordering time and finally lead to a huge cost.

In the literature, Tapiero and Venezia first considered the variability in maintenance cost and presented a replacement policy based on such variability

X.-S. Si et al., *Data-Driven Remaining Useful Life Prognosis Techniques*,
Springer Series in Reliability Engineering, DOI 10.1007/978-3-662-54030-5_16

[10]. Filar et al. in [11] discussed and compared several measures of variability for stochastic processes with rewards. Motivated by Tapiero and Venezia [10], Chen and Jin [12], Gosavi [13], and Giri and Dohi [14] investigated many maintenance policies by minimizing the cost with the variability discount. However, all these works formulated the maintenance policy in the age-based replacement framework and thus the real-time condition monitoring information is not considered. Therefore, these policies aimed to schedule the maintenance activity for a population but not for an individual. However, considering real time condition monitoring information is desirable so that the optimal replacement time is repeatedly updated during equipment operation to ensure the most recently calculated replacement time accurately reflects the current reality of the monitored equipment. Recently, Chen et al. in [15] considered a condition-based replacement model based on the variability of the maintenance cost in condition-based maintenance framework. However, this work did not incorporate the prognostic information. Si et al. in [16] proposed a cost-variability-sensitive maintenance model incorporating the prognostic information. But they did not consider the issue about the spare parts inventory decision.

In this chapter, we develop a new sequential maintenance and inventory model to consider the effects of both the expectation of the maintenance cost and its variability. Given the prognostic information obtained from condition monitoring and variable maintenance cost, we obtain a maintenance decision which is different from that of the mean maintenance cost-based model under the same setting. Moreover, we also establish an inventory decision model on the basis of the determined maintenance time. The prognostic information is obtained from a degradation process modeled as an adaptive Wiener process in real time. We demonstrate the proposed method with a practical case study. The results indicate that our method can effectively mitigate the management risk for the maintenance and spare parts inventory, but with a small cost increase.

The remainder parts are organized as follows. In Sect. 16.2, we present a Wiener-process-based degradation model for prognostics. In Sect. 16.3 we propose a parameter estimation procedure based on expectation maximization (EM) algorithm. Section 16.4 establishes the real-time variable cost-based maintenance model and the inventory decision model with the prognostic information involved. Section 16.5 provides a practical case study.

16.2 Degradation Modeling for Prognostics

In this chapter, we consider applying a Wiener-process-based model with a recursive filter algorithm developed in [3, 17] to obtain the prognostics information and then use this information to establish the replacement model and spare parts ordering model.

16.2.1 Degradation Modeling

In general, a Wiener process $\{X(t), t \geq 0\}$ can be represented as

$$X(t) = \lambda t + \sigma_B B(t), \tag{16.1}$$

where λ is a defined drift coefficient, $\sigma_B > 0$ is a diffusion coefficient, and $B(t)$ is the standard Brownian motion (BM). Without loss of generality, we assume $X(0) = 0$ in this chapter.

Considering the potential for updating knowledge of the process when new degradation observations become available, for $t > t_i$, we model the degradation process over time since t_i as

$$X(t) = x_i + \lambda(t - t_i) + \sigma_B B(t - t_i). \tag{16.2}$$

For deriving the RUL estimation according to Eq. (16.2), we employ the concept of the first hitting time (FHT) to define the lifetime, which can be interpreted as the FHT of the degradation state crossing a threshold level [18]. From the concept of FHT, when degradation $X(t)$ modeled as Eq. (16.2) reaches a pre-set critical level w, the equipment can be declared to be failed and thus there was no useful lifetime left. Then, the RUL L_i at t_i can be defined as

$$L_i = \inf\left\{l_i : X(t_i + l_i) \geq w | \mathbf{X}_{1:i}, \forall 1 \leq j \leq i, x_j < w\right\}, \tag{16.3}$$

with the probability density function (PDF) $f_{L_i|\mathbf{X}_{1:i}}(l_i|\mathbf{X}_{1:i})$ and the cumulative density function (CDF) $F_{L_i|\mathbf{X}_{1:i}}(l_i|\mathbf{X}_{1:i})$.

16.2.2 RUL Estimation

To incorporate the history of the observations of the degradation, we consider an updating procedure for the drifting parameter λ by making λ evolving as $\lambda_i = \lambda_{i-1} + \eta$ over time, where $\eta \sim N(0, Q)$. As such, the drift parameter is evolving as a random variable with a probabilistic distribution, conditional on the observed data up to time t_i. The reason to make λ time-varying with the collecting data is that the degradation trend is dominated by the drift coefficient while the variance coefficient reflects the uncertainty in degradation process. In order to establish the linkage between the drift parameter and the observation history up to date, the degradation equation can be reconstructed and taken to be a *self-organizing state-space model* as

$$\begin{cases} \lambda_i = \lambda_{i-1} + \eta \\ x_i = x_{i-1} + \lambda_{i-1}(t_i - t_{i-1}) + \sigma_B \varepsilon_i \end{cases} \tag{16.4}$$

where the error terms are distributed as $\eta \sim N(0, Q)$ and $\varepsilon_i \sim N(0, t_i - t_{i-1})$. The use of $t_i - t_{i-1}$ as the variance of ε_i is derived by the property of BM. In this chapter we assume that the initial drift λ_0 follows a normal distribution with the mean μ_0 and the variance P_0. As such, the drift parameter is considered as a latent "state" and can only be estimated from the historical information $\mathbf{X}_{1:i} = \{x_1, x_2, \ldots x_i\}$ due to its unobservable nature. In the established state-space model (16.4), the state equation and the observation equation are linear and the associated noises are Gaussian. As such, we utilize Kalman filter to estimate the latent drift parameter. The specific Kalman filter algorithm can be found in [3] and here we do not discuss this issue.

By Kalman filter, we can estimate λ_i at time t_i from the historical observations. First, define $\hat{\lambda}_{i|i} = \mathrm{E}(\lambda_i | X_{1:i})$ and $P_{i|i} = \mathrm{Var}(\lambda_i | X_{1:i})$ as the expectation and variance of λ_i which are conditional on the whole historical information, respectively. Then we can obtain that $\lambda_i \sim N(\hat{\lambda}_{i|i}, P_{i|i})$ at time t_i. Here we need to consider the impact of the uncertainty of estimated λ_i on the RUL distribution. In order to achieve this aim, according to the law of total probability, the PDF of the RUL at time t_i can be formulated with the available degradation measurement $\mathbf{X}_{1:i}$ as

$$f_{L_i|\mathbf{X}_{1:i}}(l_i|\mathbf{X}_{1:i}) = \frac{w - x_i}{\sqrt{2\pi l_i^3 \left(P_{i|i} l_i + \sigma_B^2\right)}} \exp\left(-\frac{\left(w - x_i - \hat{\lambda}_{i|i} l_i\right)^2}{2 l_i \left(P_{i|i} l_i + \sigma_B^2\right)}\right), l_i > 0 \tag{16.5}$$

and the CDF can be written as

$$F_{L_i|\mathbf{X}_{1:i}}(l_i|\mathbf{X}_{1:i}) = 1 - \Phi\left(\frac{w - x_i - \hat{\lambda}_{i|i} l_i}{\sqrt{P_{i|i} l_i^2 + \sigma_B^2 l_i}}\right) + \exp\left[\frac{2\hat{\lambda}_{i|i}(w - x_i)}{\sigma_B^2} + \frac{2P_{i|i}(w - x_i)^2}{\sigma_B^4}\right] \Phi\left[-\frac{2P_{i|i}(w - x_i) l_i + \sigma_B^2(\hat{\lambda}_{i|i} l_i + w - x_i)}{\sigma_B^2 \sqrt{P_{i|i} l_i^2 + \sigma_B^2 l_i}}\right] \tag{16.6}$$

where $\hat{\lambda}_{i|i}$ and $P_{i|i}$ can be obtained from Kalman filter, $\Phi(\cdot)$ denotes the CDF of the normal distribution.

Once the new degradation observation is available and the parameter is updated, the RUL of this monitored equipment can be computed using Eq. (16.5). In fact, the parameters in the state-space model can be estimated simultaneously in conjunction with the drifting parameter. In the next section, we will discuss the issue about parameter estimation.

16.3 Parameter Estimation

In the above section we have already obtained the PDF and CDF of the RUL. Since the model parameter vector $\boldsymbol{\theta} = [\mu_0, P_0, \sigma_B, Q]$ is unknown, we need to estimate them based on the available CM information up to the current time. It is difficult to estimate the parameters directly because of the latent "state", i.e., drifting parameter. Thus, in this chapter, the EM algorithm is used to estimate the unknown parameters.

The EM algorithm is an iterative method and its key idea is to compute the conditional expectation of one log-likelihood function which consists of the complete data set $(\lambda_{0:i}, X_{0:i})$. This implies that maximizing the conditional expectation of the complete data log-likelihood function iteratively can generate a sequence of parameter estimates which eventually converge to the MLE of the parameters. Let $\hat{\boldsymbol{\theta}}^{(j)}$ denote the estimate of the parameter from the jth iteration of the EM algorithm, then the conditional expectation of the complete data log-likelihood function denoted by $\mathscr{Q}(\boldsymbol{\theta}, \hat{\boldsymbol{\theta}}^{(j)})$ can be formulated as

$$\mathscr{Q}(\boldsymbol{\theta}, \hat{\boldsymbol{\theta}}^{(j)}) = \mathrm{E}_{\hat{\boldsymbol{\theta}}^{(j)}}\{[\log p(\lambda_{0:i}, X_{0:i}|\boldsymbol{\theta})]|X_{0:i}\} \tag{16.7}$$

where $\log p(\lambda_{0:i}, X_{0:i}|\boldsymbol{\theta})$ is the complete data log-likelihood function. The more details can be found in [19]. In sum, the parameter estimation procedure consists of the following two steps:

(1) ***E-step***

$$\text{Compute: } \mathscr{Q}(\boldsymbol{\theta}, \hat{\boldsymbol{\theta}}^{(j)});$$

(2) ***M-step***

$$\text{Compute: } \hat{\boldsymbol{\theta}}^{(j+1)} = \arg\max_{\boldsymbol{\theta}} \mathscr{Q}(\boldsymbol{\theta}, \hat{\boldsymbol{\theta}}^{(j)})$$

The above steps are iterated with an initial guess $\hat{\boldsymbol{\theta}}^{(0)}$ until the criterion of convergence is satisfied.

Specifically for our model (16.4), to calculate $\mathscr{Q}(\boldsymbol{\theta}, \hat{\boldsymbol{\theta}}^{(j)})$, for $k = 0, 1, 2, \ldots, i$, we first define the following quantities:

$$\begin{aligned} \hat{\lambda}_{k|i} = \mathrm{E}_{\hat{\boldsymbol{\theta}}^{(j)}}(\lambda_k|X_{0:i}),\ \ P_{k|i} &= \mathrm{E}_{\hat{\boldsymbol{\theta}}^{(j)}}(\lambda_k^2|X_{0:i}) - \hat{\lambda}_{k|i}^2, \\ P_{k,k-1|i} &= \mathrm{E}_{\hat{\boldsymbol{\theta}}^{(j)}}(\lambda_k\lambda_{k-1}|X_{0:i}) - \hat{\lambda}_{k|i}\hat{\lambda}_{k-1|i} \end{aligned} \tag{16.8}$$

After some algebraic manipulations, we have

$$\begin{aligned} \mathscr{Q}(\boldsymbol{\theta}, \hat{\boldsymbol{\theta}}^{(j)}) &\propto -\log P_0 - \left((P_{0|i} + \hat{\lambda}_{0|i}^2 + \mu_0^2 - 2\mu_0\hat{\lambda}_{0|i}) \Big/ P_0\right) \\ &- \sum_{k=1}^{i}\left((\log Q + P_{k|i} + \hat{\lambda}_{k|i}^2 + P_{k-1|i} + \hat{\lambda}_{k-1|i}^2 - 2(P_{k,k-1|i} + \hat{\lambda}_{k|i}\hat{\lambda}_{k-1|i}))\Big/ Q\right) \\ &- \sum_{k=1}^{i}\Big[\log\sigma_B + \Big((X_k - X_{k-1})^2 - 2\hat{\lambda}_{k-1|i}(X_k - X_{k-1})(t_k - t_{k-1}) \\ &\quad +(t_k - t_{k-1})^2(P_{k-1|i} + \hat{\lambda}_{k-1|i}^2)\Big) \Big/ (\sigma_B^2(t_k - t_{k-1}))\Big] \end{aligned} \tag{16.9}$$

Clearly, computing the conditional expectation requires evaluating the quantities defined in Eq. (16.8) with respect to the estimated parameter $\hat{\boldsymbol{\theta}}^{(j)}$ at the jth iteration. In this chapter, we utilize the Kalman Smoother to estimate these quantities. Kalman smoother includes two parts: one part is forward recursion, i.e., filtering, and the other is backward recursion, i.e., smoothing. The more details can be found in [17].

After obtaining $\mathcal{Q}(\boldsymbol{\theta}, \hat{\boldsymbol{\theta}}^{(j)})$, the unknown parameter vector $\hat{\boldsymbol{\theta}}^{(j+1)}$ at the $(j+1)$th iteration can be obtained through maximizing the conditional expectation with respect to $\boldsymbol{\theta}$. Particularly, for computing convenience, according to the first order necessary condition, we calculate the partial derivatives of $\mathcal{Q}(\boldsymbol{\theta}, \hat{\boldsymbol{\theta}}^{(j)})$ with respect to μ_0, P_0, σ_B^2, Q. Then we have

$$\begin{aligned}
&\hat{\mu}_0 = \hat{\lambda}_{0|i}, \qquad \hat{P}_0 = P_{0|i},\\
&\hat{\sigma}_B^2 = \frac{1}{i}\sum_{k=1}^{i} \frac{(X_k - X_{k-1})^2 - 2\hat{\lambda}_{k-1|i}(X_k - X_{k-1})(t_k - t_{k-1}) + (t_k - t_{k-1})^2 (P_{k-1|i} + \hat{\lambda}_{k-1|i}^2)}{t_k - t_{k-1}},\\
&\hat{Q} = \left(P_{k|i} + \hat{\lambda}_{k|i}^2 + P_{k-1|i} + \hat{\lambda}_{k-1|i}^2 - 2(P_{k,k-1|i} + \hat{\lambda}_{k|i}\hat{\lambda}_{k-1|i})\right)\Big/ i.
\end{aligned} \tag{16.10}$$

From the above derivation, we can see that once the new observation data are available, we can utilize the EM algorithm to evaluate and update the parameters until the criterion of convergence is satisfied. Additionally, in the process of parameter estimation, we can also obtain the updated drift parameter simultaneously.

16.4 Replacement and Inventory Decision Modeling

In order to study a variable cost-based maintenance policy and further to study the influence of the considered maintenance policy on the subsequent spare parts inventory, the measure of the cost variability needs to be first defined. In this chapter, the long-run variance measure recommended by [13] is used. A counting process $\{N(t),\ t \geq 0\}$ is considered. Let T_n denote the time between the $(n-1)$th and the nth replacement in this process with $n \geq 1$. If $\{T_1, T_2, \ldots\}$ is a sequence of non-negative random variables which are independent and identically distributed, the counting process is called a renewal process. Let C_n denote the cost associated with the nth renewal under policy π, then $C_\pi(t) = \sum_{n=1}^{N(t)} C_n$ is the sum of the single cost up to time t and $C_\pi^2(t) = \sum_{n=1}^{N(t)} (C_n)^2$ is the sum of the square of the single cost up to time t, according to the renewal theorem, the long-run expected cost $\mathrm{E}[C_\pi(\mathrm{t})]$, the long-run expected square cost $\mathrm{E}[C_\pi^2(t)]$ and the expected cycle $\mathrm{E}[T_\pi]$ can be written as

$$\mathrm{E}[C_\pi(\mathrm{t})]{=}\mathrm{E}[C_n], \quad \mathrm{E}[C_\pi^2(t)] = \mathrm{E}[(C_n)^2], \quad \mathrm{E}[T_\pi] = \mathrm{E}[T_n] \tag{16.11}$$

From [13], the long-run expected cost per unit time under maintenance policy π can be defined as

$$\varphi_\pi = \lim_{t\to\infty} \frac{\sum_{n=1}^{N(t)} C_n}{t} \tag{16.12}$$

The *long-run mean square* cost under maintenance policy π is defined as

$$\psi_\pi = \lim_{t\to\infty} \frac{\sum_{n=1}^{N(t)} (C_n)^2}{t} \tag{16.13}$$

The *long-run variance* of the cost under maintenance policy π is defined as

$$V_\pi = \lim_{t\to\infty} \frac{\sum_{n=1}^{N(t)} (C_n - \mathrm{E}[C_\pi(\mathrm{t})])^2}{t} \tag{16.14}$$

From the above definitions, we have the following lemma.

Lemma 16.1 *If* $|E[C_\pi(t)]| < \infty$, $E[T_\pi] < \infty$, $E[C_\pi^2(t)] < \infty$, *then with probability 1(w.p.1)*

$$\varphi_\pi = \frac{E[C_\pi(t)]}{E[T_\pi]}, \quad \psi_\pi = \frac{E[C_\pi^2(t)]}{E[T_\pi]}, \quad V_\pi = \frac{E[C_\pi^2(t)] - (E[C_\pi(t)])^2}{E[T_\pi]} \tag{16.15}$$

Proof We only give the proof of V_π , the others can be proved accordingly.

$$\begin{aligned} V_\pi &= \lim_{t\to\infty} \frac{\sum_{n=1}^{N(t)} \left(C_n - \mathrm{E}[C_\pi(\mathrm{t})]\right)^2}{t} = \lim_{t\to\infty} \left(\frac{\sum_{n=1}^{N(t)} \left(C_n - \mathrm{E}[C_\pi(\mathrm{t})]\right)^2}{N(t)}\right)\left(\frac{N(t)}{t}\right) \\ &= \frac{\mathrm{E}[C_\pi^2(\mathrm{t})] - \left(\mathrm{E}[C_\pi(\mathrm{t})]\right)^2}{\mathrm{E}[T_\pi]} \quad \text{w.p.1} \end{aligned} \tag{16.16}$$

Equation (16.16) follows from the fact that

$$\begin{aligned} \frac{\sum_{n=1}^{N(t)} (C_n - \mathrm{E}[C_\pi(\mathrm{t})])^2}{N(t)} &= \frac{\sum_{n=1}^{N(t)} (C_n)^2}{N(t)} - 2\frac{\sum_{n=1}^{N(t)} C_n \mathrm{E}[C_\pi(\mathrm{t})]}{N(t)} + \frac{\sum_{n=1}^{N(t)} (\mathrm{E}[C_\pi(\mathrm{t})])^2}{N(t)} \\ &= \mathrm{E}[C_\pi^2(\mathrm{t})] - (\mathrm{E}[C_\pi(\mathrm{t})])^2 \quad \text{w.p.1 as } t \to \infty \end{aligned}$$

and the elementary renewal theorem, which implies that $\lim_{t\to\infty} N(t)/t = 1/\mathrm{E}[T_\pi]$ w.p.1. This completes the proof.

Then, the variable cost-based maintenance optimization problem can be formulated as

$$\min_\pi \left[(\varphi_\pi) + \alpha V_\pi\right], \alpha \geq 0 \tag{16.17}$$

where π is the class of the considered maintenance polices, and α, is the cost-variability-sensitive factor. From Eq. (16.17), we can see that the objective function is equivalent to the traditional cost policy when $\alpha = 0$.

In the above equation, α can be interpreted as the relative weight of variability. The specific value of α is usually set based on either engineering expert knowledge or accepted industrial standards. The intuitive meaning of $\alpha < 1$ is that the decision maker considers that the improvement in variability has less impact than the same degree of improvement in the mean cost. In other words, the cost-variability is less significant than the mean cost. Similarly, $\alpha > 1$ means that the variability has a more significant impact than the mean cost on the decision-making. We note that the most common intuition of the decision makers is to consider the mean cost to be more significant than the cost-variability. Therefore, the decision makers should be cautious when they decide to use a α, with a value greater than one.

Under prognostic information provided in Sect. 16.2, the above optimization problem can be formulated as Variance-Penalized Decision Problem. Suppose that the degradation observation is x_i at the ith CM point t_i, the associated variable cost optimization problem can be represented as follows

$$C_\pi(i, x_i) = \min_{t^* \in [t_i, \infty)} \left[(\varphi_\pi(i, x_i)) + \alpha V_\pi(i, x_i) \right], \alpha \in [0, \infty) \tag{16.18}$$

with

$$\varphi_\pi(i, x_i) = \frac{c_p + (c_f - c_p)\Pr(L_i < t_r - t_i \,|\mathbf{X}_{1:i})}{t_i + (t_r - t_i)\left(1 - \Pr(L_i < t_r - t_i \,|\mathbf{X}_{1:i})\right) + \int_{l_i=0}^{t_r - t_i} l_i f_{L_i|\mathbf{X}_{1:i}}(l_i|\mathbf{X}_{1:i}) dl_i} \tag{16.19}$$

$$V_\pi(i, x_i) = \frac{c_p^2 + \left(c_f^2 - c_p^2\right)\Pr(L_i < t_r - t_i \,|\mathbf{X}_{1:i}) - \left(c_p + (c_f - c_p)\Pr(L_i < t_r - t_i \,|\mathbf{X}_{1:i})\right)^2}{t_i + (t_r - t_i)\left(1 - \Pr(L_i < t_r - t_i \,|\mathbf{X}_{1:i})\right) + \int_0^{t_r - t_i} l_i f_{L_i|\mathbf{X}_{1:i}}(l_i|\mathbf{X}_{1:i}) dl_i} \tag{16.20}$$

where $\Pr(L_i < t - t_i|\mathbf{X}_{1:i}) = \int_0^{t-t_i} f_{L_i|\mathbf{X}_{1:i}}(l_i|\mathbf{X}_{1:i})dl_i$ and $f_{L_i|\mathbf{X}_{1:i}}(l_i|\mathbf{X}_{1:i})$ can be obtained from Eq. (16.5), t_r is the decision variable representing the planned replacement time at the ith CM point, c_p is the mean cost of a preventive replacement, and c_f is the mean replacement cost associated with the failure; $\varphi_\pi(i, x_i)$ and $V_\pi(i, x_i)$ are the long-run average cost per unit time and *long-run variance* of the cost under maintenance policy π at the state (i, x_i), respectively.

Suppose that the optimal planned replacement time is t_r^* at time t_i, then according to the spare parts inventory policy developed in [8, 13], the long-run average inventory cost per unit time with prognostic information $C_O(i, x_i)$ can be formulated as

$$C_O(i, x_i) = \frac{k_s \int_{t_0}^{t_0+L} F_{L_i|\mathbf{X}_{1:i}}(l_i|\mathbf{X}_{1:i})dl_i + k_h \int_{t_0+L}^{t_r^*} \left(1 - F_{L_i|\mathbf{X}_{1:i}}(l_i|\mathbf{X}_{1:i})\right) dl_i}{\int_{t_0}^{t_0+L} F_{L_i|\mathbf{X}_{1:i}}(l_i|\mathbf{X}_{1:i})dl_i + \int_0^{t_r^*} \left(1 - F_{L_i|\mathbf{X}_{1:i}}(l_i|\mathbf{X}_{1:i})\right) dl_i + t_i} \tag{16.21}$$

where $F_{L_i|\mathbf{X}_{1:i}}(l_i|\mathbf{X}_{1:i})dl_i$ can be obtained from Eq. (16.6), t_0 is the decision variable representing the inventory ordering time at time t_i, k_h is the holding cost per unit time, k_s is the shortage cost per unit time, and L is the fixed leading time elapsed from the moment of placing the order up till order receipt. Note that the optimal inventory ordering time at any CM time point, t_0^* should satisfy the constraint $t_0^* + L \leq t_r^*$. Actually, by implementing the Nelder–Mead Simplex method in Matlab toolbox, Eqs. (16.18) and (16.21) can be minimized. Therefore, the optimal replacement time and spare parts ordering time can be optimized and determined sequentially.

16.5 Case Study

The inertial measurement unit (IMU) is a critical equipment of strapdown inertial navigation system (SINS), and its operating state has a direct influence on the navigation precision. The key components of IMU are three gyros and three accelerometers, which measure the angular velocity and linear acceleration, respectively. In engineering practice, the gyros operate with certain load under various environments, and the wheels of the gyros rotate at very high speed. In such case, the rotation axis wears over time and finally results in the gyros' drift. Once the drift is equal to or beyond one pre-set threshold, the IMU will be considered to be failed and has to be repaired. Thus, the drift of the gyro is often used as a performance indicator to evaluate the RUL of an IMU and schedule maintenance activity and spare parts ordering time.

Since the dr0ift along the sense axis is more sensitive to the performance degradation of IMU, in this chapter, we take the drift along the sense axis as the degradation variable. Moreover, according to the engineering experience and the request to the navigation precision of IMU, we set the failure threshold equal to 0.38 (°/hour) with respect to the drift along the sense axis. The data set is illustrated as Fig. 16.1 with a regular CM interval 3 h.

First, we use the approach developed in Sects. 16.2 and 16.3 to estimate and update the unknown parameters of model (16.4) and the RUL distribution once the new CM data are available, then we incorporate the estimated RUL into the subsequent maintenance and inventory model. The initial parameter values of model

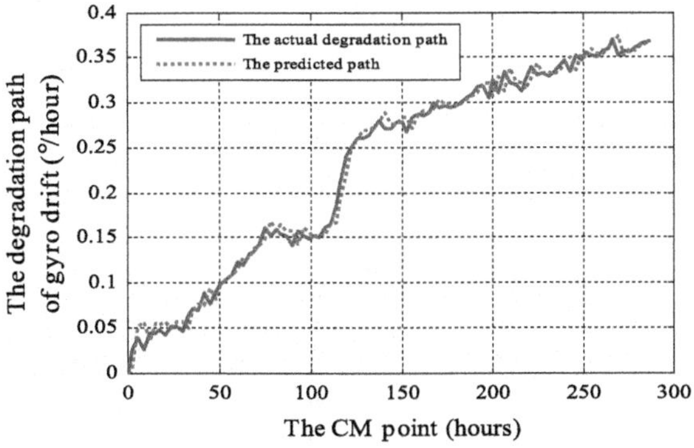

Fig. 16.1 The actual drift data and the predictions with adapted BM

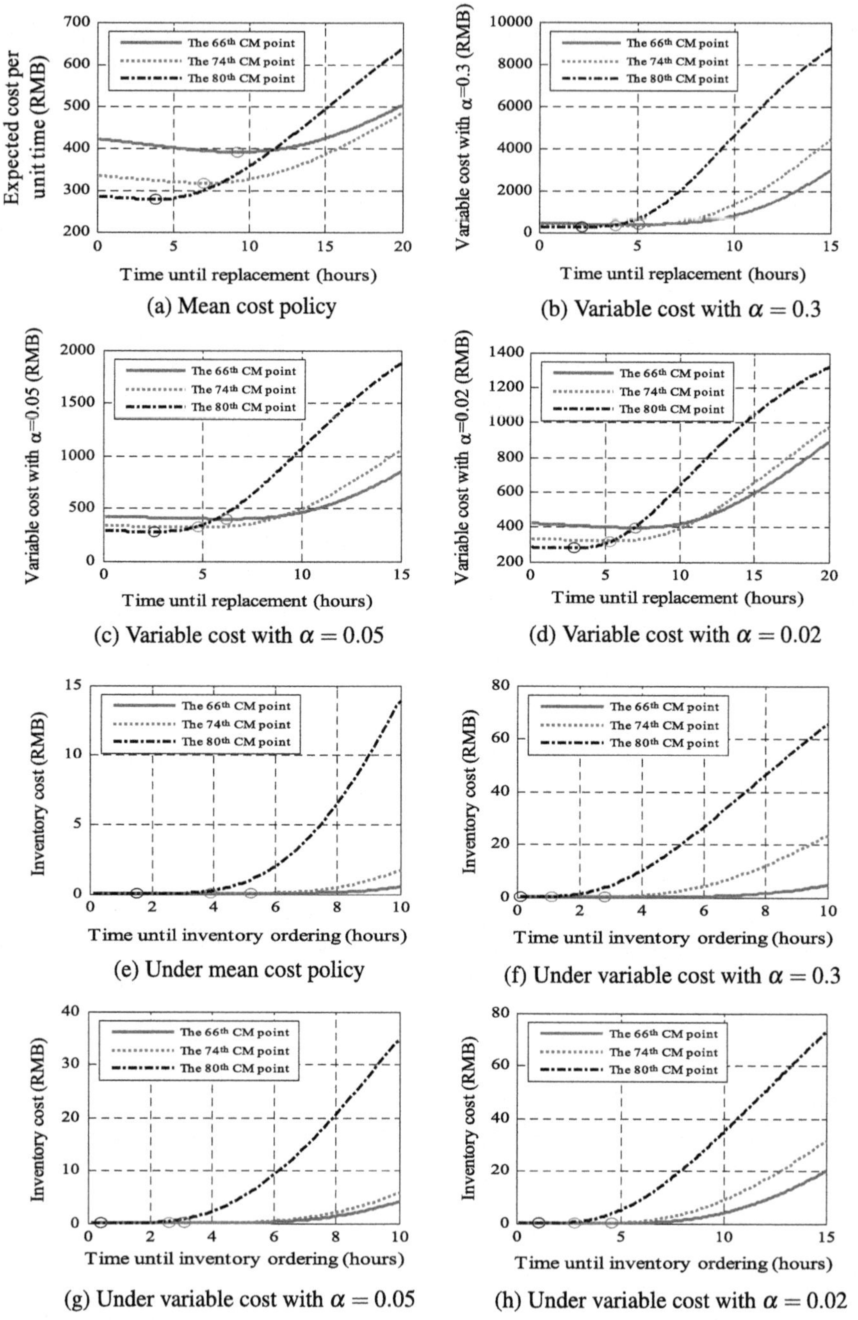

Fig. 16.2 Illustration of the condition-based replacement and inventory decisions at three CM points

(16.4) are set to $\mu_0 = 0.005\ P_0 = 0.01$, $\sigma_B = 0.1$, and $Q = 0.001$. The final estimated parameter values are $\mu_0 = 0.0066\ P_0 = 1.1089 \times 10^{-5}$, $\sigma_B = 0.0097$, and $Q = 1.1304 \times 10^{-6}$. Then, for illustrative purpose, in this case study, we set $c_f = 10000\,\text{RMB} c_p = 4000\,\text{RMB}$, the holding cost per unit time $k_h = 1\,\text{RMB}$ the shortage cost per unit time $k_s = 30000\,\text{RMB}$ and the leading time $L = 2\,\text{h}$. Figure 16.2a–d illustrate the expected cost per unit time against the associated time until replacement, and the variable cost with $\alpha = 0.3\ \alpha = 0.05$, and $\alpha = 0.02$ at three different CM points, the 66th, 74th and 80th CM points. Figure 16.2e–h illustrate the corresponding inventory ordering results with respect to the three cases.

From Fig. 16.2, it can be seen that the optimal replacement times (denoted by circles) under variable cost policy are always earlier than the corresponding mean cost policy, at different CM points. In other words, the variable cost optimal policy tends to be more conservative than the mean cost optimal policy. In addition, it seems that the larger the relative weight α, the shorter is the preventive replacement interval. This seems to be consistent with the intuition that the maintenance risk can be reduced through increasing the replacement frequency. Furthermore, we can find that the optimal inventory ordering times are different under different replacement policies. The shorter the replacement interval, the earlier the inventory ordering time is. Thus, considering the variable cost can reduce the maintenance and inventory risks simultaneously. To clearly compare the optimal maintenance and inventory cost between mean cost policy and variable cost policy, the minimized maintenance cost and long-run mean inventory cost are summarized in Table 16.1, where the maintenance cost is called Cost 1 and the inventory cost is called Cost 2.

The results in Table 16.1 indicate that our proposed method has a small maintenance cost increase as opposed to the mean cost policy. However, our method can effectively mitigate the maintenance risk by scheduling the replacement early. In addition, we can see that the inventory cost under variable cost policy is less than the one under mean cost policy. Thus, considering the variable maintenance cost has a direct influence on the subsequent inventory decision.

Table 16.1 Optimal maintenance and inventory costs with mean cost policy and variable cost policy

	Mean cost policy		$\alpha = 0.3$		$\alpha = 0.05$		$\alpha = 0.02$	
Unit: RMB	Cost 1	Cost 2	Cost 1	Cost 2	Cost 1	Cost 2	Cost 1	Cost 2
66th point	389.79	0.0423	401.94	0.0048	398.19	0.0081	395.89	0.0108
74th point	316.53	0.0321	323.26	0.0066	321.23	0.0076	319.99	0.0090
74th point	277.49	0.0197	280.38	0.0138	279.51	0.0074	278.98	0.0078

References

1. Si XS, Wang W, Hu CH, Zhou DH (2011) Remaining useful life estimation-a review on the statistical data driven approaches. Eur J Oper Res 213(1):1–14
2. Wang W (2007) A two-stage prognosis model in condition based maintenance. Eur J Oper Res 182(3):1177–1187
3. Si XS, Hu CH, Wang W, Chen MY (2011) An adaptive and nonlinear drift-based Wiener process for remaining useful life estimation. 2011 Prognostics and health management conference, PHM2011 Shenzhen, IEEE Xplore, pp 1–5
4. Si XS, Wang W, Hu CH, Zhou DH, Pecht M (2012) Remaining useful life estimation based on a nonlinear diffusion degradation process. IEEE Trans Reliab 61(1):50–67
5. Carr MJ, Wang W (2011) An approximate algorithm for prognostic modeling using condition monitoring information. Eur J Oper Res 211(1):90–96
6. Gebraeel NZ, Lawley MA, Li R, Ryan JK (2005) Residual life distribution from component degradation signals: a Bayesian approach. IIE Trans 37(6):543–557
7. Kaiser KA, Gebraeel NZ (2009) Predictive maintenance management using sensor-based degradation models. IEEE Trans Syst Man Cybern Part A Syst Hum 39(4):840–849
8. Armstrong M, Atkins D (1996) Joint optimization of maintenance and inventory policies for a simple system. IIE Trans 28(5):415–424
9. Elwany AH, Gebraeel NZ (2008) Sensor-driven prognostic models for equipment replacement and spare parts inventory. IIE Trans 40(7):629–639
10. Tapiero CS, Venezia I (1979) A mean variance approach to the optimal machine maintenance and replacement problem. J Oper Res Soc 30(5):457–466
11. Filar JA, Kallenberg LCM, Lee H-M (1989) Variance-penalized Markov decision processes. Math Oper Res 14(1):147–161
12. Chen Y, Jin JH (2003) Cost-variability-sensitive preventive maintenance considering management risk. IIE Trans 35(12):1091–1101
13. Gosavi A (2006) A risk-sensitive approach to total productive maintenance. Automatica 42(8):1321–1330
14. Giri BC, Dohi T (2010) Quantifying the risk in age and block replacement policies. J Oper Res Soc 61(7):1151–1158
15. Chen N, Chen Y, Li ZG, Zhou SY, Sievenpiper C (2011) Optimal variability sensitive condition-based maintenance with a Cox PH model. Int J Prod Res 49(7):2083–2100
16. Si XS, Hu CH, Wang W (2012) A real-time variable cost-based maintenance model from prognostic information. 2012 Prognostics and system health management conference, PHM2012 Beijing, IEEE Xplore, pp 1–6
17. Si XS, Wang WB, Hu CH, Chen MY, Zhou DH (2013) A Wiener-process-based degradation model with a recursive filter algorithm for remaining useful life estimation. Mech Syst Signal Process 35(1–2):219–237
18. Lee MLT, Whitmore GA (2006) Threshold regression for survival analysis: modeling event times by a stochastic process reaching a boundary. Stat Sci 21(4):501–513
19. Dempster A, Laird N, Rubin D (1977) Maximum likelihood from incomplete data via the EM algorithm. J R Stat Soc Ser B 39(1):1–38

CPSIA information can be obtained
at www.ICGtesting.com
Printed in the USA
LVOW02*0710120217
523967LV00002B/41/P

9 783662 540282